Recent Trends in Fracture and Damage Mechanics

Geralf Hütter · Lutz Zybell
Editors

Recent Trends in Fracture and Damage Mechanics

Springer

Editors
Geralf Hütter
Institute of Mechanics and Fluid Dynamics
TU Bergakademie Freiberg
Freiberg
Germany

Lutz Zybell
Institute of Mechanics and Fluid Dynamics
TU Bergakademie Freiberg
Freiberg
Germany

ISBN 978-3-319-21466-5 ISBN 978-3-319-21467-2 (eBook)
DOI 10.1007/978-3-319-21467-2

Library of Congress Control Number: 2015945615

Springer Cham Heidelberg New York Dordrecht London

Printed on acid-free paper

Springer International Publishing AG Switzerland is part of Springer Science+Business Media (www.springer.com)

Preface

In September 2015, an international symposium entitled "Recent Trends in Fracture and Damage Mechanics" was held at the Institute of Mechanics and Fluid Dynamics at TU Bergakademie Freiberg. The symposium was organized in honor of the scientific contributions of Meinhard Kuna to the field of fracture and damage mechanics. In parallel, celebrating his 65th birthday, Professor Kuna also officially retired at this event. All those who knew him were confident that this would in no way deter him from further research. Many of his alumni, former colleagues, scientific companions, cooperating partners as well as personal friends agreed without hesitation to contribute to the symposium. The present volume, which includes the proceedings of the symposium, provides for the interested reader an overview of the development of some fields of fracture and damage mechanics in the last decades. We are pleased to dignify thus the lifetime achievement of Meinhard Kuna in an appropriate manner.

First of all, we would like to thank Meinhard Kuna for being our scientific mentor for the last years. Through his own experiences at the IFE Halle, he also tried to provide excellent working conditions during our doctorate time in Freiberg. Personally, we became acquainted with an initially somewhat distant scientist, whose soft core, expressed as a kind-hearted, equitable, and fatherly supervisor, appeared after some time. We think that retiring—or better said, the release of responsibility in a gradual manner—will not be an easy step in the life of Meinhard Kuna. Therefore, we wish him and his family all the best for this challenging task and their future.

Furthermore, we would like to thank all authors contributing to this volume and the anonymous reviewers for their comments helping to improve the manuscripts.

The support of Nathalie Jacobs and Cynthia Feenstra of Springer Publishing during the publication process is also gratefully acknowledged. Last but not least, we highly appreciate the help of Karina Hartmann during the preparation of the symposium.

Freiberg
June 2015

Geralf Hütter
Lutz Zybell

Contents

Part I
Historical Perspective

Meinhard Kuna: Physics and Engineering at the Crack Tip—A Retrospective

Geralf Hütter and Lutz Zybell

1 A Brief Scientific Biography

Meinhard Kuna was born in January 1950 in Eckartsberga in the former GDR. He studied physics at TU Magdeburg from 1968 to 1972 with a focus on theoretical physics. Afterwards he was a scientific research assistant at the Institute of Solid State Physics and Electron Microscopy (IFE) of the Academy of Sciences of the GDR in Halle/Saale. He worked on elastic-plastic crack tip analyses and received his doctorate (Dr. rer. nat.) in theoretical physics from the University Halle-Wittenberg in 1978. During the following years he became leader of the research group "Numerical Methods in Fracture Mechanics" at the IFE in Halle. In 1990 he habilitated in theoretical physics at the Martin Luther University Halle-Wittenberg with his thesis about hybrid crack tip finite elements. From 1991 to 1994, Meinhard Kuna was the head of the department "Microstructure of Materials and Systems" of the Fraunhofer Institute of Mechanics of Materials (IWM) in Halle. During this time he also worked at the IWM headquarters in Freiburg. In 1995, he was a visiting professor for Mechanics at the Otto von Guericke University Magdeburg in the Faculty of Mechanical Engineering. From 1996 to 1997, he worked again in southern Germany and was head of the department "Numerical Simulation" at the Materials Testing Institute (MPA) of the University of Stuttgart.

In 1997 Meinhard Kuna was appointed as a full professor for Solid Mechanics at TU Bergakademie Freiberg. Since then he has served two periods as director of the Institute of Mechanics and Fluid Dynamics. During his years in Freiberg he taught

G. Hütter · L. Zybell (✉)
Institute of Mechanics and Fluid Dynamics, TU Bergakademie Freiberg,
Freiberg, Germany
e-mail: lutz.zybell@imfd.tu-freiberg.de

G. Hütter
e-mail: geralf.huetter@imfd.tu-freiberg.de

G. Hütter and L. Zybell, *Recent Trends in Fracture and Damage Mechanics*, DOI 10.1007/978-3-319-21467-2_1

basic courses in mechanics to thousands of students and inspired numerous students to do their majors or to start a PhD in mechanics (Fig. 1). In 2011, he was honored with the Julius Weisbach Price for his outstanding merits in teaching at TU Bergakademie Freiberg. Over a number of years he also served in several important functions in the university. From 2005 to 2012, he was the liaison officer of the German Research Foundation (DFG) for TU Bergakademie Freiberg. Between 2006 and 2009, he acted as the Vice Dean of the Faculty of Mechanical, Process and Energy Engineering and since 2009 he has been an elected member of the university council. In 2011, he founded an international masters course on Computational Material Science at TU Bergakademie Freiberg and since then he has served as the Dean of Study of this course (Fig. 2). Altogether, in his time as a professor in Freiberg he supervised around 30 PhD students, mentored three habilitations and hosted eight Humboldt research fellows, where eight out of all scholars later became professor. All through his time in Freiberg, Meinhard Kuna has closely identified himself with the university. He became involved in several large-scale research projects and worked successfully together for many years with other colleagues from Freiberg. Especially with Gerhard Pusch, Peter Hübner and Horst Biermann from the Institute of Materials Engineering of TU Bergakademie Freiberg he maintained very close collaboration, resulting in several research projects in the field of fracture mechanics. Moreover, Meinhard Kuna has cherished longtime fruitful cooperations with many scientists all over the world.

Meinhard Kuna is an internationally recognized expert in fracture and damage mechanics. Furthermore, his research interests comprise inelastic material modeling, as well as the development of numerical methods (particularly FEM and BEM) towards fracture applications, the modeling of smart materials with a focus on piezoelectric material behavior and the development of miniaturized test methods,

Fig. 1 Meinhard Kuna teaching mechanics of materials at TU Bergakademie Freiberg in 2012

Fig. 2 Meinhard Kuna discussing with international master students of the Computational Materials Science course at TU Bergakademie Freiberg in 2014

in particular the Small Punch Test (SPT). Since 1976 Meinhard Kuna has authored around 350 scientific publications, 200 of which are peer reviewed articles. His most cited articles are his study of ductile failure in nodular cast iron [40] by 3D cell models from 1996 and his review article on the development of FEM techniques for fracture mechanics analysis in piezoelectric materials [34] from 2010. In 2008 Meinhard Kuna published his textbook "Numerische Beanspruchungsanalyse von Rissen: Finite Elemente in der Bruchmechanik" [33], which was later also published in English [35]. This monograph is nowadays established as the standard book for the numerical treatment of cracks by FEM and was awarded in 2014 with the *TEXTY* Textbook Excellence Award by the International Text and Academic Authors Association (TAA).

Throughout his life of research and teaching, Meinhard Kuna has been very engaged in scientific societies. He is an active member of the German Association for Materials Research and Testing (DVM) and the Society of Applied Mathematics and Mechanics (GAMM), contributing regularly to their annual meetings. Within the DVM he was chairman of the working group on fracture mechanics from 2003 to 2007 and organized the annual meetings of this group numerous times. Between 2003 and 2009 he was the representative of the German National Fracture Group in the European Structural Integrity Society (ESIS) and within the International Conference on Fracture (ICF). Under the auspices of ESIS and DVM he initiated the 18th European Conference on Fracture (ECF 18) at Dresden. Moreover, in 2001 he organized the 11th International Workshop on Computational Mechanics of Materials (IWCMM11) [71] at TU Bergakademie Freiberg together with Siegfried Schmauder from MPA Stuttgart. In 2009, Andreas Ricoeur and Meinhard Kuna hosted the IUTAM Symposium on Multiscale Modelling of Fatigue, Damage and

Fracture in Smart Materials [39], again at the TU Bergakademie Freiberg. In honor of his achievements for the fracture mechanics community, the DVM awarded him the August Wöhler Medal in 2015, see Fig. 3. Due to his expertise, Meinhard Kuna is much sought after as a reviewer for scientific organizations and for numerous international journals. Since 2012 he has been a member of the Board of Reviewers of the German Research Foundation (DFG) for the field of Applied Mechanics, being responsible for the evaluation of research proposals. Also within the Alexander von Humboldt Foundation he is part of the selection committee. Moreover, he is a member of the editorial board of the *International Journal of Fracture*, *Archive of Applied Mechanics*, and *Engineering Fracture Mechanics*, those journals being most closely related to his fields of research.

Meinhard Kuna has been an important member of the international fracture mechanics community for the last 35 years. His main scientific achievements are in the development of numerical methods for treating problems of damage and fracture. He made outstanding contributions particularly in the field of the development of specialized FEM and BEM techniques for crack tip analyses and in the study of fracture mechanics of piezoelectric materials. An important characteristic in his scientific work was that he did not stick to purely academic problems. Rather, his ambition—to start research projects motivated by practical problems, to explore new fields thoroughly, and finally to lead solutions back to practice—has been characteristic of his whole academic career and resulted in many industrial research projects. His main scientific projects and some personal anecdotes are outlined in the following section.

Fig. 3 Presentation of the August Wöhler Medal of the DVM to Meinhard Kuna by Peter Hübner (*left*) and Gerhard Pusch (*right*) in 2015

2 The Making of an Engineering Physicist in Fracture and Damage Mechanics

2.1 Studies at TU Magdeburg and Working at IFE Halle Until 1990

As already mentioned, Meinhard Kuna was born in January 1950 in Eckartsberga, which is located near Naumburg in Thuringia in the former GDR. His father was a school headmaster and his mother a teacher and he was the eldest of three brothers and one sister. After 1953 Meinhard Kuna spent his childhood in Wernigerode, where he also finished primary and secondary school. The two last years of his high school time he attended a special advanced education class for maths and natural sciences at the TU Magdeburg. These special classes were introduced in the GDR in the 1950s at several universities in order to promote outstanding pupils.

After finishing school with highest honors, Meinhard Kuna began his study of physics with a focus on theoretical physics at the same university. At that time he was—next to his studies—a very enthusiastic musician. He played guitar in a beat band and served as the pianist in the student cabaret "Prolästerat für Studienungelegenheiten". During that period he also got to know his later wife Christine, who did a teacher-training study course for math and physics at TU Magdeburg.

In his diploma thesis [23], Meinhard Kuna worked on atomistic modeling of crack tip deformations in α-iron, see Fig. 4. Due to the high degree of difficulty of that topic, the thesis was issued as a cooperation of two students. So Meinhard Kuna, in the early 1970s, tried together with his longtime college friend Uwe Riemann to implement what is nowadays known as molecular dynamics. The thesis was supervised by Heribert Stroppe and Rainer Clos. For the young ambitious physicist Meinhard Kuna this time felt like a first scientific disillusion. Rather than

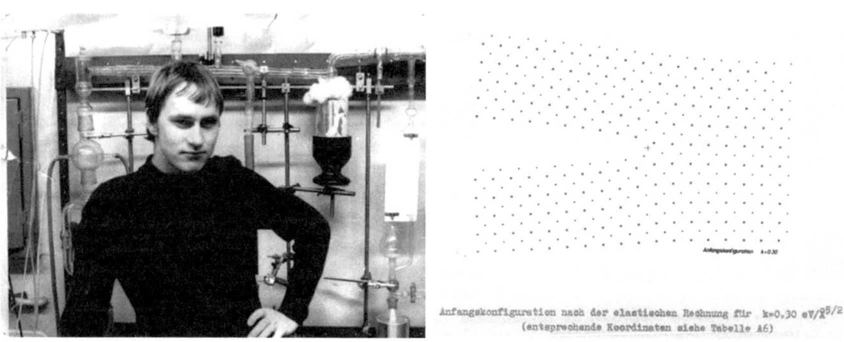

Fig. 4 Meinhard Kuna as a student of Theoretical Physics around 1970 (*left*) and sketch of atom positions at the crack tip in α-iron from his Diploma Thesis [23] (*right*)

continuing his study of Einstein's already famous theories, he had to orient himself towards engineering fracture mechanics and handling IT problems.

Due to critical political discussions with lecturers for Marxism-Leninism and undesirable activities during their study years, Uwe Riemann and Meinhard Kuna were early identified as "non-true to the line" by representatives of the socialistic system. This meant that a later career at any university was absolutely barred. By chance, Meinhard Kuna applied for a so-called "Aspirantur" (a type of PhD position in the former GDR) at the Institute of Solid State Physics and Electron Microscopy (IFE) of the Academy of Sciences of the GDR in Halle/Saale. This institute was founded and headed by Heinz Bethge, who was an outstanding scientist in the field of surface physics. Due to his international standing, the IFE had excellent equipment and offered best working conditions—even for young talented researchers being regarded as politically non-conforming.

During this time at the IFE Halle, Meinhard Kuna worked on elastic-plastic crack tip analyses. The initial plan, to apply to that problem what is nowadays known as dislocation dynamics, did not work due to limited computing resources in the middle of the 1970s. For this reason, he decided together with his supervisor Volker Schmidt to solve the problem within the framework of continuum plasticity theory by using the finite element method (FEM). This decision meant for Meinhard Kuna the final anticlimax from theoretical physics towards engineering mechanics. By initial assistance and further collaboration with the pioneering research group for FEM within GDR, Jürgen Dankert and Siegfried Koczik at TU Magdeburg, Meinhard Kuna developed an elastic-plastic FEM program for 2D crack analyses, enabling evaluation of the J-integral values along different contours [24]. At that time, the post-processing was still performed by-hand and thus very time-consuming, see Fig. 5.

During his doctorate time at IFE Halle, Meinhard Kuna learned working as an independent scientist. The excellent working conditions at IFE—including an open-minded, considerate and friendly atmosphere, many seminars including scientific discussions, regular visits of international guests and a strong support of young scientist by the superiors—led to his first outstanding scientific results in the field of elastic-plastic crack tip analyses [24]. On account of his success Meinhard

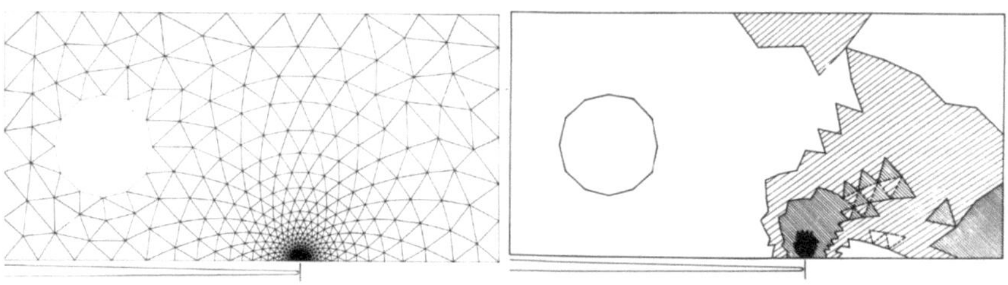

Fig. 5 FEM simulation of elastic-plastic crack tip behavior [24]: Meshing of a CT specimen with 689 TRID3 finite elements (*left*) and graphical representation of the plastic zone made by-hand (*right*)

Kuna was honored in 1977 with the Young Researchers Prize of the Academy of Sciences of the GDR. At that time, also the later very fruitful cooperation [43, 44, 76] with Zdenek Bilek and Zdenek Knesl from the Institute of Physics of Materials (IPM) in Brno started. The colleagues from IPM were specialized in experimental fracture mechanics, which complemented perfectly the simulation work of Meinhard Kuna and his supervisor Volker Schmidt. In 1978, Meinhard Kuna's first working period at IFE Halle was finished by defending his dissertation [25], see Fig. 6. Since then, Meinhard Kuna was simply called "Mr. CT" at the institute.

In the late 1970s, Meinhard Kuna became known as a formidable newcomer in the international fracture mechanics community. He regularly participated in conferences in Eastern Europe and he was also allowed to go to the "West" upon special application. Visiting the *1st International Conference on Numerical Methods in Fracture Mechanics* in Swansea/Wales, which was organized by Luxmoore and Owen, was an important experience for his later career, since there he met several leading scientists in fracture mechanics such as Rice, Turner, Kobayashi, and Knott.

To perform expensive FEM computations, the best russian super computer BESM6 was only available at the Computing Center of the Academy of Sciences of the GDR in Berlin-Adlershof, see Fig. 7. The configuration of BESM6 was based on transistor technology, magnetic drums and digital tape resulting in a clock rate of 10 MHz and 32 kilowords central memory. Meinhard Kuna traveled once a week by train from Halle to Berlin, carrying a bag full of punch cards containing input files. Including self-made post-processing, this allowed 20 jobs per day.

Fig. 6 Defending his Dissertation in 1978: Presentation of the doctoral cap by Heinz Bethge (head of IFE Halle) (*left*) and Meinhard Kuna as "Mr. CT" (*right*)

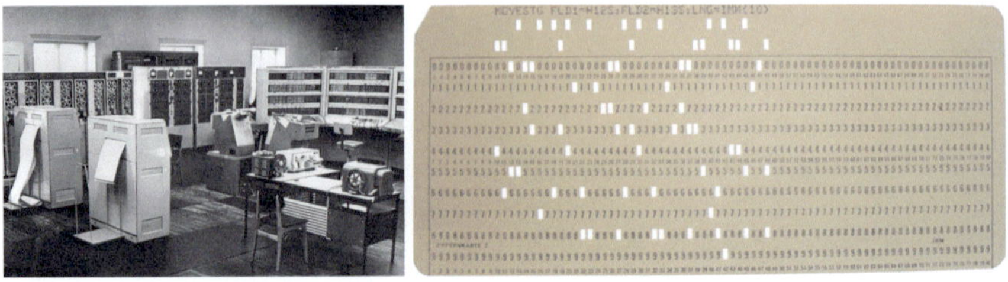

Fig. 7 FEM computations in the 1970s: russian super computer BESM6 at the Computing Center of the Academy of Sciences of the GDR in Berlin-Adlershof (*left*) and Hollerith punch card containing a part of an input file (*right*)

Due to his scientific success in the field of theoretical fracture mechanics and its industrial application towards the technical safety of chemical and nuclear facilities, Meinhard Kuna was assigned to establish the working group "Numerical Methods in Fracture Mechanics" at the IFE Halle in 1978. There was a large demand in the GDR for assessing industrial cases of damage by fracture mechanics analyses and besides the group of Holger Theilig from the School of Engineering in Zittau there was no one able to do this. Also Horst Blumenauer and Gerhard Pusch from TU Magdeburg, who comprised the leading group in experimental fracture mechanics in the GDR at that time [6], had a strong interest in numerical analyses of their experiments. So Meinhard Kuna started his group with Wolfgang Uhlmann, Günther Maschke, and Ulrich-Michael Eisentraut whereas Matthias Petersilge, Michael Busch, and Thomas Hantschel joined somewhat later.

One of the first impressive failure cases investigated by the new Kuna group was the explosion of a high pressure facility for ethylene production within the industrial complex "Leunawerke Walter Ulbricht" south of Halle, see Fig. 8. After only four months of operation a pipe elbow failed and the whole factory building was destroyed. The fractographic investigation, which was performed by Martin Möser at the IFE Halle [60], revealed that hydrogen-assisted cracking led to the growth of a fatigue crack, finally leading to failure by brittle fracture. The fracture mechanics assessment of this damage event was performed by using hybrid crack tip elements.

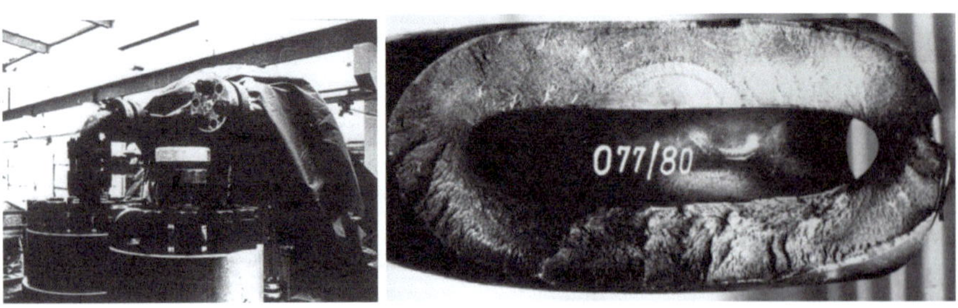

Fig. 8 Explosion of a high pressure facility for ethylene production within the industrial complex "Leunawerke Walter Ulbricht" in 1980

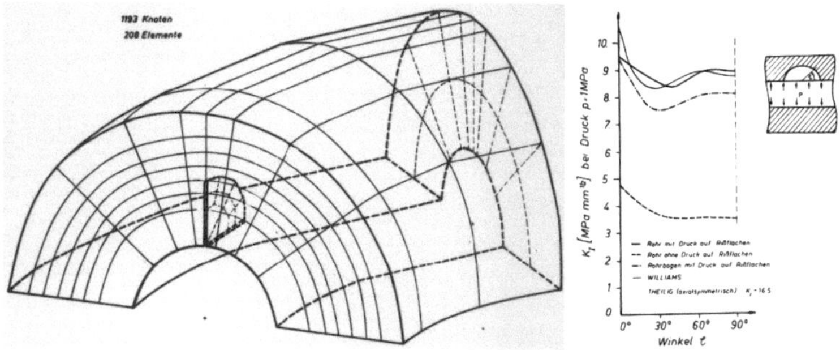

Fig. 9 Application of hybrid crack tip finite elements for the fracture mechanics assessment of the damage event at "Leunawerke Walter Ulbricht" in 1980

The results, see Fig. 9, showed that the internal pressure and the geometry of the pipe elbow had an enormous influence on the crack tip loading conditions.

The Kuna group worked on several fields of numerical fracture mechanics. Together with Lothar Wiltinger and Johannes Altenbach from TU Magdeburg a FEM software for linear-elastic fracture mechanics was developed. After the first version *CRACK2D* [45] for plane crack problems, this software was later extended towards 3D crack analyses [9, 27, 28]. This program was used for thermo-elastic crack analyses, which were important for failure analyses caused by thermo shock of the reactor pressure vessel WWER440 for the nuclear power plant in Greifswald. Regarding the safety of nuclear pressure vessels of type WWER there was a strong collaboration between all countries in the COMECON. In those years, Meinhard Kuna and his group also developed so-called hybrid crack tip finite elements [26, 29, 37, 42]. Based on hybrid variational principles of elasticity, these elements allowed the use of different shape functions for displacements and stresses, so that the stress singularity at the crack tip can be directly incorporated in one element, yet ensuring comparability of boundary displacements with standard isoparametric elements.

Due to limited computing resources, especially for 3D applications of the FEM, Günther Maschke and Meinhard Kuna decided in the mid 1980s to start using the boundary element method (BEM) for fracture mechanics applications. In addition to plane and spatial crack configurations, the BEM was also extended towards elastic-plastic material behavior [8, 14, 55, 56]. The developed BEM code was also used in collaboration with Hans-Achim Bahr and Herbert Balke from the Central Institute of Solid Mechanics and Materials Science ZFW Dresden for investigating crack patterns generated by thermal shock [4, 46].

At the end of the 1980s a frequent and intensive collaboration started between the Kuna group and research institutes in the western part of Germany. First contacts between Meinhard Kuna and scientists from the FRG had already happened at international conferences and afterwards these scientists were invited for a stay at the IFE Halle. For example, Hans-Georg Blauel and Erwin Sommer from the Fraunhofer Institute for Mechanics of Materials (IWM) in Freiburg visited the IFE

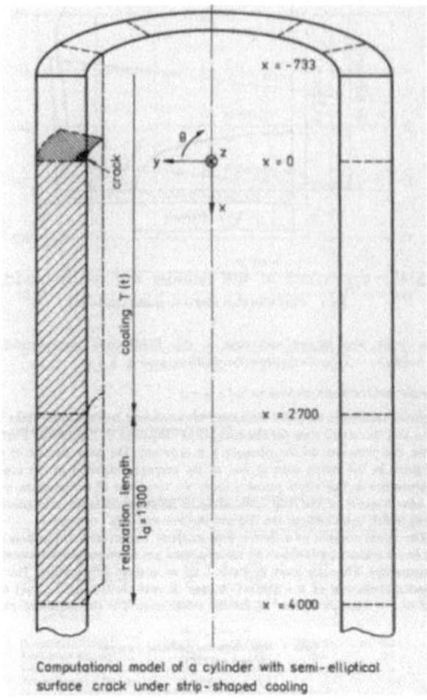

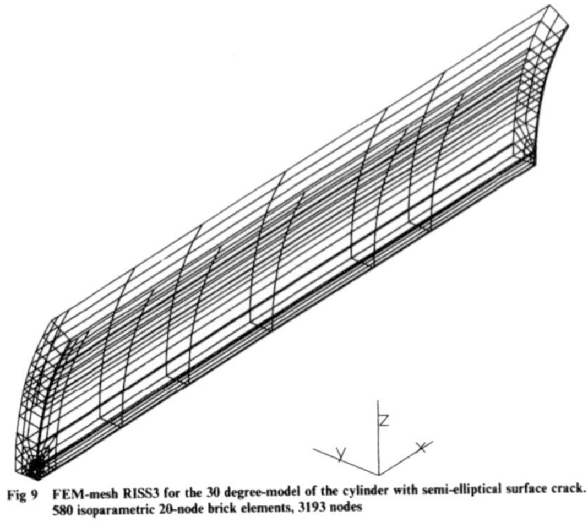

Fig. 10 Thermoelastic-plastic analysis of a nuclear pressure vessel: Sketch of the WWER 440 (*left*) and finite element mesh of 1/6 of the vessel (*right*) (from [48])

Halle in the late 1980s. In this period there were also first contacts with Dietmar Klingbeil and Wolfgang Brocks from the Federal Institute for Materials Research and Testing (BAM) in Berlin. During a fracture mechanics conference in Freiburg in the 1980s, Meinhard Kuna met also Dietmar Gross from TU Darmstadt for the first time, which was the beginning of an initially distant but later close friendship. Although trips to western Germany were strongly regulated and policed, in 1988 Meinhard Kuna was allowed to visit together with Volker Schmidt the IWM in Freiburg for four weeks. During this stay he investigated the behavior of a nuclear pressure vessel under a loss of coolant accident (LOCA) scenario, see Fig. 10. Together with Horst Kordisch and Andrea Ockewitz the *J*-integral for a surface crack during a pressurized thermal shock was analyzed with the FEM code ADINA [48].

Other important intra-German contacts had been established with Karl-Heinz Schwalbe from the GKSS in Geesthacht and Hans-Georg Hahn and Hans Albert Richard from the TU Kaiserslautern. The latter led to a very fruitful scientific cooperation on mixed-mode fracture [68]. Meinhard Kuna calculated the stress intensity factors for the so-called AFM specimen, see Fig. 11, which was developed by Hans Albert Richard to investigate all three fracture modes. The results of this study showed that modes II and III can not be decoupled. Finally, a couple of months before the fall of the Berlin Wall, Meinhard Kuna visited Satya N. Atluri at the Georgia Institute of Technology in the US for some weeks. They worked together on superposition techniques for obtaining weight functions by the BEM [47].

Summarizing Meinhard Kuna's scientific period from its beginning to the end of the GDR, it can be stated that due to happy circumstances, exceptional working conditions and strong promotion by Heinz Bethge and Volker Schmidt at the IFE

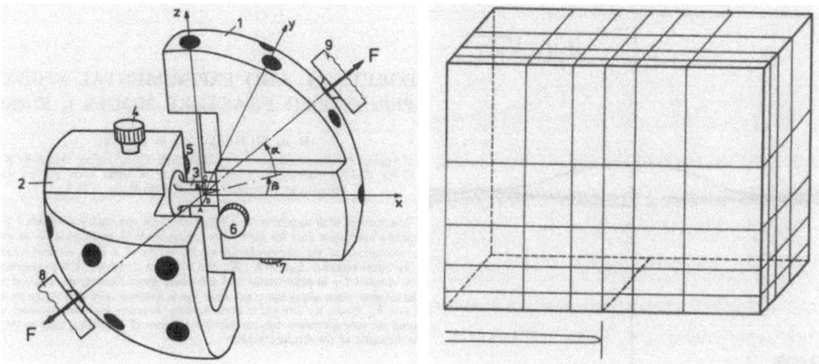

Fig. 11 German-German scientific cooperation with Hans Albert Richard using hybrid crack tip finite elements: sketch of the AFM specimen for mixed-mode fracture experiments (*left*) and finite element mesh of the crack tip region (*right*) (from [68])

Halle, and on account of his high personal engagement, Meinhard Kuna had become an internationally recognized expert in numerical fracture mechanics and a respected scientific leader of his working group at IFE Halle.

2.2 *The Time After the Reunification of Germany*

The severe changes encompassing the Reunification of Germany in 1990 also had a strong effect on the career of Meinhard Kuna. According to the Unification Treaty, the Academy of Sciences of the GDR was disbanded and its institutes had to be integrated into the scientific landscape of the Federal Republic of Germany. After a first evaluation, the IFE in Halle was intended to join the Max Planck Society. Since this largest German research organization is dedicated to basic research, Meinhard Kuna and some colleagues looked for more seminal perspectives for the applied research groups of the IFE. Due to their contacts with the Fraunhofer IWM in Freiburg and on account of their very similar research areas, a common proposal, to found the branch department "Microstructure of Materials and Systems" in Halle within the Fraunhofer Society, was elaborated and finally approved by the German Science Council.

The department in Halle, headed by Meinhard Kuna, included 20 scientists who continued to work successfully in three groups: numerical fracture mechanics led by Meinhard Kuna, the field of electron microscopy analysis of materials under the auspices of Jürgen Hopfe and investigation of the mechanics of microelectronic materials and devices, which was managed by Dieter Katzer and Matthias Petzold. The latter research topic was the basis for Meinhard Kuna's continuing interest in fracture and damage of silicon and solders as well as on miniaturized testing.

During this time of severe political changes, researchers from the former socialistic GDR had to adapt to the capitalistic mechanisms of research, the necessity of formulating research proposals and acquiring third-party funding from industry and public partners. In parallel, doors to new areas were opened. In 1991, Meinhard Kuna habilitated with a thesis on hybrid crack tip elements [30], which was not allowed to him in the GDR for political reasons.

Due to several conflicts between the Fraunhofer IWM headquarters in Freiburg and the department in Halle, Meinhard Kuna was transferred in 1994 to the headquarter in Freiburg. There, he came closely into touch with damage mechanics. Together with Dong-Zhi Sun he investigated the damage behavior of nodular cast iron by cell model simulations of representative volume elements [40], see Fig. 12. Both, damage mechanics and the material called nodular cast iron, remained relevant topics of research during Meinhard Kuna's further career. The time at IWM Freiburg was also the starting point for a long-term friendship with Wolfgang Brocks, Reinhold Kienzler, and Igor Varfolomeev, who also worked there at that time.

In 1995, upon invitation of Horst Blumenauer, Meinhard Kuna became a visiting professor for damage mechanics at the Otto von Guericke University in Magdeburg. There at TU Magdeburg, an Innovative Research Center for "Adaptive Mechanical Structures" was initiated, where Meinhard Kuna got involved. This marks the beginning of his work on numerical methods and criteria of fracture for smart materials [31, 32].

After his visiting stay in Magdeburg, in 1996 Meinhard Kuna changed to the Materials Testing Institute (MPA) of the University of Stuttgart, where he became head of the department "Numerical Simulation". There, in addition to industrial research projects in the field of nuclear safety, he continued working on ductile damage models (in particular the Rousselier model) in collaboration with Michael Seidenfuß and Ludwig Stumpfrock. Within a cooperative EU-project with Ivo Dlouhý from the Czech Academy of Sciences, Meinhard Kuna started simulating the small punch test (SPT), bringing together his interests on miniaturized testing

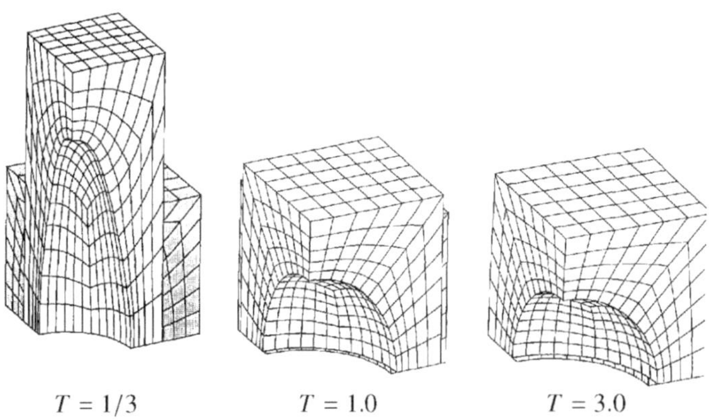

Fig. 12 Deformed cubic primitive unit cells for nodular cast iron under loading with different stress triaxialities (from [40])

and damage models—a combination that inspired several future research projects. Furthermore, together with Fritz Aldinger from the MPI Stuttgart and Bernd Kröplin from the University of Stuttgart, Meinhard Kuna continued investigating the fracture of smart materials at that time.

2.3 Professorship at TU Bergakademie Freiberg

In 1997, Meinhard Kuna was appointed as a full professor for "Applied Mechanics and Solid Mechanics" at TU Bergakademie Freiberg. Within the almost 20 following years in Freiberg he continued working in manifold fields, which had attracted his interest during his former stations. They were numerical fracture mechanics and safety assessment of structures, damage mechanics and micromechanics of materials, smart materials, miniaturized testing of materials as well as the development of corresponding numerical methods. From the many projects, amounting to about one million Euro per year, which he initiated in this period, only a short abstract can be given here.

In the field of numerical fracture mechanics, Meinhard Kuna's scientific origin, he worked on formulations of energy balance integrals [11] and numerical methods allowing efficient computations of these quantities [12, 78]. Based on former cooperations of his coworker Matthias Scherzer with Reiner Kreißig and Arnd Meyer from the TU Chemnitz, Meinhard Kuna was also involved in the development of adaptive FEM methods within the collaborative research center on "Parallel Numerical Simulation for Physics and Mechanics" (SFB 393) [59].

However, Meinhard Kuna did not stick to purely academic research but was always looking for potential practical applications of his theories and methods. That is why the programs *J-POST* [20] for the computation of energy balance integrals and *PROCRACK* [65] for FEM-simulations of fatigue crack growth in complex 3D structures were developed at his chair. Both programs met a kind reception in industry and found many applications in the safety assessment of structures, as shown for instance in Fig. 13.

Via collaborations with industry, authorizing agencies and the Institute of Materials Engineering of the TU Bergakademie Freiberg (Gerhard Pusch, Peter Hübner, Horst Biermann), Meinhard Kuna came again in touch with nodular cast iron. Safety assessment studies were carried out for industrial components made from this material like railway wheels [50], gas pipelines [49], and transportation casks for spent nuclear fuel rods [35]. In addition to industrial applications of fracture mechanics, combined experimental and numerical investigations including macroscopic and micromechanical models for damage, fracture and crack propagation in nodular cast iron were developed for fatigue [61, 64, 79], dynamic [10] and ductile failure [80] and for the ductile-brittle transition regime [17].

Meinhard Kuna's research on damage mechanics and micromechanics of materials was of course not restricted to nodular cast iron. Strongly increasing computing capacities allowed more detailed micromechanics models. In parallel,

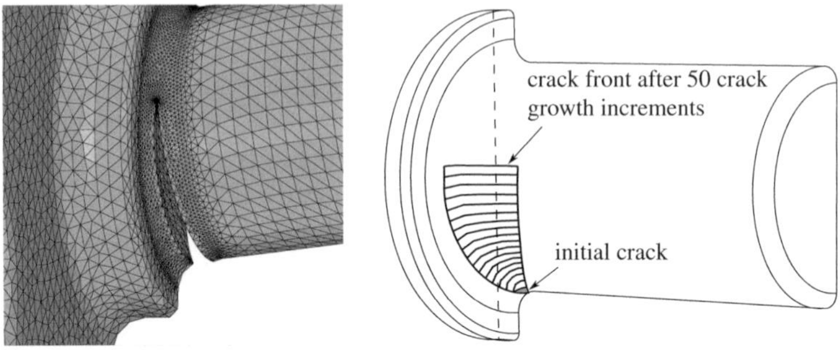

Fig. 13 FE simulation of fatigue crack growth in a steering knuckle (from [65])

non-linear FEM codes called for improved material models of damage in order to overcome the limitations of classical elastic-plastic fracture mechanics for all relevant damage mechanisms of engineering materials. For ductile failure, nonlocal extensions of the Gurson model were formulated and implemented in FEM software [18, 53]. Furthermore, simulations with a discretely resolved microstructure in the process zone at the crack tip allowed a deeper insight into the micromechanical processes [15]. Also improved models were developed to describe the ductile-brittle transition of engineering metals [16, 52]. The mechanisms of fatigue were addressed at the microscale [64] and a cyclic cohesive zone model was developed [69] to simulate initiation and propagation of fatigue cracks under arbitrary load sequences. Returning to microelectronic components, the creep damage of lead-free solders was investigated, too [41].

Applying sophisticated damage models, it turned out that the determination of the often large number of parameters of these models is a challenging task itself. Appropriate strategies of parameter identification like neural networks [1, 52] and measurement of local deformation fields [74] were developed in collaboration with Martin Abendroth, Marcel Springmann, and Thomas Linse. In particular, the small punch test (SPT) was employed for this purpose since this miniaturized test requires only a small amount of testing material. In the meantime, due to activities together with Martin Abendroth and Stefan Rasche the SPT was also extended towards ceramics and high temperature applications [2, 67].

However, the work of Meinhard Kuna was not limited to classical metallic materials. Rather, if required in collaborations with industry or institutes, he did not hesitate to deal with new materials or new methods. So the collaborative research centers on "TRIP-Matrix-Composites" (SFB 799 coordinated by Horst Biermann) and on "Multi-Functional Filters for Metal Melt Filtration" (SFB 920 coordinated by Christos Aneziris) and the Centre of Excellence *ADDE* (coordinated by David Rafaja), initiated at the TU Bergakademie Freiberg, required the development of respective material models. In particular, models for phase transformations in zirconia ceramics, TRIP steel, and silicon had to be developed, implemented and verified [7, 57, 63]. The filters investigated within the SFBs 799 and 920 are open-cell ceramic foams. This required modeling of the failure of individual rods

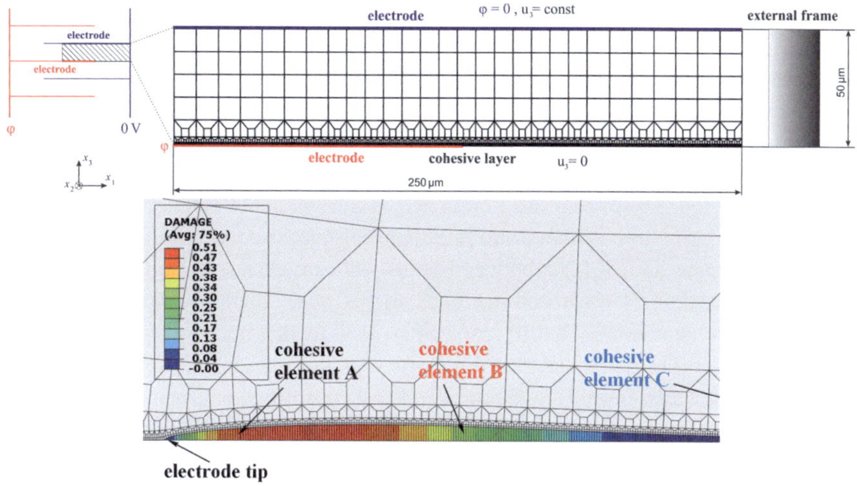

Fig. 14 FE simulation of damage accumulation at the electrode tip of a piezoelectric stack actuator under cyclic electric loading (from [21]): model (*top*) and resulting cohesive damage at electrode tip after 100 cycles (*bottom*)

and the implications to integrity of the complete filter structure under complex thermomechanical loading conditions of a melt filtration [73, 75]. Collaboration with the growing solar industry in Freiberg required simulations of technological processes like wire sawing and polishing [54, 62].

Furthermore, Meinhard Kuna continued to work on fracture of smart materials in Freiberg together with his co-workers Andreas Ricoeur, Matthias Scherzer and Marco Enderlein. Energy balance integrals and configurational forces were employed and numerical techniques for their effective computation were developed and implemented [3, 12, 51, 66, 72]. In addition, analytical solutions for interface crack problems in piezoelectric materials were investigated [70]. Since Meinhard Kuna was always looking for beneficial links between his various research activities, under his guidance also numerical methods and modeling techniques like BEM [13, 77], X-FEM [5], adaptive FEM [19] or cyclic cohesive zone models [21, 22] found their application to fracture of piezoelectric materials, see Fig. 14. Simultaneously, Meinhard Kuna looked for novel applications of piezoelectric sensors for in situ monitoring of crack propagation and damage [36, 58]. Due to his expertise in numerical modeling of fracture in smart materials, numerous foreign researchers joined the research group in Freiberg for visiting stays, thus contributing significantly to its success, among them Fulin Shang, Eric Béchet, Qun Li, and Sergii Kozinov.

Despite his manifold research activities, Meinhard Kuna was always very engaged in teaching. Together with his coworkers Eckehard Kullig, Uwe Mühlich, Monika Müller, Andreas Ricoeur, and Matthias Scherzer he offered in his chair a wide range of lectures in the field of mechanics, going far beyond what was demanded by study regulations. Meinhard Kuna wanted to provide profound basics in applied mechanics and fracture mechanics to students for their future careers both in industry and academia. His demanding but likable style of teaching enjoyed great popularity by the students. Due to his engagement, in 2011 he received the Weisbach award of the TU Bergakademie Freiberg for excellent teaching. Regarding the staff for his many research projects, he did not rely on external graduates. Rather, he always aimed to recruit the best of his own students for PhD theses at his chair. Observing the trends and requirements in science and industry, in 2011 Meinhard Kuna initiated the international master course on Computational Materials Science as a bridge between solid state physics, materials science and engineering mechanics, thus coming back to his own scientific roots. Moreover, Meinhard Kuna was not only fully committed to students, but he identified himself closely with the TU Bergakademie Freiberg. For this reason he accepted responsibility for the university in different official positions as mentioned before. Meinhard Kuna also campaigned for the historical research at TU Bergakademie Freiberg. Together with Norman Pohl he published a volume about the work of the polymath Julius Weisbach, appreciating his merits for the development of engineering sciences at the TU Bergakademie Freiberg, on occasion of Weisbach's 200th birthday in 2006 [38].

Upon his retirement in September 2015, Meinhard Kuna handed over his chair to his successor. Within the almost 20 years at the TU Bergakademie Freiberg he had built up a highly productive research group, with more than 20 young scientist belonging to it in peak times. The loyal atmosphere within the group was also supported by the biannual retreats, see Fig. 15, which were introduced by Meinhard Kuna in 2009. Recapitulating his activity at TU Bergakademie Freiberg and in national and international scientific organizations since 1997, Meinhard Kuna

Fig. 15 The Kuna research group at a retreat in 2014

promoted fracture mechanics in engineering as well as material sciences remarkably —especially by bringing together researchers from different scientific fields. Also in the near future, he will surely remain an active member of the fracture mechanics community.

References

1. Abendroth M, Kuna M (2006) Identification of ductile damage and fracture parameters from the small punch test using neural networks. Eng Frac Mech 73(6):710–725
2. Abendroth M, Soltysiak S (2015) Assessment of material properties by means of the small punch test. In: Hütter G, Zybell L (eds) Recent Trends in Fracture and Damage Mechanics. Springer, Berlin
3. Abendroth M, Groh U, Kuna M, Ricoeur A (2002) Finite element-computation of the electromechanical J-integral for 2-D and 3-D crack analysis. Int J Fracture 114(4):359–378
4. Bahr H, Balke H, Kuna M, Liesk H (1987) Fracture analysis of a single edge cracked strip under thermal shock. Theor Appl Fract Mech 8(1):33–39
5. Béchet E, Scherzer M, Kuna M (2009) Application of the X-FEM to the fracture of piezoelectric materials. Int J Numer Meth Eng 77(11):1535–1565
6. Blumenauer H, Pusch G (1987) Technische Bruchmechanik. Deutscher Verlag für Grundstoffindustrie
7. Budnitzki M, Kuna M (2012) A thermomechanical constitutive model for phase transformations in silicon under pressure and contact loading conditions. Int J Sol Struct 49 (11–12):1316–1324
8. Busch M (1989) Anwendung des BEM-Programms OREAS zur Lösung bruchmechanischer dreidimensionaler Aufgaben der Elastostatik. Technische Mechanik 10:100–105
9. Eisentraut U, Kuna M (1986) Ein FEM-Programm zur Lösung ebener, axialsymmetrischer und räumlicher Riss-, Festigkeits-und Wärmeleitprobleme. Technische Mechanik 7:51–58
10. Emrich A, Mühlich U, Kuna M, Ludwig A, Trubitz P (2007) Indirect measuring of crack growth by means of a key-curve-method in pre-cracked Charpy specimens made of nodular cast iron. Int J Fract 145(1):47–61
11. Enderlein M, Ricoeur A, Kuna M (2003) Comparison of finite element techniques for 2D and 3D crack analysis under impact loading. Int J Sol Struct 40(13):3425–3437
12. Enderlein M, Ricoeur A, Kuna M (2005) Finite element techniques for dynamic crack analysis in piezoelectrics. Int J Fract 134(3–4):191–208
13. Groh U, Kuna M (2005) Efficient boundary element analysis of cracks in 2D piezoelectric structures. Int J Sol Struct 42(8):2399–2416
14. Hantschel T, Busch M, Kuna M, Maschke H (1990) Solution of elastic-plastic crack problems by an advanced boundary element method. In: Luxmoore A, Owen D (eds) Proceedings of the 5th international conference on numerical methods in fracture mechanics, pp 29–40
15. Hütter G, Zybell L, Mühlich U, Kuna M (2012) Ductile crack propagation by plastic collapse of the intervoid ligaments. Int J Fract 176(1):81–96
16. Hütter G, Linse T, Roth S, Mühlich U, Kuna M (2014) A modeling approach for the complete ductile-brittle transition region: Cohesive zone in combination with a non-local Gurson-model. Int J Fract 185:129–153
17. Hütter G, Zybell L, Kuna M (2014) Micromechanical modeling of crack propagation with competing ductile and cleavage failure. Proc Mat Sci 3:428–433
18. Jackiewicz J, Kuna M (2003) Non-local regularization for FE simulation of damage in ductile materials. Comp Mater Sci 28(3):684–695

19. Jański Ł, Scherzer M, Steinhorst P, Kuna M (2010) Adaptive finite element computation of dielectric and mechanical intensity factors in piezoelectrics with impermeable cracks. Int J Numer Meth Eng 81(12):1492–1513
20. Kiyak Y, Klingbeil D, Enderlein M, Kuna M (2009) Vergleichende Untersuchungen von J-Integral Postprozessoren bei statischen und dynamischen Rissbeanspruchungen. DVM-Bericht 241:293–302
21. Kozinov S, Kuna M (2015) Numerical analysis of fracture of pre-stressed ferroelectric actuator taking into account cohesive zone for damage accumulation. In: Araúo A, Mota Soares C (eds) 7th ECCOMAS thematic conference on smart structures and materials (SMART 2015)
22. Kozinov S, Kuna M, Roth S (2014) A cohesive zone model for the electromechanical damage of piezoelectric/ferroelectric materials. Smart Mat Struct 23(5):055, 024
23. Kuna M (1972) Berechnung der Atompositionen an der Rissspitze in α-Eisen. Diploma thesis, TU Magdeburg
24. Kuna M (1976) An application of the finite element method to elastic-plastic analysis of the compact tension fracture test specimen. Int J Fract 12(1):175–177
25. Kuna M (1978) Zur Anwendung der Methode der finiten Elemente in der Bruchmechanik elastischer und elastisch-plastischer Körper. Dissertation, University Halle-Wittenberg
26. Kuna M (1982) Konstruktion und Anwendung hybrider Rißspitzenelemente für dreidimensionale bruchmechanische Aufgaben. Technische Mechanik 3(2):37–43
27. Kuna M (1982) Three-dimensional elastic analysis of ct specimen with straight and curved crack fronts. Int J Fract 19(3):R63–R67
28. Kuna M (1984) Behandlung räumlicher Rissprobleme mit der Methode der finiten Elemente. Tech Mech 5:23–26
29. Kuna M (1984) Hybrid crack tip elements for three dimensional fracture problems. In: Sih GC, Sommer E, Dahl W (eds) Application of fracture mechanics to materials and structures. Springer, Berlin, pp 607–617
30. Kuna M (1991) Entwicklung und Anwendung effizienter numerischer Verfahren zur bruchmechanischen Beanspruchungsanalyse am Beispiel hybrider finiter Rissspitzenelemente. Habilitation Thesis, Universität Halle-Wittenberg
31. Kuna M (1995) Energiebilanzintegrale für Risse in piezolelektrischen Werkstoffen unter elektrischen und mechanischen Beanspruchungen. Technische Mechanik 15:195–204
32. Kuna M (1998) Finite element analyses of crack problems in piezoelectric structures. Comp Mater Sci 13(1):67–80
33. Kuna M (2008) Numerische Beanspruchungsanalyse von Rissen—Finite Elemente in der Bruchmechanik. Vieweg Teubner, Wiesbaden
34. Kuna M (2010) Fracture mechanics of piezoelectric materials—Where are we right now? Eng Frac Mech 77(2):309–326
35. Kuna M (2013) Finite elements in fracture mechanics: theory-numerics-applications. Solid mechanics and its applications, vol 201. Springer, Berlin
36. Kuna M, Bäcker D (2014) A PVDF sensor for the in-situ measurement of stress intensity factors during fatigue crack growth. Proc Mat Sci 3:473–478
37. Kuna M, Khanh D (1978) Ein spezielles Hybrid-Element für die Spannungsanalyse ebener Körper mit Rissen. In: Berichte des VIII. Int. Kongresses über Anwendungen der Mathematik in den Ingenieurwissenschaften, Weimar, pp 71–76
38. Kuna M, Pohl N (2006) Julius L. Weisbach (1806–1871): Gedenkschrift zu seinem 200. Geburtstag. TU Bergakademie Freiberg
39. Kuna M, Ricoeur A (2010) Proceedings of the IUTAM symposium on multiscale modelling of fatigue, damage and fracture in smart materials, IUTAM Bookseries, vol 24. Springer, Berlin
40. Kuna M, Sun D (1996) Three-dimensional cell model analyses of void growth in ductile materials. Int J Fract 81(3):235–258
41. Kuna M, Wippler S (2010) A cyclic viscoplastic and creep damage model for lead free solder alloys. Eng Frac Mech 77(18):3635–3647
42. Kuna M, Zwicke M (1990) A mixed hybrid finite element for three-dimensional elastic crack analysis. Int J Fract 45(1):65–79

43. Kuna M, Bilek Z, Knésl Z, Schmidt V (1978) The study of crack tip stress and strain field in elasto-plastic materials. Czech J Phys B 28(1):88–107
44. Kuna M, Bilek Z, Knesl Z, Schmidt V (1980) Theoretical and experimental investigation of the behavior of plane bodies with a crack in the elastic and elastic-plastic regions. Part 1—Analysis of stresses and strains by the finite element method. Strength Mater 12(11):1366–1372
45. Kuna M, Wiltinger L, Altenbach J (1981) CRACK2D—Ein Finite-Element-Programm zur bruchmechanischen Analyse ebener elastostatischer Rissprobleme. Maschinenbautechnik 30:75–78
46. Kuna M, Maschke H, Bahr H, Weiss H (1987) Analytical and numerical study of thermo-shock induced fracture by single and multiple crack growth. In: Transactions of the 9th international conference on structural mechanics in reactor technology. Vol A
47. Kuna M, Rajiyah H, Atluri S (1990) A new approach to determine weight functions from Bueckner's fundamental field by the superposition technique. Int J Fract 44(4):R57–R63
48. Kuna M, Kordisch H, Ockewitz A (1991) Thermoelastic-plastic FEM-analysis of a semi-elliptical surface crack in a cylinder under non-axisymmetric cooling. In: Blauel J, Schwalbe KH (eds) Defect assessment in components—fundamentals and applications. ESIS/EGF, Mechanical Engineering Publications, London, pp 101–115
49. Kuna M, Wulf H, Rusakov A, Pusch G, Hübner P (2003) Entwicklung und Verifikation eines bruchmechanischen Bewertungssystems für Hochdruck-Ferngasleitungen. DVM-Bericht 235:153–162
50. Kuna M, Springmann M, Mädler K, Hübner P, Pusch G (2005) Fracture mechanics based design of a railway wheel made of austempered ductile iron. Eng Frac Mech 72(2):241–253
51. Li Q, Kuna M (2012) Evaluation of electromechanical fracture behavior by configurational forces in cracked ferroelectric polycrystals. Comp Mater Sci 57:94–101
52. Linse T, Kuna M, Schuhknecht J, Viehrig HW (2008) Usage of the small-punch-test for the characterisation of reactor vessel steels in the brittle-ductile transition region. Eng Frac Mech 75(11):3520–3533
53. Linse T, Hütter G, Kuna M (2012) Simulation of crack propagation using a gradient-enriched ductile damage model based on dilatational strain. Eng Frac Mech 95:13–28
54. Ludwig C, Kuna M (2012) An analytical approach to determine the pressure distribution during chemical mechanical polishing. J Electr Mat 41(9):2606–2612
55. Maschke H (1984) Anwendung der Randintegralmethode auf bruchmechanische Aufgaben. Technische Mechanik 5:20–22
56. Maschke H, Kuna M (1985) A review of boundary and finite element methods in fracture mechanics. Theor Appl Fract Mech 4(3):181–189
57. Mehlhorn L, Mühlich U, Kuna M (2013) A material model of particle size dependent transformation plasticity of PSZ ceramics under thermomechanical loading. Adv Eng Mat 15 (7):638–645
58. Menzer M, Kuna M (2014) Damage identification in two-dimensional structures using Lamb waves. In: Sensors and smart structures technologies for civil, mechanical, and aerospace systems, vol 9061, pp 906,105–906,113
59. Meyer A, Rabold F, Scherzer M (2006) Efficient finite element simulation of crack propagation using adaptive iterative solvers. Comm Num Meth Eng 22(2):93–108
60. Möser M (1991) Schäden am Rohrbogen einer Äthylenanlage. In: VDI-Bericht Nr. 902. VDI-Verlag Düsseldorf
61. Mottitschka T, Pusch G, Biermann H, Zybell L, Kuna M (2012) Influence of graphite spherical size on fatigue behaviour and fracture toughness of ductile cast iron EN-GJS-400-18LT. Int J Mat Res 103(1):87–96
62. Nassauer B, Liedke T, Kuna M (2013) Polyhedral particles for the discrete element method. Gran Matt 15(1):85–93
63. Prüger S, Seupel A, Kuna M (2014) A thermomechanically coupled material model for trip-steel. Int J Plast 55:182–197

64. Rabold F, Kuna M (2005) Cell model simulation of void growth in nodular cast iron under cyclic loading. Comp Mater Sci 32(3):489–497
65. Rabold F, Kuna M, Leibelt T (2013) PROCRACK: a software for simulating three-dimensional fatigue crack growth. In: Advanced finite element methods and applications, Springer, Berlin, pp 355–374
66. Rao B, Kuna M (2008) Interaction integrals for fracture analysis of functionally graded piezoelectric materials. Int J Sol Struct 45(20):5237–5257
67. Rasche S, Strobl S, Kuna M, Bermejo R, Lube T (2014) Determination of strength and fracture toughness of small ceramic discs using the small punch test and the ball-on-three-balls test. Proc Mat Sci 3:961–966
68. Richard H, Kuna M (1990) Theoretical and experimental study of superimposed fracture modes I, II and III. Eng Frac Mech 35(6):949–960
69. Roth S, Hütter G, Kuna M (2014) Simulation of fatigue crack growth with a cyclic cohesive zone model. Int J Fract 188(1):23–45
70. Scherzer M, Kuna M (2001) Asymptotic analysis of interface problems in piezoelectric composite materials. In: Smart materials—proceedings of the 1st cesarean. Springer, Bonn, pp 137–148
71. Schmauder S, Kuna M (2003) Special issue of the 11th international workshop on computational mechanics of materials. Comp Mat Sci 26
72. Shang F, Kuna M, Abendroth M (2003) Finite element analyses of three-dimensional crack problems in piezoelectric structures. Eng Frac Mech 70(2):143–160
73. Sieber T, Mühlich U, Liedke T, Ballaschk U, Berek H, Aneziris CG, Ehinger D, Wolf S, Krüger L (2011) Deformation and failure of open-cell foams made of trip-steel-ZrO_2-composite materials: experimental observations versus numerical simulations. Steel Res Int 82 (9):1004–1016
74. Springmann M, Kuna M (2003) Identification of material parameters of the Rousselier model by non-linear optimization. Comp Mater Sci 26:202–209
75. Storm J, Abendroth M, Emmel M, Liedke T, Ballaschk U, Voigt C, Sieber T, Kuna M (2013) Geometrical modelling of foam structures using implicit functions. Int J Sol Struct 50(3):548–555
76. Uhlmann W, Knésl Z, Kuna M, Bilek Z (1976) Approximate representation of elastic-plastic small scale yielding solution for crack problems. Int J Fract 12(3):507–509
77. Wippler K, Kuna M (2007) Crack analyses in three-dimensional piezoelectric structures by the BEM. Comp Mater Sci 39(1):261–266
78. Wünsche M, Zhang C, Kuna M, Hirose S, Sladek J, Sladek V (2009) A hypersingular time-domain BEM for 2D dynamic crack analysis in anisotropic solids. Int J Numer Meth Eng 78(2):127–150
79. Zybell L, Chaves H, Kuna M, Mottitschka T, Pusch G, Biermann H (2012) Optical in-situ investigations of overload effects during fatigue crack growth in nodular cast iron. Eng Frac Mech 95:45–56
80. Zybell L, Hütter G, Linse T, Mühlich U, Kuna M (2014) Size effects in ductile failure of porous materials containing two populations of voids. Eur J Mech A/Sol 45:8–19

Experimental and Numerical Fracture Mechanics—An Individually Dyed History

Wolfgang Brocks and Karl-Heinz Schwalbe

Abstract Almost half a century ago, fracture mechanics started in Germany with the foundation of the *DVM Working Group Fracture Mechanics* in 1969. The present authors have been partly involved in the further development of fracture and damage mechanics, one with particular interest in elastic-plastic fracture and modelling, the other in thin-walled structures, fatigue and assessment. They take the colloquium in honour of the 65th birthday of Professor Meinhard Kuna as occasion to highlight some significant achievements on the background of personal experience. In particular, they intend to show that both fracture and damage mechanics started with paradigm changes which were partly looked at with distrust in the beginning but turned out to be seminal.

1 Introduction

Today, in the early 21st century, both fracture and damage mechanics appear as established and acknowledged domains of science in the continuity of continuum mechanics. This has not at all been the case at their respective implementation. As the American physicist and philosopher of science Kuhn [46] described in his fundamental book on "The Structure of Scientific Revolutions", which first appeared in 1962, science does not progress via an accumulation of new knowledge, but undergoes periodic "paradigm shifts", in which scientific exploration within a particular field is abruptly transformed. This is most obvious for the great

W. Brocks (✉)
Christian Albrecht University, Kiel, Germany
e-mail: wbrocks@kabelmail.de

K.-H. Schwalbe
Hamburg University of Technology, Hamburg, and Ruhr-Universität Bochum, Bochum, Germany
e-mail: karl-heinz.schwalbe@web.de

G. Hütter and L. Zybell, *Recent Trends in Fracture and Damage Mechanics*, DOI 10.1007/978-3-319-21467-2_2

scientific revolutions like the change of a geocentric to a heliocentric system or from Newtonian to relativistic mechanics.

At a much smaller scale, fracture and damage mechanics required paradigm shifts as well and had to fight against deadlocked concepts of what "mechanics" was regarded to be. It additionally interfered with an obsolete (though still existing) understanding of physics as a science not only dealing with real objects but presenting "genuine reality", which antagonises with the idea that science can do nothing but develop "models" of reality.

Classical fracture mechanics is a direct application of classical continuum mechanics. But whereas the traditional science of strength of materials did not know anything about "defects", fracture mechanics came up with this unfamiliar term in order to explain and predict failure of structures which had been designed properly according to the state of engineering science as established in the 19th and beginning of 20th century. This first and foremost challenged engineers and companies to admit that their products were not "defect-free", a demand which particularly people from the nuclear industry refused to comply to, even after applying (more or less of necessity) fracture mechanics concepts.

Moreover, fracture mechanics introduced a so far unknown length parameter into structural assessment, namely the size of a presumed or existing defect, the crack length. How was it to be defined, particularly since it affected the load bearing capacity and lifetime of a structure significantly? Furthermore, new physical quantities and material parameters of seemingly obscure dimensions emerged with the new theory, which was used as argument to discredit its reputation.

First-hand examples from the personal experience and history of the authors will be highlighted not only as witty contributions to an anniversary but as "writing on the wall" dedicated to the present generation of scientists that similar unreasonableness can and will recur in presence and future.

By looking at the number of publications, it may be interesting to note that during the first two or three decades the majority of papers and hence of research work was done in the US. During the following decades a shift occurred towards Europe and Asia, where now substantially more publications than in the US have their origin.

2 Linear Elastic Fracture Mechanics (LEFM)

2.1 Fundamentals

2.1.1 Historic Development

Whereas the theoretical background of linear elastic fracture mechanics, that is the mathematical description of stress and strain fields at stress concentrators [37] and an assessment of the energy balance in cracked bodies [28], dates back a century ago, its significance for structural integrity was spotted not until some spectacular accidents had occurred and Irwin published his seminal paper [40] in the mid 20th century. Remarkably, Irwin did not only depict the relationship between the stress intensity and the energy approach but also introduced a two-parameter description of the stress field, long before similar concepts were discussed intensively within the fracture-mechanics community: "The influence of the test configuration, loads and crack length upon the stresses near an end of the crack may be expressed in terms of two parameters. One of these is an adjustable uniform stress parallel to the direction of a crack extension. … The other parameter, called the stress intensity factor, is proportional to the square root of the force tending to cause crack extension". Actually, the first parameter is Rice's T-stress [64], and the crack driving force is Griffith's strain-energy release rate [28].

It took another ten years to launch fracture mechanics in Germany with the foundation of the DVM Working Group Fracture Mechanics in 1969, eight years after a DVM meeting, where "studies on failure mechanics and fracture research had been reported to an international audience in the presence of Dr. George Irwin" [52]. By no means, fracture mechanics had been finally established, then. Acknowledged scientists and engineers fought the new concepts as non-scientific. The "father" of notch mechanics, Heinz Neuber, attacked fracture mechanics in the 3rd edition of his seminal book [58] as follows: "The present new edition provides evidence for the various deficiencies of fracture mechanics. This is primarily about the violation of the stress distribution in the vicinity of the notch or crack tip. Furthermore the lateral dimensions of the crack and the radius of surface curvature are disregarded. Moreover, all effects related to deviations from linear elasticity are ignored."[1] Likewise in the 1980s, the retired president of the Federal Institute of Materials Testing (BAM) in Berlin regarded fracture mechanics as pseudo-science, since it introduced a surface energy (Griffith's energy release rate) and a material parameter with the weird dimension of $\text{MPa}\sqrt{\text{m}}$.

These arguments may sound quite amusing today but culminated in fierce disputes not that long ago. They were not very honest, of course, as field singularities are common in physical theories from Newtonian to relativistic mechanics. Singularities arise as limit cases of mathematical equations, but nobody expects a quantity actually to become infinite, and their existence does not devaluate the significance of a theory provided the latter describes the surrounding neighbourhood correctly. The crack tip itself, $r \to 0$, is a mathematical artefact where continuum mechanics is not applicable anyway. Likewise, boundary layer theories are known from fluid mechanics, for instance, in which the thickness of the boundary layer is negligibly small but nevertheless embodies a finite energy.

Removing the "unphysical" stress singularity remained an issue, however, and resulted in Irwin's small-scale yielding approach [41], Dugdale's strip yield model [21], Fig. 1a, and Barenblatt's cohesive zone [4], Fig. 1b, ancestor of modern cohesive models (see Sect. 6.2).

[1]Translation from the preface of the German edition [57] by the present authors who made an effort to meet Neuber's particular terminology as authentically as possible.

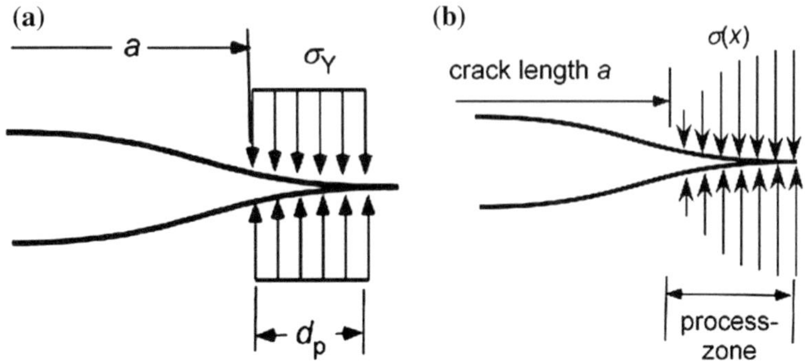

Fig. 1 Models avoiding the stress singularity at the crack tip: **a** Dugdale [21], **b** Barenblatt [4]

2.1.2 Non-singular Terms

A series expansion of the stress state near a crack tip has been presented by Williams [100] in 1957 showing that beside the terms with a $1/\sqrt{r}$ singularity there is a constant term of the normal stresses parallel to the crack face,

$$\sigma_{xx} = \frac{A_{-1}}{\sqrt{r}}\cos\frac{\theta}{2}\left(1 - \sin\frac{\theta}{2}\sin\frac{3\theta}{2}\right) - \frac{C_{-1}}{\sqrt{r}}\sin\frac{\theta}{2}\left(2 + \cos\frac{\theta}{2}\cos\frac{3\theta}{2}\right) + 2A_0, \quad (1)$$

which does not vanish for $r \to 0$. It depends on the biaxiality of the external loading. For instance, in a cracked infinite panel under biaxial loading by tensile stresses, $\sigma_{xx} = \lambda\sigma_\infty$, $\sigma_{yy} = \sigma_\infty$, this term is

$$A_0 = \sigma_\infty(1 - \lambda). \quad (2)$$

It becomes a maximum for $\lambda = 0$ and vanishes for $\lambda = 1$. Since the singularity of stresses appeared so dominant, the constant term had been neglected and forgotten for a long time until the "geometry dependence" of fracture parameters alerted the community.

Larsson and Carlsson [49] investigated the influence of non-singular stress terms and specimen geometry on small scale yielding at crack-tips in elastic-plastic materials. They found that the size and shape of the plastic zone, which is important for the definition of valid K_{Ic} values, was significantly affected by the biaxiality of loading and the specimen geometry.

On this background, Rice [64] introduced the T-stress,

$$\sigma_{ij}(r,\theta) = \frac{K_I}{\sqrt{2\pi r}}f_{ij}(\theta) + T\,\delta_{1i}\delta_{1j}, \quad (3)$$

which characterises the “inherent stress biaxiality in fracture specimens” [50] in small scale yielding and became the starting point for all considerations on a “second parameter” in fracture mechanics affecting the fracture behaviour [6, 20].

2.2 Crack Extension by Fatigue

In the 1960s the fast growing aerospace industries faced challenges towards reducing structural mass and increasing reliability, thus calling for sound methods and codes for quantifying the reliability of their products. The major item to be dealt with was—and still is—structural fatigue. It was in that decade when pioneers such as Paul C. Paris, Richard Hertzberg, Art J. McEvily, Jaap Schijve and many others laid the ground for quantifying life expectancy of structural components containing already some fatigue damage in the form of cracks.

A major break-through was achieved by Paris and Erdogan [62] who correlated the rate of crack extension, *da/dN* with the cyclic stress intensity factor, ΔK. It turned out that there is a power law relationship between these two parameters:

$$da/dN = C\Delta K^m, \quad (4)$$

where the exponent, m, for many metallic materials is typically between 2 and 4. However, when experiments were conducted at very low and very high values of ΔK, researchers found that an S-shaped curve resulted, where Eq. (4) is valid only in the intermediate section of the curve, Fig. 2a. It is worth noting that the fractographical appearance of a fatigue fracture surface changes with the rate of crack propagation, thus demonstrating that different mechanisms are at work, Fig. 2b.

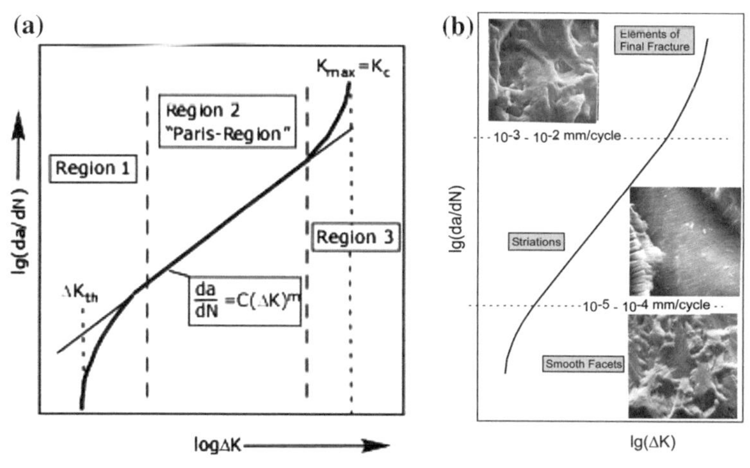

Fig. 2 Fatigue crack extension diagrams: **a** Schematic **b** Fractographic features of the aluminium alloy AlZnMgCu0.5 F46 in correlation with the crack propagation curve, after Schwalbe [73]

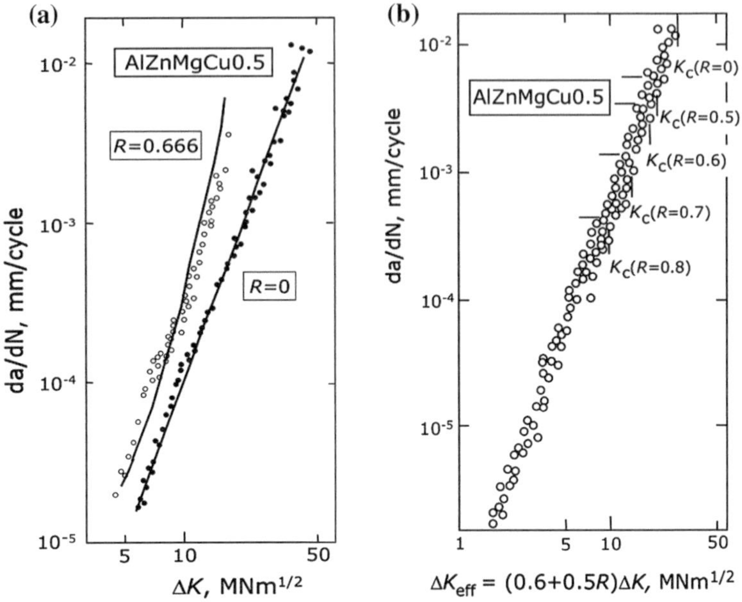

Fig. 3 Fatigue crack propagation in AlZnMgCu0.5 for two stress ratios, Schwalbe [73]: **a** plotted versus ΔK, **b** plotted versus ΔK_{eff}

Furthermore, the unique correlation with ΔK was soon questioned when it was found that crack propagation was also dependent on the R-ratio of the applied stress. A large number of equations have been developed to describe the S-shape and the stress ratio effect. With the following two modifications the S-shape can be modelled:

$$\frac{da}{dN} = \frac{C(\varDelta K - \varDelta K_{\text{th}})^m}{(1 - R)(K_{\text{c}} - \varDelta K)}. \tag{5}$$

However, this dilemma was soon solved by Elber [22] who found that the R-ratio effect was due to partial crack closure during unloading the specimen. If this effect is quantitatively included in the horizontal axis—by using only that cyclic stress intensity factor, $\Delta K_{eff,}$ describing the load range during which the crack is open, and thus effective for stresses and strains at the crack tip—then the stress ratio effect disappears in the graph. This effect is demonstrated for an aluminium alloy in Fig. 3.

With a simple analytical model, the crack extension rate can be estimated, Schwalbe [72]. It is based on the assumption that under each load cycle a crack extends by that amount which corresponds to the distance from the crack tip where the true fracture strain is reached

$$da/dN = \frac{(1 - 2\nu)^2}{4\pi\sigma_Y^2(1 + n)} \left[\frac{2\sigma_Y}{E\varepsilon_f}\right]^{1+n} \Delta K^2 \tag{6}$$

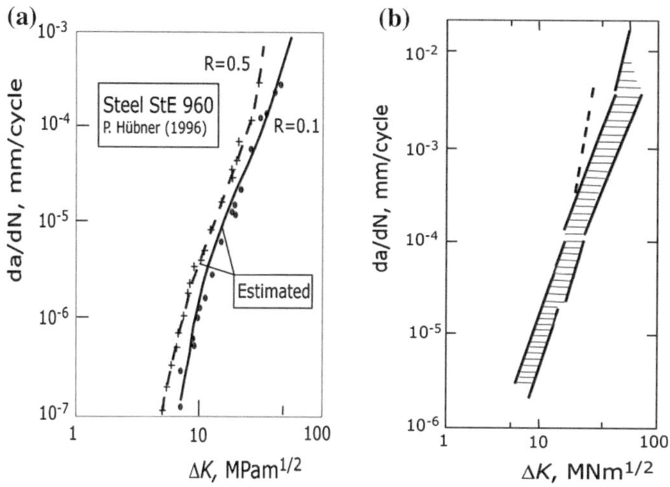

Fig. 4 Fatigue crack propagation diagrams: **a** Experimentally determined crack propagation rates of a high strength steel at two *R*-ratios with Eq. (6), Hübner [35]; **b** Variety of structural steels, with yield strengths between 250 and 1660 MPa, Schwalbe [73]

where σ_Y is yield strength, *n* strain hardening exponent, *E* modulus of elasticity, ν Poisson's number and ε_f the true fracture strain. Several comparisons with experiments on various metallic materials yielded very good results. An example is shown in Fig. 4a where Hübner [35] compares his experiments on a high strength steel tested at two *R*-ratios.

Interestingly, as opposed to popular opinion, the material's strength has no substantial effect on the propagation rate. Figure 4b shows a compilation of data obtained on 17 steels with a wide range of yield strengths. The only outlier is a very brittle steel (broken line) with low fracture toughness.

It may be worth noting that there was substantial resistance in the service load fatigue community against using fracture mechanics. The reason was that there were two communities—fracture mechanics and fatigue—with little contact with each other. Fracture mechanics was still looked at with reservation as something esoterical.

However, the activities in the application of fracture mechanics methods to the behaviour of cracks under cyclic loading—primarily in the US—gave rise to the quantitative predictability of the fatigue life of structural components by simply integrating the crack extension equation. Prerequisite is either the existence of a crack found by inspection or of an assumed crack. Due to the pioneering developments at NASA, such predictions have even become possible for variable amplitude loading, Newman [59]. This has been the break-through for general application of fracture mechanics to fatigue problems: It is now an unquestioned standard tool for qualifying structures with crack-like flaws, being used world-wide.

3 Elastic-Plastic Fracture Mechanics (EPFM)

LEFM based structural assessment finally was accepted and required for the life assessment of "high-risk" structures, particularly in the aerospace industry and in nuclear engineering.[2] It became increasingly obsolete, however, with the increasing ductility of structural steels, for example in the nuclear industry which became a major driving force in the development of EPFM. "Valid" data for "plane-strain fracture toughness" [1] at room temperature require specimen dimensions at the meter scale, which are often beyond structural dimensions and raise the question of the significance of the data. Structural engineers were encountered by a new demand: plasticity.

The advantage of LEFM is that due to the linearity of the constitutive equations and backed by the assumption of small strains, closed form solutions for stress and strain fields at a crack tip could be obtained. This is generally impossible in incremental plasticity, as the constitutive equations are not only non-linear but the current stress-strain state depends on the loading history. People tend to keep to the well known, and Griffith's concept of an (elastic) strain-energy release rate had become familiar by now. Its mathematical equivalent was a "path-independent integral", the J-integral, which independently Cherepanov [17] and Rice [63] introduced. Its significance as an intensity parameter of the crack-tip fields was demonstrated by Hutchinson [36] and Rice and Rosengren [67], hence named HRR singularity, and formulas for its experimental determination as energy release rate were provided by Rice et al. [66].

Thus, a perfect analogy to LEFM had been finally established, and those disliking integrals could recall its physical property as a plastic energy release rate. Aside from this, it was and still is accepted in some sectors of industry to convert J-values to K-values by $K_J = \sqrt{JE'}$ as in elasticity even under large scale plastic conditions. Specimen size conditions for measuring valid J_{Ic} data [2] are much less restrictive than for K_{Ic} [1].

One basic limitation of this theory should be kept in mind, namely the underlying assumption of "deformation theory of plasticity", which is a theory of hyper-elasticity, rather, requiring the existence of a strain-energy density as potential of stresses,

$$\sigma_{ij} = \frac{\partial w}{\partial \varepsilon_{ij}} \quad \text{with} \quad w = \int_0^t \sigma_{ij} \dot{\varepsilon}_{ij} \mathrm{d}t. \tag{7}$$

This assumption forbids not only global unloading of a structure but also any local re-arrangement of stresses due to yielding. Numerous numerical analyses based on incremental plasticity and large strains, e.g. [12, 54], have shown that

[2]Where it has later been replaced by the J-integral, see below, or by K_J-values calculated from J according to LEFM.

under plane-strain conditions and in sufficiently thick 3D structures the stress states are approximately J-dominated as long as the crack does not extend, and criteria for J-dominance [55, 86] were established.

3.1 Crack Extension: JR-Curves

Since the J-concept worked satisfactorily for predicting crack initiation, at least in thick-walled components, its limitation that any crack growth causes local stress redistributions which violate Eq. (7) sank into oblivion. Regardless of this, the fracture mechanics community started extending the concept to growing cracks and developed the concept of resistance curves [2] in terms of $J(\Delta a)$.

What followed was, adopting Kuhn's conception [46], a period of "normal science", when scientists attempt to enlarge the central paradigm by "puzzle-solving", which is extremely productive. Journals and conferences were flooded with R-curves. Anomalies were found: R-curves depending on specimen size and shape [26], though by definition, a material parameter must be geometry independent. These findings further increased the number of measured R-curves. It was not even clear whether all of them were measured and evaluated correctly, which opened up the chance of intensive discussions whether or not the measured effects were "real". Evaluation formulas for J were actually still controversial beyond the turn of the millennium [8].

During a period of normal science, the failure of a result to conform to the paradigm is not seen as refuting the paradigm, but as the mistake of the researcher. When the present author published a numerical study [15] on the path dependence of J for large crack extension in 1989, showing that J became zero at the crack tip, $r \to 0$, thus questioning the significance of J as a parameter governing crack growth, he was blamed for false FE analyses. However, Rice et al. [65] had shown in 1980 already, that the singularity of the strain energy density at a moving crack is $\ln(r^{-1})$ in incremental plasticity, whereas in order that J remains finite for $r \to 0$ it has to be an r^{-1} singularity as in elasticity and "deformation theory" of plasticity. The same authors also proposed a "far-field" value, J_{f}, which is different from the deformation theory value of J.

Likewise, geometry dependent R-curves were attributed to faulty testing. The necessity of introducing a "second parameter" characterising the "constraint" of a structure was fiercely fought by the US apologists of a "one parameter characterisation" [86] until finally one of them came up with his own "two-parameter approach" [60, 61]

$$\sigma_{ij}(r,\theta) = \sigma_{ij}^{\mathrm{HRR}}(\theta,n) + Q\sigma_0\delta_{ij} \quad \text{for} \quad |\theta| < \frac{\pi}{2}, \tag{8}$$

mimicking Eq. (3) and unobtrusively ignoring other people's prior suggestions of triaxiality parameters based on the hydrostatic stress [11, 13]. Q is no constant second term of an analytical series expansion like Rice's T-stress, however, but a phenomenological approximation of several higher order terms [85].

Two ASTM conferences on "Constraint Effects" were held in the United States in 1993 and 1995, and in Germany, a Priority Programme (*Schwerpunktprogramm*) on "Ductile Fracture Mechanics" (*Fließbruchmechanik*) was funded by the German Research Foundation (DFG) from 1989–1996 with a budget of 12.8 million DM. The existence of "constraint effects" had been accepted, finally, and the "conventional" along with the "anomalous" results, i.e. the geometry dependent R-curves, were subsumed into one framework. It remained a phenomenological patchwork after all, as a global quantity, J, was combined with a local field parameter, the crack-tip triaxiality, which was not even uniquely defined, and any evident and physical background was lacking.

From today's point of view, the fracture-mechanics community missed two essential points:

- The cumulative quantity J, which increases with crack length, ceases to be an energy-release rate, as soon as the crack starts extending, and an incremental quantity would be required instead, as Turner [94] pointed out. What has been understood to be an extension of Griffith's theory was not, in the end.
- Although physically meaningful models of the failure processes occurring at the crack tip were available [68, 93], the purely phenomenological J concept did not consider them. Respective local models came up as the "local approach" [69], "micromechanical models" [92] and "damage mechanics" [10, 56] but were eyed with distrust by the J-community, in the beginning.

3.2 Energy Dissipation Rate

In the middle of the discussions on the geometry dependence of J_R curves, Turner [94] introduced the energy dissipation rate as an alternative measure of tearing resistance,

$$R = \frac{\partial U_{\text{dis}}}{B\partial a} = \frac{\partial W_{\text{ext}}}{B\partial a} - \frac{\partial U_{\text{el}}}{B\partial a}, \tag{9}$$

where W_{ext} is the external work and U_{el} the (recoverable) elastic strain energy. This definition is a straight transfer of Griffith's elastic energy release rate [28] to plastic processes which is consistent with the incremental theory of plasticity. The dissipation rate has the same dimension as J and characterises the increment of irreversible work per incremental crack extension, da. It falls with increasing crack length in gross plasticity and consists of two contributions, namely work of remote plastic deformation and local work of separation,

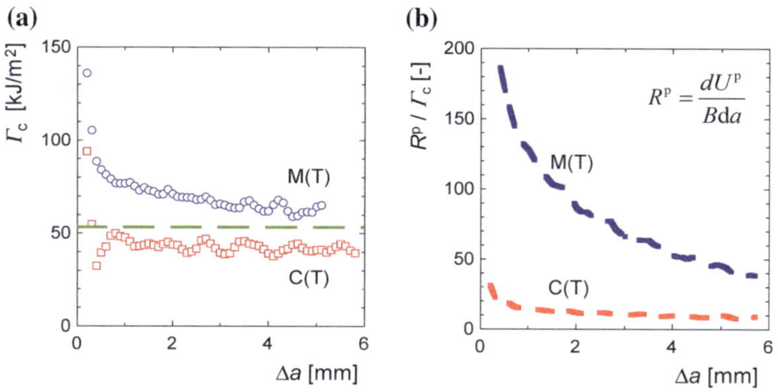

Fig. 5 Energy dissipation rates in M(T) and C(T) specimens calculated from FE analyses with the GTN model [89]: **a** local work of separation, **b** (normalized) remote plastic work

$$R = \frac{\partial U_{\text{pl}}}{B\partial a} + \frac{\partial U_{\text{sep}}}{B\partial a} = R_{\text{pl}} + \Gamma_{\text{c}}. \tag{10}$$

A simple energy balance put the dilemma of R-curves straight: What people measure as "fracture resistance" results to a great deal from remote plasticity and not from local material separation [95], and the problem of geometry dependence is hence inherent and unsolvable. Turner's approach was enlightening but showed no way out. There was no possibility based on continuum mechanics to split the two contributions in Eq. (10), and attempts to establish "an alternative view of R-curve testing" [90] based on the dissipation rate did not become accepted. Only understanding of the energy dissipation mechanisms in the process zone at the crack tip is precisely what is necessary to identify "fracture toughness" as a material property.

A numerical analysis of crack extension in a C(T) and an M(T) specimen with the model of Gurson et al. [29, 56] brought additional quantitative evidence [89]. There is a minor geometry effect on the local work of separation, Γ_c, Fig. 5a, but a major effect on the global plastic work, R_{pl}, Fig. 5b. Note that R_{pl} is normalized by Γ_c.

Regardless of Turner's arguments, J_R-curve testing continued, and damage mechanics developed in parallel, suspiciously eyeballed by the mainstream fracture community.

3.3 The CTOD Concept

Historically the first elastic-plastic fracture mechanics concept was developed in the 1960s at the Welding Institute, Cambridge, U.K. Wells [99] started with the idea that the spot where fracture initiates—the crack tip—should be looked at, and that the deformation there at the moment of fracture should be taken as a property characteristic of the material tested. In the experimental method developed at The

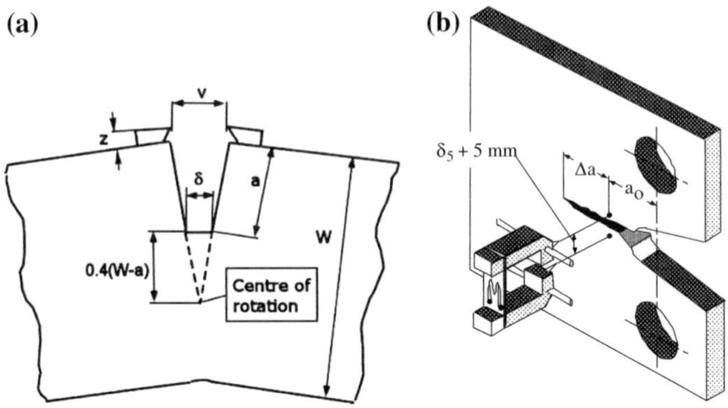

Fig. 6 Crack tip opening displacement (CTOD) **a** Determination according to BS 5762 [16]; **b** δ_5 according to Schwalbe [74]

Welding Institute (TWI) the crack tip opening displacement (CTOD), δ, is determined indirectly from measurements of the displacement, v, at the specimen's front face, Fig. 6a,

$$\delta = \frac{K^2(1-\nu)}{2\sigma_Y E} + \frac{0.4(W-a)}{0.4W + 0.6a + z} v_{pl}, \tag{11}$$

where v_{pl} is the plastic part of v.

The CTOD test method became the British Standard BS 5762 in 1979 and was later also integrated in ISO 12135 [38] and ESIS [23] methods.

Another method for determining the crack tip opening displacement was developed in the authors' group, the δ_5 method. It is particularly suited for determining the crack extension properties of a material and is measured at the specimen's side face, Fig. 6b, Schwalbe [75] and Schwalbe et al. [82]. This experimental technique found also its way into standards: ISO 22889 [39] and ASTM E 2472 [3].

A further application of the δ_5 technique has been demonstrated by Hellmann and Schwalbe [33], where it was shown that both definitions of the CTOD yield practically identical results. The main difference between both method consists in the applicability to test piece geometry: Whereas the BS 5762 technique can only be used with C(T) and SE(B) specimens, the δ_5 technique can be applied to any geometry with a surface breaking crack, including structural components.

It turned out that the δ_5 method is particularly suited for testing and analysing thin sections, and it is for these cases that criteria for the validity of CTOD crack extension resistance curves have been established, Heerens and Schödel [32]. Such a curve is independent of the specimen's width, W, if

- $\Delta a \leq 0.25(W - a_0)$ for a C(T) specimen,
- $\Delta a \leq W - a_0 - 4B$ for an M(T) specimen, $W - a_0 > 4B$,

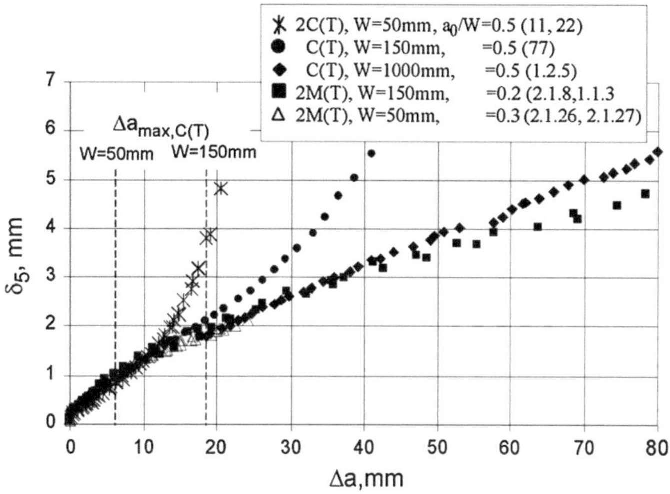

Fig. 7 δ_5 R-curves obtained on M(T) and C(T) test pieces with different width dimensions, Heerens and Schödel [32]

where B is the thickness of the specimen, ISO 22889 [39]. An example is shown in Fig. 7. It is clearly seen that at an amount of 25 % of the original net section width, $W - a_0$, the R-curves become width dependent, hence, are no longer valid. In contrast to this finding, an M(T) specimen exhibits a much longer curve. The δ_5 R-curve method is particularly suited for thin-walled materials used in light-weight structures.

At very high loads, leading to full plasticity of the remaining net section, the cyclic CTOD, $\Delta\delta_5$, can successfully be used to correlate the fatigue crack propagation rate, da/dN, Hellman and Schwalbe [34].

The two elastic-plastic fracture mechanics concepts are not independent of each other, they are compatible. Otherwise, at least one of these concepts would be wrong. A very simple correlation is

- $J \approx \delta\sigma_Y$ for plane stress,
- $J \approx 2\delta\sigma_Y$ for plane strain,

A more rigorous correlation is given in Sect. 5 on assessment procedures.

Interestingly, a further fracture parameter, the crack tip opening angle, CTOA, whose theoretical background can be found in the analysis of the near-tip field at a growing crack by Rice et al. [65], is practically identical with the stabilised slope of the δ_5 R-curve, Heerens and Schödel [32]. Its direct measurement on a specimen surface causes difficulties insofar as the tip of the crack is often not detectable due to the amount of plastic deformation causing a blackish area which obscures the crack

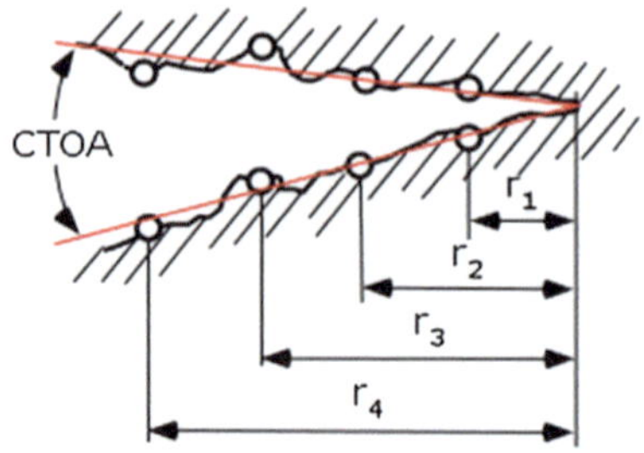

Fig. 8 Determination of the crack tip opening angle using four pairs of reference points, Schödel [71]

tip. A solution was found by Schödel in his PhD Thesis [71], defining a few pairs of reference points along the crack contour which allow easy determination of the CTOA, Fig. 8. The CTOA test method is included in the test standard ISO 22889 [39].

4 Test Procedures

4.1 Some Test Techniques

The new concept of fracture mechanics, of course, required adequate test techniques to take account of the effect of a crack on material properties. The explosion of a Polaris rocket in the US due to brittle fracture gave rise to the foundation of the first committee dealing with fracture mechanics, the ASTM Committee E 24. Now it is Committee E 8. This committee created in the 1960s the first test standard, E 399 [1], which has become a world-wide used method. This method deals with materials and test conditions for linear elastic behaviour and plane strain conditions. The resulting material parameter was dubbed K_{Ic}. This test method has been in use with only minor modifications until now. Key researchers involved in this area were John Srawley, Bill Brown, Gil Kaufman and John Shannon.

In the beginning, all kinds of fracture mechanics tests were very unpopular in industry because of their high cost. Much time and money go into the pre-cracking procedure, making fracture mechanics testing expensive.

Fracture mechanics tests owe a remarkable attribute: the values determined are dependent on the size and geometry of the test piece; however, in the regime of linear elastic behaviour, beyond a certain size plain strain conditions prevail and the value of the fracture property remains fairly constant. And beyond that size, the values are regarded as "valid". As a consequence, the "validity", in other words, size independence, of such values can only be checked after the test has been done. This is unique to the world of fracture mechanics and can make the determination of useful parameters quite cumbersome.

The minimum size requirement for achieving a size independent value is given by

$$B, (W - a) \geq 2.5 \left(\frac{K_{\text{Ic}}}{\sigma_{\text{Y}}} \right)^2, \tag{12}$$

where W designates the width of the specimen, B the thickness and a the crack length. Usually, materials with relatively moderate strength levels have high toughness and therefore require enormous specimen sizes and hence very big and expensive test equipment. This is e.g. the area of pressure vessels where materials with high toughness are preferred. Consequently, German institutions, e.g. the *Bundesanstalt für Materialprüfung*, *University of Stuttgart* and also the *University of Aachen* installed test equipment for performing tests on large-scale specimens [48].

This dilemma was solved when the emerging nuclear industry in the U.S. was in need of information about the fracture properties of their high toughness materials. This led to the developments of elastic-plastic methods as described in Sect. 3.

The test alone is only a part of the problem when fracture mechanics is to be applied to an actual structural component. It had to be demonstrated that the component would fail under the same conditions as were present in the test. Therefore, in the earlier years of fracture mechanics, huge components such as thick-walled pressure vessels were tested, requiring appropriate budgets.

The development of elastic-plastic fracture mechanics and the utmost exploitation of the structural mass and the increasing necessity to define safety margins brought along the need to know quantitatively the material resistance beyond the initiation of crack extension. To his end, appropriate experimental techniques for the determination of the most important quantity—the increasing crack size during the test—had to be developed.

After having played around unsuccessfully with standard ultrasonic equipment, the very first usable technique was developed by the Westinghouse research group in Monroeville near Pittsburgh, PA. It was dubbed the "unloading compliance method" since it consists of periodic unloading during the loading path of the specimen. The unloading traces are elastic, and their slope is a measure of the actual crack size. Garth Clarke from the Westinghouse group worked on the computerisation of this method [18]. It is the first method that was ever standardised.

However, the first technical realisation of this method had a problem: In the initial part of the loading the crack seemed to decrease its length because the unloading slopes got steeper than the initial loading slope of the specimen. It turned out that the friction between the loading bolts and the loading holes of the C(T) specimen prevented proper unloading. The solution was the introduction of flat-bottomed holes in the loading clevises. Still the same artefact of "negative crack growth" can be found in rather recent publications.

The second technique—the electrical potential drop method—exploits the effect of crack size on the electrical resistance of specimens made of metallic materials. AC and DC methods were developed. Its standardised form which makes use of direct current was developed in the authors' research group, Schwalbe and Hellmann [79]. It is based on the disturbance of a DC electrical field by the presence of a crack. Johnson [42] developed a closed form relationship between the crack size, a, and the potential drop measured across two well defined points. Solving that equation for the crack length yields [79].

$$a = \frac{2W}{\pi}\cos^{-1}\left(\frac{\cosh\left(\frac{\pi y}{2W}\right)}{\cosh\left[\left(\frac{\varphi}{\varphi_0}\right)\cosh^{-1}\left\{\frac{\cosh\left(\frac{\pi y}{2W}\right)}{\cos\left(\frac{\pi a_0}{2W}\right)}\right\}\right]}\right), \tag{13}$$

Here

$2y$ signifies the distance between the potential pick-up points,
φ_0 is the potential related to the starting crack length, a_0,
φ is the current potential drop related to the increasing crack length.

Although this equation was derived for an M(T) specimen geometry it was demonstrated that it is applicable to C(T) and SE(B) geometries as well, whereas compliance based techniques have to be calibrated for each specimen geometry. On the other hand, it can be used for metallic materials only.

Initially, arguments were raised against this method such that it should depend on the material and on the test temperature due to the variation of resistivity. However, it is the normalisation of the current potential by the starting value: φ/φ_0 ruling out these arguments. This technique is being used in a number of test methods.

4.2 Harmonisation of Test Procedures

Whereas in the beginning of standardisation of fracture mechanics tests, a method for each fracture parameter, K, J, CTOD and for plane strain and plane stress was developed, over the time a harmonisation was attempted. The idea was that a single method should be sufficient for all fracture parameters and that the specimen response should tell whether the evaluation is to be done according to either linear elastic or elastic-plastic procedures. Furthermore, the test interpretation either in terms of J or CTOD should be given in the test method to be developed.

A first attempt was undertaken by the European Group on Fracture (EGF) by creating its first test method, EGF P1-87D, Schwalbe et al. [81]. This procedure describes the determination of crack extension resistance curves in terms of J and δ where δ had to be determined according to BS 5762 [16]. It allows also the determination of J and δ at initiation of stable crack extension. Later, this procedure was superseded by ESIS P2-92 [23], which includes the stress intensity factor as a fracture parameter and an additional method for determining initiation of stable crack extension using measurement of the critical stretch zone width as outlined by Heerens et el. [31]. A further initiative by ISO led to the standard ISO 12135 [38] very similar to ESIS P2-92 [23].

The handbook EFAM GTP02 [78] developed at GKSS Research Centre is probably the most comprehensive procedure. It is based on ESIS P2-92 and includes additional features such as M(T) specimens, the δ_5 technique, determination of K-based fracture parameters, crack tip opening angle, rate of dissipated energy, testing of weldments, statistical treatment of scatter and special validity criteria for tests on specimen with low constraint. In addition to material parameters related to stable crack extension, the determination of parameters for unstable fracture is also described.

5 Assessment Procedures

A number of engineering assessment methods have been developed in different institutions from various countries. Due to the number of procedures and the complexity of the subject, only a very coarse overview can be given here.

Methods for assessing the integrity of structures with crack-like defects have to be standardised in order to make assessments independent of individual methods and persons using them.

Industry specific methods such as for pipelines, pressure vessels and aircrafts have been developed, however, basically they all have to satisfy the inequality

$$\text{Crack Driving Force} < \text{Material Resistance}.$$

The difference lies in the details used for expressing either side of this inequality. The left hand side poses most of the problems, e.g. using finite element analysis or one of several analytical expressions, whereas on the right hand side it has to be decided which material properties have to be used and how this has to be done.

The most important information from such an assessment are maximum load a component is able to carry, critical crack size and residual life time. From this information inspection intervals can be quantified in order to enable the operator to find a crack before it becomes critical.

As long as the structural behaviour can be characterised by the framework of LEFM, the problem of the left hand side, as given by $K = \sigma(\pi a)^{0.5}Y(a/W)$, is relatively easy to solve. The stress, σ, acting in the structural cross section containing the crack can be determined by linear-elastic stress analysis. The dimensionless function, $Y(a/W)$, depending on geometrical and loading conditions, can be found in handbooks for numerous cases. Otherwise, modern computational methods such as finite element analyses make it possible to generate solutions for unusual problems.

When it comes, however, to elastic-plastic conditions in the cross section under consideration, then J or δ have to be determined using elastic-plastic analyses. In order to facilitate the task of determining the elastic-plastic crack driving force parameter, several "engineering" assessment schemes have been developed, the first one again at TWI, Harrison et al. [30]. They called it the *Design Curve*. In this method, the CTOD is expressed in normalised form such that for two degrees of yielding the normalised critical CTOD, $\bar{\delta}_c$, is given by

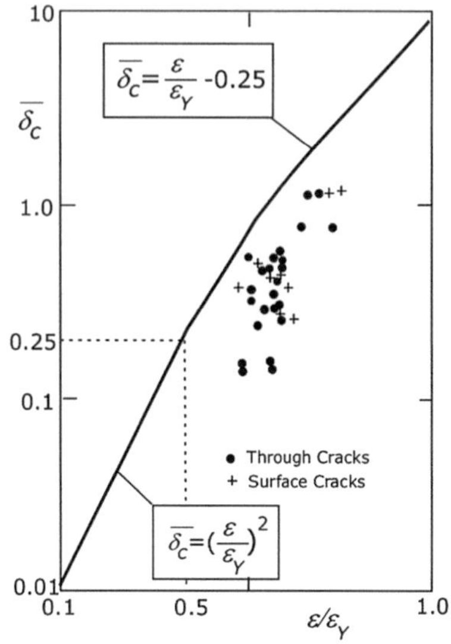

Fig. 9 Experimental data compared with the Design Curve, data from [30]

$$\begin{aligned} \bar{\delta}_c &= \frac{\delta_c}{2\pi\varepsilon_Y a} = \left(\frac{\varepsilon}{\varepsilon_Y}\right)^2 && \text{for } \frac{\varepsilon}{\varepsilon_Y} < 0.5 \\ \bar{\delta}_c &= \frac{\delta_c}{2\pi\varepsilon_Y a} = \left(\frac{\varepsilon}{\varepsilon_Y}\right) - 0.25 && \text{for } \frac{\varepsilon}{\varepsilon_Y} > 0.5 \end{aligned} \quad . \tag{14}$$

These equations are supposed to be used for short cracks where the CTOD is proportional to the local strain. The advantage is that no solution for the stress intensity factor is needed. Similar equations were derived for J as a driving force parameter. Design curve approaches do not attempt to provide more or less accurate failure conditions; they rather give the user conditions which are supposed to be "safe". The example shown in Fig. 9 demonstrates that all experimental values are below the Design Curve and can hence be regarded as safe.

A further method, the "*Failure Assessment Diagram*" (FAD) was developed in the U.K. by the then nationalised Central Electricity Generating Board (CEGB).

In its early version, a simple curve was proposed which—similarly to the COD Design Curve—divides an area of safety and one of unsafe conditions; that is to say, this curve is not a curve for predicting failure. Its vertical axis is given by the applied stress intensity factor, K, normalised by the critical value, K_c: $K_r = K/K_c$. The horizontal axis is again a normalised quantity, a quantity representing the applied load related to a plastic limit load: $S_r = F/F_L$.

The FAD line was formulated as

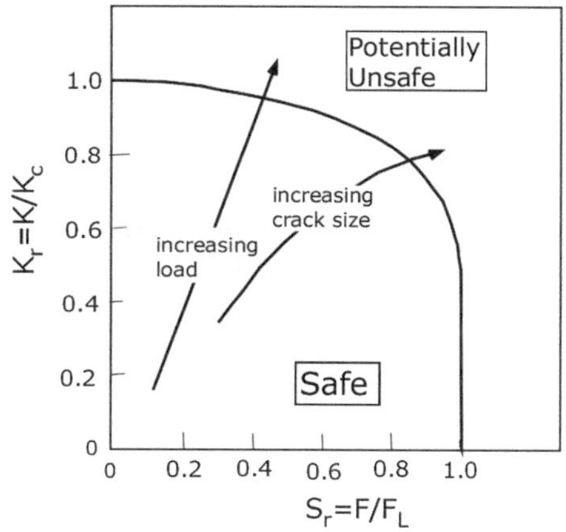

Fig. 10 Failure Assessment Diagram, as cited by [102]

$$K_r = S_r \left[\left(\frac{8}{\pi^2} \right) \ln \sec \left(\frac{\pi S_r}{2} \right) \right]^{-1/2}, \tag{15}$$

see Fig. 10. The square bracket is almost identical to the expression for the crack tip opening displacement based on the Dugdale model [21] shown in Fig. 1a. This expression—although frequently related to Dugdale—in fact was derived by Goodier and Field [27].

Several additional formulations followed to adjust the FAD to the development in fracture mechanics, see e.g. [103].

In the US, the following developments can be observed: In the 1970s/1980s, driven by the needs of the developing nuclear industry and by the contemporary development of the *J*-integral theory, an attempt was initiated to develop a hand book for *J* as a driving force parameter, similar to the stress intensity factor handbooks. This way the *EPRI*[3]*Handbook* emerged [47, 87].

It is based on partitioning the *J* integral into an elastic and a plastic component:

$$J = J_{el} + J_{pl}, \tag{16}$$

with

$$J_{el} = \frac{K_{eff}^2}{E}, \tag{17}$$

$J_{pl} = 0$ for contained yielding conditions ($F \leq F_Y$), where F_Y is the applied force at the attainment of net section yielding,
and

[3]EPRI is the acronym for Electric Power Research Institute which is financed by American power generating companies.

$$J_{\rm pl} = \alpha g_1(a/W,n)\left(\frac{F}{F_{\rm Y}}\right)^{(1+n)/n} \tag{18}$$

for fully plastic conditions ($F \geq F_Y$), where α is the coefficient and n the strain hardening exponent, respectively in the Ramberg-Osgood strain hardening law, and $g_1(a/W, n)$ is a function providing the effects of the geometry of the component and of strain hardening. Functions $g_1(a/W, n)$ have been determined for a number of configurations from finite element analyses.

Whereas the stress intensity factor is only a function of geometry variables, elastic-plastic parameters such as J are also dependent on the deformation properties of the material considered. Although the originators of the handbook simplified the procedure by the above mentioned method, the enormous efforts needed for a suitable handbook led then to a stop, simply because of the parameter explosion following from the interaction of geometrical parameters with the deformation properties of the material. Nevertheless, the handbook has set a landmark, and its way to develop J-expressions has influenced other authors.

It should be kept in mind that the EPRI Handbook provides only the left hand side of the assessment equation, whereas the COD Design Curve and FAD methods deal with both sides of the equation.

The *Engineering Failure Assessment Method* (EFAM) [76] is the only comprehensive assessment scheme which includes both the experimental determination of the relevant material properties including corrosive environments, strength mismatched welded joints and high temperature behaviour, and driving force estimating schemes, including creep conditions and mismatched welded joints. All elements are written in procedural form. The formal document for this assessment scheme can be found under [74].

The *Engineering Treatment Model* (ETM) [83] is one out of six elements of the EFAM. It describes the determination of driving force parameters in analytical form. The detailed procedure is given in [77].

Below the yield load, when $F < F_{\rm Y}$, the crack driving force expression for CTOD is

$$\delta_5 = \frac{\beta_1}{E}K + \frac{1}{mE\sigma_{\rm Y}}\frac{F}{F_{\rm Y}}K_{\rm eff}^2. \tag{19}$$

For fully plastic conditions ($F \geq F_{\rm Y}$), the ETM driving force formulations is obtained by transferring the stress—strain curve of the material under consideration to the yielding net section. For this purpose, the stress strain curve is represented by a power law for stresses beyond the yield strength, $\sigma_{\rm Y}$

$$\frac{\sigma}{\sigma_{\rm Y}} = \left(\frac{\varepsilon}{\varepsilon_{\rm Y}}\right)^N, \tag{20}$$

where $0 < N < 1$.

The resulting ETM equation reads

$$\frac{\delta_5}{\delta_{5\mathrm{Y}}} = \left(\frac{F}{F_{\mathrm{Y}}}\right)^{\frac{1}{N}} = \left(\frac{J}{J_{\mathrm{Y}}}\right)^{\frac{1}{1+N}}. \tag{21}$$

This expression shows a clear correlation between δ_5 and J. It can be easily used for sensitivity analyses, because the analytical expression shows clearly how the various parameters affect the result. The relationship in Eq. (21) has been verified in numerous experimental and numerical investigations.

For very short cracks, i.e. for $a \ll W$,

$$\frac{\delta_5}{\delta_{5\mathrm{Y}}} = \frac{\varepsilon_{\mathrm{a}}}{\varepsilon_{\mathrm{Y}}}. \tag{22}$$

δ_5 is directly proportional to the applied strain, ε_{a}

An application of the ETM to mixed mode cracks was reported by Dalle Donne and Döker [19], Fig. 11. For their tests on cruciform specimens with mixed-mode loading a modified δ_5 technique was developed, namely a vector defined by the Mode I and Mode II components. More validations see e.g. in [84].

The ETM was extended for analysing yield-strength mismatched welded joints [80, 102]. To this end, a welded joint is characterised by the base metal, BM, and a strip of weld metal, WM, between two pieces of base metal. Thus, a bi-materiel model emerged where both components are given piece-wise power law

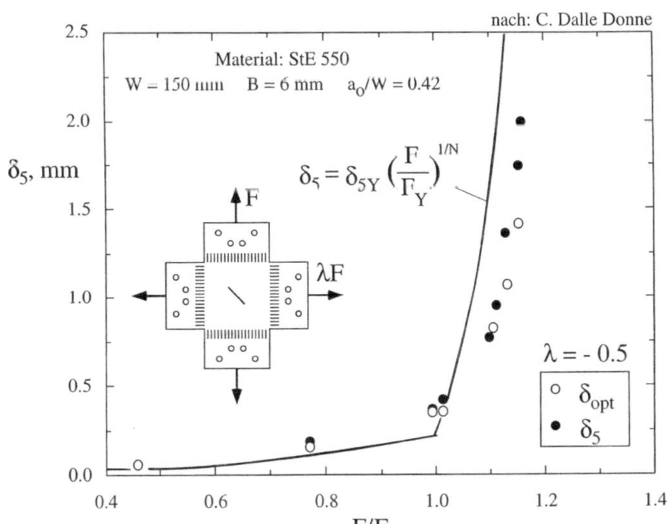

Fig. 11 Experimental data obtained for a crack under mixed-mode loading in comparison with the ETM prediction [19]

deformation behaviour. The model is for the δ_5 route only. The δ_5 driving force is given by the strain applied to the base metal. The procedure is quite complex, nevertheless, all conditions can be described with analytical expressions. For the details, the document EFAM ETM-MM 96 [80] should be consulted which includes also yield load solutions for some standard cases.

In the framework of a European Brite-Euram project, the "Structural Integrity Assessment Procedure for European Industry (SINTAP)" was developed. Seventeen institutions from nine European countries contributed to this project [103] SINTAP offers the FAD routine as well as the crack driving force (CDF) routines which deliver identical results; it is a matter of personal preference which method is going to be used. And also, the user has the option to use the J-integral or the crack tip opening displacement.

The principle of the FAD routine is shown in Fig. 10, however, the ordinate is based on J or δ instead of K. Several levels of analysis are offered [103]:

- The *Basic Option* requires only two basic material properties, namely the yield strength and the Charpy energy.
- The *Standard Option* needs the fracture toughness and yield and tensile strength. It distinguishes between materials with and without Lüders plateau. Elements of the ETM and of the British R6 procedure are included.
- The *Mismatch Option* deals with strength mismatched welded joints are treated, again based on R6 and the ETM.
- The *Stress-Strain Defined Option* requires the complete stress-strain curve and the fracture toughness of the material. Strength mismatched situations can also be analysed.
- The *J-Integral Analysis* includes also the use of the CTOD and is based on FE analyses of J and δ.
- Finally, the *Constraint Option* deals specifically with low-constraint cases with special reference to the δ_5 technique.

A second European project, the *Fitness-for Service Network* (FITNET) comprised about 50 organisations. It includes several modules: a fracture module based on the above mentioned SINTAP options and modules for fatigue crack extension, fatigue life, corrosion and creep [103].

6 Models of the Process Zone

The idea of a "process zone" at the crack tip, where material degradation and separation occur, is quite old and does not come into conflict with classical fracture mechanics. The concept is that a continuum field exists around this zone which may be K- or J-dominated. Dugdale [21] assumed that the stresses cannot exceed the yield strength, σ_0, at the crack tip, see Fig. 1a. Barenblatt [4] considered a zone of material degradation, where the stresses $\sigma(x)$ become zero at the crack tip, see

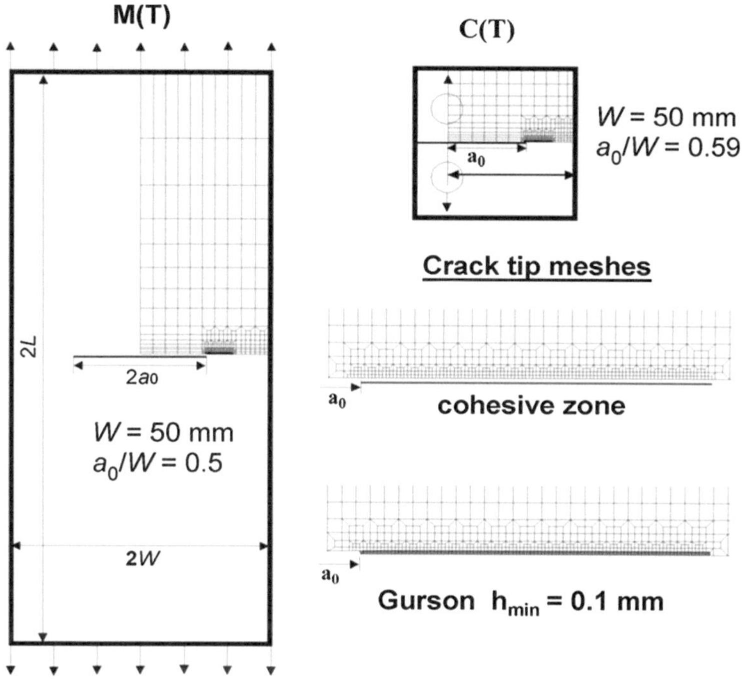

Fig. 12 FE-models of M(T) and C(T) with cohesive zone and damage zone (GTN model [29, 56]), respectively, in the ligament

Fig. 1b. This stress distribution could not be measured, however, and it needed models of damage to calculate it.

With the concept of a "process zone" ahead of the crack tip, uncoupling of remote plastic work, R_{pl}, and local separation energy, Γ_c, in Eq. (10) can be realised in numerical models, if according to Barenblatt's idea [4] specific elements are introduced, where material degradation and separation occur [88, 89], Fig. 12. The simulations require a constitutive description of the material behaviour in the process zone, which can mirror the local loss of stress carrying capacity. In general, two alternatives are used: Micromechanically based damage models or phenomenological cohesive models.

6.1 Damage Models

The development of damage mechanics began in 1958 when Kachanov [43] published the first paper introducing a damage variable for creep failure of metals, which is nearly the same year as the birth of fracture mechanics identified with Irwin's paper [40]. The concept of "continuum damage mechanics" (CDM) was picked up again in the eighties [44], particularly in France [51], extended to fatigue and ductile fracture [69] and generalised within the framework of thermodynamics of irreversible processes.

The attractiveness of CDM is its unified framework. Constitutive equations of a damaged material are derived from the same formalism as for a non-damaged material except that the stresses, σ_{ij}, are replaced by the effective stresses", $\tilde{\sigma}_{ij}$, which is called the principle of strain equivalence. In the simplest case of isotropic damage, effective stresses result from

$$\tilde{\sigma}_{ij} = \frac{\sigma_{ij}}{1 - D}, \quad 0 \le D < 1, \tag{23}$$

where D is a scalar damage variable. In analogy to the plastic strain rates as derived from a yield potential (normality rule), the evolution of damage is derived from a generalised dissipation potential, Φ,

$$\dot{D} = \lambda \frac{\partial \Phi}{\partial Y}, \tag{24}$$

which is a convex function of the conjugate stress variables σ_{ij}, R, ξ_{ij}, Y. $R(\bar{\varepsilon}_p)$ is the uniaxial flow stress of isotropic hardening, ξ_{ij} are the back stresses of kinematic hardening, and Y is an equivalent "damage stress" as dual state variable to D, also called energy density release rate [51]. The latter is capable of being misunderstood and confused with Griffith's elastic energy release rate, however.

This is a nice theoretical framework consistent with thermodynamics but does not answer the question wherefrom to get the dissipation potential. The latter requires a physically based description of the micromechanical damage processes in a material. The apparent theoretical consistency of CDM attempts mechanics people, in particular, to permanently develop new damage models, sometimes without any precise perception of "damage" and apparently without ever having seen a "real" material.

Alternatively, damage models based on the micro-mechanisms of ductile rupture [93, 96], namely the nucleation, growth and coalescence of voids (Fig. 13), were developed. The mechanism of void growth in a plastic material had been analysed in the late 60s, already [53, 68], and the essential influence of the hydrostatic stress was well known. Remarkably however, this physical understanding [13] had little effect on the discussions on "constraint effects" within the fracture mechanics community, which by the majority and particularly in the USA preferred "non-singular stresses" like T or Q as triaxiality parameters [6, 20, 60, 61] and bashed damage models as poor mechanics.

This is all the more incomprehensible since a fundamental model of ductile damage based on a yield potential for dilatant plasticity by Gurson [29] originates from the USA and was applied to simulations of ductile rupture [56, 57] around the same time as Rousselier's "local approach" [69] in France. The model of Gurson, Tvergaard and Needleman (GTN model) has been favoured in Europe predominantly [10, 92] and particularly promoted by the Fraunhofer Institute in Freiburg. It needed a second conversion of the apologist of a one-parameter J-approach [86],

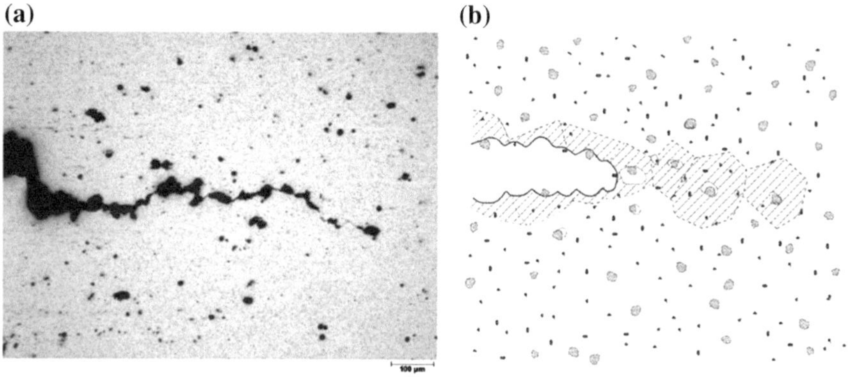

Fig. 13 Ductile crack extension by void nucleation, growth and coalescence: **a** experimental observation for an Al alloy, **b** schematic of a process zone with “cells”

then a two-parameter *J*-*Q*-approach [60, 61] to popularise the GTN model in the USA, which necessarily had to come along with a special new name “computational cell” [101] or “cell model” [24, 25].

Numerical models can employ the constitutive equations of continuum damage either for the complete structure, thus allowing for arbitrary directions of crack growth and not even needing an initial crack [56, 98] or just in one row of elements along the ligament considered as “process zone” as in Fig. 12 [88, 89], thus resembling a cohesive zone model. Significant void growth is commonly restricted to one row of elements.

A serious problem of damage models as established in the constitutive framework of “simple materials” is their feature of localisation, which gave reason to severe objections against the significance of the respective numerical results from the viewpoint of numerical mathematics. Actually, the results of FE simulations are mesh dependent, and the bearing load of a structure decreases with decreasing element height in the ligament. At the least, it is argued, that the postulation of convergence with reducing mesh size is violated, but some people deny simulations with softening materials any significance at all. Again, it is a question of the basic perception of models one has, and whether models and numerical methods are confused with “reality”. Continuum mechanics in general is an approximate description of real matter, since the assumption $(\Delta x, \Delta y, \Delta z) \rightarrow 0$ conflicts with the micro-structure of materials. This has no impact on the solution for hardening materials but becomes an issue for softening. The constitutive equations are relations between stresses and strains in *solid elements*, representing the micro-structure of the material in an *average sense*. In order to obtain physically meaningful results for the dissipation rate, Γ_c, in the process zone, a length parameter, h_0, has to be introduced, which depends on the average spacing of void nucleating particles and the hardening behaviour of the metallic matrix. The respective relation is established by an energy equivalence for a representative volume element (RVE) or “cell” (see Fig. 13b). The work of separation, ΔU_{sep}, per incremental crack extension, Δa, is

$$\Gamma_c = \frac{\Delta U_{\mathrm{sep}}}{B\Delta a} = \frac{1}{B\Delta a} \int\limits_{V_0} \left(\int\limits_{t}^{t+\Delta t} (1-f)\bar{\sigma}\, \mathrm{d}\bar{\varepsilon}^{\mathrm{pl}} \right) \mathrm{d}V = \bar{u}_{\mathrm{sep}}\, h_0, \tag{25}$$

with f and $\bar{u}_{\mathrm{sep}}$ being the average void volume fraction and the average strain energy density, respectively, in an RVE of volume $V_0 = h_0 B\Delta a$ [7]. If "local" constitutive equations are applied, which do not contain an intrinsic length scale, the height of the finite elements, h_0, in the ligament has to be considered as a characteristic material parameter [5, 91]—which is devil's notion for some FE experts. If a "finite" element is considered as a mathematical entity for solving boundary value problems in continuum mechanics, the perception of an element size as material parameter sounds weird, indeed. But if it is regarded as a representation of an RVE it appears more natural. In order to point out this difference, Shih introduced the term "computational cell" [24, 25, 101] instead of finite element. There are other concepts for mending the "pathological" mesh dependence like non-local approaches or gradient theories which please the theoretical requirements better than introducing the finite element size as material parameter. These approaches need special subroutines exceeding the capabilities of commercial FE codes, however, and they encounter new problems.

An additional problem with the application of damage models is their commonly large number of parameters, which are supposed to represent micromechanical properties but nevertheless difficult to identify [5] and to verify their uniqueness. An initial euphoria that all parameters of the GTN model can be determined from tensile test data turned out to be unrealistic since stress triaxiality plays an important role and is too low in a tensile specimen. Altogether, the application of damage models is still mostly a preserve of experts.

6.2 *Cohesive Models*

Cohesive models (CM) describe various kinds of decohesion processes, see Fig. 14, by a relation between (normal) surface tractions, σ_n, and respective material separation, δ_n, i.e. the traction-separation law (TSL) or cohesive law. For this, particular surface elements are introduced at the boundaries of solid elements along a pre-defined crack path as shown in Fig. 12. The constitutive relation of the interface elements represents the effective mechanical behaviour due to the physical processes, for instance micro-void nucleation, growth and coalescence in a ductile material. Commonly, the cohesive law is defined by two parameters, a cohesive strength (CS), σ_c, and a critical separation, δ_c, Fig. 15, or, alternatively, a separation energy (SE), Γ_c, which simply represents the area under the traction-separation law.

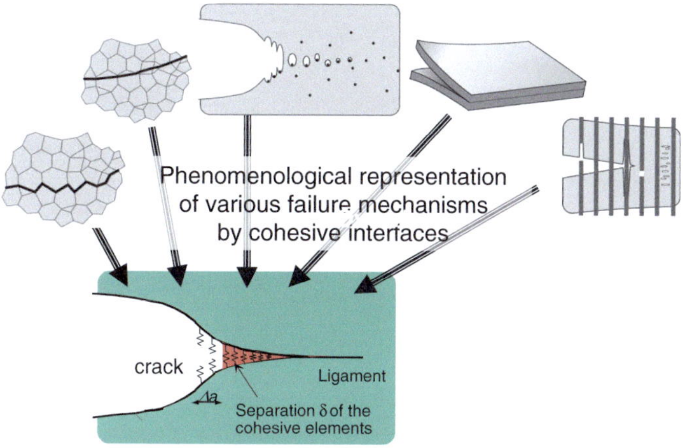

Fig. 14 Cohesive model as phenomenological representation of various decohesion processes in materials

$$\Gamma_c = \int_0^{\delta_c} \sigma_n(\delta_n) d\delta_n. \tag{26}$$

The cohesive model can be regarded as a renascence of Griffith's concept of a surface energy. The significant differences, however, are that

- though Γ_c is supposed to be a "surface" energy, the respective physical separation process occurs in a volume of finite, though commonly small thickness, in reality, Eq. (25), and
- the CS, σ_c, is an additional independent, phenomenological parameter, representing the maximum tensile stress which can be sustained by the material microstructure.

Cohesive laws can also be established for mixed mode separation processes, which will require an additional assumption on the interaction of tensile and shear modes [70]. The TSL requires significantly fewer material parameters than damage models, and numerical simulations based on cohesive models are less susceptible to convergence problems. A vital advantage compared to the continuum models of damage is that they do not show pathological mesh dependence and do not require the introduction of an additional length parameter via the FE mesh, since they are established as a relation between stresses, σ, and displacements, δ, instead of stresses and strains, ε. A major drawback is their restriction to pre-defined crack paths along the boundaries of solid elements.

The TSL is purely phenomenological and cannot be measured directly, in general. Various relations have been proposed in the literature, see Fig. 15 and overview in [9]. Since it represents micro-mechanical processes of material

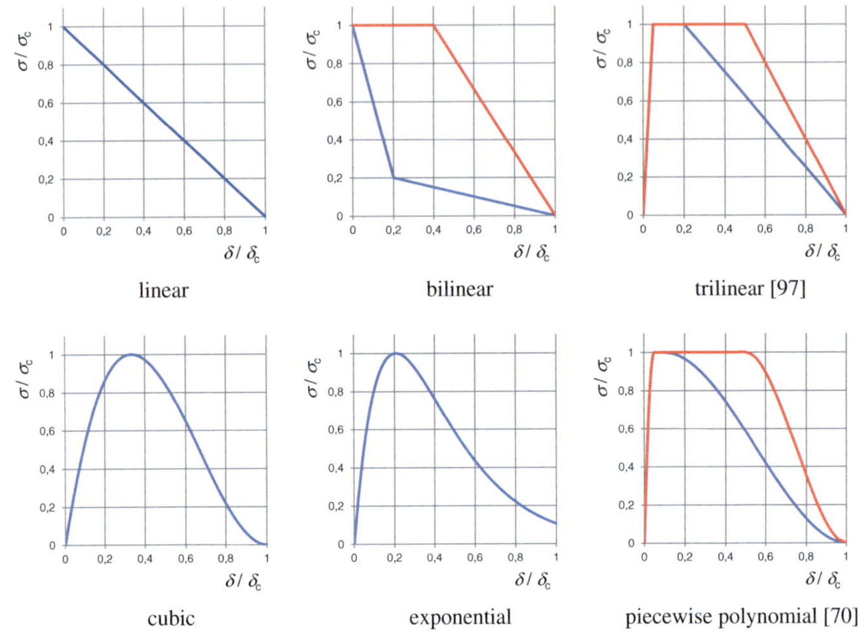

Fig. 15 Various traction-separation laws in the literature (overview from [9])

degradation and damage, respective micro-mechanical models may help to identify it. Processes of void growth have been investigated by numerical simulations of RVEs or "unit cells" containing a void [14, 45], which are assumed to represent a typical periodic microstructure of ductile materials. The TSL for ductile rupture by Scheider [70], which is, among others, depicted in Fig. 15, has been supported by micromechanical analyses [7, 88].

Numerical studies have also demonstrated that material separation based on void growth and coalescence depends on the stress triaxiality. This observed local "constraint" effect is quite evident: higher triaxiality causes an increase of the "fracture stress", i.e. a higher CS, and a decrease of ductility, i.e. a lower SE as in Fig. 5a, see [7, 88]. The particular micromechanical process of void growth and coalescence governing ductile rupture thus exhibits a local constraint effect, which adds to global constraint effects on the overall plastification of the structure. As the (local) SE is very small compared to the (global) plastic work per crack extension, however, Fig. 5b, the effect of triaxiality on the cohesive parameters is commonly negligible and CS and SE can be regarded as material constants from an engineering point of view [89]. The geometry dependence of J_R curves can be accurately predicted by FE simulations employing cohesive elements, Fig. 16.

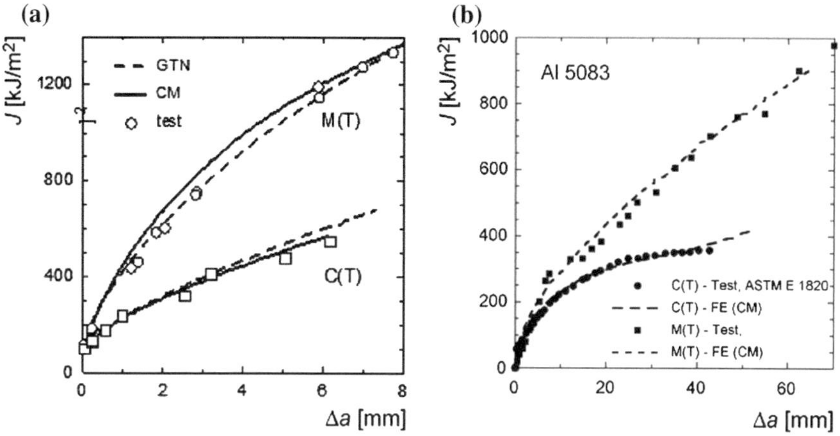

Fig. 16 J_R curves of C(T) and M(T) specimens, tests and FE simulations with cohesive model: **a** steel DIN StE460, side-grooved specimens, plane strain [89], **b** Al 5083, 3 mm thick panels, plain stress [8]

7 Conclusions

The history of fracture mechanics is an example of a more esoterical science looked at with scepticism in some parts of the world. Special resistance was observed in the service load fatigue community which relied on tests only. However, the new paradigm finally transformed into a versatile tool box for the development, design and operation of modern engineering structures. Pressure on high exploitation of structural mass and on economical operation without compromising safety led to wide acceptance of fracture mechanics. The concept of fracture mechanics has become an indispensable ingredient in numerous industrial codes. After several decades of development, it is now well established, demonstrated by standardisation of test methods and establishment of assessment procedures, sometimes in industry-specific form.

An interesting aspect in the development of fracture mechanics is the local distribution of the most important activities. In the beginning, the major driving forces were observed in the US. LEFM was driven in that country by failures in the aerospace area, where safety is of utmost importance. A very important aspect in the development of LEFM is again its application to the aerospace area: Fatigue crack propagation analysed with fracture mechanics made it possible to quantify the residual life time of a cracked component. A similar technological pressure emerged from the upcoming nuclear industry, driving the J-integral methods in terms of test procedures and assessment schemes. Westinghouse and some university groups for theoretical basics were key players.

The activities in the UK were historically earlier, here devoted primarily to welded joints, in particular for the developing offshore industry. The Welding

Institute and CEGB were the most prominent driving forces. CEGB was unique insofar as here the research institutes were immediately responsible for safe operation of their power plants. This way, experience in operation had a short way into the research groups, and vice versa. Whereas in the US the focus was on the *J*-integral, the UK worked on the CTOD. Both worlds were ultimately united as alternatives of equal rights in test standards and in assessment methods.

In Germany, major contributions have been provided by research centres such as Bundesanstalt für Materialprüfung (BAM) in Berlin, Fraunhofer-Institut für Werkstoffmechanik in Freiburg and GKSS Research Centre in Geesthacht, and University institutes in Karlsruhe, Magdeburg and Freiberg.

It should be noted that the present mature position of classical fracture mechanics would not have been possible without the contribution of numerical methods which have been gradually developed to high standards to meet the demands of understanding the experimentally observed phenomena of fracture and to transfer fracture mechanics material parameters to structural behavior. However, the heydays of classical fracture mechanics are now over, the major problems and their solutions are laid down in standards and codes, and the amount of research has substantially decreased. Only a few activities are still visible, such as application to new materials or material combinations. Even teaching fracture mechanics is about to vanish. Mechanical modelling in terms of damage models and cohesive zone models emerges as a more demanding tool, capable of dealing with problems inaccessible by classical fracture mechanics.

In conclusion, the history of fracture mechanics is an example of technology driven progress in science which finally found its way into wide-spread practical application. It also shows that being open to new ideas and working on them with enthusiasm does not only provide scientific merits, more importantly, it pays in terms of technological leadership, in contrast to more conventional traits.

References

1. ASTM E 399, Standard test method for linear-elastic plane-strain fracture toughness K_{Ic} of metallic materials. Annual book of ASTM Standards, vol 03.01. American Society for Testing and Materials, Philadelphia
2. ASTM E 1820-06, Standard test method for measurement of fracture toughness. Annual book of ASTM Standards, vol 03.01. American Society for Testing and Materials, Philadelphia
3. ASTM E 2472, Standard test method for determination of resistance to stable crack extension under low-constraint conditions, ASTM Book of Standards, vol 03.01
4. Barenblatt GI (1959) The formation of equilibrium cracks during brittle fracture: general ideas and hypothesis, axially symmetric cracks. Appl Math Mech 23:623–636
5. Bernauer G, Brocks W (2002) Micro-mechanical modelling of ductile damage and tearing—Results of a European numerical round robin. Fatigue Fract. Engng. Mater Struct 25:363–384
6. Betegón C, Hancock JW (1991) Two-parameter characterization of elastic-plastic crack tip fields. J Appl Mech 58:104–110

7. Brocks W (2005) Cohesive strength and separation energy as characteristic parameters of fracture toughness and their relation to micromechanics. Struct Integr Durab 1:233–241
8. Brocks W, Anuschewski P, Scheider I (2010) Ductile tearing resistance of metal sheets. Eng Fail Anal 17:607–616
9. Brocks W, Cornec A, Scheider I (2003) Computational aspects of nonlinear fracture mechanics. In: Milne I, Ritchie RO, Karihaloo B (eds) Comprehensive structural integrity—Numerical and computational methods, vol 3. Elsevier, Oxford, 127–209. ISBN: 0-08-043749-4
10. Brocks W, Klingbeil D, Künecke G, Sun DZ (1995) Application of the Gurson model to ductile tearing resistance. In: Kirk M, Bakker A (eds) Constraint effects in fracture—Theory and applications: ASTM STP 1244, vol Second. American Society for Testing and Materials, Philadelphia, pp 232–252
11. Brocks W, Künecke G, Noack HD, Veith H (1989) On the transferability of fracture mechanics parameters to structures using FEM. Nucl Eng Design 112:1–14
12. Brocks W, Olschewski J (1986) On J-dominance of crack-tip fields in largely yielded 3D structures. Int J Solids Struct 22:693–708
13. Brocks W, Schmitt W (1993) Quantitative assessment of the role of crack tip constraint on ductile tearing. In: Hackett EM, Schwalbe KH, Dodds RH (eds) Constraint effects in fracture, ASTM STP 1171. American Society for Testing and Materials, pp 64–78
14. Brocks W, Sun DZ, Hönig A (1995) Verification of the transferability of micromechanical parameters by cell model calculations for visco-plastic materials. Int J Plast 11:971–989
15. Brocks W, Yuan H (1989) Numerical investigations on the significance of J for large stable crack growth. Eng Fract Mech 32:459–468
16. BS 5762 (1979) Methods for crack opening displacement (COD) testing. British Standards Institution
17. Cherepanov CP (1967) Crack propagation in continuous media. Appl Math Mech 31:476–488
18. Clarke GA, Brown GM (1980) Computerized methods for J_{Ic} determination using unloading compliance techniques. ASTM STP 710:110–126
19. Dalle Donne C, Döker H (1994) Biaxial load effects on plane stress J-Δa and δ5-Δa curves. In: Schwalbe KH, Berger C (eds) Proceedings of 10th Biennial European conference on fracture, EMAS, Cradley Heath
20. Du ZZ, Hancock JW (1991) The effect of non-singular stresses on crack-tip constraint. J Mech Phys Solids 39:555–567
21. Dugdale DS (1960) Yielding of steel sheets containing slits. J Mech Phys Solids 8:100–104
22. Elber W (1970) Fatigue crack closure under cyclic tension. Eng Fract Mech 2:37–44
23. ESIS P2-92 (1992) ESIS Procedure for determining the fracture behaviour of materials. European Structural Integrity Society
24. Faleskog J, Gao X, Shih CF (1998) Cell model for nonlinear fracture analysis—I. Micromechanics calibration. Int J Fract 89:355–373
25. Gao X, Faleskog J, Shih CF (1998) Cell model for nonlinear fracture analysis—II. Fracture-process calibration and verification. Int J Fract 89:375–398
26. Garwood SJ (1979) Effect of specimen geometry on crack growth resistance. In: Smith CW (ed) ASTM STP 677, American Society for Testing and Materials, pp 511–532
27. Goodier JN, Field FA (1963) Plastic energy dissipation in crack propagation. In: Drucker DC, Gilman JJ (eds) Fracture of solids. Wiley, New York, pp 103–118
28. Griffith AA (1920) The phenomena of rupture and flow in solids. Phil Trans Roy Soc Lond A211:163–198
29. Gurson AL (1977) Continuum theory of ductile rupture by void nucleation and growth—Part I: Yield criteria and flow rules for porous ductile media. J Eng Mater Technol 99:2–15
30. Harrison JD, Dawes MG, Archer GL, Kamath KD (1978) The COD approach and its application to welded structures. Research Report 55/1978/E, The Welding Institute, Cambridge

31. Heerens J, Cornec A, Schwalbe KH (1988) Results of a round robin on stretch zone width determination. Fatigue Fract Eng Mater Struct 11:19–29
32. Heerens J, Schödel M (2009) Characterization of stable crack extension in aluminium sheet material using the crack tip opening angle determined optically and by the δ5 clip gauge technique. Eng Fract Mech 76:101–113
33. Hellmann D, Schwalbe KH (1986) On the experimental determination of CTOD based R-curves. In: Schwalbe KH (ed) The crack tip opening displacement in elastic-plastic fracture mechanics, Springer, Berlin, pp 115–132
34. Hellmann D, Schwalbe KH (1992) Methods for material characterisation in the range of high crack growth rates, In: Rie KT (ed) Proceedings of third international conference on LCF and elasto-plastic behaviour of materials. Elsevier Applied Sciences Publishers, pp 521–526
35. Hübner P (1996) Schwingfestigkeit der hochfesten schweißbaren Baustähle StE 885 und StE 960. PhD Dissertation, Technische Universität Bergakademie Freiberg
36. Hutchinson JW (1968) Singular behaviour at the end of a tensile crack in a hardening material. J Mech Phys Solids 16:13–31
37. Inglis CE (1913) Stresses in a plate due to the presence of cracks und sharp corners. Trans Inst Naval Arch 60:219–230
38. International Standard, ISO 12135 (2002) Metallic materials—Unified method of test for the determination of quasistatic fracture toughness. International Standards Organisation, Geneva
39. International Standard, ISO 22889 (2007) Metallic materials—method of test for the determination of resistance to stable crack extension using specimens of low constraint. International Standards Organisation, Geneva
40. Irwin GR (1957) Analysis of stresses and strains near the end of a crack traversing a plate. J Appl Mech 24:361–364
41. Irwin GR (1964) Structural aspects of brittle fracture. Appl Mater Res 3:65–81
42. Johnson HH (1965) Calibrating the electrical potential method for studying slow stable crack growth. Mater Res Stand 5:442–445
43. Kachanov LM (1958) On the creep rupture time (in Russian). Izv AN SSSR Otd Tehn Nauk 8:26–31
44. Kachanov LM (1986) Introduction to continuum damage mechanics. Kluwer Academic Publishers, Berlin
45. Koplik J, Needleman A (1988) Void growth and coalescence in porous plastic solids. Int J Solids Struct 24:835–853
46. Kuhn TS (2012) The structure of scientific revolutions, 4th edn. University of Chicago Press, Chicago ISBN 9780226458113
47. Kumar V, German MD, Shih CF (1981) An engineering approach for elastic-plastic fracture analysis. EPRI-Report NP-1931. EPRI, Palo Alto
48. Kussmaul, K (ed) (1991) Fracture mechanics verification by large-scale testing, EGF/ESIS Publication 8. Mechanical Engineering Publications Limited, London
49. Larsson SG, Carlsson AJ (1973) Influence of non-singular stress terms and specimen geometry on small scale yielding at crack-tips in elastic plastic materials. J Mech Phys Solids 21:263–278
50. Leevers PS, Radon JC (1982) Inherent stress biaxiality in various fracture specimens geometries. Int J Fract 19:311–325
51. Lemaitre J, Chaboche JL (1985) Mércanique des Matériaux Solids. Dunod, Paris
52. Luke M (2013) Das Fraunhofer-Institut für Werkstoffmechanik IWM. In: DVM-Nachrichten 57. Deutscher Verband für Materialforschung und –prüfung, Berlin
53. Clintock Mc (1968) A criterion of ductile failure by the growth of holes. J Appl Mech 35:63–371
54. McMeeking RM (1977) Finite deformation analyses of crack-tip opening in elastic-plastic materials and implications for fracture. J Mech Phys Solids 25:357–381
55. McMeeking RM, Parks DM (1979) On criteria for J-dominance of crack-tip fields in large scale yielding. In: ASTM STP 668. American Society for Testing and Materials, pp 175–194

56. Needleman A, Tvergaard V (1984) An analysis of ductile rupture in notched bars. J Mech Phys Solids 32:461–490
57. Needleman A, Tvergaard V (1987) An analysis of ductile rupture at a crack tip. J Mech Phys Solids 35:151–183
58. Neuber H (1985) Kerbspannungslehre, 3rd edn. Springer, Berlin
59. Newman JC (1997) Prediction of crack growth under variable-amplitude loading in various materials. Technical Report. NASA Langley, USA
60. O'Dowd NP, Shih CF (1991) Family of crack-tip fields characterized by a triaxiality parameter: I Structure of fields. J Mech Phys Solids 39:989–1015
61. O'Dowd NP, Shih CF (1992) Family of crack-tip fields characterized by a triaxiality parameter: II Fracture applications. J Mech Phys Solids 40:939–963
62. Paris PC, Erdogan F (1963) A critical analysis of crack propagation laws. J Basic Eng Trans Am Soc Mech Eng D85:528–534
63. Rice JR (1968) A path independent integral and the approximate analysis of strain concentrations by notches and cracks. J Appl Mech 35:379–386
64. Rice JR (1974) Limitations to the small scale yielding approximation for crack tip plasticity. J Mech Phys Solids 22:17–26
65. Rice JR, Drugan WJ, Sham TL (1980) Elastic-plastic analysis of growing cracks. In: Fracture mechanics 12th conference ASTM STP 700. American Society for Testing and Materials, Philadelphia, pp 189–221
66. Rice JR, Paris PC, Merkle JG (1973) Some further results of J-integral analysis and estimates. In: Progress in flaw growth and fracture toughness testing, ASTM STP 536. American Society for Testing and Materials, pp 231–245
67. Rice JR, Rosengren GF (1968) Plane strain deformation near a crack-tip in a power-law hardening material. J Mech Phys Solids 16:1–12
68. Rice JR, Tracey DM (1969) On the ductile enlargement of voids in triaxial stress fields. J Mech Phys Solids 17:201–217
69. Rousselier G (1987) Ductile fracture models and their potential in local approach of fracture. Nucl Eng Des 105:97–111
70. Scheider I, Brocks W (2003) Simulation of cup-cone fracture using the cohesive model. Eng Fract Mech 70:1943–1961
71. Schödel M (2006) Bruchmechanische Untersuchungen der Rissöffnung bei stabilem Risswachstum in dünnem Blech aus Al 5083. PhD thesis, Technische Universität Hamburg, Harburg
72. Schwalbe KH (1978) Mechanik und Mechanismen des stabilen Risswachstums. Habilitation thesis, Ruhr-Universität Bochum
73. Schwalbe KH (1980) Bruchmechanik metallischer Werkstoffe. Hanser, Germany
74. Schwalbe KH (1996) The engineering flaw assessment method (EFAM). Document EFAM 96. GKSS Research Centre, Geesthacht
75. Schwalbe KH (1997) Introduction of δ5 as an Operational Definition of the CTOD and its practical application. In: Reuter WG, Underwood JH, Newman JC (eds) Fracture mechanics: ASTM STP 1256, vol 26. ASTM International, West Conshohocken, pp 763–778
76. Schwalbe KH (1998) The engineering flaw assessment method (EFAM). Fatigue Fract Eng Mater Struct 21:1203–1213
77. Schwalbe KH (1998) EFAM ETM 97—the ETM method for assessing the significance of crack-like defects in engineering structures, comprising the versions ETM 97/1 and ETM 97/2. GKSS Research Centre, Geesthacht
78. Schwalbe K-H, Heerens J, Zerbst U, Pisarski H, Kocak M (2002) EFAM GTP 02—The GKSS test procedure for determining the fracture behaviour of materials. GKSS Research Centre, Geesthacht
79. Schwalbe KH, Hellmann D (1981) Application of the electrical potential method to crack length measurements using Johnson's formula. J Test Eval 9:218–221

80. Schwalbe KH, Kim YJ, Hao S, Cornec A, Kocak M (1997) EFAM ETM-MM 96—the ETM method for assessing the significance of crack-like defects in joints with mechanical heterogeneity (strength mismatch). GKSS Report 97/E/9. GKSS Research Centre, Geesthacht
81. Schwalbe KH, Neale B, Ingham T (1988) Draft EGF recommendations for determining the fracture resistance of ductile materials—EGF procedure EGF P1-87D. Fatigue Fract Eng Mater Struct 11:409–420
82. Schwalbe KH, Shannon J, Newman J (2005) Fracture mechanics testing on specimens with low constraint—standardisation activities within ISO and ASTM. Eng Fract Mech 72:557–576
83. Schwalbe KH, Zerbst U (2000) The engineering treatment model. Int J Press Vess Piping 77:905–918
84. Schwalbe KH, Zerbst U, Brocks W, Cornec A, Heerens J, Amstutz H (1998) The ETM method for assessing the significance of crack-like defects in engineering structures. Fatigue Fract Eng Mater Struct 21:1215–1231
85. Sharma SM, Aravas N (1991) Determination of higher order terms in asymptotic elastoplastic crack tip solutions. J Mech Phys Solids 39:1043–1072
86. Shih CF, German MD (1985) Requirements for a one parameter characterization of crack tip fields by the HRR-singularity. Int J Fracture 29:73–84
87. Shih CF, Hutchinson JW (1976) Fully plastic solutions and large-scale yielding estimates for plane stress crack problems. J Eng Mat Tech 98:289–295
88. Siegmund T, Brocks W (1999) Prediction of the work of separation and implications to modeling. Int J Fracture 99:97–116
89. Siegmund T, Brocks W (2000) A numerical study on the correlation between the work of separation and the dissipation rate in ductile fracture. Eng Fract Mech 67:139–154
90. Sumpter JDG (1999) An alternative view of R-curve testing. Eng Fract Mech 64:161–176
91. Sun DZ, Hönig A (1994) Significance of the characteristic length for micromechanical modelling of ductile fracture. In: Aliabadi MH, Carpinteri A, Kalisky S, Cartwright DJ (eds) Proceedings of third international conference on localized damage, Computational Mechanics Publications, Southhampton, pp 287–296
92. Sun DZ, Kienzler R, Voss B, Schmitt W (1992) Application of micro-mechanical models to the prediction of ductile fracture. In: Atluri SN, Newman JC, Raju IS, Epstein JS (eds) Fracture mechanics: twenty-second symposium ASTM STP 1131. American Society for Testing and Materials, Philadelphia, pp 368–378
93. Thomason PF (1985) A three-dimensional model for ductile fracture by the growth and coalescence of microvoids. Acta Metall 33:1087–1095
94. Turner CE (1990) A re-assessment of ductile tearing resistance, Part I: The geometry dependence of J-R curves in fully plastic bending, Part II: Energy dissipation rate and associated R-curves on normalized axes. In: Firrao D (ed) Fracture behaviour and design of materials and structures, vol II. EMAS, Warley, pp 933–949, 951–968
95. Turner CE, Kolednik O (1994) Application of energy dissipation rate arguments to stable crack growth. Fatigue Fract Eng Mater Struct 20:1109–1127
96. Tvergaard V (1982) On localization in ductile materials containing spherical voids. Int J Fracture 18:237–252
97. Tvergaard V, Hutchinson JW (1992) The relation between crack growth resistance and fracture process parameters in elastic-plastic solids. J Mech Phys Solids 40:1377–1397
98. Tvergaard V, Needleman A (1984) Analysis of the cup-cone fracture in a round tensile bar. Acta Metall 32:157–169
99. Wells AA (1961) Unstable crack propagation in metals: cleavage and fast fracture. In: Proceedings of the crack propagation symposium, vol 2. Cranfield, pp 210–230
100. Williams ML (1957) On the stress distribution at the base of a stationary crack. J Appl Mech 24:109–114

101. Xia L, Shih FC (1995) Ductile crack growth—I. A numerical study using computational cells with microstructurally-based length scales. J Mech Phys Solids 43:223–259
102. Zerbst U, Ainsworth RA, Schwalbe KH (2000) Basic principles of analytical flaw assessment methods. Int J Press Vess Pip 77:855–867
103. Zerbst U, Schödel M, Webster S, Ainsworth R (2007) Fitness-for-service fracture assessment of structures containing cracks. Elsevier, Oxford

Part II
Applications

Fracture Mechanics Assessment of Welded Components at Static Loading

Peter Hübner and Uwe Zerbst

Abstract Special aspects of welded components such as geometrical imperfections, inhomogeneity, strength mismatch and welding residual stresses have to be taken into account in a fracture mechanics analysis since they affect the crack driving force in the component as well as the fracture resistance of the material. The treatment of components subjected to static loading is based on methods such as the European SINTAP procedure, which has been validated in a number of case studies. In the present paper the influence of strength mismatch and residual stresses on the fracture resistance and the assessment of a component are discussed.

1 Introduction

The thermal cycle in welding differs from a common heat treatment in that its local pattern is significantly inhomogeneous and in the very short time scale of the welding process. The result of these specific characteristics consists of a number of effects with respect to the crack driving force in the components as well as to the fracture resistance of the material (Fig. 1). Essential effects are

- Modifications of the microstructure which cause strength mismatch and inhomogeneous fracture toughness across the heat affected zone and the weld metal,
- Generation of welding residual stresses due to prevented shrinking and phase transformations in the weld, and
- Deviations from the design geometry of the weldment, so-called misalignment.

P. Hübner (✉)
Hochschule Mittweida, Mittweida, Germany
e-mail: peter.huebner@hs-mittweida.de

U. Zerbst
Bundesanstalt für Materialprüfung, Berlin, Germany
e-mail: uwe.zerbst@bam.de

G. Hütter and L. Zybell, *Recent Trends in Fracture and Damage Mechanics*, DOI 10.1007/978-3-319-21467-2_3

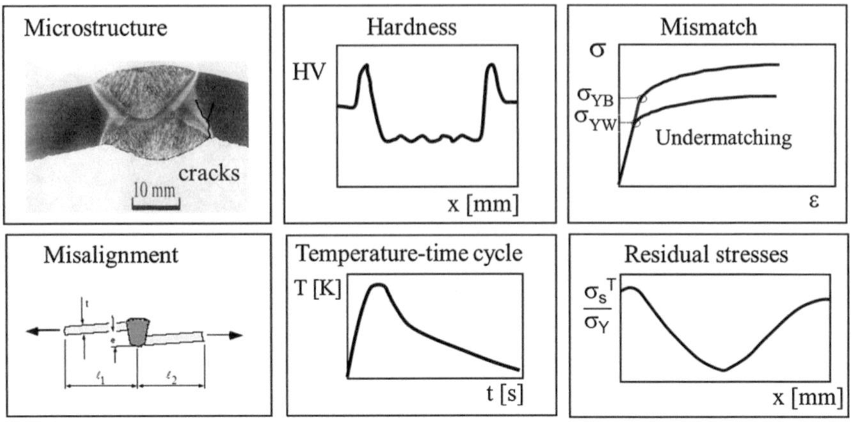

Fig. 1 Thermic cycle during welding and effects on the weldment

The inhomogeneity of the microstructure and the residual stress field cause a significant scatter in the fracture toughness. Furthermore, the combination of zones of locally reduced toughness and welding residual stresses might prompt unexpected crack initiation with the final consequence of component failure. Attributed to the specific technological conditions of the welding process, cracks or crack-like defects such as hot or solidification cracks, cold cracks or lack of fusion can occur. Besides this, surface notches and volumetric defects, porosity or slag inclusions can be the initiation points of crack extension at cyclic loading. Fracture mechanics assessment is significantly complicated by such complex conditions. In the present paper the effect of strength mismatch and welding residual stresses on the structural integrity of weldments shall be discussed in more detail. With respect to further influence quantities such as misalignment, crack initiation at notches and statistical aspects of the fracture toughness the reader is also referred to the review paper [1].

2 Inhomogeneity of the Microstructure and Strength Mismatch

Depending on the material type, the effects of the temperature-time cycle during welding are different. E.g., in dispersion hardened and cold worked or strain-hardened aluminum alloys significant softening in the weldment and heat affected zone are observed. The conditions in steels are more complex since the transition behavior depends on the cooling rate such that the resulting microstructure can be martensitic or of mixed nature but in any case deviating from that of the base metal. That's why increased as well as decreased strength values may be possible.

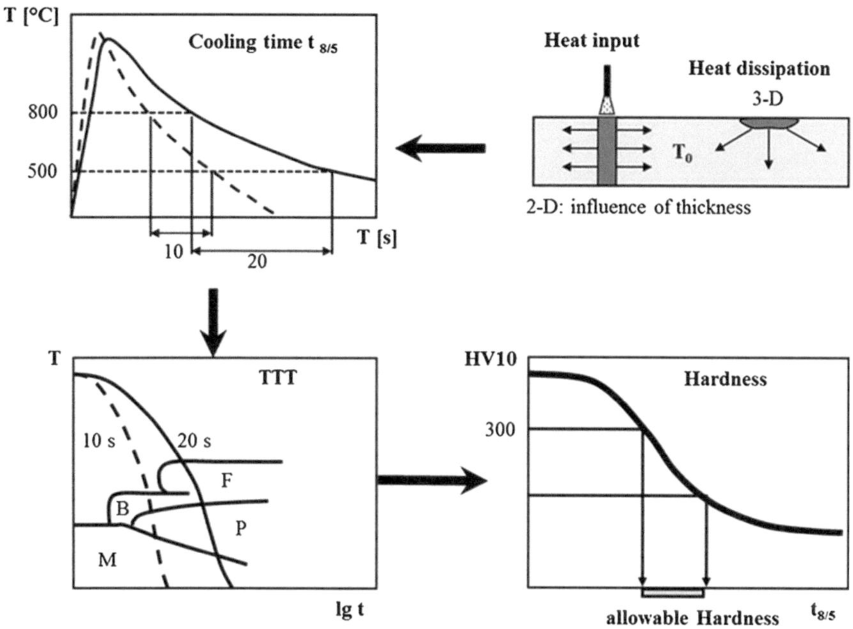

Fig. 2 $t_{8/5}$ concept for predicting the phase volumes in steel hardening. *TTT* Time Temperature Transformation diagram, *F* ferrite, *P* perlite, *B* bainite, *M* martensite

With respect to structural steels, the $t_{8/5}$ concept is proven to be a suitable measure of the technological conditions of the welding process. The welding technology is characterized by the energy input per unit weld length, a potential preheat temperature T_0 and the cooling conditions (2D, 3D) [2, 3]. The $t_{8/5}$ time is the time over which the weld cools down from 800 to 500 °C (Fig. 2). Weldment specific time-temperature-transformation (TTT) diagrams can be used for specifying the volume of structural phases which, for their part, determine the hardness and strength in the heat affected zone. Single-pass welds can be described very well by this concept. Note that TTT diagrams, which allow for routine assessments, exist for various structural steels dependent on their chemical composition [4, 5].

The situation is more complex for multi-pass weldments, since the exact distribution of the phase volumes at each position of the heat affected zone depends on the complete thermal history. It can only be determined by simulation (Fig. 3) [6]. Based on microstructure-strength correlations a hardness distribution along the weld can be obtained which is in line with strength differences across it, at least as long as residual stress effects do not play a major role [7]. It has to be emphasized that particularly coarse grain areas of high hardness tend to embrittlement. Therefore, it is important to identify such zones.

The difference in the strength properties between weld and base metal caused by variations in the microstructure is designated as strength mismatch. Its effect on potential cracks consists in strain localization in the lower strength zone of the

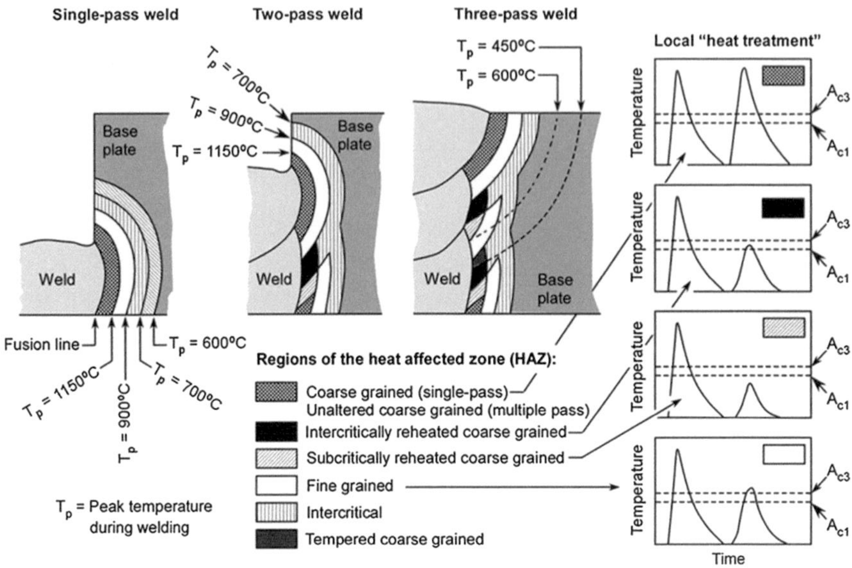

Fig. 3 Formation of the heat affected zone in multi-pass welds. Most dangerous are coarse grain zones, since they can be locations of embrittlement [8]

weldment (Fig. 4) which is detrimental when the crack is located near such an area. Usually strength mismatch is characterized by the so-called strength mismatch ratio M

$$M = \frac{\sigma_{YW}}{\sigma_{YB}}, \tag{1}$$

which is simplify based on the yield strengths of the weld metal σ_{YW} and the base metal σ_{YB}. *Overmatching* (OM) characterizes the case of a weld metal strength higher than those of the base metal. The weld is mechanically shielded, i.e. it will still be in an elastic state when the base metal already plastically deforms. In contrast, *Undermatching* (UM) exists when the yield strength of the weld metal is lower than that of the base metal. In this case, strain localization takes place in the weld. Note that the maximum local strain in mismatch components exceeds that of all base or weld metal components (Fig. 4). Dependent on the location of the crack with respect to this zone, a significant increase in the crack driving force can be the consequence. Thus, with respect to the danger of component failure, the location of the crack with respect to the strain concentration zone is essential. The risk is especially high for weld metal cracks in undermatched and heat affected zone or fusion line cracks in overmatched weldments where the location of the highest crack driving force is frequently identical to those of lowest toughness, an effect which additionally intensifies the detrimental influence of the strength mismatch effect.

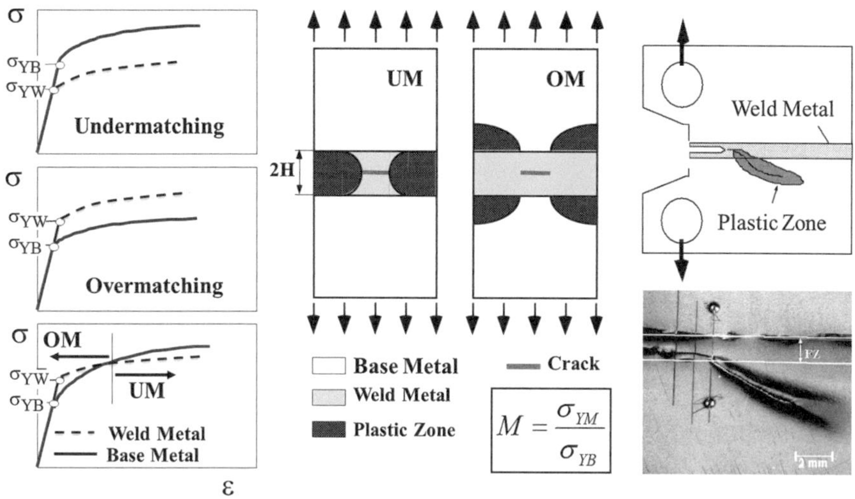

Fig. 4 Definition of strength mismatch. The plastic strain concentration causes deviating of the crack to the lower strength region; photo [9]

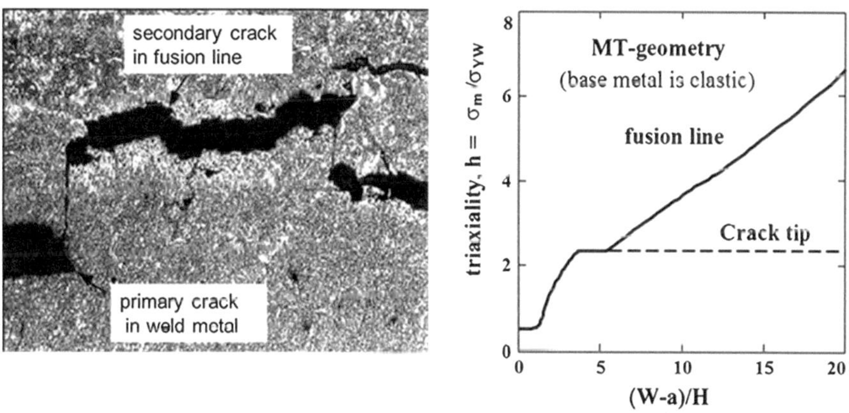

Fig. 5 Secondary crack at a secondary stress peak at the weld line away from the main crack tip (Figure: M. Koçak). *Right side* Secondary maximum of stress triaxiality [10]

Vice versa, there might be cases where a heat affected zone crack in an overmatch joint can be harmless despite its poor fracture resistance because the crack tends to deviate to the region of lower strength but higher toughness (Fig. 4, right). Occasionally it is even observed that cracks initiate away from the main crack (Fig. 5). This can be explained by the mismatch-caused formation of a locally constraint stress state. In the Fig. 5 an example of a tension plate with a middle crack is shown in which a secondary stress peak larger than that at the crack tip exists due to this effect. The strength mismatch effect is particularly distinct in laser and electron beam welds due to the deep welding effect.

Particularly in overmatching steel welds, crack deviation is often observed, what makes the determination of toughness sometimes a tricky business. In contrast, strength undermatching is typical for aluminum alloys due to local softening during the thermal cycle.

3 Welding Residual Stresses

The thermal cycle during welding causes thermal expansion during heating and shrinking during cooling. Since the heating is locally restricted, free shrinkage is hindered. It is simply restrained by the surrounding colder material. As a consequence, type I residual stresses develop. In addition, phase transformations in conjunction with volume changes overlay the effect of thermal deformation. Temperature dependent values of the elastic modulus and yield strength as well as the global clamping and restraint conditions of the component during welding make things even more complicated. The pattern of the residual stresses in longitudinal and transverse direction can be understood by imagining a hot beam which is fitted into the weld, linked up with the base metal and subsequently cooled down. This makes immediately clear that the welding technique is of essential influence for the formation of the residual stress state. A large molten bath and fewer weld passes cause higher residual stresses than multi-pass welds or stringer bead technique. The magnitude of the residual stresses is reduced by preheating and so-called back-step welding. This illustrates why residual stress fields can be different even when parameters such as energy input per unit length and welding speed are identical. This is particularly the case in manual welding.

A characteristic feature of welding residual stresses is that the corresponding forces and moments are in internal equilibrium, i.e., they integrate to zero across the volume. That means the tension and compression stresses exist at the same time but at different locations across the weldment, see (Fig. 6). With respect to fracture mechanics the location of the crack tip with respect to the local stress is of great importance.

The stresses at surface change in thickness direction, see (Fig. 7). Note that transverse residual stresses are most dangerous in butt welds loaded in transverse direction since they can act as crack openers in cases of incomplete welding, at root defects and at undercuts.

The magnitude of the residual stresses also depends on the clamping conditions during welding. The resulting restraint stresses superimpose the residual stress profiles and cause disequilibrium of forces and moments in the section potentially containing the crack, see Fig. 8.

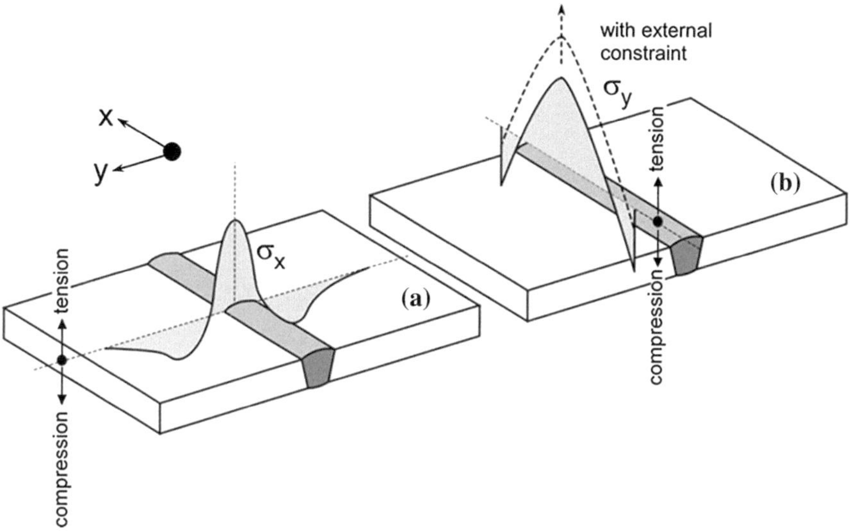

Fig. 6 Schematic welding residual stress distributions in a butt joint; **a** longitudinal, σ_x; **b** transverse, σ_y (according to [11], p. 308, modified). Note that the stress profile is different for external restraint during welding

Fig. 7 Idealised schematic through-thickness profile of transverse welding residual stresses in a butt joint

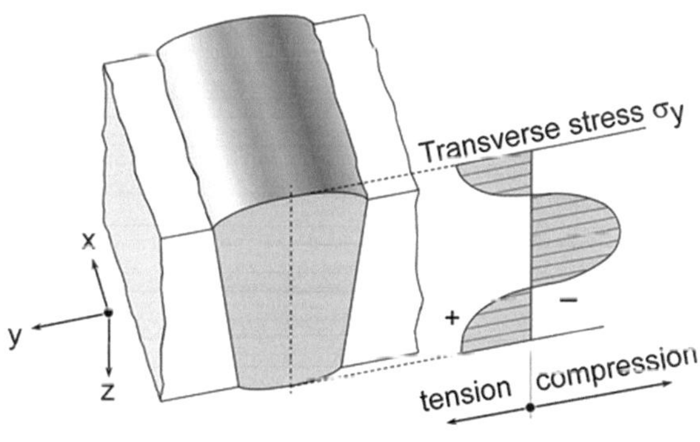

Fig. 8 Effect of membrane restraint on transverse residual stresses; multi-pass welded joint of austenitic steel plates, according to [12]

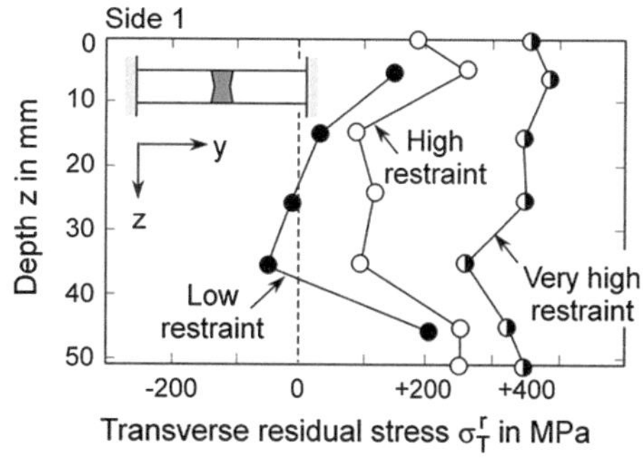

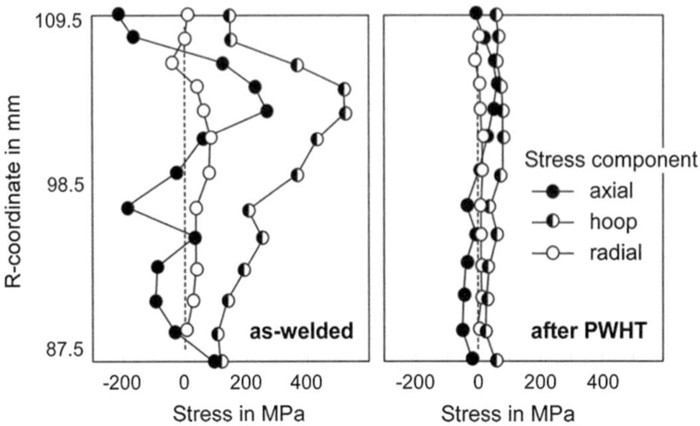

Fig. 9 Through-wall variations of residual stresses at a circumferential butt weld in a pipe in the as-welded state and after thermal stress relief by post weld heat treatment (PWHT); according to Josephson (1983), cit. in [12]

Welding residual stresses can be reduced by stress relief annealing, provided that the clamping conditions allow for plastic deformation and the yield strength at the annealing temperature is sufficiently reduced (Fig. 9). Note that the residual stresses cannot be reduced below that level. In the case of steels it has to be checked that no phase transformations occur, which could level off the effect of the heat treatment.

An essential problem is how to quantify the residual stress profiles. Although there exists a number of experimental techniques such as the hole drilling method, X-ray and neutron refraction most of them yield only limited information usually at the surface. In addition, particularly with respect to stress gradients across the wall, they sometimes show quite divergent results compared to each other. Finite element results will generate satisfying results only when beside the temperature-time record phase transformations are modeled. This requires appropriate simulation strategies by which in a first step the portions of the respective phases and the corresponding transformation-caused strains have to be determined from the temperature-time characteristics. Only then the resulting stresses can be determined taking into account the temperature dependent modulus of elasticity as well as the temperature dependent yield strength. The results of a round robin exercise in Fig. 10, concerning experimental as well as numerical data, illustrate that the scatter in such an analysis can be considerably.

Although the residual stress distribution frequently is qualitatively known, its magnitude, however, is uncertain. In such cases the generation of upper bound curves might be an option. These curves can be used as conservative input information to fracture mechanics analyses, see Fig. 11. The maximum value at the surface is usually assumed as identical to the yield strength. Upper bound residual stress profiles exist for plate and pipe butt and seam welds and for repair welds. They are summarized in various compendia of fracture mechanics assessment procedures, including those of SINTAP [14]. A problem is that these solutions

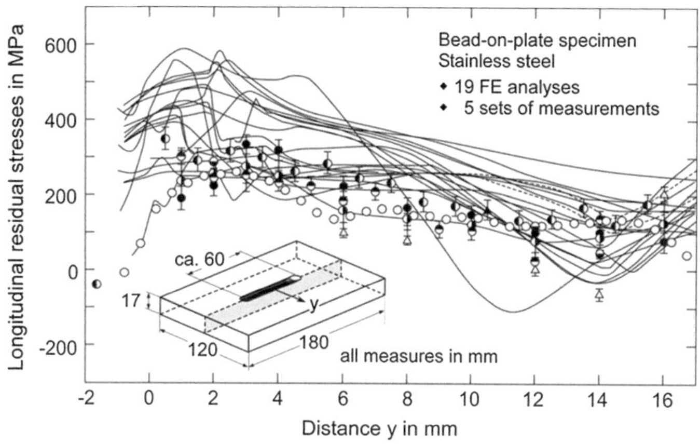

Fig. 10 Results of a Round Robin for the determination of welding residual stresses (according to Bouchard [13])

Fig. 11 Through thickness residual stress profile for tubular T-butt welds, experimental data and upper bound curve, according to [15]. σ_Y^* is the smaller of the yield strengths of the weld metal and base metal

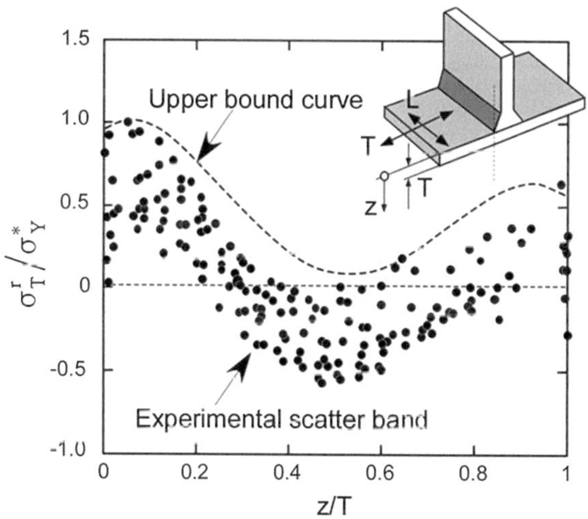

violate the equilibrium conditions in that they are shifted to the tension range of stresses. It is also known that the residual stresses become rapidly smaller in transverse direction compared to the center of the weld. Frequently they will be very small at the weld toe transition between the weld and the heat affected zone/base metal [1]. The consequence is a danger of high conservatism when the upper bound curves are used in fracture mechanics analyses, e.g., for fatigue crack propagation. A way out could be sensitivity analyses for judging the effect of the residual stresses in a specific fracture mechanics analysis.

4 Strength Mismatch and Residual Stresses in Determining the Fracture Toughness

A number of test recommendations exist for the fracture toughness testing of weldments [16–18]. These deal with specific aspects of weldments compared to non-welded specimens. Features which are addressed are

- The notch position (with respect to the weld, the heat affected zone and the fusion line),
- The direction of crack extension (longitudinal or transverse with respect to the axial direction),
- Questions of pre-cracking for instance with respect to the residual stresses which frequently do not allow for straight crack fronts,
- Post-test verification that the crack tip has really been positioned in the microstructure of interest, and
- The number of specimens needed for statistical safety.

Both, strength mismatch and welding residual stresses can not only have a significant effect on the fracture toughness and but can also cause serious experimental problems. One example was already mentioned: cracks deviating from its original planes due to mismatch. Another effect is irregular crack fronts.

In addition, the crack tip parameters might be affected and have sometimes to be corrected. E.g., the determination of strength mismatched weld specimens requires a specific shape function for the J integral solution, depending from weld geometry, see Fig. 12. For a more detailed discussion see [1].

For the case of SE(B) specimens without or with limited strength mismatch the following J integral solution according to [18] is commonly used:

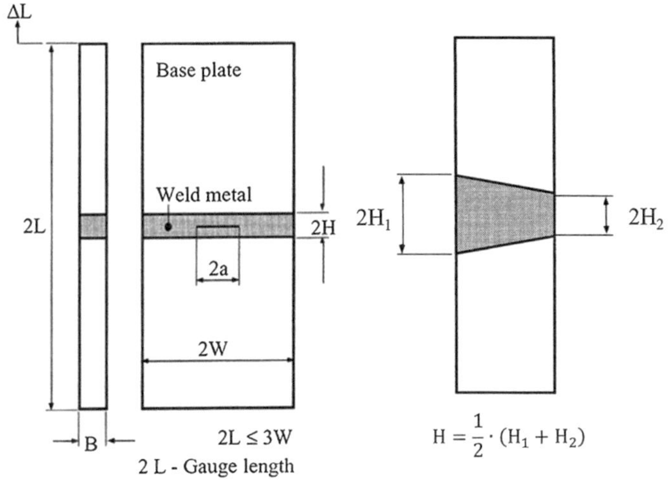

Fig. 12 Definition of the weld width height 2H of a V weld. [18]

$$J_0 = \frac{K^2}{E} + \frac{\eta_{pl}^{CMOD}}{B \cdot (W-a)} \cdot A \quad \text{with} \quad \eta_{pl}^{CMOD} = 3.724 - 2.244 \cdot \frac{a}{W} + 0.408 \cdot \left(\frac{a}{W}\right)^2 \tag{2}$$

A is the plastic area of the force-displacement plot. For weld cracks, the deviation from the exact mismatch solution is less than 10 % for mismatch ratios between $M = 0.9$ and 1.25 and $a/W = 0.1$–0.5. With respect to heat affected zone cracks the same is true for $M < 0.9$ and $a/W = 0.1$–0.7. Outside of these ranges a strength mismatch corrected η_{pl} has to be determined such as shown in Fig. 13.

For middle crack tension specimens M(T) without or with limited strengths mismatch J is obtained by [18]:

$$J_0 = \frac{K^2}{E} + \frac{\eta_{pl}}{B \cdot (W-a)} \cdot A \tag{3}$$

with $\eta_{pl} = 1$. Equation (3) is valid for $a/W = 0.125$–0.5.

Within a range of $M = 0.9$–1.1 the deviation from the mismatch corrected solution is less than 10 % for weld cracks and plane strain as well as plain stress conditions and this is also correct for heat affected zone cracks and $M = 0.9$–2. Outside of this range η_{pl} has to be determined according to Fig. 14.

The effect of a strength mismatch correction on the η factor in Eq. (2) is shown in Fig. 16. The example refers to a dissimilar electron beam weld of aluminum alloys Al 2017 and Al 7075 [19]. The fracture resistance has been obtained for the weld metal. The test specimen was an SE(B) specimen with a weld width $2H = 2.4$ mm, a specimen width $W = 20$ mm and a specimen thickness $B = 10$ mm. The initial crack

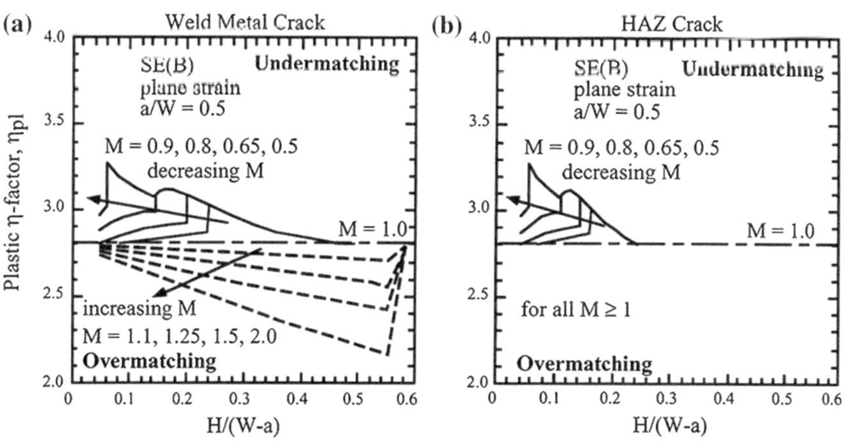

Fig. 13 Strength mismatch corrected function η_{pl} for J integral determination using SE(B) specimens with weld or heat affected zone cracks (plane strain) [18]

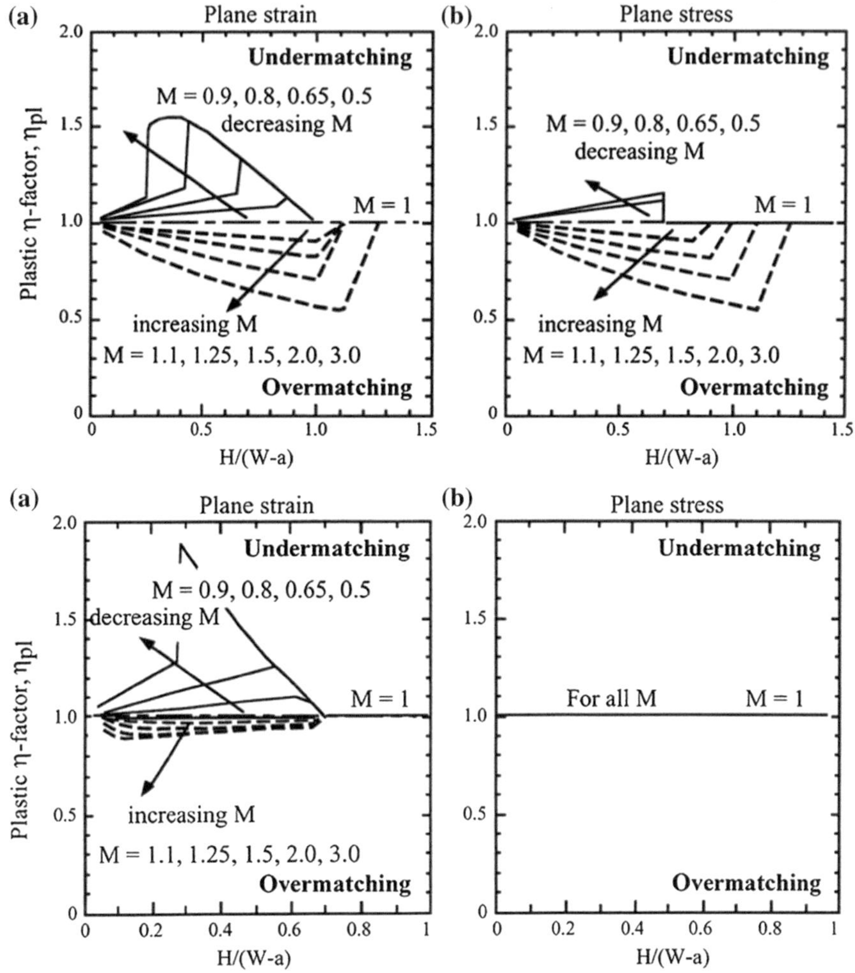

Fig. 14 Strength mismatch corrected function η_{pl} for J integral determination using M(T) specimens with weld or heat affected zone cracks (plane stress and plane strain); *Upper figure* M (T)-specimens crack in weld metal; *Lower figure* M(T)-specimen crack in HAZ [18]

depth was a_0 = 10 mm and the mismatch ratio M = 0.54. According to Fig. 15 the corrected η_{pl} factor for the weld crack was determined as η_{pl} = 3.2.

Using this corrected η_{pl} factor, a significant strength mismatch effect on the crack resistance curve is obtained as it is shown in Fig. 16. The analysis has been carried out based on the test guideline EFAM GTP 02 [18]. The all base metal resistance against stable crack initiation, $J_{0.2}$, would be underestimated by 27 % and also the R curves show a difference of this order. What is, however, uncertain is whether the physical crack initiation resistance, J_i, would behave the same way since no stretch zone measurement was carried out in the present case. Note that Fig. 16 is just an

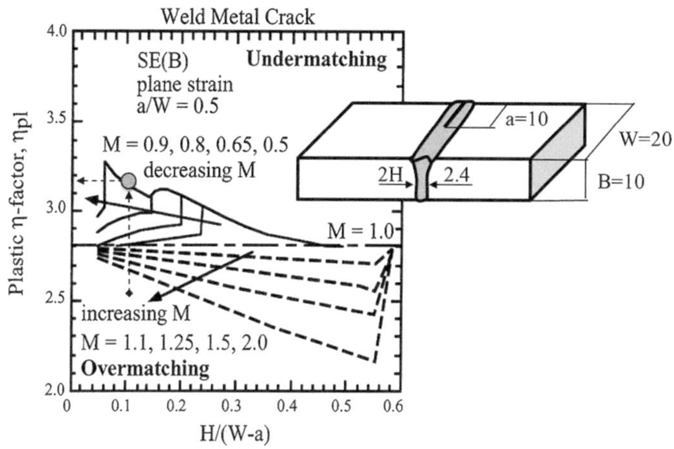

Fig. 15 Mismatch correction for the η_{pl} factor of the J integral solution (Eq. 3), according to [18]

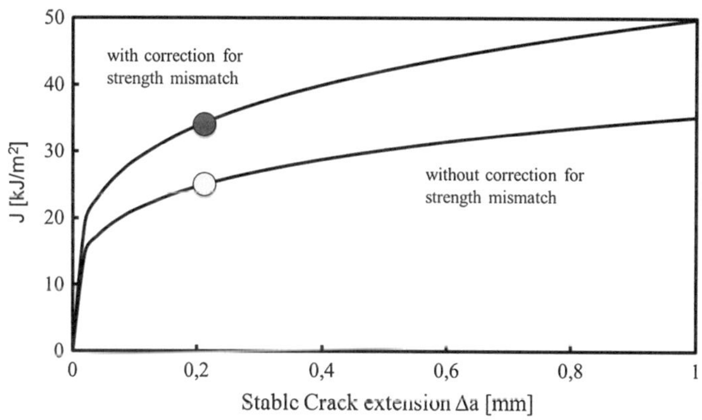

Fig. 16 Effect of strength mismatch correction on the crack resistance curve and the "engineering" resistance against stable crack initiation, $J_{0.2}$, dissimilar joint of Al 2017 and Al 7075, strength undermatching

example. As can be concluded from Fig. 15, strength mismatch is of benefit for the R curve only in the case of undermatching whilst overmatching would cause a detrimental effect. Unfortunately the latter is the more representative case in practical application.

Figure 17 illustrates the effect of strength mismatch on the statistical distribution of the critical crack tip opening CTOD of a bi-metal joint. When the crack is located in the higher strength material the crack resistance is shifted to lower values. In the case of undermatching this is quite opposite [20].

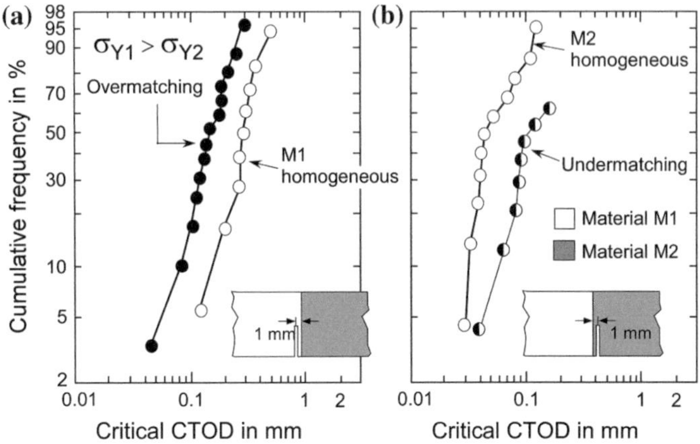

Fig. 17 Effect of strength mismatch in the vicinity of the notch tip on the distribution of fracture toughness in terms of the critical CTOD, **a** crack in the higher strength material; **b** crack in the lower strength material; according to [20]

5 Fracture Mechanics Assessment of Components

5.1 General

The assessment of a welded component for fracture may be based on the SINTAP procedure [14], which in turn forms the basis for the DVS Merkblatt (leaflet) [21]. Further developments are found in FITNET [22] the updated R6 procedure [23] and the British standard BS 7910 [24] (see also [25–29]). The key element of these procedures is a "plasticity correction" of the formally determined stress intensity factor which is based on a ligament yielding factor L_r identical to the ratio of the applied load and the yield or limit load. The latter roughly refers to the load at which the general yielding state of the ligament ahead of the crack is reached. The analysis can be carried out in so-called FAD and CDF formats.

In the FAD format (Fig. 18, FAD stands for Failure Assessment Diagram) the applied K factor is referred to the fracture resistance which, in general terms, is designated by K_{mat}. A failure line, the so called FAD, hereafter designated by $f(L_r)$, separates the region of safe operation from that of potential non-safety. The analysis is then based on the relative position of a design point with respect to the failure line. The component is regarded as safe as long as this point lies inside the area circumscribed by $f(L_r)$. Note that the FAD line is dependent on the stress-strain curve of the material but regarded as independent of the component geometry and the loading type. In contrast, the design point characterizes the component under consideration including the fracture resistance and the parameters of the stress-strain curve of the material, the crack size and its geometry and the loading type, i.e., bending, tension etc. This is illustrated in Fig. 18a, b for shifting the design point towards the failure line by increasing load and crack size.

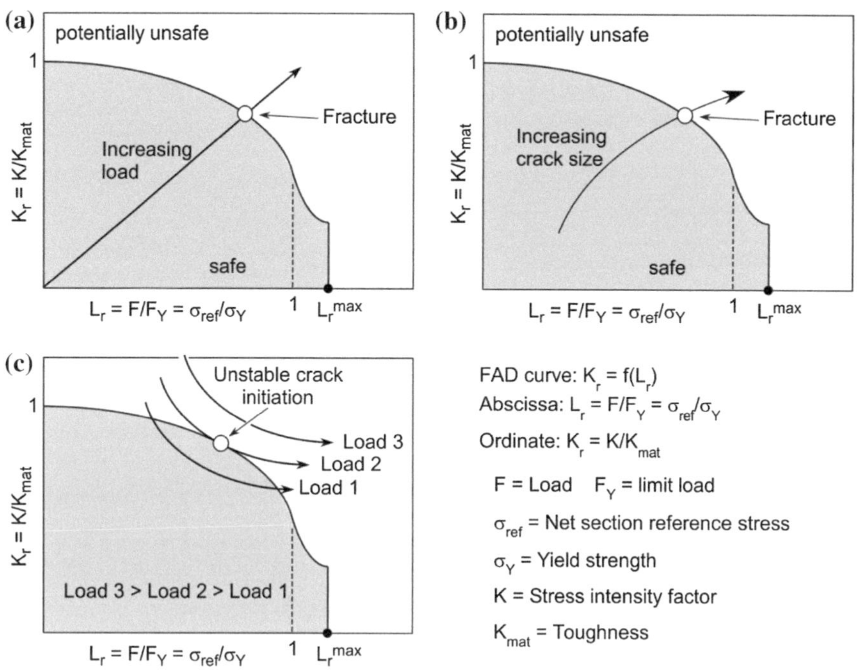

Fig. 18 FAD format of fracture assessment using the SINTAP or similar procedures. **a** for increasing load, **b** for increasing crack size, **c** for ductile instability

Figure 18c illustrates an assessment for unstable failure. Whereas (a) and (b) are based on a single toughness value, K_{mat}, (c) uses an R curve such as shown in (Fig. 16). Since in that case not only K and L_r are crack size dependent but also J_{max}, the design path differs from those in (a) and (b) and failure is defined for the assessment curve which tangentially touches the FAD line.

Following the CDF format (CDF stands for Crack Driving Force) the results are numerically identical with those of the FAD approach. What is different is the assessment philosophy which consists of two separate steps:

(a) The crack driving force is determined in terms of the J integral or CTOD *(δ)* by

$$J = J_e \cdot [f(L_r)]^{-2} \quad \text{or} \quad \delta = \delta_e \cdot [f(L_r)]^{-2} \tag{4}$$

with

$$J_e = \frac{K^2}{E'} \quad \text{or} \quad \delta_e = \frac{K^2}{\sigma_Y \cdot E'} \tag{5}$$

(b) The obtained value is then compared with the fracture resistance J_{mat} or δ_{mat}.

The ligament yielding parameter L_r

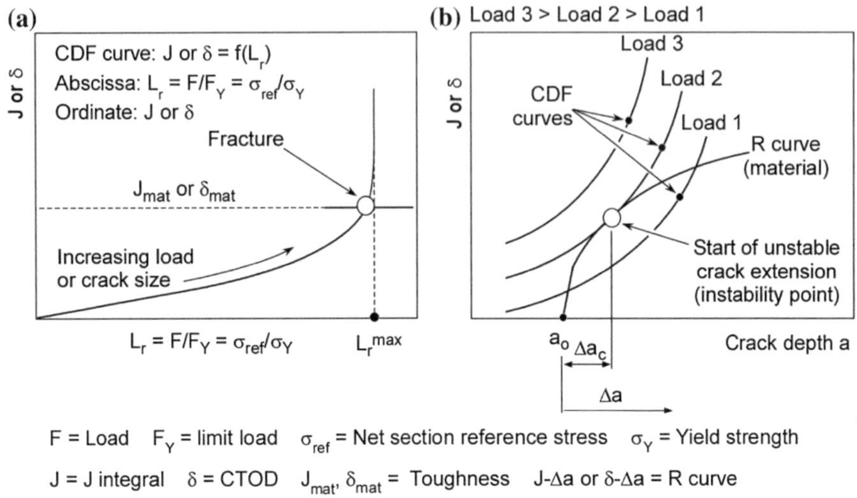

Fig. 19 Determination of the critical condition of a component by the CDF approach **a** Toughness given by a single value (e.g., resistance against stable crack initiation); **b** Toughness given by an R-curve (assessment for crack instability)

$$L_r = F/F_Y \tag{6}$$

is given in compendia such as [14, 21–24] or [30–32]. Figure 19a illustrates the CDF format determination of the critical load for stable and Fig. 19b for unstable crack initiation. In the latter case, instability is given at the load for which the CDF curve is tangential to the J- or δ-Δa curve of the material.

5.2 *Consideration of Strength Mismatch*

Beyond a mismatch ratio $M > 1.1$ or below $M < 0.9$ both, a mismatch modified yield or limit load and a mismatch modified FAD line have to be used. For weldments with neither the base nor the weld metal showing a Lüders' plateau [14] the latter is given by

$$f(L_r) = \left[1 + \frac{1}{2} \cdot L_r^2\right]^{-1/2} \cdot \left[0.3 + 0.7 \cdot \exp\left(-\mu_M \cdot L_r^6\right)\right] \tag{7}$$

for $0 \le L_r < 1$

and

$$f(L_r) = f(L_r = 1) \cdot L_r^{\left(N_M - \frac{1}{2 \cdot N_M}\right)} \tag{8}$$

for $1 \leq L_r < L_{r\,max}$

For $L_r > L_{r\,max}$, i.e. plastic collapse, $f(L_r)$ is set to zero.

The mismatch corrected coefficients μ_M and N_M with the latter being a mismatch-equivalent strain hardening coefficient are determined by

$$\mu_M = (M-1) \cdot \left[\frac{\left(\frac{F_{YM}}{F_{YB}} - 1 \right)}{\mu_W} + \frac{\left(M - \frac{F_{YM}}{F_{YB}} \right)}{\mu_B} \right]^{-1}, \quad \mu_M \leq 0.6 \tag{9}$$

with

$$\mu_B = 0.001 \cdot \frac{E_B}{R_{p0.2,B}}, \; E_B \quad \text{and} \quad R_{p0.2,B} \text{ in MPa}, \; \mu_B \leq 0.6 \tag{10}$$

and

$$\mu_W = 0.001 \cdot \frac{E_W}{R_{p0.2,W}}, \; E_W \quad \text{and} \quad R_{p0.2,W} \text{ in MPa}, \; \mu_W \leq 0.6 \tag{11}$$

$$N_M = (M-1) \cdot \left[\frac{\left(\frac{F_{YM}}{F_{YB}} - 1 \right)}{N_W} + \frac{\frac{(M - F_{YM})}{F_{YB}}}{N_W} \right]^{-1} \tag{12}$$

$$N_B = 0.3 \cdot \left[1 - \frac{R_{p0.2}}{R_m} \right]_B \tag{13}$$

$$N_W = 0.3 \cdot \left[1 - \frac{R_{p0.2}}{R_m} \right]_W \tag{14}$$

$$L_{r\,max} = 0.5 \cdot \left(1 + \frac{0.3}{0.3 - N_M} \right) \tag{15}$$

In the equations above, the index B describes the base metal and the index W the weld metal. F_{YB} is the yield or limit load of the all-base metal component with crack and F_{YW} those of its all-weld metal counterpart. F_{YM} characterizes the welded component with strength mismatch. With respect to the latter, the compendia mentioned comprise solutions for plates subjected to tension and bending and hollow cylinders with axial and circumferential cracks with the cracks being located in the weld metal and the heat affected zone. All solutions require the knowledge of the yield strengths of the base and weld metal. An example is given in Fig. 20, further ones can be found in [33].

Component geometry:

Welded plate with through wall center crack, tension loading

$a/W = 0.5$, plain stress

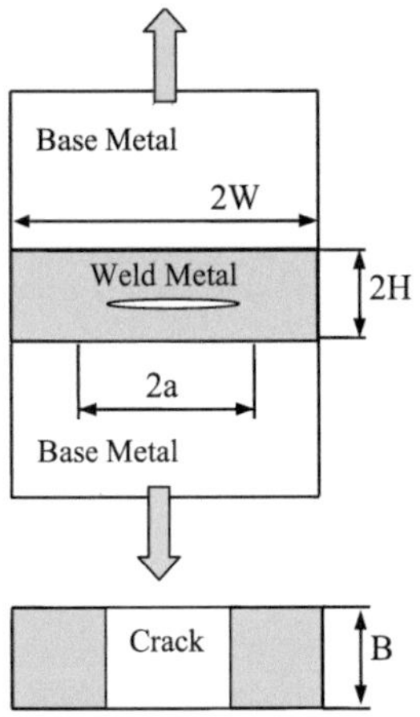

Yield load of the joined component F_{YM}:

Undermatching (UM): $M = \sigma_{YW}/\sigma_{YB} < 1$

$$\frac{F^{*}_{YM}}{F_{YB}} = \begin{cases} M & \text{for } 0 \le \psi \le 1.43 \\ M \cdot \left[\frac{2}{\sqrt{3}} - \left(\frac{2-\sqrt{3}}{\sqrt{3}}\right) \cdot \frac{1.43}{\psi}\right] & \text{for } 1.43 \le \psi \end{cases}$$

$$\frac{F^{**}_{YM}}{F_{YB}} = \begin{cases} M & \text{for } 0 \le \psi \le 1.43 \\ 1 - (1 - M) \cdot \frac{1.43}{\psi} & \text{for } 1.43 \le \psi \end{cases}$$

$$F_{YM} = \min.(F^{*}_{YM}, F^{**}_{YM})$$

Overmatching (OM): $M = \sigma_{YW}/\sigma_{YB} > 1$

$$\frac{F^{*}_{YM}}{F_{YB}} = \begin{cases} M & \text{for } \psi \le \psi_1 \\ \frac{24(M-1)}{25} \cdot \frac{\psi_1}{\psi} + \frac{M+24}{25} & \text{for } \psi > \psi_1 \end{cases}$$

$$\psi_1 = \left(1 + 0.43 \cdot e^{-5(M-1)}\right) \cdot e^{-(M-1)/5}$$

$$F^{**}_{YM} = (2 \cdot \sigma_{YB} \cdot B \cdot W) \text{ for all } \psi$$

$$F_{YM} = \max.(F^{*}_{YM}, F^{**}_{YM})$$

$$\psi = \frac{1}{H} \cdot (W - a)$$

Reference yield load (base metal) F_{YB}:

$$F_{YB} = 2 \cdot \sigma_{YB} \cdot B \cdot (W - a)$$
$$F_{YW} = M \cdot F_{YB}$$

Fig. 20 Strength mismatch corrected yield load of a centre cracked welded plate with a crack in the weld metal (over- and under-matching). Solution for plain stress conditions from [33]

5.3 Considering Welding Residual Stresses

In a fracture mechanics assessment it is essential to distinguish between primary and secondary stresses. Primary stresses are defined as stresses due to applied loading, i.e., forces, moments, pressure, dead weight, etc. They can cause plastic collapse because of which both, the K factor and the yield or limit load or L_r respectively have to be determined for the primary stresses. In contrast, secondary stresses are the result of locally suppressed shrinking or expansion in the component. Since they are in internal equilibrium across the section of interest they do not contribute to plastic collapse with the consequence that they are considered only in K factor determination but not for the yield load or L_r.

Usually, welding residual stresses are considered as secondary stresses. However, care is advised since residual stresses (respectively the forces and moments behind them) which are self-equilibrating across the component volume may not self-equilibrate across the section containing the crack. This is definitely the case for reaction stresses which, e.g., are generated when a beam is welded into a larger structure. What counts is the cross section containing the crack. In such cases even welding residual stresses have to be classified as primary stresses.

For mode I crack opening and small scale yielding conditions the overall stress intensity factor can be determined by simple superposition of its primary and secondary components. These are marked by the upper indices "p" and "s".

$$K_I = K_I^p + K_I^s \tag{16}$$

However, in general, when the boundary conditions of the linear elastic fracture mechanics are no valid any longer, interaction effects between the primary and secondary stresses have to be taken into account. As a consequence the real crack driving force will be larger than the sum of K_I^p and K_I^s in Eq. (16). This happens in the contained yielding range, i.e. at L_r values roughly in the range between 0 and 1. Above that value, and actually starting even below this, relaxation effects become active and the overall crack driving force becomes smaller than the sum of K_I^p and K_I^s. This interaction characteristics is described by a factor V, which is used as a correction function to K_I^s

$$K_I = K_I^p + V \cdot K_I^s \tag{17}$$

The parameter V, see Fig. 21, depends on the magnitude of the primary and secondary stresses, on the crack depth and on the ligament yielding L_r.

It is determined by the equation:

$$V = \frac{K_p^s}{K_I^s} \cdot \xi \tag{18}$$

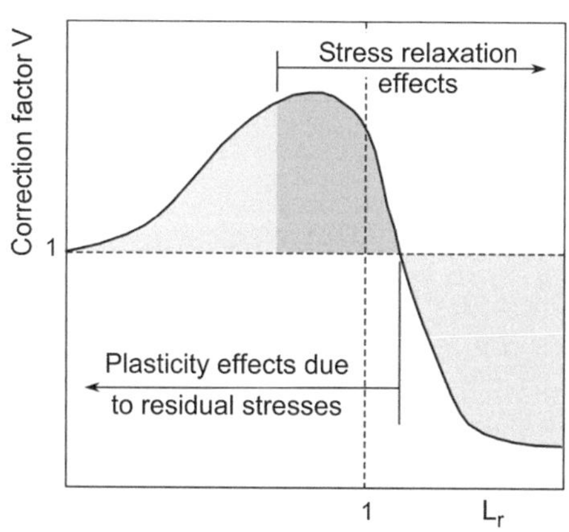

Fig. 21 V correction term as a function of the ligament yielding parameter L_r (schematic)

with K_I^s being the stress intensity factor due to the secondary stresses and K_p^s being a plastic zone-corrected value of K_I^s according to

$$K_p^s(a) = \left[\frac{a_{eff}}{a}\right]^{1/2} \cdot K_I^s(a) \tag{19}$$

K_p^s may be based on an effective crack depth, a_{eff}, which is given as the original crack depth plus the radius of the plastic zone ahead of it:

$$a_{eff} = a + \frac{1}{2\pi\beta} \cdot \left[\frac{K_I^s(a)}{\sigma_Y}\right]^2 \tag{20}$$

(β = 1 in plane stress; β = 3 in plane strain).

The function ξ is provided in Tables in the compendia mentioned where it is determined as a function of the parameters L_r and $K_p^s/(K_I^p/L_r)$.

For $K_p^s/(K_I^p/L_r) \leq 4$ a simplified procedure may be used for V, which does not need the determination of K_p^s. This is given by

$$V = \begin{cases} 1 + 0.2L_r + 0.02 \cdot [K_I^s/K_I^p/L_r] & \text{for } L_r < L_r^* \\ 3.1 - 2L_r & \text{for } L_r^* < L_r < 1.05 \\ 1 & \text{for } L_r > 1.05 \end{cases} \tag{21}$$

L_r^* is determined by the intersection of the two first equations in Eq. (21).

5.4 Validation of the SINTAP Strength Mismatch Option

The various mismatch options of SINTAP have been successfully validated [34, 35]. An example is given in Fig. 23, which is based on the mismatch-corrected yield load according to Fig. 22. As can be seen, the failure load of a friction stir weld of Al 6013 T6 was conservatively predicted, i.e., the analysis underestimated the experimental value by 16 %. Note that this is a rather poor result. Particularly the results for strength overmatched components are usually much less conservative.

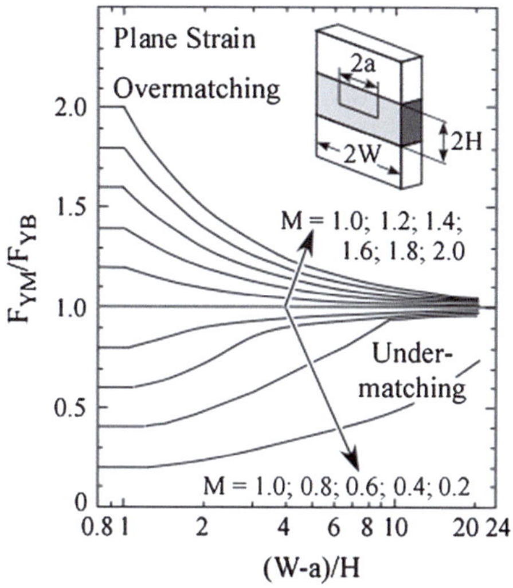

Fig. 22 Strength mismatch-corrected yield or limit load for a tension loaded plate with a central crack in the weld metal, plane strain, from [33]

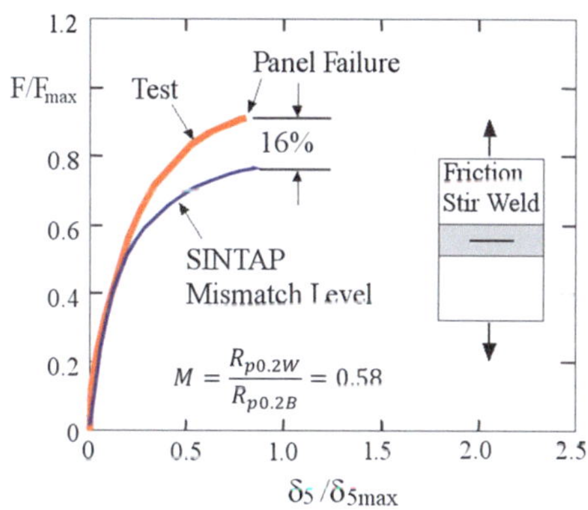

Fig. 23 Comparison of SINTAP prediction and experiment. Tension loaded plate of Al 6013 T6, friction stir weldment, Mismatch ratio M = 0.58 [36]

6 Application to a Component with Strength Mismatch

The example is an electron beam weld of an Al 7075 alloy. The analysis is performed for an M(T) specimen with a crack in the weld metal (MM = mismatch) (Table 1). It does not only include a mismatch correction for the crack driving force but considers the effect also at the material crack resistance side. The target information of the analysis is the critical load, i.e., in the present case the critical tension force for stable crack initiation.

Table 1 Input data for failure assessment

Dimensions	Material data
B = 5 mm 2 W = 160 mm $2a$ = 80 mm $2H$ = 2.4 mm $\psi = \frac{1}{H} \cdot (W - a) = 33.3$	$R_{p0.2W}$ = 276 MPa R_{mW} = 345 MPa $R_{p0.2B}$ = 507 MPa R_{mB} = 580 MPa M = 0.54 $K_{mat,k}$ = 51.0 MPa$\sqrt{m}$ from $J_{0.2}$ (MM corrected) $K_{mat,u}$ = 43.6 MPa$\sqrt{m}$ from $J_{0.2}$ (not corrected) K_{Ic} = 23.9 MPa$\sqrt{m}$ base metal E = 70,000 MPa

The yield load is determined according to Fig. 20:

$$F_{YB} = 2\sigma_{YB} \cdot B \cdot (W - a) = 202800\,\text{N}$$
$$F_{YW} = M \cdot F_{YB} = 110400\,\text{N}$$
$$F_{YM} = M \cdot \left[\frac{2}{\sqrt{3}} - \left(\frac{2 - \sqrt{3}}{\sqrt{3}}\right) \cdot \frac{1.43}{\psi}\right] = 126746\,\text{N}$$

The mismatch corrected factors, μ_M und N_M, needed for specifying the FAD line are determined by:

$$\mu_M = (M - 1) \cdot \left[\frac{\left(\frac{F_{YM}}{F_{YB}} - 1\right)}{\mu_W} + \frac{\left(M - \frac{F_{YM}}{F_{YB}}\right)}{\mu_B}\right]^{-1} = 0.221$$

with

$$\mu_B = 0.138$$
$$\mu_W = 0.254$$

and

$$N_M = (M - 1) \cdot \left[\frac{\left(\frac{F_{YM}}{F_{YB}} - 1\right)}{N_W} + \frac{\frac{(M - F_{YM})}{F_{YB}}}{N_B}\right]^{-1} = 0.05433$$

with

$$N_B = 0.0378$$
$$N_W = 0.06$$

The plastic collapse limit is determined as:

$$L_{r\,max} = 0.5 \cdot \left(1 + \frac{0.3}{0.3 - N_m}\right) = 1.11$$

The FAD failure line is given by:

$$\begin{array}{ll} f(L_r) = \left[1 + \frac{1}{2} \cdot L_r^2\right]^{-1/2} \cdot \left[0.3 + 0.7 \cdot \exp\left(-0.221 \cdot L_r^6\right)\right] & 0 \le L_r < 1 \\ f(L_r) = f(L_r = 1) \cdot L_r^{\left(N_M - \frac{1}{2 \cdot N_M}\right)} = 0.703 \cdot L_r^{-9.15} & 1 \le L_r < L_{r\,max} \\ f(L_r) = 0 & L_r < L_{r\,max} \end{array}$$

For specifying the assessment point (K_r, L_r), stress intensity factors and yield loads are needed for the geometry, crack size and loading of the component. Using the K factor solution of the M(T) plate

$$K = \frac{F}{B \cdot \sqrt{W}} \cdot \frac{1}{2} \cdot \sqrt{\frac{\pi \cdot a}{W} \cdot \left[\cos\left(\frac{\pi \cdot a}{2 \cdot W}\right)\right]^{-1}}$$

K_r is obtained as

$$K_r = \frac{K_r}{K_{mat}}$$

and

$$L_r = \frac{F}{F_{YM}}$$

Figure 24 illustrates the determination of the applied load at which the failure line FAD is met. The analysis follows the same route when instead all base metal or all weld metal conditions are assumed, i.e. without mismatch correction. All results are compared in Fig. 25.

As can be seen, the highest crack initiation load of 86 kN is obtained for the mismatch option using the mismatch corrected $J_{0.2}$ crack resistance. In comparison, the all weld metal analyses yields a crack initiation load of 73 kN which is about 15 % smaller than those based on the full mismatch analysis. The deviation becomes even more pronounced for the all base metal analysis. Caused by the lower crack resistance of the base metal the failure load drops to 52 % of the mismatch value. Note that this is also due to the fact that the base metal failure occurs as cleavage fracture such that a K_{IC} value has to be used as the input information.

The example illustrates that only a failure analysis taking into account the strength mismatch effect provides an adequate assessment and allows for the exploitation of the real loading capacity of the component. This benefit justifies the increased effort of this kind of analysis.

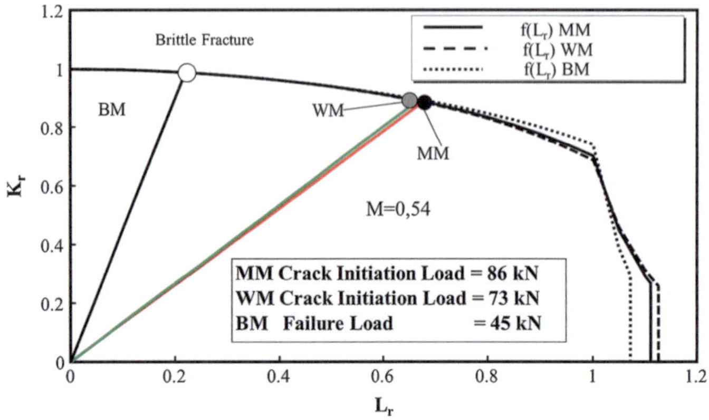

Fig. 24 FAD analysis of an M(T) specimen of Al 707, electron beam weld, analysis with mismatch correction. *BM* Base Metal, *WM* Weld Metal, *MM* mismatch

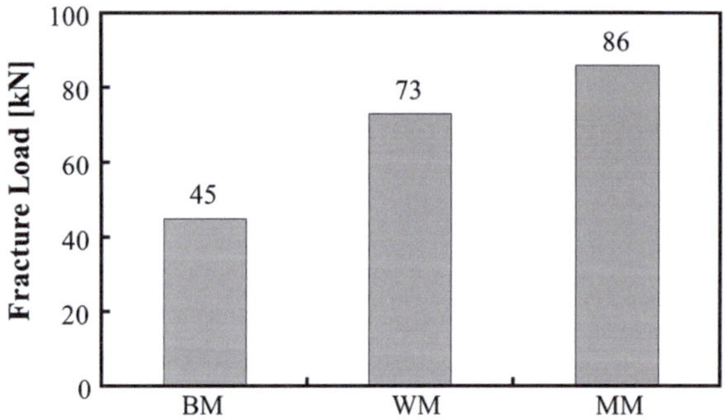

Fig. 25 Failure loads of the Al 7076 M(T) specimen of the electron beam weld analyzed in Fig. 24. Comparison with the results of all base metal (BM) and all weld metal (WM) analyses. MM stands for strength mismatch

7 Summary

With respect to fracture mechanics assessment, weldments are characterised by a number of specific features which affect the fracture toughness as well as the crack driving force in the component. These comprise geometrical imperfections (misalignment), non-homogeneous microstructures, welding residual stresses and strength mismatch. The present paper highlights two of these items: strength mismatch and residual stresses. It is demonstrated how they have to be taken into account in fracture mechanics assessment under static loading. An example is provided for the influence of strength mismatch on fracture toughness. As shown, non-consideration would have significantly affected the result. The same statement is true with respect to component behavior. This shows why strength mismatch has to be considered at both sides of fracture assessment. Laser and electron beam welds are much stronger affected by strength mismatch as conventional welds.

References

1. Zerbst U, Ainsworth RA, Beier HTh, Pisarski H, Zhang ZL, Nikbin T, Nitschke- Pagel T, Münstermann S, Kucharczyk P, Klingbeil D (2014) Review on fracture and crack propagation in weldments—A fracture mechanics perspective. Eng Fract Mech 132:200–276
2. Degenkolbe J, Uwer D, Wegman H (1985) Kennzeichnung von Schweißtemperaturzyklen hinsichtlich ihrer Auswirkungen auf die mechanischen Eigenschaften von Schweißverbindungen durch die Abkühlzeit t8/5 und deren Ermittlung. Technical Report, Thyssen
3. Werkstoffblatt, SEW 088 (1993) Schweißgeeignete Feinkornbaustähle, Richtlinien für die Verarbeitung, besonders für das Schmelzschweißen
4. Seyffarth P, Meyer B, Scharff A (1992) «Großer Atlas Schweiß-ZTU-Schaubilder». DVS-Verlag, Düsseldorf. Fachbuchreihe Schweißtechnik Band 110
5. Buchmayr B (1991) Computer in der Werkstoff und Schweißtechnik/Anwendung von mathematischen Modellen. DVS-Verlag, Düsseldorf
6. Koch F, Enderlein M, Pietrzyk M (2013) Simulation of the temperature field and the microstructure evolution during multi-pass welding of L485 MB pipeline steel. Comput Methods Mater Sci 13:173–180
7. Yorioka N, Okumura M, Kasuya T, Cotton HJ (1987) Prediction of HAZ Hardness of transformable steels. Metal Constr 19:217–223
8. Toyoda M (1989) Significance of procedure/evaluation of CTOD test of weldments International M; Document X-1192-89. Institute of Welding (IIW)
9. Dos Santos J, Çam G, Torster F, Isfan A, Riekehr S, Ventzke V, Koçak M (2000) Properties of power beam welded steels, Al- and Ti-alloys: significance of strength mismatch. Weld World 44:42–64
10. Schwalbe K-H, Kim Y-J, Hao S, Cornec A, Koçak M (1997) EFAM ETM-MM 96: the ETM method for assessing the significance of crack-like defects in joints with mechanical heterogeneity (strength mismatch), GKSS Forschungszentrum, GKSS Bericht 97/E/9
11. Masubuchi K, Blodgett OW, Matsui AS, Ruud CO, Tsai CI, Hesse TD (2001) Residual stresses and distortion. In: Welding handbook, 9th edn, vol 1. Welding Science and Technology, Miami, FL. American Welding Society, pp 297–357
12. Leggatt RH (2008) Residual stresses in welded structures. Int J Pres Ves Piping 85:144–151
13. Bouchard PJ (2008) Combined use of FE-simulations and neutron/x-ray experiments, VDI-Expert Forum 10th April 2008. http://ebookbrowse.com/vdi-08-06-bouchard-pdf-d299890187
14. SINTAP (1999) Structural integrity assessment procedure. Final Revision. EU-Project BE 95-1462. Brite Euram Programme, Brüssel
15. Bate SK, Green D, Buttle D (1997) A review of residual stress distributions in welded joints for the defect assessment of offshore structures. In: Offshore Technology Report (OTH) Series 482, Health and Safety Executive (HSE)
16. ISO 15653 (2010) Metallic materials—Method for the determination of quasistatic fracture toughness of welds. International Organisation for Standardization (ISO)
17. BS 7448 (1997) Fracture mechanics toughness tests. Part 2: Method for determination of KIc, critical CTOD and critical J values of welds in metallic materials. British Standards Institution, London

18. Schwalbe K-H, Heerens J, Zerbst U, Koçak M (2001) EFAM-GTP 02. The GKSS procedure for determining the fracture behaviour of materials. GKSS-For-schungszentrum Geesthacht. GKSS-Bericht 2002/2024
19. Peter S (2005) Untersuchung der mechanischen und bruchmechanischen Eigenschaften von reibrühr- und elektronenstrahlgeschweißten Al-Legierungen TU Bergakademie Freiberg Diplomarbeit
20. Toyoda M (2002) Transferability of fracture mechanics parameters to fracture performance evaluation of welds with mismatching. Prog Struct Eng Mater 4:117–125
21. Zerbst U, Hübner P (2004) Bruchmechanische Bewertung von Fehlern in Schweißverbindungen. DVS-Merkblatt 2401, DVS-Verlag Düsseldorf, Fachbuchreihe Schweißtechnik, Bd. 101
22. Koçak M, Webster S, Janosch JJ, Ainsworth RA, Koers R (2006) Fitness for service procedure (FITNET), Annex C, Brussels
23. R6, Revision 4 (2009) Assessment of the integrity of structures containing defects. EDF Energy, Barnwood, Gloucester
24. BS7910 (2005) Guide on methods for assessing the acceptability of flaws in metallic structures. British Standards Institution (BSI), London
25. Zerbst U, Schödel M, Webster S, Ainsworth RA (2007) Fitness-for-service fracture assessment of structures containing cracks. A workbook based on the European SINTAP/FITNET Procedure. Elsevier, Amsterdam
26. Ainsworth RA (2003) Failure assessment diagram methods. In: Practical failure assessment methods. Comprehensive structural integrity (CSI), vol 7, Chapter 7.03. Elsevier, Amsterdam, pp 89–132
27. Zerbst U, Ainsworth RA, Schwalbe K (2000) Basic principles of analytical flaw assessment methods. Int J Press Vess Pip 77:855–867
28. Zerbst U, Schwalbe K, Ainsworth RA (2003) An overview of failure assessment methods in codes and standards. In: Practical failure assessment methods. Comprehensive structural integrity (CSI), vol 7, Chapter 7.01. Elsevier, Amsterdam, pp. 1–48
29. Schwalbe K-H, Zerbst U (2003) Crack driving force estimation methods. In: Practical failure assessment methods. comprehensive structural integrity (CSI), vol 7, Chapter 7.04. Elsevier, Amsterdam, pp 133–1756
30. Miller AG (1988) Review of limit loads of structures containing defects. Int J Press Vessel Pip 32:197–327
31. Laham S (1999) SINTAP: Stress intensity factor and limit load handbook, vols Report EPD/GEN/REP/0316/98, Issue 1. British Energy Generation Ltd. (BEGL)
32. Schwalbe K-H, Zerbst U, Kim Y-J, Brocks W, Cornec A, Heerens J, Amstutz H (1998) EFAM ETM 97: the ETM method for assessing crack-like defects in engineering structures, GKSS Report 98/E/6, Appendix 1, Geesthacht. GKSS Research Centre, Germany
33. Schwalbe K-H, Kim Y-J, Hao S, Cornec A, Koçak M (1997) EFAM ETM-MM 96: The ETM method for assessing the significance of crack-like defects in joints with mechanical heterogeneity (strength mismatch). GKSS-Forschungszentrum, GKSS Bericht 97/E/9
34. Zerbst U (2005) Erfahrungen mit der europäischen Fehlerbewertungsprozedur SINTAP. Materialprüfung 1/2005
35. Zerbst U, Brocks W, Heerens J, Schödel M, Scheider I, Steglich D, Seib E, Cornec A, Schwalbe K-H (2005) Failure assessment concepts for thin-walled structures containing crack-like defects ICAF 2005. Hamburg, 6–10 Jun 2005
36. Seib E, Koçak M, Assler H (2004) Fracture assessment of welded aerospace aluminium alloys using SINTAP route. Weld World 48:2–8

Application of Fracture Mechanics for the Life Prediction of Critical Rotating Parts for Aero Engines

M. Springmann, M. Schurig, D. Hünert, W. Rothkegel and H. Schlums

Abstract The application of fracture mechanics based methods forms an integral part in the prediction of safe life and damage tolerant behaviour of critical parts in aero engines. For compressor and turbine discs, this comprises the calculation of cyclic life to burst, integrity under extreme loading conditions and damage tolerant behaviour for different kinds of anomalies encountered in manufacturing and/or in-service application. Taking the example of the introduction of the new nickel-based disc alloy ATI 718Plus®, the typical methods, the extent of material and component testing, and some limitations of current methods are presented. This project is a joint effort between Rolls-Royce Deutschland and several universities and research institutions.

Keywords ATI 718Plus · Material testing · Fracture mechanics · Damage tolerance

1 Introduction

The requirements for jet engines are increasing every year. In the development process of a jet engine compressor and turbine architecture the relatively massive discs play an important role in considering mass reduction and improved safety concepts. Precise knowledge of the crack initiation resistance and crack propagation behaviour in highly loaded discs is a precondition for achieving an optimal design in terms of life time prediction and ultimate load failure.

M. Springmann (✉)
Beuth Hochschule für Technik Berlin, Luxemburger Straße 10, 13353 Berlin, Germany
e-mail: springmann@beuth-hochschule.de

M. Schurig · D. Hünert · W. Rothkegel · H. Schlums
Rolls-Royce Deutschland Ltd & Co KG, Eschenweg 11, 15827 Blankenfelde-Mahlow, Germany

G. Hütter and L. Zybell, *Recent Trends in Fracture and Damage Mechanics*, DOI 10.1007/978-3-319-21467-2_4

For the introduction of a new material in the context of the certification process of engine discs a comprehensive laboratory specimen test programme is required that allows adequate characterisation of the disc material under various loading and environmental conditions such as high temperature, large local deformation and cyclic events. The specimen test data are used to quantify material models for creep, plasticity, crack growth and crack initiation behaviour as well as correlations for lifetime predictions. A component test programme is required in order to validate the material behaviour on the component level under more realistic complex loading conditions. This allows the representation and consideration of bulk residual stresses, of high material volume accounting for scatter in material properties, as well as of machining and surface treatment effects such as shot-peening and their influences on crack initiation. For certain disc areas such as air holes or disc fir trees, so-called subcomponent tests are utilized reflecting the geometric features and associated typical loading conditions.

The complex loading of aero engine discs is the result of different flight phases (generally for civil application: taxi, take-off, climb, cruise, decent and approach), involving varying disc speeds and corresponding temperature profiles. Every material point of an engine disc is subjected to a specific combination of stress and temperature over the flight cycle. An example is shown in Fig. 1 [1].

Engine discs are designed for finite fatigue life. Hence, the number of released flight cycles is limited and could possibly determine the shop visit intervals of an aero engine. Therefore, the precise knowledge of the finite low cycle fatigue (LCF) life is a huge cost and safety issue. To fulfil the safety requirements every disc loading and perturbation with a certain probability must be assessed and

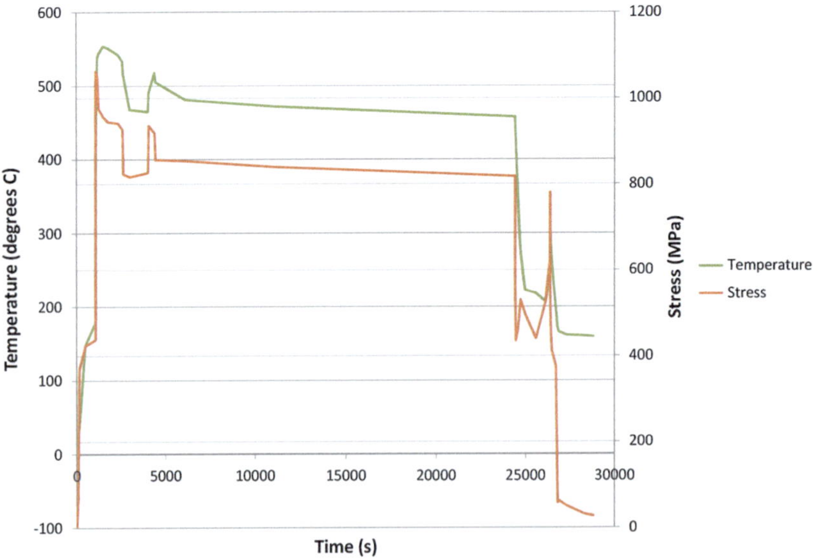

Fig. 1 Typical stress and temperature time histories [1]

included in the calculation of the safe cyclic life. Any damage or material anomaly potentially reduces the LCF initiation life. Damage tolerance is an engineering approach to account for those damages and anomalies using fracture mechanics methods.

2 Material A718Plus

ATI 718Plus® (A718Plus) is a new nickel-based superalloy for potential future implementation for compressor and turbine aero engine discs developed and supplied as billet material by ATI Allvac. A718Plus is currently being evaluated at Rolls-Royce Deutschland (RRD) as a low-cost alternative to Waspaloy or Udimet 720Li or to replace IN718 at higher operating temperatures.

Historically, IN718 has been limited by its maximum operating temperature. A718Plus combines the desirable processability and weldability of IN718 with the higher temperature capability and thermal stability of Waspaloy or Udimet 720Li but at lower costs.

The alloy is strengthened by precipitations of Gamma-Prime and Delta/Eta-phase, which ensure the high thermal stability of the alloy. The high chromium and cobalt content guarantees a superior oxidation resistance. The exact composition of A718Plus is shown in Table 1 and the effects of the microstructure on the material properties can be consulted in [2].

A new forging route and heat treatment procedure was developed by Rolls-Royce Deutschland and Otto Fuchs (forging house) to optimise the mechanical properties of the alloy in terms of integrity, LCF and damage tolerance requirements. A main focus of the multi-stage forging and heat treatment process was set on the adjustment of the microstructure to improve the resistance of the alloy against dwell crack propagation, see [3].

The mechanical properties, especially the crack propagation resistance, are highly dependent on the amount and orientation of the Delta-phase within the forging as well as the size of the Gamma-Prime precipitates. Both precipitates were adjusted during forging and heat treatment in such a way, that they oppose the main stress during operation. The orientation and size of the Delta-phase can be influenced by adapting the multistep forging procedure and by applying additional heat treatments prior to and during forging. The size of the Gamma-Prime precipitates is controlled by the heat treatment. The microstructure of A718Plus after forging and heat treatment is shown in Fig. 2. Here, Delta-phase precipitates are visible as white particles concentrated at the grain boundaries.

Table 1 Detailed composition of A718Plus (in weight %)

Ni	Cr	Fe	Co	Nb	Mo	W	Al	Ti	C	P	B
Bal.	18	10	9	5.45	2.8	1	1.45	0.7	0.025	0.014	0.006

Fig. 2 Microstructure of A718Plus after forging and heat treatment, by courtesy of RRD

A comprehensive material development programme is running to assess and qualify A718Plus. This material is being developed for engine use as part of the German-funded LuFoIV Aerospace Research Programme.

3 Material Testing

An extensive material specimen test programme has been set up in order to qualify the new nickel-based disc alloy A718Plus. An overview about this test programme is given in [4].

To understand the crack propagation resistance of A718Plus, Corner Crack (CC) as well as Compact Tension (CT) specimens are used for testing, see Fig. 3. A wide range of test conditions (e.g. temperature, different environments, R-ratio and different dwell times at maximum load condition) was investigated, also to understand when time- and temperature-dependent effects become important.

In order to underpin the growing importance on damage tolerance of disc materials, the crack propagation tests are accompanied by crack initiation LCF tests,

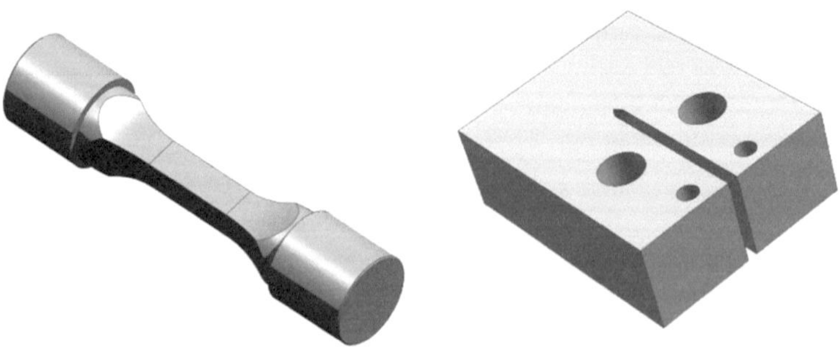

Fig. 3 *Left* CC specimen, *right* CT specimen

where the influence of surface damages like dents or scratches on the crack initiation behaviour are studied. Specimens with different notch geometries (notch factors) were damaged by scratching the surface in a highly stressed area using a defined procedure; an example is shown in Fig. 4. The scratches were measured precisely by replica technique to obtain the geometry of the induced damage. Thus, the artificial damages could be compared to real disc damages and a strict process quality control could be applied to the scratch insertion. In principle the number of cycles for crack initiation plus the number of cycles for stable crack growth represent the lifetime of the considered damaged specimen or component, respectively.

Apart from the specimen test programme as described above, further testing is performed in order to qualify full-scale disc forgings. For this purpose disc forgings are cut up not only to characterise the microstructure of the material (e.g. grain size and grain size distribution or content of Delta-phase and its orientation and distribution) but also to investigate how homogeneous material properties are distributed in different areas of the forging. For this purpose specimens are machined from different areas of the forging and also in different orientation and tested at a subset of conditions as described before. These tests consist again of standard tensile tests, LCF tests, creep tests and crack propagation tests for defined test conditions.

The aim of the different specimen test programmes is not only to gain a general insight into the material behaviour but also to do sufficient testing to enable the evaluation of material scatter and to derive "minimum" material properties.

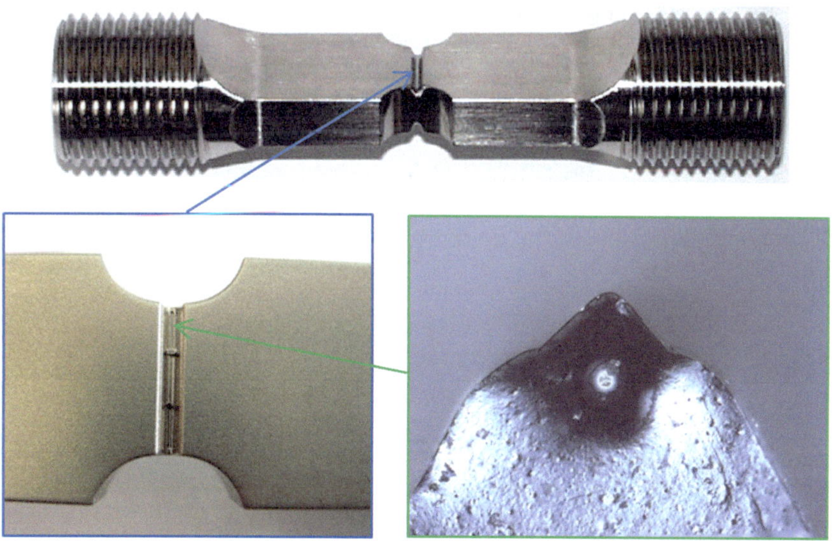

Fig. 4 *High* K_t notched specimen, scratched surface in the notch area, section of replica, by courtesy of RRD

4 Validation

For the certification of engine discs as critical parts, the understanding of the material behaviour derived from specimen testing needs to be transferred to the component level. The certification requirements comprise among others the proof of rotor integrity in an overspeed event, the declaration of a safe cyclic life and appropriate damage tolerance assessments to address the potential failure from material, manufacturing and service-induced anomalies within the approved life of the critical part.

The results of the specimen test programme are the basis for the characterisation of the fundamental material behaviour of the new alloy A718Plus under relevant loading conditions. However, on engine component level additional influences, like multi-axial stress fields, bulk residual stresses, stress redistribution processes due to plastic deformation, volume effects and surface conditions from manufacturing, determine substantially the potential failure of discs.

For some typical disc features the conventional spin tests are not suitable to achieve local loading conditions that are comparable to engine conditions. On this account subcomponent tests were set up to validate features such as disc rim air holes and disc fir trees, respectively.

The disc rim air holes are subjected to high tensile and high compressive loads due to the varying disc speed in a flight cycle in combination with the transient temperature gradient. In a conventional spin test the transient and stabilised engine temperature gradient could not be achieved. For that reason a subcomponent air holes test was developed. A test disc with a reduced number of air holes with manufacturing process parameters equivalent to those used to produce the engine discs was manufactured. The subcomponents are cut out of the disc including one air hole each, see Fig. 5. Using a conventional test machine, tensile and compressive loads were applied to the subcomponent in order to achieve the required local air hole loading.

The test definition was supported by numerical studies using material data generated by simple specimen tests. Assuming an engineering crack (with a defined crack length) occurred at the position of highest air hole stress, a crack propagation analysis was accomplished to obtain the number of cycles from crack initiation until total failure of the subcomponent specimen. Thus, the appropriate number of cycles for crack initiation can be determined based on the measured total number of cycles to subcomponent specimen failure and the calculated number of stable crack growth. Here, the crack initiation positions of the failed subcomponent test specimen and the analysis assumption must be aligned and were validated by the test outcome. However, the experimental validation also included the time to crack initiation that relies on the loading history. For turbine discs, this includes cyclic out-of-phase thermo-mechanical loading that could not be applied to a subcomponent test specimen. Therefore, the subcomponent tests were augmented by thermo-mechanical fatigue testing that proved conservatism of the isothermal test

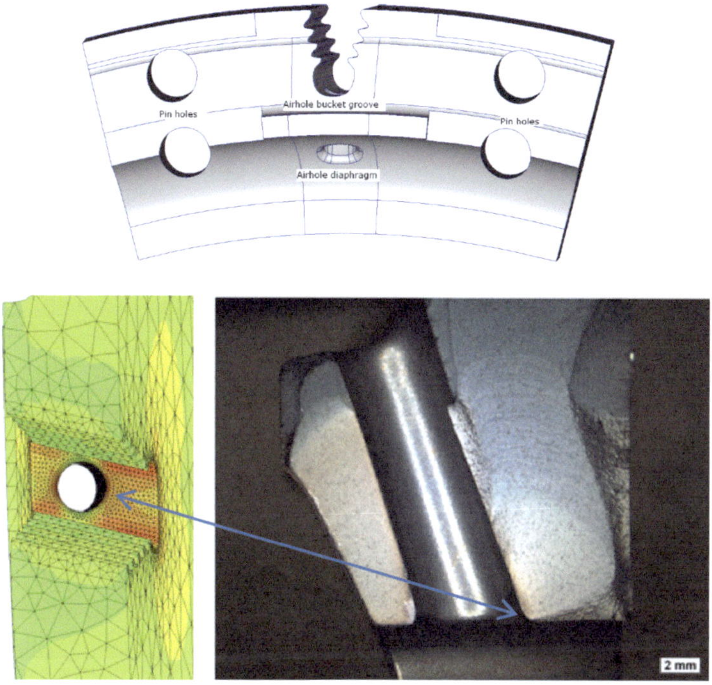

Fig. 5 Subcomponent air hole test specimen, FEM model, fracture surface, by courtesy of RRD

setup [5]. A typical fracture surface and FEM predicted initiation site is shown in Fig. 5.

Within bladed engine discs the fir tree is often one of the areas with the highest loading. Under engine conditions, a glazing layer often forms at high temperature between disc and blade contact faces for components made from nickel-based alloys. Spin tests are usually run under vacuum conditions where the glazing layer cannot be formed. This would lead to significantly higher friction between the mating parts and hence to unrepresentative fir tree loading. For that reason a fir tree bi-axial test was set up (see Fig. 6). The tensile load reproduces the blade centrifugal load and the compressive load represents the neighbouring blades. The load path of the compressive load was optimised to achieve the same work of friction as in the engine over a flight cycle.

By cyclic testing, crack initiation and propagation were optically monitored, see Fig. 6 (right). In some tests, the test conditions were changed after crack initiation and some cycles of crack propagation in order to study the influence of different parameters on further crack propagation in a subcomponent. Figure 7 shows the fracture surface of a lab-opened bi-axial fir tree test specimen run with short and subsequently long dwell times. It was shown that the appearance of the fracture surface and the propagation rate is influenced, while other possible mechanisms (e.g. out-of-plane crack growth due to complex creep fatigue interaction) could be

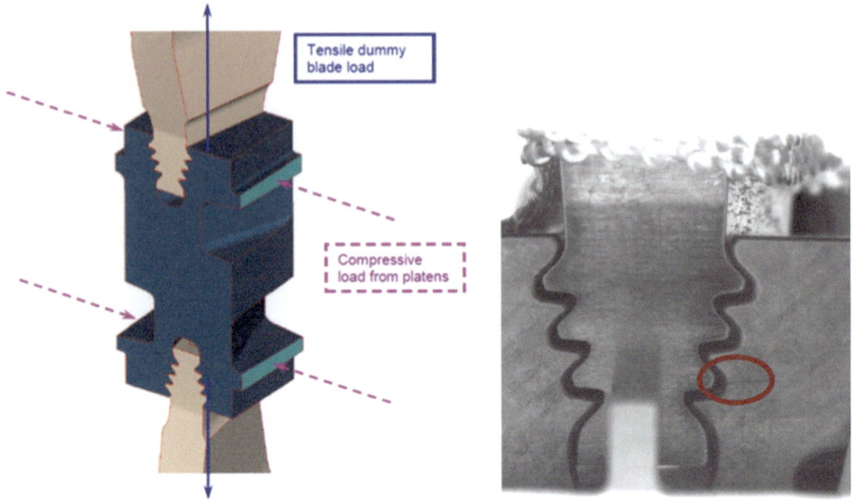

Fig. 6 *Left* Fir tree bi-axial subcomponent test setup, *right* cracked subcomponent test specimen, by courtesy of RRD

excluded. In Fig. 7 superimposed red ellipsoidals show results of simulation that predicted the crack shape and size.

Besides subcomponent LCF tests, crack initiation and propagation tests with a scratched surface were performed, see Fig. 8. The aim was to validate the results from simple specimen tests on the component level. A main focus was on the orientation of the proceeding crack. Again, the simulation resulted in a prediction of the crack propagation phase after initiation of a crack from artificial surface damage (a scratch) at the same location as shown in Fig. 6 (right). At this stage a generalised statement is not yet possible, because only a few experiments have been analysed and further investigations are ongoing.

Fig. 7 Crack propagation test results of bi-axial fir tree test specimen and superimposed crack shape prediction, by courtesy of RRD

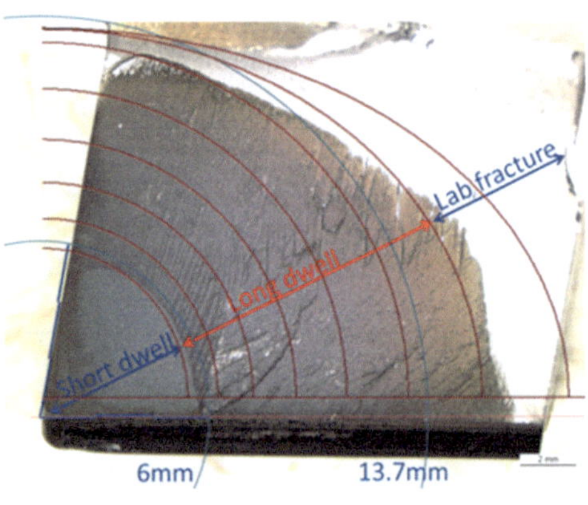

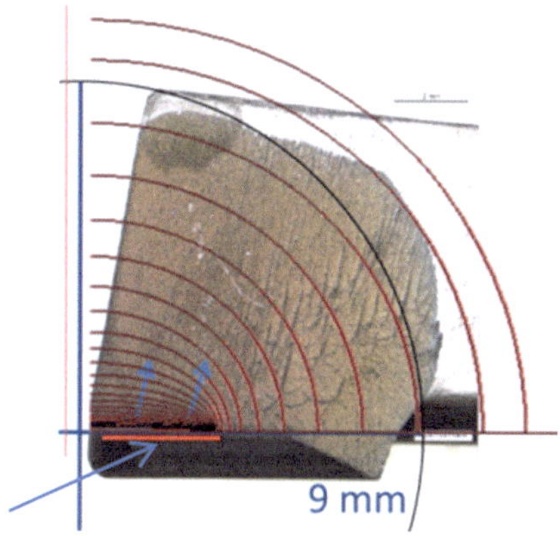

Fig. 8 Crack propagation test results of scratched bi-axial fir tree test specimen and superimposed crack shape prediction, by courtesy of RRD

5 Modelling

The material testing results are the basis for the selection of appropriate material models and for the identification of the material parameters.

The results and the processing of various A718Plus crack propagation experimental data indicated that the application of the widely used Paris law is reasonable:

$$\frac{da}{dN} = C \cdot \Delta K^n \quad (1)$$

Here, da/dN is the crack growth rate and C as well as n are material parameters depending on temperature, R-ratio and dwell time.

The parameter identification was done initially for every valid experiment. Subsequently, the adjusted curves were compared for the same test conditions. In this context, dependencies in terms of location and orientation of the specimen within the forging could be investigated.

To achieve a certain internal material quality level required for critical parts a defined number of experiments at given test conditions must be available. From the available data a "typical" and "minimum" -3σ (i.e. describing the fastest propagation rate) curve were derived for every test condition.

The previous investigations revealed material parameters for explicit test conditions, see Fig. 9. For the application regarding engine conditions the crack propagation parameters must be continuously available over the relevant temperature and R-ratio range.

The temperature dependence of the Paris law parameters C and n is considered from approximately 20 °C until 700 °C. In this temperature range the crack growth rate (da/dN) over temperature is outlined for a defined $\Delta K_{\min}$ and $\Delta K_{\max}$,

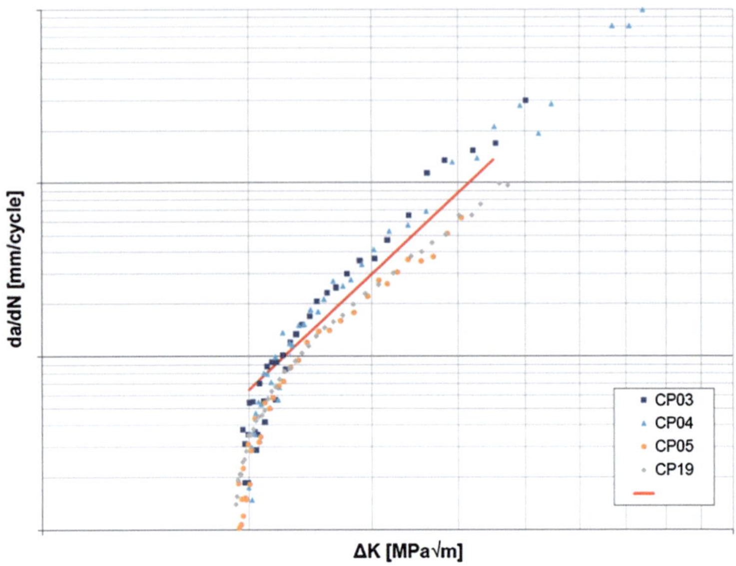

Fig. 9 Experimental crack propagation results and a "typical" Paris law based approximation for a defined test condition

representing the used minimum and maximum cyclic stress intensities for a test condition. The measured data points are approximated by a polynomial of the kind:

$$\log\left(\frac{da}{dN_{\max/\min}}\right)(T) = A_0 + A_1 T + A_2 T^2 + \cdots + A_p T^p \qquad (2)$$

In Figs. 10 and 11 it can be noticed that A718Plus has a special behaviour around 400 °C. The crack growth rate at 400 °C is on average lower compared to the rate at 20 °C. In tensile tests the so-called Portevin-Le Chatelier Effect can be observed at around 400 °C test temperature [2]. Microstructural investigations are ongoing to find the root cause of these effects.

The temperature-dependent material parameter $n(T)$ of the Paris law can be calculated as follows:

$$n(T) = \frac{\log\left(\frac{da}{dN_{\max}}(T)\right) - \log\left(\frac{da}{dN_{\min}}(T)\right)}{\log \Delta K_{\max} - \log \Delta K_{\min}} \qquad (3)$$

The parameter $C(T)$ is then derived from the Paris law:

$$C(T) = \frac{da}{dN}(T) \cdot \Delta K^{-n(T)} \qquad (4)$$

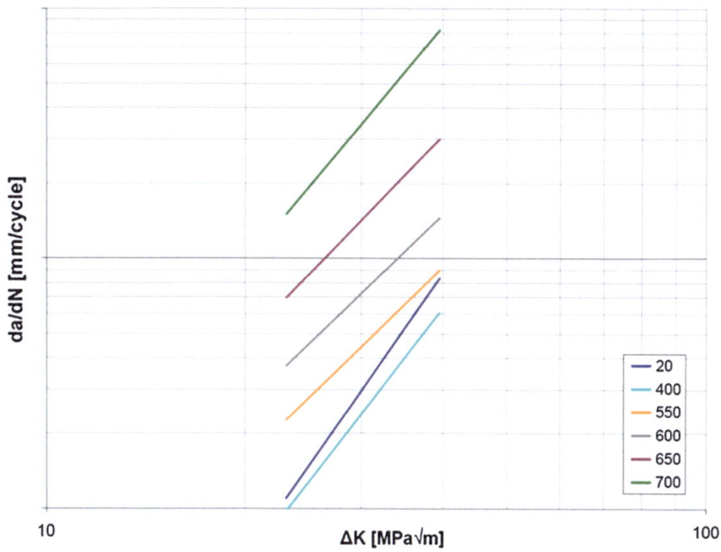

Fig. 10 Adjusted crack propagation curves for a fixed R-ratio but pending on temperature in (°C)

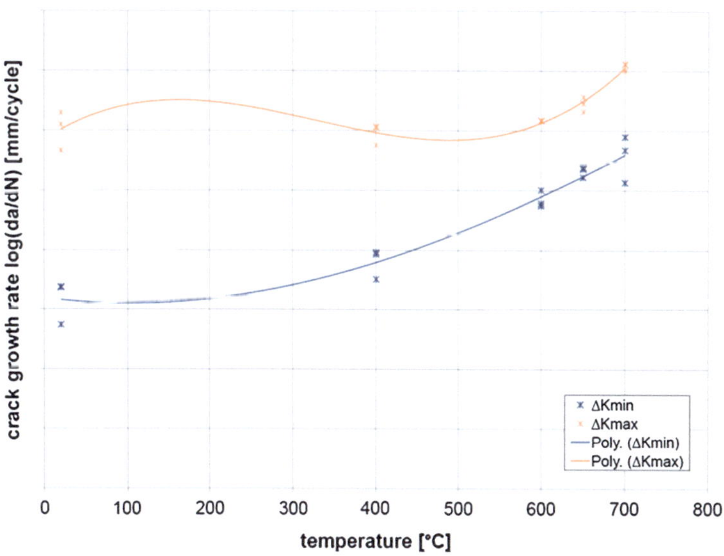

Fig. 11 Selected results for temperature dependent crack growth rates

During an aero engine flight cycle, complex stress scenarios occur within an engine disc and result in different R-ratios at different locations on the disc. The crack propagation test programme covers selected R-ratios. Intermediate points are calculated by using the cyclic stress intensity $\Delta K_{R=0}$ at $R = 0$ and the so-called Walker exponent m:

$$\Delta K_R = \frac{\Delta K_{R=0}}{(1-R)^{m-1}} \tag{5}$$

The Walker exponent is a specific material constant and determined by various experiments with different R-ratios, see Fig. 12. If the condition is fulfilled that in a log-log diagram the crack propagation curves are parallel for different R-ratios, then the Walker exponent m can be calculated by results of experiments with two different R-ratios (a and b) as follows:

$$m = 1 + \frac{\log(\Delta K_{R_b}) - \log(\Delta K_{R_a})}{\log(1-R_a) - \log(1-R_b)} \tag{6}$$

Experimental results for A718Plus showed that a specific Walker exponent does not cover the whole range of conditions. Therefore, the adjustment of m was accomplished on a sequence of intervals.

Engine discs are subjected to severe stresses and high temperatures over a relatively long period of time. Experimental investigations with Corner Crack specimens have shown that the crack growth rate is strongly dependent on the dwell time at maximum load (see Fig. 13), when the test temperature exceeds a material-specific threshold value. That means that, besides the time-independent

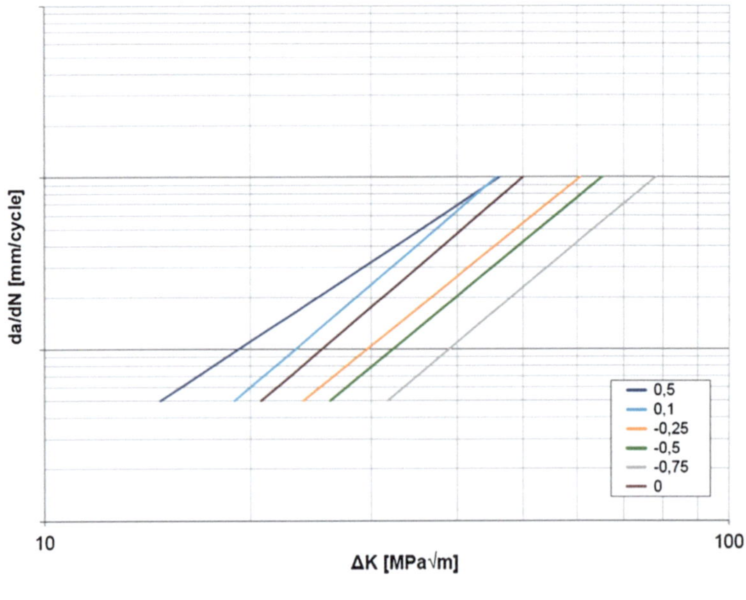

Fig. 12 Adjusted crack propagation curves at a specific temperature for different R-ratios

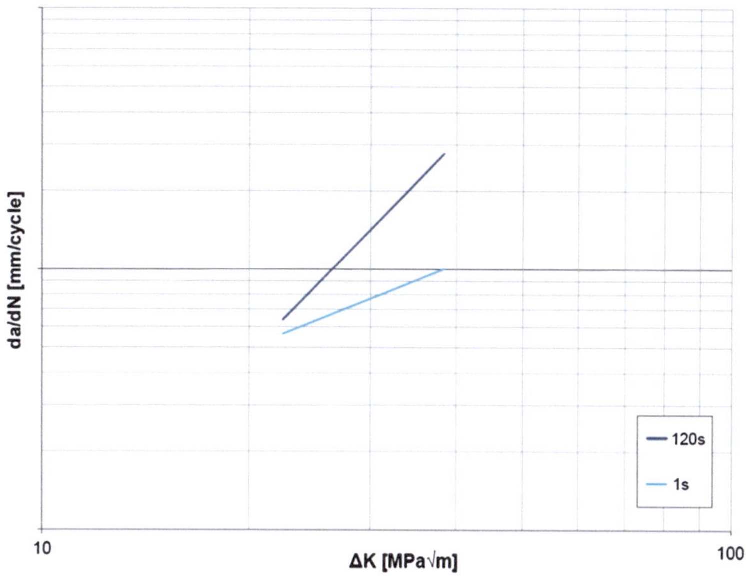

Fig. 13 Crack propagation at elevated temperature for different dwell times at maximum load

crack growth behaviour, the time-dependent crack growth behaviour needs to be taken into consideration. The total crack growth rate is composed by cycle-dependent (da/dN) and time-dependent (da/dt) portions:

$$\left(\frac{da}{dN}\right)_{total} = C \cdot \Delta K^n + \sum_{cycle} D' \cdot K^n \tag{7}$$

with

$$D' = A \cdot \exp\left(-\frac{B}{T}\right) \tag{8}$$

as a function of temperature T and using the temperature-independent material parameters A and B [6]. K denotes the instantaneous stress intensity factor.

Using the latest experimental evidence (more that 300 specimens at different temperatures, R-ratios and dwell times) the parameter fit for $\{C(T), n(T), m(T), A, B\}$ was updated. Experimental results of crack propagation rates are compared to the model predictions in Fig. 14. Fitting is generally performed on averaged results as shown above, the quality is judged by using several points from each test, replicating different stress intensities. While the scatter is higher in the tests with long dwell time, the qualitative shape of the curve is better predicted in these tests. The main reason is that near-threshold effects are not accounted for in the model.

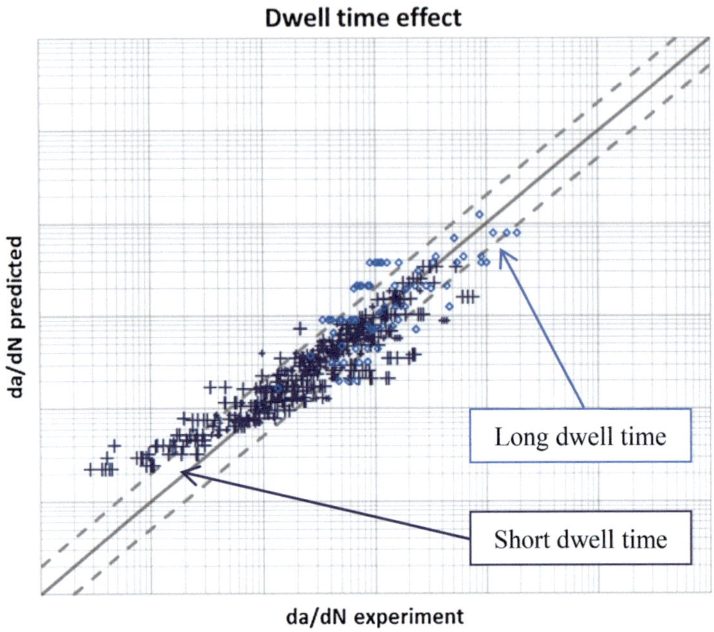

Fig. 14 Comparison of predicted and experimental crack propagation rates for different dwell times

6 Summary and Outlook

The introduction of a new material for critical parts in aero engines requires a comprehensive test programme and analyses to maintain the disc integrity requirements and to provide the basis for the determination of safe cyclic life. In this contribution the fracture mechanical material test programme and crack propagation modelling of the new nickel-based alloy A718Plus was briefly presented.

Besides the introduction of the material for future applications, the specimen and component test programme will be completed to provide data for the full designated application range. The test results will also be used to develop material models which describe the crack initiation behaviour and damage mechanisms of the alloy. Special focus is put on damage tolerance assessments with regard to various surface impacts. The consideration of the damage initiation test results based on testing of specimens with artificially introduced scratches at the peak stress position for a general damage tolerance approach is still ongoing.

Acknowledgments The work was conducted as a part of the German-funded research programme RoKoTec (LuFoIV) and supported by the Bundesministerium für Wirtschaft und Energie (BMWi) under grant number 20T0813. Special thanks to Bastian Kähler for the extensive processing of the test data and performance of numerical studies.

References

1. Timbrell C, Chandwani R, Jacques S, Waterhouse L, Wisbey A, Williams S (2013) Comparing crack growth testing and simulation results under thermo-mechanical fatigue conditions. In: 13th international conference on fracture, Beijing, China, June 16–21
2. Löhnert K. Pyczak F (2010) Microstructure evolution in the Nickel base superalloy Allvac®718Plus™, superalloy 718 and derivatives, Pittsburgh, USA, pp 876–891
3. Löhnert K (2011) Einfluss der thermomechanischen Vorbehandlung auf die Eigenschaften der Nickelbasissuperlegierung A718Plus, Dissertation, University Erlangen-Nürnberg, Germany
4. Springmann M, Rothkegel W, Hünert D, Schlums H (2013) Introduction of the new nickelbase alloy ATI 718Plus® for critical parts. In: 7th International conference on low cycle fatigue, Aachen, Germany
5. Schurig M, Huenert D, Rothkegel W, Schlums H (2014) LCF specimen and subcomponent tests for the introduction of the new Nickel-based turbine disc material ATI 718Plus. In: Grellmann W, Frenz H (eds) Tagungsband Werkstoffprüfung, Berlin, Germany, pp 91–96
6. Timbrell C, Chandwani R, MacLachlan D, Williams S (2012) A time dependent crack growth law for high temperature conditions. In: NAFEMS European conference: multiphysics simulation, Frankfurt, Germany, Oct 16–17

Consideration of Fatigue Crack Growth Aspects in the Design and Assessment of Railway Axles

Igor Varfolomeev and Michael Luke

Abstract Current design rules for railway wheelsets do not directly address issues related to fatigue crack propagation. Nevertheless, the latter topic is a part of the revised safety concept for passenger trains recently adopted in German railway applications. Numerous research activities, including international cooperative projects, have been conducted in the past decade aiming at quantifying fatigue crack growth rates in railway axles and estimating their inspection intervals based on the fracture mechanics methodology. This paper summarizes some experience and findings obtained by the authors within several studies dealing with the assessment of fatigue crack propagation in railway steels. Particular aspects highlighted in the paper include material characterization, effects of the specimen geometry and crack tip constraint on fatigue crack growth rates, stress analyses of axles and wheelsets, the derivation of stress intensity factor solutions applicable to specific conditions achieved in railway axles, considerations of the variability and scatter of geometrical parameters and material data in fatigue crack growth calculations.

1 Introduction

Railway axles are traditionally designed to meet criteria of "infinite" life, i.e. to achieve a sufficient number of load cycles which exceeds the conventional endurance range of about 10^7 cycles by at least two orders of magnitude. Relevant design rules developed during the course of several decades are currently incorporated in a number of industrial standards, e.g. [1, 2]. Besides the requirement of safe design, regular non-destructive examinations (NDE) of wheelsets are included

I. Varfolomeev (✉) · M. Luke
Fraunhofer IWM, Wöhlerstr. 11, 79108 Freiburg, Germany
e-mail: igor.varfolomeev@iwm.fraunhofer.de

M. Luke
e-mail: michael.luke@iwm.fraunhofer.de

G. Hütter and L. Zybell, *Recent Trends in Fracture and Damage Mechanics*, DOI 10.1007/978-3-319-21467-2_5

in maintenance procedures of railway operators in order to prevent that possible fatigue, corrosion or ballast impact damage of axle parts may lead to crack initiation and growth with a potential hazard of train derailment, see e.g. [3, 4]. In this context, the fracture mechanics methodology may help to combine information obtained by NDE with quantitative estimates of fatigue crack propagation, starting from a reasonably postulated initial crack size, and thus establish a basis for specifying inspection intervals. The way from safe design towards safe operation of railway axles as well as developments of the assessment methodology and tools are reflected in numerous publications, e.g. [5–13]. Moreover, several international cooperative projects [14–16] have been devoted to particular aspects of the design and fatigue assessment of railway wheelsets. Despite extensive research conducted in the past, needs for further improving the analysis methods, extending material knowledge and better understanding of various effects, such as those due to residual stresses, short crack behaviour, crack tip constraint, uncertainties in the definition of input data for crack growth analyses, are some principal targets of on-going research activities [17].

This paper summarizes some experience gained by the authors within several studies dealing with the assessment of fatigue crack propagation in railway axles. Particular aspects addressed in the paper are material characterization, axle stress analysis, accurate determination of stress intensity factors for loading conditions achieved in railway axles, effects of the specimen geometry on fatigue crack growth rates, incorporation of the variability and scatter of geometrical parameters and material data in the fatigue life assessment procedure.

2 Review of Fatigue Crack Growth Data for Railway Axle Materials

2.1 Material Characterisation

Among the materials employed in railway axle manufacture, two mild steel grades EA1N (material designation C35) and EA4T (25CrMo4) are particularly referred to in the European standards EN 13103 [1] and EN 13104 [2]. Further high-strength steel grades, 34CrNiMo6+QT and 30NiCrMoV12, are common for high-speed train design and applications. In [5], fatigue endurance limits were experimentally determined for the EA1N, EA4T and 34CrNiMo6+QT steel grades on both small-scale laboratory specimens and full-scale axles. Besides characterising fatigue strength for the axle body, a number of tests were carried out in [5] focusing on crack initiation in wheel seats. Additional fatigue strength data for the above three steel grades are available in [9, 14–16].

Fatigue crack growth (FCG) data for railway axle materials can be found e.g. in [7–10, 15, 18–23]. Beretta et al. [8, 18–20, 23] performed a series of FCG tests using different specimen geometries. In most of their studies, the single-edge bend specimen, SE(B), was employed at negative stress ratios, $R < 0$, whereas a modified compact tension specimen, C(T), and a middle-cracked tension specimen, M(T), were used at both positive and negative stress ratios. Additionally, a single-edge cracked tension, SE(T), specimen design was adopted in [19]. When comparing test results for standard specimens and full-scale axles, a good agreement was found between FCG rates for cracks in axles, on the one hand, and SE(B) and SE(T) specimens, on the other hand [18–20]. A particular observation made by Beretta and Carboni [20] for the EA1N material was a large scatter of the experimental results at $R = -1$, especially in the near-threshold regime. Furthermore, those authors concluded a considerable effect of the fatigue pre-cracking procedure (standard approach [24] vs. compression pre-cracking) and testing method (constant load vs. load shedding) on the resulting FCG rates.

The high-strength steel grade 30NiCrMoV12 was investigated by Beretta et al. [7, 18], while similar specimen types as for the EA1N material, including full-scale axle tests, were used. In particular, FCG data are provided in [18] for the near-threshold regime at $-2 \leq R \leq 0$ and for the Paris regime at positive stress ratios of $R = 0.05,\ 0.3$.

Steel grades EA4T and 34CrNiMo6+QT were extensively studied in [9, 10] using both M(T) and axle-like specimens. In those investigations, basic FCG data were derived on M(T) specimens at two stress ratios of $R = -1$ and $R = 0.1$, as shown in Fig. 1. The curves plotted in the diagram represent a smooth fit of the experimental data for EA4T using the Mettu-Forman equation following its description in [25]. The above equation is particularly employed in Sect. 5 for crack propagation analyses. According to Fig. 1, at equal stress intensity factor ranges, ΔK, the material 34CrNiMo6+QT exhibits somewhat higher FCG rates and, at least

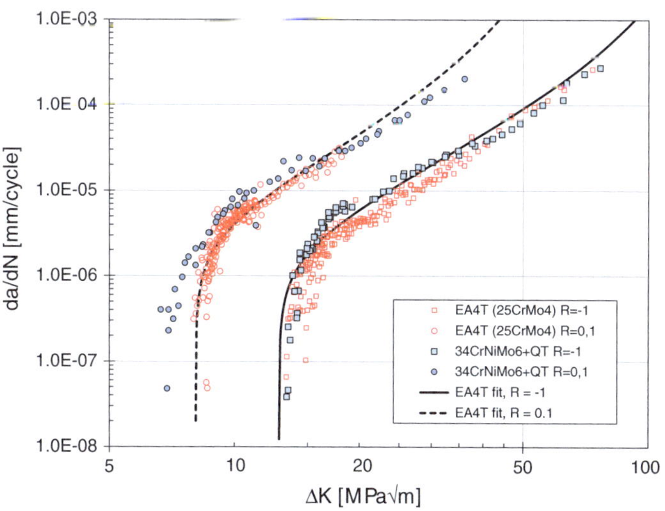

Fig. 1 FCG rates for EA4T and 34CrNiMo6+QT according to [9, 10]

Table 1 Threshold value, ΔK_{th}, and parameters of the Paris-Erdogan equation for EA4T and 34CrNiMo6+QT according to [9, 10]

Steel grade	$R = -1$			$R = 0.1$		
	ΔK_{th}	C	m	ΔK_{th}	C	m
EA4T	13.0	2.74×10^{-10}	3.2	7.5	2.65×10^{-9}	3.2
34CrNiMo6+QT	13.0	4.32×10^{-9}	2.5	6.5	2.42×10^{-8}	2.5

Units: MPa$\sqrt{\text{m}}$ for ΔK, mm/cycle for da/dN

at $R = 0.1$, lower threshold values, ΔK_{th}, as compared to EA4T. The latter aspect is of special interest for the fracture mechanics assessment of railway axles, since the major stage of crack propagation is expected to occur in the near-threshold regime.

The threshold values and parameters C and m of the Paris-Erdogan equation

$$\frac{da}{dN} = C\Delta K^m \tag{1}$$

estimated in [9, 10] for both materials are listed in Table 1. Note that in view of data scatter, which is especially considerable for the material EA4T, the reported ΔK_{th} values represent respective lower bound estimates, whereas the parameters C and m provide an average fit of the experimental data.

Further experimental investigations of fatigue crack growth in steel grades EA4T and 34CrNiMo6 are available in [19, 21], respectively. Note that all FCG data mentioned above refer to constant amplitude loading (CAL) tests, whereas railway axles are generally subjected to variable amplitude loading (VAL) representative of in-service load spectra, which examples are provided in [9, 14, 16, 20, 26]. To quantify the effect of VAL on fatigue crack propagation, a limited amount of test data on both standard specimens and full-scale axles are available [9, 10, 14, 20, 26], most of which obtained under block loading conditions. Exceptions are full-scale axle tests reported in [26], where a complex test load sequence was applied according to respective service load measurements.

2.2 *Effect of Specimen Geometry*

When applying the fracture mechanics methodology to the assessment of railway axles, one of the key issues is the transferability of material data derived in tests on standard specimens to a component containing a surface crack. For the latter, the initial size is typically assumed just below the detection limit of a specific NDE technique, whereas the final crack size considered in the assessment can be selected to reasonably bound FCG rates at the final stage of crack propagation. For instance, a final crack depth of about 10 mm [26] can be regarded as a rational limit, above which the remaining fatigue life of an axle until fracture is usually a small fraction of the total lifetime. Taking into account that the axle outer diameter is normally within 150–200 mm, crack sizes relevant for the axle assessment are essentially smaller than those in standard fracture mechanics specimens. As a consequence, the local stress-strain field achieved along the crack front in an axle may significantly differ from that in a standard specimen.

The above effect is often considered in connection with the crack tip constraint (e.g. expressed in terms of the T-stress) and can influence FCG rates in different fracture mechanics specimens [22, 27–29]. As a part of the experimental program in [9], a large number of axle-like specimens made of both EA4T and 34CrNiMo6 +QT were tested under in-plane bending, including both CAL and VAL block programmes. The respective specimens were manufactured as full-scale or 1:3 scaled solid axles containing initial fatigue pre-cracks generated from electrically discharge machined (EDM) notches of semi-elliptical shape with depths of about 2–5 mm. The test evaluation revealed considerably lower FCG rates for cracks in the axle-like specimens as compared to the standard M(T) geometry. It is worth mentioning that stress levels applied in the respective axle tests [9], both at CAL and VAL, produced ΔK and da/dN values within the Paris regime, so that the analytical FCG curve, Eq. (1) along with the parameters according to Table 1, could directly be used to estimate fatigue crack propagation in the axle specimens. As a result, the calculated remaining fatigue lives were approximately 3–5 times smaller than those experimentally observed, see [9, 10]. Though this finding may be positively appraised, concluding that the use of the material data and the analytical approach adopted in [9, 10] provides conservative estimates of fatigue lives for cracked axles, the need to better understand the geometry (or constraint) effects in fatigue crack propagation analyses becomes obvious.

The influence of the specimen geometry on FCG rates is demonstrated in Fig. 2 using experimental data from [22] derived for the material EA4T. In that study, two

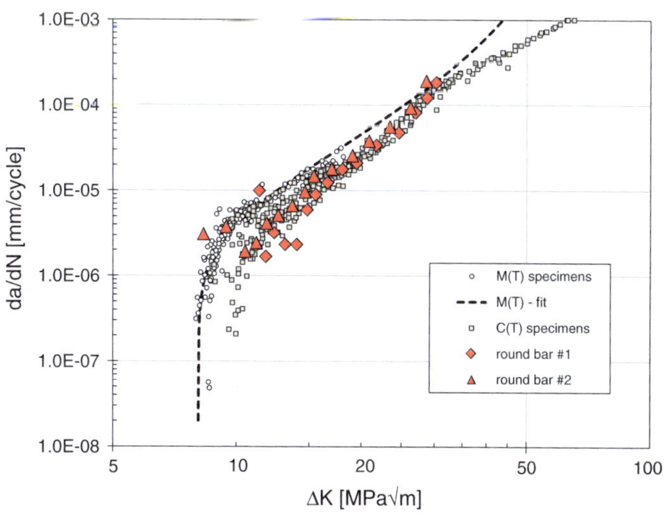

Fig. 2 FCG rates for the steel grade EA4T using different specimen geometries, $R = 0.1$ [22]

standard specimen types, M(T) and C(T), as well as two round bars with a diameter of 50 mm were tested at $R = 0.1$. The round bar specimens contained semi-elliptical fatigue pre-cracks which extension was monitored by means of beach marks. To minimize crack closure effects, tests on the round bar specimens were performed at a constant maximum load. Note that due to a narrow spacing between the initial beach marks and their fuzzy contours, the evaluation of the beginning stage of crack propagation in the round bar specimens was rather ambiguous. For this reason, few initial points for both respective specimens exhibit a considerable scatter and deviate from the anticipated trend line. On the whole, the results in Fig. 2 reveal similar FCG rates for the C(T) and round bar specimens, whereas the M(T) specimen produces faster crack growth, especially in the near-threshold regime. When comparing the M(T) and C(T) geometries, similar observations were made by Vecchio et al. [27] and Hutar et al. [28], whereas Tong [29] concluded the opposite trend.

3 Stress Analysis for Railway Axles

The axle body, geometric transitions and wheel seats are often considered as most critical locations with respect to crack initiation and potential propagation, see e.g. [6, 9, 12, 13, 16, 30]. As a prerequisite for a fracture mechanics based assessment, detailed stress calculations for the respective axle zones are required which are usually performed using the finite-element method. A particular emphasis is on an accurate description of stress gradients in geometric transitions. Besides bending stresses resulting from loads acting on the axle [1, 2], residual stresses due to press fittings need to be considered [30], which may significantly affect crack propagation near wheel, gear or brake seats.

3.1 Example of Stress Calculations

In this section, examples of stress calculations are presented which refer to an axle geometry considered within the project [9]. The respective axle design is representative of a solid leading trailer axle of a commuter train with speeds up to 160 km/h. Figure 3 shows the finite-element model of the axle and wheel assembly, as well as an indication of two cross-sections in the transition zone from the wheel seat to the shaft selected for fracture mechanics analyses. Characteristic dimensions in the assessment part of the axle are as follows: the shaft diameter $d = 160$ mm, the wheel seat diameter $D = 185$ mm, the transition radii in the fillet $R_1 = 15$ mm and $R_2 = 75$ mm. The assessment cross-section #1 is located at a distance of 10.3 mm from the wheel seat edge and corresponds to the transition from the fillet radius R_1–R_2. The assessment cross-section #2 is located within the fillet radius R_2, at an axial

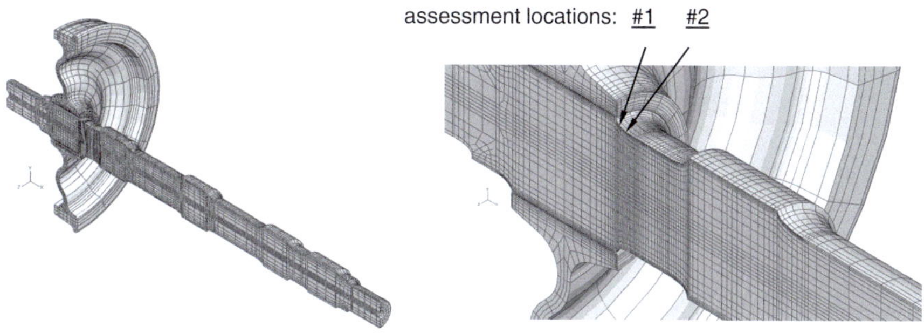

Fig. 3 Finite-element model of the reference axle [9] with indication of assessment locations between the wheel and gear seats

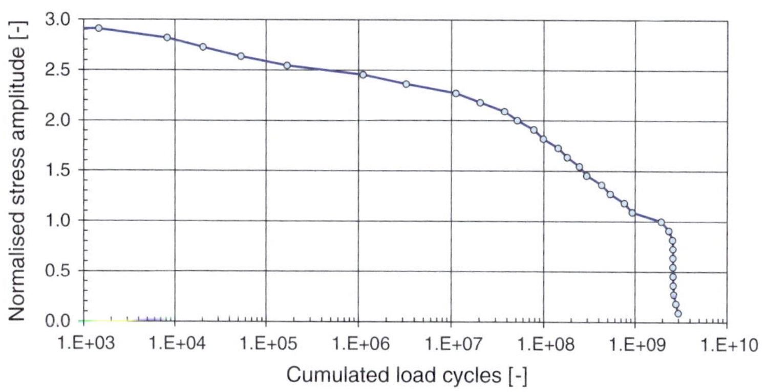

Fig. 4 Load spectrum determined for the reference axle geometry in [9]

distance of 22 mm from the wheel seat edge. The axle diameter at the cross-sections #1 and #2 is approximately 165.2 and 160.8 mm, respectively.

The service load spectrum used in the calculations is given in Fig. 4 as a histogram of the axial stress amplitudes normalised by the value corresponding to the straight-track regime (vehicle's weight as the only axle load). The data were derived in [9] from strain gauge measurements at the axle free surface, at a characteristic position referred to a cylindrical part with a nominal diameter of 160 mm. The load spectrum is based on data records performed over a mileage of about 5300 km, including representative track sections, and subsequently extrapolated to cover an overall design service time corresponding to 3×10^9 cycles. Details of the load spectrum measurement procedure and further results can be found in [9].

Axle stress calculations were performed in two steps. First, press fitting was modelled assuming the maximum allowable interference of $\Delta D = 0.386$ mm between the wheel seat and wheel bore diameters. Although not indicated in Fig. 3, press fitting of the gear was also modelled. In the second step, the bending moment corresponding to a nominal stress of 100 MPa in the axle shaft (d = 160 mm) was

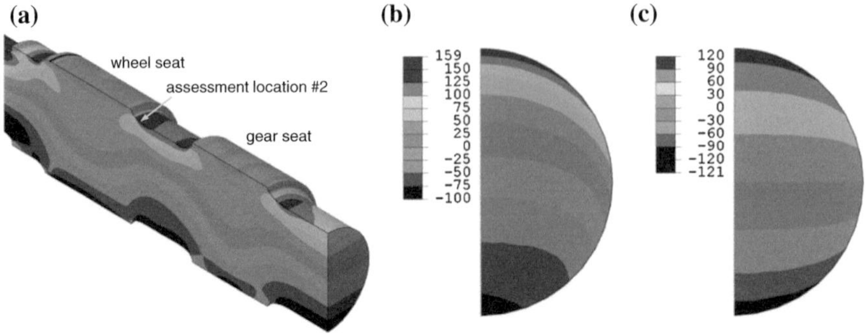

Fig. 5 Axial stress in the axle at superimposed press fitting and bending at a nominal bending stress of 100 MPa (**a**). Details (**b**) and (**c**): stress distributions in the assessment section #2 at superimposed press fitting and bending and at pure bending load, respectively. Results for ΔD = 0.386 mm, COF = 0.12

additionally applied. Besides the interference, ΔD, the coefficient of friction (COF) is a further parameter influencing the stress magnitude and distribution due to both press fitting and bending [31]. In the example considered here, this was assumed to be $COF = 0.12$ which is characteristic of a value at the beginning of the service life for a new wheelset assembly.

The resulting axial stress distribution due to superimposed press fitting and bending is shown in Fig. 5a for an axle part including both seats as well as the adjacent transitions. Figure 5b is a detailed stress plot in the assessment cross-section #2 demonstrating a complex two-dimensional (2D) stress distribution with gradients in both the radial and the circumferential direction. This feature is explained, on the one hand, by the superposition of the bending stresses with the axisymmetric stress field due to press fitting. On the other hand, as shown in Fig. 5c, pure bending load results in 2D stress pattern in cross-sections close to geometric transitions where the classical beam bending theory does not apply. This important feature requires appropriate solutions for stress intensity factors which are able of taking into account 2D stress gradients in a prospective crack plane.

3.2 *Effect of Press Fit Conditions*

As mentioned above, both the interference of press fitted parts and the associated friction coefficient influence the stress magnitude and distribution. To quantify the respective effects, analyses were performed assuming the above parameters to vary within reasonable ranges.

According to the wheelset design specification [9], the minimum and the maximum values of the allowance for interference are equal to $\Delta D_{min} = 0.2$ mm and $\Delta D_{max} = 0.386$ mm, respectively. To cover the whole range of the interference

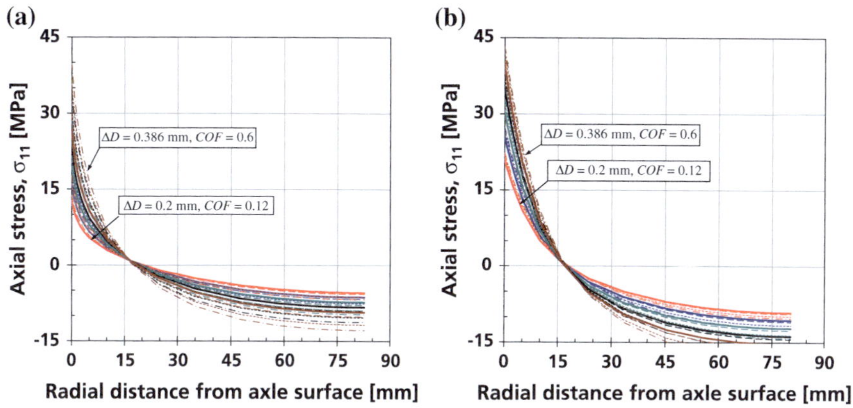

Fig. 6 Axial stress due to press fitting at assessment cross-sections #1 (**a**) and #2 (**b**)

achieved for axles in operation, the following discrete values are considered below: $\Delta D = 0.2, 0.246, 0.294, 0.34$ and 0.386 mm. The lower bound of the coefficient of friction is estimated to be $COF = 0.12$, being representative of a new (as manufactured) wheelset assembly. Further values of $COF = 0.2$, 0.4 and 0.6 are additionally considered to take into account presumable changes of the contact interaction conditions during the service life. Altogether, twenty parameter combinations of ΔD and COF are analysed.

The calculated axial residual stress due to press fitting is shown in Fig. 6 as a function of the radial distance in two assessment cross sections, for different pairs $(\Delta D,\ COF)$ considered. The curves in the diagrams are arranged according to ΔD and COF values: the smaller the allowance for interference and the friction coefficient, the lower residual stresses are achieved at the axle surface. Note also that the assessment section #2 is subjected to higher press fit stresses with a more pronounced in-depth penetration, as compared to the section #1.

The latter conclusion is confirmed in Fig. 7a which is a plot of the axial stress component due to press fitting along the axle surface. The abscissa in the latter diagram represents the axial distance measured from the wheel seat edge. Accordingly, the stress at the surface continuously increases with increasing both the interference and the friction coefficient. At the same time, the location of the stress maximum is not uniquely defined, shifting towards the wheel seat edge with increasing ΔD and COF. At superimposed press fit and bending loading, two rather distinct stress maxima can be identified at the axle surface (see Fig. 7b where the nominal bending stress in the shaft equals to 100 MPa) corresponding to the assessment cross-sections #1 and #2, as previously defined in Sect. 3.1.

Inter alia, these results demonstrate that both fatigue crack initiation and subsequent crack growth in railway axles should generally be regarded as probabilistic phenomena by taking into account the scatter, variability and uncertainties of all relevant input parameters, including material data, axle geometry and loading.

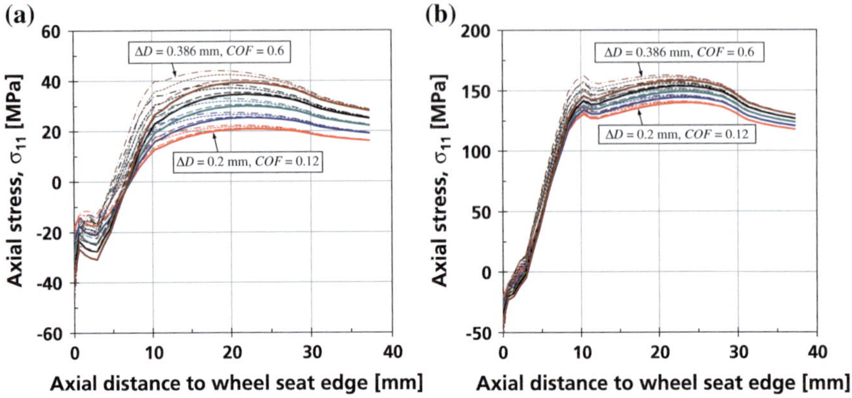

Fig. 7 Axial stress along the axle surface due to press fitting (**a**) and at superimposed press fit and bending loading with the nominal bending stress magnitude of 100 MPa (**b**)

4 Stress Intensity Factors for Surface Cracks in Axles

A number of analytical models and associated stress intensity factor (Mode I) solutions have been developed in the past for analyses of surface crack growth in cylindrical bars. Some of the models developed for this purpose consider the crack front described by a circular segment [32–35], whereas a more universal approximation is achieved assuming a semi-elliptical crack shape [36–41], see Fig. 8. In the latter crack model, a denotes the crack depth or the minor ellipse half-axis, the parameter $2c$ represents the major axis of an ellipse describing the crack front, and r is the cylinder radius. The physical crack length, $2s$, is measured along the cylinder circumference between the intersection points of the ellipse with a circle of the radius r.

The corresponding part-elliptical crack model is usually parameterised by two independent dimensionless geometrical parameters, a/c and a/r, whereas the

Fig. 8 Fracture surface of a round bar specimen with beach marks (*left*) and the model of a semi-elliptical surface crack (*right*)

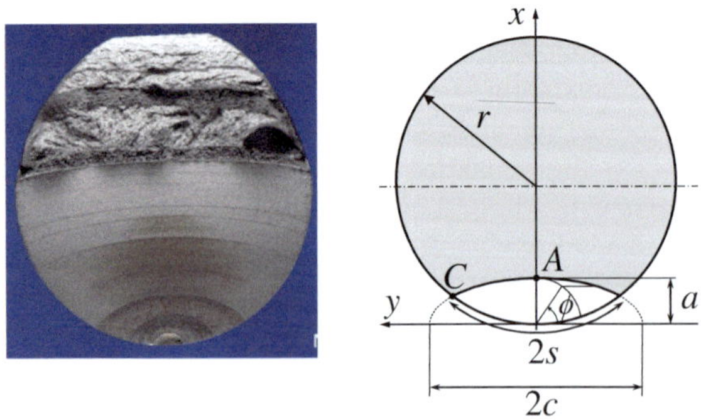

evolution of the crack shape, *a*/*c*, can be estimated by independently calculating crack growth increments in the depth and length directions, respectively. In such a case, the stress intensity factor, *K*, should be available at least for two characteristic locations on the crack front (Fig. 8) representative for the crack growth in the radial (crack centre or deepest point *A*) and circumferential (surface point *C*) directions [36, 39–41]. Some simplified solutions provide the stress intensity factor for the deepest point only, while assuming the crack shape evolution, *a*/*c* versus *a*/*r*, to follow a predefined curve according to particular experimental observations [32, 38].

Most of the *K* solutions mentioned above have been derived for cylindrical rods subjected to tension and bending loads, so that their application to railway axles is generally limited to crack locations in cylindrical parts not affected by stress concentrators or press fits. An exception is likely the solution after Shiratori et al. [36] referred to in [37], although the relevant data reproduced in the handbook [37] are given only for a combined tension and bending loading (linear stress distribution in the prospective crack plane).

Several authors point out that, except for the crack centre point *A*, rotary bending produces a higher stress intensity factor range as compared to the case of plane bending [36, 37, 39, 40]. Accordingly, stress intensity factors for both plane and rotary bending are provided in those papers. However, the effect of rotary bending is rather negligible for small cracks and becomes pronounced only for relatively large cracks sizes which are usually not tolerable in railway axles under service conditions. For this reason, distinguishing between the plane and rotary bending is regarded to be of minor practical importance in the assessment of lifetime and inspection intervals for railway axles [9, 30]. Nonetheless, appropriate *K* solutions for cracks subjected to rotary bending may help to better understand crack behaviour in test axles where substantial amount of crack growth is achieved.

A further summary of stress intensity factor solutions applicable to railway axles can be found in [42]. A special emphasis in that paper is on accurate stress intensity factor calculations taking into account loading conditions accomplished in critical axle parts, such as geometric transitions and zones affected by press fits. A particular analysis method included in [42], Chap. 5, was originally derived by the present authors within the research project [9]. The respective solution is based on the polynomial influence functions technique and applies to a general case of continuous stress distribution in the prospective crack plane (see e.g. Figure 5) which can be approximated by a two-dimensional 4th order polynomial

$$\sigma(x,y) = \sum_{m=0}^{4}\sum_{n=0}^{4}\left[D_{mn}^{(1)}\left(\frac{x}{r}\right)^{m}\left|\frac{y}{r}\right|^{n} + D_{mn}^{(2)}\left(\frac{x}{r}\right)^{m}\operatorname{sgn}(y)\left|\frac{y}{r}\right|^{n}\right] \tag{2}$$

with geometric notations according to Fig. 8. The two polynomial functions in the brackets of Eq. (2), associated with the coefficients $D_{mn}^{(1)}$ and $D_{mn}^{(2)}$, represent the even and the odd terms of the stress field about the symmetry plane $y = 0$, respectively. Both cases need to be considered at rotary bending, whereas the odd term can be omitted in case of plane bending. The stress approximation by a two-dimensional polynomial of the coordinates x and y is essential to assure that the stress gradients in both radial and circumferential directions are taken into consideration (see Fig. 5b, c).

Two sets of stress intensity factors, $K_{mn}^{(1)}$ and $K_{mn}^{(2)}$, were numerically calculated in [9] for 50 basic load cases of type

$$\sigma_{mn}^{(1)} = \left(\frac{x}{a}\right)^m \left|\frac{y}{c}\right|^n, \quad \sigma_{mn}^{(2)} = \mathrm{sgn}(y)\left(\frac{x}{a}\right)^m \left|\frac{y}{c}\right|^n, \qquad m, n = 0, \ldots, 4 \tag{3}$$

where the superscripts $^{(1)}$ and $^{(2)}$ refer to the even and odd stress distributions, respectively. Subsequently, the polynomial influence functions were derived as dimensionless coefficients

$$f_{mn}^{(1)} = \frac{K_{mn}^{(1)}}{\sqrt{\pi a}}, \quad f_{mn}^{(2)} = \frac{K_{mn}^{(2)}}{\sqrt{\pi a}} \tag{4}$$

The latter are tabulated in [9, 42] for the surface and the deepest point of the crack front and various crack geometries covering parameter ranges of $0.2 \le a/c \le 1$, $0.05 \le a/r \le 1$. Note that the corresponding results compilation (Table 12 in [9] or Table 20 in [42]) contain a *typing error*: instead of $a/c = 0.8$, the correct value of $a/c = 0.75$ should be used whenever indicated.

Given the coefficients $D_{mn}^{(1)}$ and $D_{mn}^{(2)}$ in Eq. (2) determined via a stress field approximation, e.g. using the least-square fit procedure, the calculation of the stress intensity factor for a particular crack geometry reduces to

$$K\left(\frac{a}{c}, \frac{a}{r}, \phi\right) = \sqrt{\pi a} \sum_{m=0}^{4} \sum_{n=0}^{4} \left[D_{mn}^{(1)} f_{mn}^{(1)}\left(\frac{a}{c}, \frac{a}{r}, \phi\right) + D_{mn}^{(2)} f_{mn}^{(2)}\left(\frac{a}{c}, \frac{a}{r}, \phi\right)\right] \left(\frac{a}{r}\right)^{m+n} \left(\frac{a}{c}\right)^{-n} \tag{5}$$

In the latter equation, ϕ denotes the angular location along the crack front, as indicated in Fig. 8.

5 Fatigue Crack Growth Calculations

In this section, crack growth calculations are performed for the reference axle geometry containing a fatigue crack initiated at the location #1 or #2, as specified in Sect. 3 (see Fig. 3). The initial crack is assumed to be of the size $a \times 2c = 2 \times 5\ \mathrm{mm}^2$ (crack aspect ratio $a/c = 0.8$), which is considered as an appropriate postulate within the wheelset specification requirements currently adopted by the German railway operator Deutsche Bahn [26]. The final crack size in FCG calculations is limited to $a = 10$ mm as discussed in Sect. 2.2.

Among factors affecting the estimated residual fatigue life of the axle with a crack, the following issues are specifically addressed:

- consideration of residual stresses due to press fitting;
- evaluation of stress amplitudes and stress ratios relevant for crack propagation;
- probabilistic aspects of the FCG analysis.

As already mentioned, both the interference of press fitted parts and the associated coefficient of friction may significantly affect the stress state at prospective crack location and thus the crack propagation behaviour [9, 30, 31]. To demonstrate the effect of press fitting on the residual lifetime, the reference axle example is considered assuming a crack located in the cross-section #2, the maximum interference of $\Delta D = 0.386$ mm, and a friction coefficient of $COF = 0.12$. Furthermore, the load spectrum shape according to Fig. 4 is used, whereas the vehicle weight (equivalently, the basic load level used in the normalisation) is selected to achieve a remaining lifetime corresponding to a mileage of about 300,000 km. The so defined number of load cycles is denoted by N_{ref} and used as a normalisation factor for the computed lifetime throughout the analyses presented below.

Figure 9 compares the residual lifetime estimates obtained with and without considering residual stresses due to press fitting. The results demonstrate that ignoring the effect of press fitting leads to a considerable overestimation of the fatigue life, which is not acceptable for safety reasons.

A tensile residual stress due to press fitting (Fig. 6) acts as a mean stress shifting the stress intensity ratio, $R_K = K_{\min}/K_{\max}$, from $R_K = -1$ corresponding to a pure rotary bending load towards higher values. The latter depend on the particular stress amplitude of the load spectrum, the residual stress magnitude and the current crack size. Thereby, a field of the threshold values, ΔK_{th}, and FCG curves, $\Delta K - da/dN$, both being functions of R_K, should be considered in the crack propagation analysis.

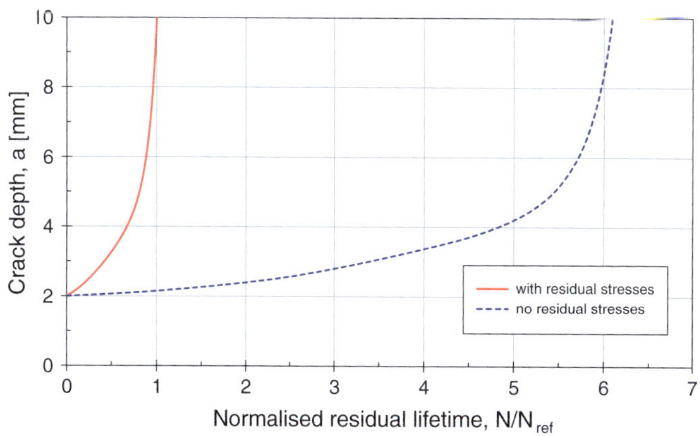

Fig. 9 Effect of stresses due to press fitting on the estimated residual lifetime: crack location at cross-section #2, $\Delta D = 0.386$ mm, $COF = 0.12$

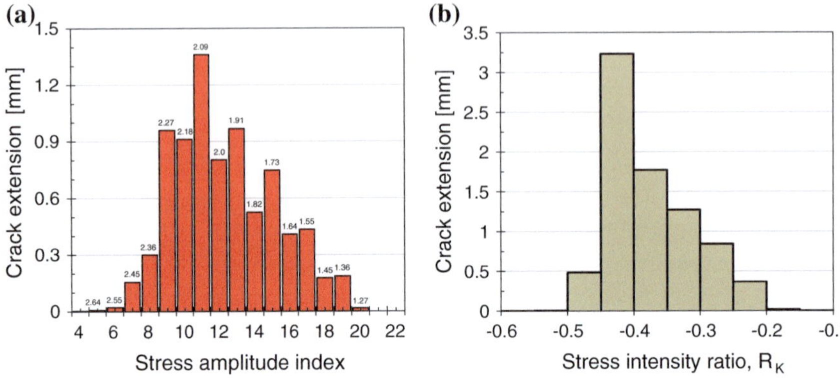

Fig. 10 Contribution of different stress amplitudes of the load spectrum (**a**) and respective stress intensity ratios (**b**) to fatigue crack propagation. Example data: crack location at cross-section #2, $\Delta D = 0.386$ mm, $COF = 0.12$. Numbers in the diagram area in (a) represent the normalised values of stress amplitudes according to Fig. 4

As a result, the shape of the load spectrum and its magnitude, as defined through the vehicle weight, being interacted with the residual stress distribution produce a complex dependency of the amount of crack propagation upon the relevant load parameters.

For the axle example and load conditions considered above (see the solid line in Fig. 9), Fig. 10a shows the contribution of different steps of the load spectrum to the total crack extension from the initial depth of 2 mm until the final depth of 10 mm. Accordingly, the load steps with normalised amplitudes both exceeding the value of about 2.5 and those below about 1.3 have a negligible effect on the remaining lifetime. This is due to the fact that those high stress amplitudes occur very rarely, as compared to the total number of cycles in the spectrum, whereas the very low stress amplitudes with a considerable amount of cycles produce stress intensity factor ranges below the respective threshold. The R_K values relevant for the FCG analysis can be estimated from Fig. 10b. This shows that the major part of crack growth occurs within the range of $-0.5 \le R_K \le -0.2$ which may be regarded as a suggestion for a rational material testing programme.

As a matter of fact, most FCG analyses for railway axles contain various uncertainties. These can be handled by e.g. adopting upper bound $\Delta K - da/dN$ curves for a particular material, including conservatively estimated ΔK_{th} values, assuming the most unfavourable combination of the press fit parameters (maximum allowable interference and a very high COF value), and eventually using conservative assumptions regarding the load spectrum shape and associated stress magnitudes. Although such an approach is principally acceptable, it may considerably underestimate the axle fatigue life, on the one hand, and provides no insight into the large scatter of axle test results, on the other hand.

In the following example, only a part of the parameter spectrum relevant for the scatter and uncertainties in fracture mechanics calculations for railway axles is addressed. This concerns an inherent variability of the interference for the axle-wheel pair as well as changes of the friction coefficient during the service time. The respective parameter variations were already defined and considered in the axle stress calculations, Sect. 3.2. To derive a statistical representation of the fracture mechanics based results, the interference is considered as a distributed quantity following a normal distribution. The distribution mean is set to $\mu_{\Delta D} = 0.294$ mm which is the arithmetic mean of the maximum and minimum interference values according to the wheelset design specification. The standard deviation is defined as $\sigma_{\Delta D} = 0.046$ mm, which corresponds to the assumption that the specified lower and upper ΔD values of 0.2 mm and 0.386 mm, respectively, cover the range of $\mu_{\Delta D} \pm 2\sigma_{\Delta D}$, i.e. the allowance for interference is complied with the wheelset design specification in 95.4 % cases.

FCG analyses are then carried out using a Monte-Carlo approach to generate the interference value, whereas respective residual stress distributions are estimated by interpolating the calculation results for discrete ΔD values previously derived in Sect. 3.2 (see Fig. 6). Each individual simulation is performed at a constant friction coefficient ($COF = 0.12$, 0.2, 0.4 or 0.6) and for the crack location in the cross-section #1 or #2. Hence, the lifetime estimates obtained in this way are believed to be representative for the whole fillet zone of the axle.

The results of this exercise are statistical distributions of the remaining axle lifetime plotted in Fig. 11 as a dimensionless quantity normalized by the reference lifetime N_{ref} (solid line in Fig. 9). Different curves are grouped according to the assessment location within the axle fillet. All data series in Fig. 11 are approximated by a three-parameter Weibull distribution, whereas the normal distribution function also provides a good data fit for the majority of cases.

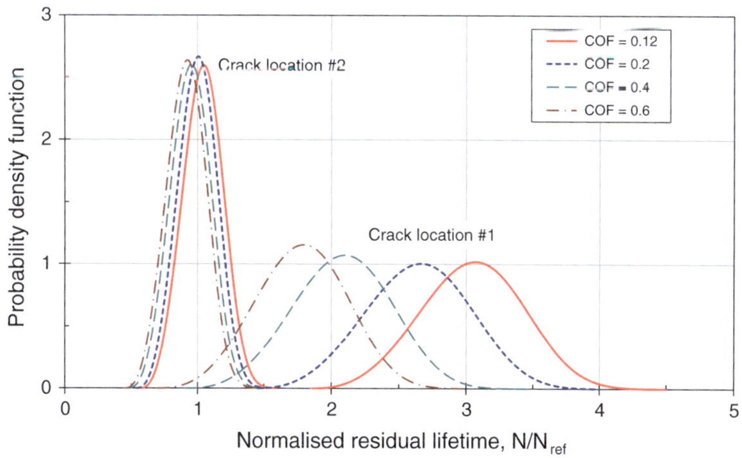

Fig. 11 Distribution functions for residual lifetime. Results at varying press fit parameters, two assessment locations in the axle, and $a_0/c_0 = 0.8$

A considerable variation in the calculated fatigue lives is predicted as a result of varying manufacturing tolerances, while assuming the other influencing factors constant. The results scatter is quantified by the ratio of the fatigue lives estimated at 97.7 and 2.3 % probability levels, the latter being in accordance with the assumed distribution parameters for ΔD. Apparently, a pronounced effect (scatter factor of 1.6–2.2) is achieved by varying the interference alone. Additionally including potential changes of the friction coefficient into consideration produces an increase of the scatter factor up to 3–4. When comparing curves for both assessment cross-sections at equal values of the friction coefficient, a lower scatter can be deduced for the location #1.

At the same time, due to the variation of the friction coefficient, the median values of fatigue lives reveal a larger scatter at the location #1 (factor of about 1.7). For comparison, the results for the location #2 differ by only 10 %. This implies that a possible increase of the friction coefficient during the service life (e.g. due to a loss of lubricant, wear, heating-up) would not lead to a considerable increase of crack propagation rates at the location #2. On the whole, the results for the assessment cross-section #2 can be regarded as conservative lower bounds for the whole fillet zone of the axle considered in this study. However, the criticality of the location #1 with respect to crack propagation rates increases with increasing the friction coefficient. This is confirmed by a decreasing distance between respective distribution curves in Fig. 11.

The scatter factors estimated above are representative for particular press fitting conditions, an assessment cross-section and an initial flaw size and shape. Considering that the crack initiation sites may be arbitrarily distributed within the fillet zone and the initial crack shape cannot distinctly be defined, an even larger variation in residual lifetime is achieved for axles with postulated defects under service conditions.

6 Notes on Crack Growth Behaviour in Press Fits

Depending on the axle design, material and bending stress magnitude, crack initiation may occur in press fitted parts, rather than in the axle body. Numerous examples of cracks initiated in wheel seats both in test rigs and under operation conditions can be found in [5, 13, 14, 16]. In those studies, long cracks over almost the whole seat circumference are often observed, whereas most of them exhibit a limited propagation in the depth direction. The respective cracks likely nucleate at multiple sites, usually located near the seat edge, and subsequently coalesce forming a long surface crack. Different from the observations in [5, 14, 16], multiple crack patterns of small surface defects distributed over a larger surface area are reported for the axle design of the Shinkansen train [13].

Note that most studies of crack behaviour in press fits confine to large scale tests, whereas no models for a reliable assessment of crack propagation, comparable to those established for FCG analyses in the axle body, are available yet. This is mainly due to a complex phenomenon of the crack growth in areas affected by the contact interaction, where the conventional solutions of the linear-elastic fracture mechanics are generally not applicable. Additionally, a mixed deformation mode for cracks in press fits and plastic deformations due to press fitting are further aspects which need to be considered in the assessment.

In the analysis presented in this section, two types of surface cracks are considered in the wheel seat of the reference axle geometry:

- semi-elliptical crack of the size $a \times 2c = 4 \times 16\ \text{mm}^2$;
- completely circumferential crack with the depth $a = 4$ mm.

Both cracks are assumed to be located at a distance of 4 mm from the seat edge in a plane normal to the axle surface. The principal expectation is to explore differences achieved in the crack driving force for the two crack geometries which may help to explain some of the experimental findings mentioned above.

In the finite-element calculations, the contact interaction is modelled for both the wheel-axle pair and the opposite crack faces. After press fitting, a reverse in-plane bending load with the net stress amplitudes of 200 MPa (referred to the axle shaft with $d = 160$ mm) is applied, which value is close to the stress level achieved in axle fatigue tests with the emphasis on crack initiation in press fits [5]. The resulting stress intensity factor ranges, ΔK, versus the net stress amplitude are compared in Fig. 12a for the two crack geometries considered.

Note a substantially higher crack driving force calculated for the semi-elliptical crack as compared to a completely circumferential crack of equal depth. This result,

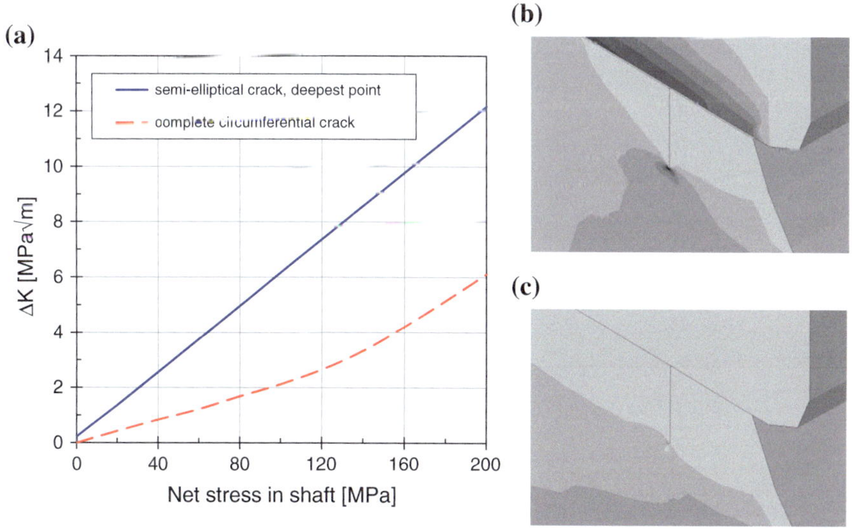

Fig. 12 Stress intensity factor ranges for two crack geometries in the wheel seat (**a**). Details (**b**) and (**c**): axial stress patterns on deformed models of the axle-wheel assembly with the crack location in the wheel seat: **b** semi-elliptical crack, $a \times 2c = 4 \times 16\ \text{mm}^2$; **c** completely circumferential crack, $a = 4$ mm. Load case: press fitting and bending

which may be regarded as implausible at first glance, can be explained by examining deformed shapes of the two models shown in Fig. 12b, c, respectively. Accordingly, local deformations around the relatively small semi-elliptical crack lead to the loss of contact in the crack vicinity (Fig. 12b) and thus promote its opening at both press fit and bending loads. In contrast, in presence of a completely circumferential crack, the wheel bore and the seat remain fully in contact at both press fitting and superimposed bending (Fig. 12c). This effect counteracts crack opening due to bending and reduces the effective crack driving force. The latter feature can also be observed by comparing the axial stress distributions in Fig. 12b, c (note the same scale used in both plots), in particularly at the respective crack tips.

Figure 13 shows contact stress patterns (radial stress component) in the wheel seat for the two crack models, at both press fitting and superimposed press fitting

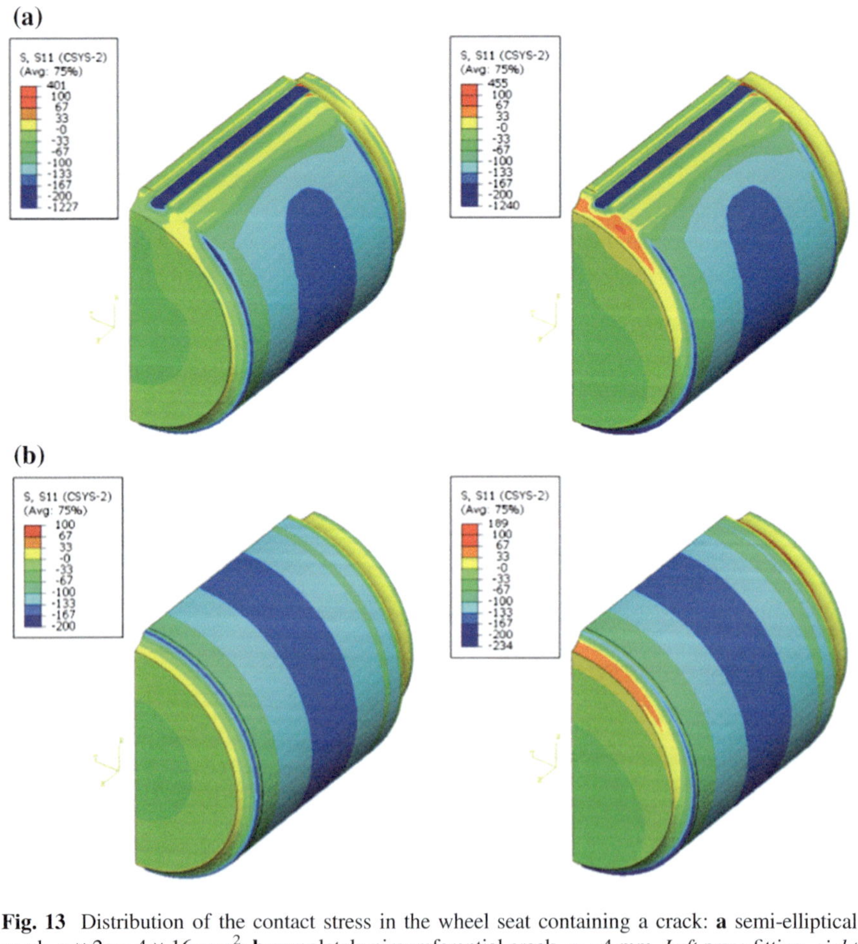

Fig. 13 Distribution of the contact stress in the wheel seat containing a crack: **a** semi-elliptical crack, $a \times 2c = 4 \times 16$ mm^2; **b** completely circumferential crack, $a = 4$ mm. *Left* press fitting, *right* press fitting and bending

and bending with a nominal bending stress of 200 MPa. These results demonstrate that a small semi-elliptical crack alters the contact interaction within a stripe area along the whole seat length, whereas no noticeable stress redistribution occurs due to the presence of a completely circumferential flaw.

The above features along with a comparison of the crack driving force parameters in Fig. 12a suggest that, being initiated at the surface, a small semi-elliptical crack may propagate in both depth and length directions, at least within the stripe area with the loss of contact. Since multiple cracks can nucleate in a press fit zone, the coalescence of several cracks located approximately at an equal distance to the wheel seat edge can be expected leading to a formation of a completely circumferential crack. As a consequence, the crack driving force for the resulting crack geometry is likely to decrease, so that a retardation of crack growth or even crack arrest may occur. Though the latter conclusion can be influenced by different factors, e.g. material properties, bending stress amplitude and press fit parameters, it seems to reflect available experimental findings of crack patterns in press fits of wheelsets [5, 13, 14, 16].

7 Conclusions

This paper gives a brief overview of different aspects relevant for the fracture mechanics based assessment of railway axles. The respective topics have frequently been addressed in the past, whereas numerous studies performed in the last two decades have especially contributed to better understanding the crack growth phenomenon in railway axles. Although the assessment methodology can be overall regarded as well established, its practical application is a subject of ongoing research activities aiming at providing generally acceptable tools for both vehicle manufacturers and operators.

One of the principal problems in the fracture mechanics based assessment of railway axles arises from the fact that the respective components have to be available for operation during a very long time. For instance, to meet economic demands regarding inspection intervals, the axle with a crack should be proven to endure a number of load cycles of the order of 10^8. This prerequisite requires highly accurate models and solutions describing crack propagation in axles, on the one hand. Some of such solutions with examples demonstrating their application are provided in the paper. On the other hand, an accurate crack growth analysis is strongly dependent on the availability of comprehensive material data and in-depth understanding of material behaviour at the crack tip. In this context, crack growth behaviour in the threshold regime under load conditions representative of those at railway axle operation is of special concern.

Summarising the current status in the field, following topics can be regarded as such that may potentially contribute to establishing a generally acceptable assessment approach:

- Comprehensive material characterisation for steel grades utilised in railway axle manufacturing. The respective investigations should focus on the determination of FCG curves at relevant stress ratios, an accurate determination of the threshold parameter ΔK_{th}, effects due to VAL, behaviour of small cracks, and an appropriate description of the scatter of material data.
- Transferability of material data from standard test geometries to cracks in full-scale axles.
- Estimation of residual stresses due to manufacturing (heat treatment, surface finishing) and their incorporation in the assessment concept.
- Use of NDE data in combination with a fracture mechanics based approach.
- Development of a fully probabilistic approach incorporating uncertainties, variability and scatter of all relevant input data.

Note that most of the issues mentioned above are subjects of the ongoing research project EBFW3 [17].

Acknowledgments The results reported in this paper were partly derived within the framework of the research project *Safe and Economic Operation of Running Gears* (Sicherer und wirtschaftlicher Betrieb von Eisenbahnfahrwerken) [9] funded by the Federal Ministry of Economic Affairs and Energy (BMWi) under research grant no. 19P4021 A-F. This financial support as well as a fruitful cooperation with all research and industrial companies involved in the above project is gratefully acknowledged.

References

1. EN 13103:2012-10 (2012) Railway applications—wheelsets and bogies—non powered axles —design method. European Committee for Standardization
2. EN 13104:2013-3 (2013) Railway applications—wheelsets and bogies—powered axles—design method. European Committee for Standardization
3. Klinger C, Bettge D, Häcker R, Heckel T, Gohlke D, Klingbeil D (2011) Schadensanalyse zum Radsatzwellenbruch ICE3. In: 43rd Conference of DVM working group fracture mechanics. Berlin, p 1–6, 22–23 February 2011
4. Manca D (2014) New insights into the Viareggio railway accident. Chem Eng Trans 36:13–18
5. Traupe M, Meinen H, Zenner H (2004) Sichere und wirtschaftliche Auslegung von Eisenbahnfahrwerken. Abschlussbericht BMBF-Projekt 19 P 0061 A bis F
6. Zerbst U, Vormwald M, Andersch C, Mädler K, Pfuff M (2005) The development of a damage tolerance concept for railway components and its demonstration for a railway axle. Eng Fract Mech 72:209–239
7. Beretta S, Carboni M (2005) Simulation of fatigue crack propagation in railway axles. J ASTM Int 2:368–381
8. Beretta S, Carboni M (2006) Experiments and stochastic model for propagation lifetime of railway axles. Eng Fract Mech 73:2627–2641
9. Lütkepohl K, Esderts A, Luke M, Varfolomeev I (2009) Sicherer und wirtschaftlicher Betrieb von Eisenbahnfahrwerken. Abschlussbericht BMWi-Projekt 19 P 4021 A bis F. http://edok01.tib.uni-hannover.de/edoks/e01fb09/614268583.pdf
10. Luke M, Varfolomeev I, Lütkepohl K, Esderts A (2011) Fatigue crack growth in railway axles: assessment concept and validation tests. Eng Fract Mech 78:714–730

11. Sander M, Richard HA (2011) Investigations on fatigue crack growth under variable amplitude loading in wheelset axles. Eng Fract Mech 78:754–763
12. Zerbst U, Schödel M, Beier HTh (2011) Parameters affecting the damage tolerance behaviour of railway axles. Eng Fract Mech 78:793–809
13. Makino T, Kato T, Hirakawa K (2011) Review of the fatigue damage tolerance of high-speed railway axles in Japan. Eng Fract Mech 78:810–825
14. WIDEM (2008) Wheelset integrated design and effective maintenance. EU project, 6th framework programme: sustainable development, global change and ecosystem. Project No. TST-CT-2005-516196. http://www.widem.org
15. MARAXIL (2012) Manufacturing railway axles with improved lifetime. Research project, Politecnico di Milano. http://maraxil.mecc.polimi.it
16. EURAXLES (2013) Minimizing the risk of fatigue failure of railway axles. EU project, 7th Framework programme for research and development. http://www.euraxles.eu
17. Deisl A, Gänser H-P, Jenne S, Pippan R (2014) Eisenbahnfahrwerke 3 – EBFW3. Description and aims of the new project. In: ESIS TC-24 Meeting, 1–2 October 2014, Politecnico di Milano. http://esistc24.mecc.polimi.it/Milan_2014/EBFW3_final.pdf
18. Beretta S, Carboni M, Cantini S, Ghidini A (2004) Application of fatigue crack growth algorithms to railway axles and comparison of two steel grades. J Rail Rapid Transit 218: 317–326
19. Beretta S, Carboni M, Martinelli E (2009) Variable amplitude crack growth in railway axles: influence on inspection intervals. In: Sonsino CM, McKeighan PC (eds) 2nd International conference on material and component performance under variable amplitude loading, Darmstadt, p 1125–1134, 23–26 March 2009
20. Beretta S, Carboni M (2011) Variable amplitude fatigue crack growth in a mild steel for railway axles: experiments and predictive models. Eng Fract Mech 78:848–862
21. Kloster V, Richard HA, Kullmer G (2012) Experimental investigations of the mean stress effect on the fatigue crack growth. In: 44th Conference of DVM working group fracture mechanics, Darmstadt, pp 121–130, 14–15 February 2012
22. Varfolomeev I, Luke M, Burdack M (2011) Effect of specimen geometry on fatigue crack growth rates for the railway axle material EA4T. Eng Fract Mech 78:742–753
23. Beretta S, Carboni M (2008) Validation of crack growth model in full-scale axles. EU project WIDEM, deliverable D7.1.6. http://www.widem.org/news.php
24. ASTM E 647-13a (2014) Standard test method for measurement of fatigue crack growth rates. ASTM International, West Conshohocken
25. NASGRO (2010) Fracture mechanics and fatigue crack growth analysis software. Reference Manual, Version 6.1, Southwest Research Institute
26. Traupe M, Jenne S, Lütkepohl K (2015) Experimental validation during the development of wheelsets for assessing the NDT periodicity of axles. In: 47th Conference of DVM working group fracture mechanics, Freiberg, pp 285–294, 10–11 Feb 2015
27. Vecchio RS, Crompton JS, Hertzberg RW (1987) The influence of specimen geometry on near threshold fatigue crack growth. Fatigue Fract Eng Mater Struct 10:333–342
28. Hutar P, Seitl S, Knésl Z (2006) Effect of constraint on fatigue crack propagation near threshold in medium carbon steel. Comput Mater Sci 37:51–57
29. Tong J (2002) T-stress and its implications for crack growth. Eng Fract Mech 69:1325–1337
30. Madia M, Beretta S, Zerbst U (2008) An investigation on the influence of rotary bending and press fitting on stress intensity factors and fatigue crack growth in railway axles. Eng Fract Mech 75:1906–1920
31. Varfolomeev I, Ivanov D, Luke M (2010) Effect of press fitting conditions on crack propagation behaviour in railway axles. In: Proceedings of the 18th European conference on fracture. Fracture of materials and structures from micro to macro scale, Dresden, 30 August to 3 September 2010
32. Forman RG, Shivakumar V (1986) Growth behavior of surface cracks in the circumferential plane of solid and hollow cylinders. In: Underwood JH, Chait R, Smith CW et al

(eds) Fracture mechanics, vol 17. ASTM STP 905. American Society for Testing and Materials, Philadelphia, pp 59–74
33. Raju IS, Newman JC (1986) Stress-intensity factors for circumferential surface cracks in pipes and rods under tension and bending loads. In: Underwood JH, Chait R, Smith CW et al (eds) Fracture mechanics, vol 17. ASTM STP 905, American Society for Testing and Materials, Philadelphia, pp 789–805
34. Caspers M, Mattheck C (1987) Weighted averaged stress intensity factors of circular-fronted cracks in cylindrical bars. Fatigue Fract Eng Mater Struct 9:329–341
35. Levan A, Royer J (1993) Part-circular cracks in round bars under tension, bending and twisting. Int J Fract 61:71–99
36. Shiratori M, Miyoshi T, Sakai Y, Zhang GR (1986) Analysis of stress intensity factors for surface cracks subjected to arbitrarily distributed surface stresses. Trans Japan Soc Mech Eng 52:390–398
37. Murakami Y (ed) (1988) Stress intensity factors handbook, vol 2. Pergamon Press, Oxford, pp 654–667
38. de Freitas M, Francois D (1995) Analysis of fatigue crack growth in rotary bend specimens and railway axles. Fatigue Fract Eng Mater Struct 18:171–178
39. Carpinteri A, Brighenti R (1996) Fatigue propagation of surface flaws in round bars: a three-parameter theoretical model. Fatigue Fract Eng Mater Struct 19:1471–1480
40. Carpinteri A, Brighenti R, Spagnoli A (1998) Surface flaws in cylindrical shafts under rotary bending. Fatigue Fract Eng Mater Struct 21:1027–1035
41. Couroneau N, Royer J (1998) Simplified model for the fatigue growth analysis of surface cracks in round bars under mode I. Int J Fatigue 20:711–718
42. Madia M, Beretta S, Schödel M, Zerbst U, Luke M, Varfolomeev I (2011) Stress intensity factor solutions for cracks in railway axles. Eng Fract Mech 78:764–792

Part III
Fracture Testing

Assessment of Material Properties by Means of the Small Punch Test

Martin Abendroth and Stefan Soltysiak

Abstract In recent years the small punch test (SPT) method has become an attractive alternative compared to traditional material testing procedures, especially in cases where only small amounts of material are available. We provide a literature review with focus on the history and application of the method. The main difficulty using the SPT is the fact that relevant material parameters cannot be as simply obtained by SPTs as by standard test methods, because of its non-uniform stress and deformation state. However, this can be achieved by comparing the experimental SPT results with those obtained by finite element computations of SPTs using advanced material models. Then the task is to determine the parameters of the material models using special optimization techniques. This paper presents SPT techniques for a broad temperature range. Work done on both ductile and brittle materials is presented. The analysis will focus on different advanced methods for determining parameters of state of the art material models for elastic-plastic, ductile damage and brittle failure behaviour. Results are provided for a weld line of a pressure gas pipe and brittle ceramic refractory materials.

Keywords Small punch test · Finite element method · Damage mechanics

1 Introduction

The small punch test (SPT) has been used now for more than thirty years to determine mechanical properties from miniaturized samples. The use of small samples becomes necessary if a sufficient amount of material for the production of standard sized samples is not available.

M. Abendroth (✉) · S. Soltysiak
Institute of Mechanics and Fluid Dynamics, TU Bergakademie Freiberg, Freiberg, Germany
e-mail: Martin.Abendroth@imfd.tu-freiberg.de

S. Soltysiak
e-mail: Stefan.Soltysiak@imfd.tu-freiberg.de

G. Hütter and L. Zybell, *Recent Trends in Fracture and Damage Mechanics*, DOI 10.1007/978-3-319-21467-2_6

The SPT is used in different sizes and different types. The smallest specimens are standard TEM sized specimen [44, 59]. Other authors use specimens cut from remnants of Charpy specimens, which are square shaped [6, 7, 53]. There has been a lot of effort to standardize the SPT and its usage [16], but this is still a running process. At least, there is a common understanding about the important features of the test.

A disk or square shaped specimen with a diameter or length D and a thickness t is placed on a circular die with a receiving hole of diameter d. This receiving die can have a rounded or straight chamfer edge of size r. The specimen can be clamped between the receiving die and a down-holder. There are also cases where the specimen is not clamped, usually for testing very brittle materials to avoid initial deformations during clamping. The specimen is loaded centrical using a punch with a spherical tip of radius R. The loading can be a constant displacement rate (CDR) of the punch, a constant force (CF) applied to the punch or an initial constant displacement (CD) of the punch followed by a holding (relaxation) time. The experimental results of the test are usually the punch displacement and/or the specimen deflection u and the punch force F. In case of time dependent material behaviour these values are stored together with the time t after starting the test. Figure 1 shows the typical test set-up and a resulting load deflection curve (LDC) for a CDR-SPT of a ductile metallic material.

Additional results from the SPT can be very useful like time and punch displacement at specimen failure. Furthermore, strain field measurements can be performed using optical devices [1, 78].

The typical results for the different types of the SPT are shown in Fig. 2 considering an elastic, visco-plastic material with a strain induced damage evolution.

The CDR-SPT can be performed at different punch velocities or specimen deflection rates. The result (see Fig. 1 right) is a load deflection curve (LDC), which starts with a short linear section (I) related to the elastic response of the specimen, followed by a typical knee (II) whose height is related to the initial yield strength of the tested material. The next part of the curve (III) corresponds to the ductile and strain rate related hardening. The decreasing slope (IV) just ahead of the force

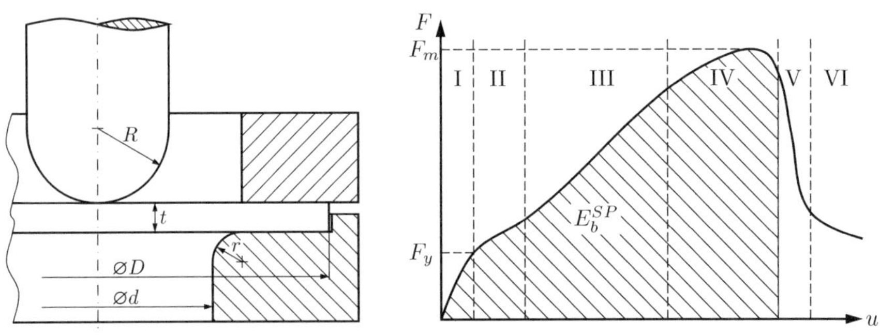

Fig. 1 Scheme of a SPT loading device and a typical resulting load deflection curve for a ductile metallic material

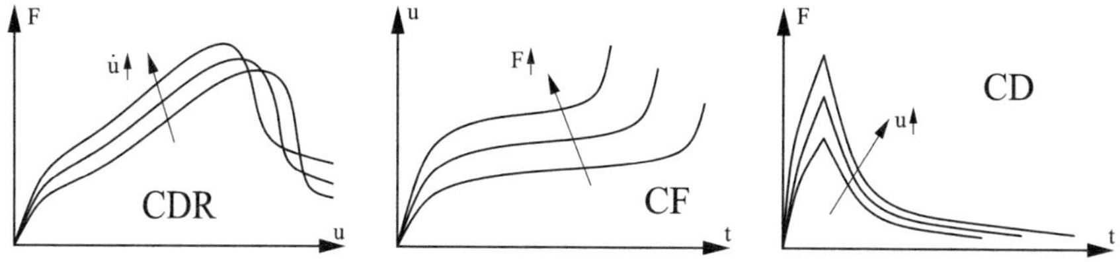

Fig. 2 Typical results for the different types of the SPT (CDR, CF, CD) for a visco-plastic metallic material with a strain induced damage evolution

maximum indicates a localization of deformation and the onset of material damage. The increasing damage leads finally to a decreasing load bearing capacity (V) of the material due to the failure of the specimen. There is a remaining force at the end of the test (VI), which is due to the frictional force between the penetrating punch and the specimen. If the punch speed is increased the curve is shifted to higher forces due to the strain rate sensitivity and the onset of damage happens at smaller deflections [3].

The CF-SPT is used to determine the material creep behaviour at different loads (stresses). As in uniaxial tensile creep tests we distinguish three parts of the curve, which are related to primary, secondary and tertiary creep, see Fig. 2 middle. The primary part of the curve is also influenced by some initial plastic deformation at rather high strain rates. The tertiary part is of course also influenced by a localization of deformation and increasing creep damage. For increasing test forces a higher mean specimen deflection will be observed together with decreasing failure times [26]. In contrast to tensile creep tests there is an initial plastic deformation within the specimen, which might have an influence on the subsequent creep behaviour.

The CD-SPT is not as common as the two other types of the test. But nevertheless it is a rather fast test to determine visco-plastic material behaviour. The test starts with a predefined deflection, which is applied in a short time, followed by a longer time period where the deflection is kept constant. The main result is the decreasing part of the curve, which depends mainly on the visco-plastic (creep) material properties, see Fig. 2 right. The advantage of this test type is that very small creep rates can be reached in a rather short time and that creep rates for a wide stress range can be determined.

2 Literature Review

This section presents the interested reader an overview of the history of the SPT as well as certain applications.

A miniaturized disk bend test (MDBT) was developed 1981 by Manahan et al. [59] for the analysis of irradiated materials used in nuclear power plants. The main reason for this development were the high costs associated with investigations arising on samples with standard size. In this early stage of development sample dimensions of $d = 3$ and $t = 0.25$ mm were used, which is similar to sample sizes used for transmission electron microscopy (TEM) [57]. The samples were so designed that no mechanical buckling occurs. This experiment was patented in the US [58]. Therein it is stated that the receiving die should have a fillet at the inner bore and the sample has to be clamped with a blank holder and loaded with a punch having a spherical tip. For the determination of material properties (e.g. $R_{p0.2}$) the finite element method (FEM) was used.

Starting in 1983 Baik et al. [6, 7] used samples having a rectangular cross-section with the dimensions $10 \times 10 \times 0.5$ mm^3 to investigate the ductile to brittle transition temperature (DBTT) of neutron irradiated steels. This test was then called small punch test, because the specimen is not only bent but also heavily stretched and partially sheared until macroscopic failure occurs. The load was applied by a steel ball having a diameter of 2.4 mm at a displacement rate of 0.02 mm s^{-1}. The measured DBTT from the SPT could be correlated with the one obtained from tests on Charpy V-notch specimens. It was found that the correlations depend on the strain rate and the stress state within the sample. In [6] the small punch work or energy E_{SP} was introduced as the integral value of the punch force over the specimen deflection for the determination of the transition temperature.

Li et al. [49] conducted experiments to determine the contact radius between punch and specimen. This value is used for analytical calculations of the tensile stresses at the specimen side opposite to the punch. Also analytical equations based on linear elastic plate theories are provided to determine the initial yield stress. Furthermore, the influence of surface defects on the failure or yield stress was investigated.

Zhang and Ardell [87] investigated the fracture toughness of ceramic materials (ZnS). Therefore, using a Vickers indenter cracks were produced on the tension side of the specimen. Two different failure patterns were observed. On one hand the expected sudden brittle fracture of the specimen was observed, but on the other hand also load drops were detected followed by an increasing load with increased deflection. Such a behaviour was explained with the initiation of small cracks or flaws which arrest in regions with lower stresses.

Norris and Parker [68] studied the influence of geometrical parameters of the SPT. Systematically, the sample thickness t, the punch radius R and the diameter of the receiving hole d have been varied. The sample diameter D was hold constant for all tests. Three different metallic materials (Al, Cu, 2.25Cr-1Mo-Steel) were investigated and correlations between the results of SPTs and standardized tensile tests examined. These correlations for yield and ultimate tensile strength have the form

$$\sigma_Y = \alpha \frac{F_y}{t^2} \quad \text{and} \quad \sigma_{UTS} = \beta \frac{F_m}{t},$$

where α and β depend on the geometrical parameters of the SPT set-up. For the definition of F_y and F_m see Fig. 1 (right). Garcia et al. [36] studied several correlations and criteria for the determination of σ_Y and evaluated their reliability.

Fleury and Ha [31] presented an analytical approach for the modelling of SPT LDCs. The experiments were performed on rectangular samples ($10 \times 10 \times 0.5$ mm^3). The load was applied using a tungsten carbide ball with a diameter of 2.4 mm at a displacement rate of 0.25 mm min^{-1}. The aim of the work was to evaluate the life time of austenitic components in steam power plants. Correlations between SPT results and material data from uniaxial tensile tests could be established. The approach for the analytical description of the SPT LDCs considers elastic-plastic plate bending and plastic membrane stretching theories. A further work of this research group [32] deals with the experimental determination of the DBTT for the same class of materials. It has been shown that a DBTT can be found using the SPT and that it correlates linearly with the fracture appearance transition temperature (FATT) obtained from conventional Charpy specimens.

A considerable number of papers is dealing with the identification of properties of ductile materials. Abendroth and Kuna [1–3] used neural networks, which are trained with results from finite element simulations to determine the parameters of the Gurson-Tvergaard-Needleman damage model (GTN-model) from small punch tests. Three different steels were examined and validated with data for the same materials obtained from tensile tests. Furthermore, the identified model parameters were applied to simulations of fracture toughness specimens (CT, SENB) and J-Δa curves could be predicted.

Finarelli et al. [30] examined the effect of neutron irradiation in austenitic and martensitic steels using analytical solutions for the determination of yield stress, equivalent fracture strain and fracture energy.

A lot of research work is related to materials used in nuclear or steam power plants. One example is P91 material which is a 9 % Chromium steel for creep resistant high temperature applications. Milička and Dobeš [64] found linear correlations between the SPT maximum force and the tensile strength. The specimen used there were obtained from remnants of surveillance specimens, which allowed the determination of additional material data without the need of additional irradiation specimens.

With the aid of the SPT the ductility properties of chromium, molybdenum and tungsten were studied [61]. The ductility of these body-centred cubic generally brittle materials can be significantly enhanced through targeted pre-deformation by the associated introduction of dislocations.

Of equally great interest such as the determination of plastic material properties is the determination of properties of brittle materials. Primarily the effects of neutron or ion irradiation on the properties of different metallic materials were studied. Ardell [5] presented investigations for a variety of irradiated inter-metallic materials, which are used in high-temperature applications.

Linse et al. [51, 52, 54] used the SPT to investigate the neutron embrittlement which shifts the DBTT towards higher temperatures. Furthermore, the tests were performed within a temperature range from −185 to +70 °C to cover the brittle, transition and ductile region. The ductile behaviour was modelled using a non-local extension of the GTN-model to overcome the mesh dependence, which often occurs in damage mechanical simulations. For covering the transition and brittle failure behaviour the Beremin model [8] was used. The parameter identification for the models was done using neural networks and non-linear optimization algorithms.

In addition to the previously described studies the SPT is often used to determine fracture mechanical parameters. Misawa et al. [66] determined the equivalent fracture strain for austenitic steels at low temperatures from SPT experiments and found a relationship to the fracture toughness $J_{\mathrm{Ic}} = k\bar{\varepsilon}_{\mathrm{qf}}$, where k is a material and geometry dependent parameter.

Saito et al. [73] made an investigation of the temperature dependence of the failure stress for different brittle ceramic materials (SiC, PSZ, Al_2O_3, Si_3N_4). To determine the onset of cracking an acoustic emission detection method was used.

Matsushita et al. [62] investigated various ferritic steels and could not confirm the linear relationship between FATT and DBTT for those materials. Due to the large scatter in the results of the DBTT they propose a multi linear regression to relate FATT and DBTT.

An approach often used is the correlation of the applied work in SPT with standardized fracture mechanics parameters (J_{Ic}, K_{Ic}). Foulds and Viswanathan [33–35] investigated this method and found that the difference between the K_{Ic} values from SPT and standard tests can be up to ±25 %.

Misawa et al. [65] used this method in combination with a statistical analysis based on the Weibull distribution. Kim et al. [46] also used the small punch work E_{SP} to determine the change of the transition temperature of a boiler tube of a power plant after use under operating conditions which caused material changes.

Bulloch [14, 15] used the small punch energy to determine both the transition temperature and the fracture toughness J_{Ic} for low alloy steels in an indirect way. First the biaxial stress state within the SP specimen is related to the SP energy and than the fracture toughness is related to the fracture strain according to the small scale yielding condition (see Rice and Johnson [72]).

Mao et al. [60] studied both brittle and ductile ceramics and metals used in power plant applications. They also used the biaxial strain at failure for the determination of J_{Ic} for the metallic materials, and the failure stress for the determination of K_{Ic} for the ceramic materials. For the test series empirical equations were derived, which permit the determination of the fracture toughness directly from SPT results.

Chi et al. [18] used the method explained above for determining the effect of neutron irradiation on the fracture toughness of nuclear reactor materials. They found that the fracture toughness is significantly reduced by neutron irradiation and that the fracture toughness J_{Ic} determined from the SPT was 20–30 % underestimated. These results were obtained by comparison of J_{Ic} values from the small punch tests with those from Charpy-V samples.

Bulloch [13] also conducted detailed fractographic investigations by scanning electron microscopy (SEM) to determine the portions of brittle and ductile fracture surface patterns within the transition and brittle region. Even at very low temperatures (−100 °C) and for macroscopic brittle failure behaviour ductile fracture patterns were found on the crack surfaces.

Bulloch [14] published an overview about the existing literature regarding CrMo-steels and found that the linear relationship between the small punch transition temperature T_{SP} and FATT proposed by Baik et al. [6] only holds for a small portion of these materials. He proposed instead a non-linear relationship as $\mathrm{FATT} = C/T_{SP}^2$, where C is a scaling factor. In addition, based on this data the relation between T_{SP} and grain size, which was proposed by Matsushita et al. [62] could be supported.

Shekhter et al. [75, 76] have also investigated a CrMo steel using SPT, Charpy-V notch test and three point bending test (3 PB). They found that the K_{Ic} values determined by SPT and Charpy-V were higher than those determined from the 3 PB test. Regarding the correlation of fracture toughness values determined by SPTs and those by conventional tests a lot of additional publications can be found [12, 17, 21, 36, 37, 47, 56, 74, 77, 83–85].

In all previously cited publications fracture mechanical parameters were determined from samples without macroscopic cracks. But there are efforts to prepare samples with defined cracks for small punch testing to improve the fracture mechanical behaviour. Ju and Kwon [43] placed a through-crack in SP samples having a rectangular cross-section and measured the true crack geometry for each single specimen. The onset of cracking was detected by acoustic emission. To determine fracture mechanical properties they used analytical solutions of elastic plate bending theory and linear elastic fracture mechanics.

Another method to produce well defined cracks in SP samples is the use of a Vickers indenter [87]. The geometry of the cracks were made visible by means of fluorescent liquid penetrating the cracks. The determined fracture toughness of Al_2O_3 measured by Rasche [69] with the SPT lay about 20 % below the reference value. Turba et al. [79] presented SP specimens with a circular notch with a diameter of 2.5 mm around the center of the specimen. These cracks were produced by electrical discharge machining into P91 material. This particular crack geometry was chosen in order to produce a plain stress state around the crack tip. A disadvantage of this crack configuration is the fact that the crack is not under pure mode I loading, but rather under a mixed-mode.

Cuesta and Alegre [19, 20] used for their investigations samples with a rectangular cross section and a thickness $t = 1$ mm. The depth of the straight crack was varied in several steps. For the evaluation of the experiments and the determination of the crack initiation point a finite element model was used. It has been demonstrated that K_{Ic} values can be determined within the range of those from Charpy tests results if the crack has a sufficient depth.

Another interesting test, which was derived from the SPT is the ball on three ball (B3B) test. Here, the specimen is supported by three balls instead by a circular die, which leads to a well defined support, which is especially necessary for very brittle materials. Rasche [69] and Rasche et al. [70] performed a study on 96 % pure alumina (Al_2O_3) to determine the parameters for a Weibull model describing the brittle failure probability. It became obvious that a size effect needs to be considered accounting for the stressed volume of different specimens.

Another application of the SPT is the investigation of time dependent material behaviour. Therefore a constant load (force) is applied, whereas this type of test is called constant force SPT (CF-SPT). To investigate the repeatability and accuracy of CF-SPTs a round robin within an European project was performed [82]. It could be shown, that the repeatability of CF-SPT experiments is as well secured as the repeatability of conventional creep tests. It could be also shown that for both a low as well as for a high-alloy steel a modified Arrhenius function for describing the temperature dependence for the time to failure can be used.

Yang and Wang [86] used a combined approach, where experiments as well as the finite element method were used to investigate the dependence between the creep strain in the center of the sample and the punch displacement. It was found that the creep strain at the sample center is independent of the applied load, the temperature and the elastic-plastic behaviour of the material.

Milička and Dobeš [63, 64] conducted studies about the creep strength of P91 steel using conventional tensile creep tests as well as CF-SPTs. They found a linear relation between the force used for CF-SPTs and the stresses within tensile creep specimen for tests with a similar time to failure. This relationship was examined later in more detail and confirmed by means of finite element analyses [25]. The same research group has shown that a Monkman-Grant relationship can be used to correlate the time to failure in the CF-SPT with the minimum creep rate [22]. In addition, the linear correlation between the CF-SPT punch force and stress in the tensile creep test, leading to the same failure time for particle reinforced aluminium alloys could be confirmed [23].

Li and Šturm studied the creep rupture strength of P91 welds [50]. Due to the small sample volume, which is required for the SPT specimens, a detailed study of the material property variations in the individual zones (primary material, heat affected zone and weld metal) of the weld was possible. A similar study on a P91 weld was done by Blagoeva and Hurst [11]. However, the focus here was on the exact application and thus confirmation of the 2006 established code of practice for small punch creep testing [16].

Kato et al. [45] used again much smaller samples $d = 3 \times 0.35$ mm to get a better resolution of the material properties of the individual zones of the weld. Dobeš and Milička [24] have shown that different approaches for the prediction of residual lifetimes of conventional experimental data (Larson-Miller, Fisher-spine) can be also applied to small punch tests.

Dymáček and Milička [26, 27] investigated the influence of friction using FEM. They came to the conclusion that a direct determination of the friction is not possible. However, by post mortem examinations on samples and comparative simulations conclusions can be drawn on the friction.

Evans and Evans [29] presented a viscoplastic constitutive law for the description of the material behaviour of a Cr-Mo steel. The material parameters were determined by means of SPTs and then validated with results from uniaxial tensile tests. Later, Evans and Wang [28] presented a validation of the model based on literature data. Based on this investigation SPT creep data could be converted into uniaxial tensile test creep data.

Holmstrom et al. [41] presented an analytical approach for the determination of uniaxial creep curves based on SPT results and the Wilshire model for creep damage. This study was related to a P91 weld.

Besterci et al. [9, 10] have shown that by means of CF-SPT anisotropic material properties can be also examined. A particle reinforced aluminium showed for a load transverse to the pressing direction the lowest creep strength.

A comprehensive investigation of CF-SPTs was performed by Hyde et al. [42]. Analytical relations between CF-SPTs and conventional creep tests have been derived. Furthermore, possibilities for the evaluation of SPTs by FEM were investigated. Conversion factors between the uniaxial creep strain or stress and the biaxial creep strains or stress could be confirmed from SPT tests. The material law used within the FEM analysis contained a Norton creep law and a continuum damage model according to Kachanov [40].

Alegre et al. [4] used a modified form of the Monkman-Grant relationship for the prediction of the time to failure of small punch tests. The modification is based on the introduction of a minimum relative punch velocity. This reformulation also allows the prediction of time to failure from interrupted experiments if the secondary creep stage has been reached.

After reviewing all the literature it can be concluded that the suitability of the SPT to determine mechanical material properties has been demonstrated for a broad range of materials. This is possible for statistical properties of brittle materials as well as for complex damage mechanical behaviour as described by the GTN model. Different materials such as martensitic steels (P91), ferritic Steels (C-0.5Mo), semiconductors (ZnS), ceramics (Al_2O_3) and plastics like ultra-high-molecular-weight polyethylene (UHMWPE) [48] were investigated. Much work has been done regarding the characterization of steels used in nuclear or steam power plants. Of particular importance is the SPT to investigate material changes due to the conditions of use. Comparative studies often provided results for correlations between SPT and conventional tests. The determination of fracture mechanical parameters on notched and unnotched specimens as well as the determination of creep data from creep and relaxation tests were done. The SPT is often accompanied by additional examinations (e.g. grain boundary etching, SEM, TEM, computer tomography) to verify or to get better interpretations of the results and to get correlations to material parameters from conventional standard experiments.

An important tool to determine material properties using the SPT is the FEM, because it allows the detailed analysis of the non-homogeneous strain and stress fields, which depend in general on the material properties. All empirical correlations inherit some material dependence and must be used therefore with great care.

3 Material Modelling

Ductile metallic materials show an elastic-plastic behaviour, where the elastic strains can be considered small in comparison to inelastic strains. Prior to yield, the material response is assumed to be linear elastic. The strain tensor is split into an elastic, a plastic and a creep part.

$$\varepsilon_{ij} = \varepsilon_{ij}^{\mathrm{el}} + \varepsilon_{ij}^{\mathrm{pl}} + \varepsilon_{ij}^{\mathrm{cr}} \tag{1}$$

Due to the axisymmetric geometry and loading of the SPT only isotropic material behaviour can be identified. Furthermore, within the SPT a monotonic loading is applied. Thus, no kinematic hardening, only isotropic hardening can be identified from experimental results.

The stresses in terms of the elastic strains are expressed by the multi axial Hooke's law, which for the isotropic case reads as

$$\sigma_{ij} = \frac{E}{1+\nu}\left(\varepsilon_{ij}^{\mathrm{el}} + \frac{\nu}{1-2\nu}\varepsilon_{kk}^{\mathrm{el}}\delta_{ij}\right), \tag{2}$$

where E denotes the Young's modulus, ν Poisson's ratio and δ_{ij} the Kronecker delta.

The plastic strain rate is derived from a flow potential Φ_{pl}.

$$\dot{\varepsilon}_{ij}^{\mathrm{pl}} = \lambda \frac{\partial \Phi_{\mathrm{pl}}}{\partial \sigma_{ij}}, \tag{3}$$

with the plastic multiplier λ. The flow rule is given by

$$\Phi_{\mathrm{pl}} = \sigma_{\mathrm{q}} - \sigma_{\mathrm{y}}\left(\varepsilon_{\mathrm{q}}^{\mathrm{pl}}\right) = 0, \tag{4}$$

where σ_{q} denotes the equivalent (von Mises) stress and $\sigma_{\mathrm{y}}(\varepsilon_{\mathrm{q}}^{\mathrm{pl}})$ an isotropic hardening function e.g. the Voce law

$$\sigma_{\mathrm{y}}\left(\varepsilon_{\mathrm{q}}^{\mathrm{pl}}\right) = \begin{cases} \sigma_0 & \text{if} \quad \varepsilon_{\mathrm{q}}^{\mathrm{pl}} \le \varepsilon_{\mathrm{lu}} \\ \sigma_0 + \sigma_1\left(\varepsilon_{\mathrm{q}}^{\mathrm{pl}} - \varepsilon_{\mathrm{lu}}\right) + \sigma_2\left\{1 - \exp\left[-n\left(\varepsilon_{\mathrm{q}}^{\mathrm{pl}} - \varepsilon_{\mathrm{lu}}\right)\right]\right\} & \text{if} \quad \varepsilon_{\mathrm{q}}^{\mathrm{pl}} > \varepsilon_{\mathrm{lu}} \end{cases} \tag{5}$$

or the well known Ramberg-Osgood law

$$\sigma_{\mathrm{y}}\left(\varepsilon_{\mathrm{q}}^{\mathrm{pl}}\right) = \begin{cases} \sigma_0 & \text{if} \quad \varepsilon_{\mathrm{q}}^{\mathrm{pl}} \le \varepsilon_{\mathrm{lu}} \\ \sigma_0\left(\frac{\varepsilon_{\mathrm{q}}^{\mathrm{pl}} - \varepsilon_{\mathrm{lu}}}{\varepsilon_0}\right)^{\frac{1}{n}} & \text{if} \quad \varepsilon_{\mathrm{q}}^{\mathrm{pl}} > \varepsilon_{\mathrm{lu}} \end{cases} \tag{6}$$

Within the two equations above σ_0 denotes the initial yield stress, σ_1 a linear and σ_2 a non-linear hardening parameter, n a hardening exponent and $\varepsilon_{\mathrm{lu}}$ a Luders strain. For a rate dependent plasticity (creep) model we define the creep strain rate as a combination of

$$\dot{\varepsilon}_{ij}^{\mathrm{cr}} = \dot{\varepsilon}_{\mathrm{p}}^{\mathrm{cr}} \frac{1}{3} \delta_{ij} + \dot{\varepsilon}_{\mathrm{q}}^{\mathrm{cr}} n_{ij} \tag{7}$$

where $\dot{\varepsilon}_{\mathrm{p}}^{\mathrm{cr}}$ is the pressure dependent volumetric strain rate and $\dot{\varepsilon}_{\mathrm{q}}^{\mathrm{cr}}$ the equivalent deviatoric creep strain rate. n_{ij} defines the direction of the creep strain derived from the equivalent stress potential.

$$n_{ij} = \frac{\partial \Phi_{\mathrm{cr}}}{\partial \sigma_{ij}} \tag{8}$$

The volumetric and deviatoric strain rates have evolution laws like:

$$\dot{\varepsilon}_{\mathrm{p}}^{\mathrm{cr}} = h_{\sigma_{\mathrm{p}}} \left(\sigma_{\mathrm{p}}, \sigma_{\mathrm{q}}, \varepsilon_{\mathrm{p}}^{\mathrm{cr}}, \varepsilon_{\mathrm{q}}^{\mathrm{cr}}, T, \ldots \right) \tag{9}$$

and

$$\dot{\varepsilon}_{\mathrm{q}}^{\mathrm{cr}} = h_{\sigma_{\mathrm{q}}} \left(\sigma_{\mathrm{p}}, \sigma_{\mathrm{q}}, \varepsilon_{\mathrm{q}}^{\mathrm{cr}}, \varepsilon_{\mathrm{q}}^{\mathrm{cr}}, T, \ldots \right). \tag{10}$$

depending on the equivalent stress σ_{q}, the hydrostatic pressure $\sigma_{\mathrm{p}} = -\frac{1}{3}\sigma_{ij}\delta_{ij}$, the volumetric and deviatoric creep strain and the temperature T. A simple example is the Norton creep law where $\dot{\varepsilon}_{\mathrm{p}}^{\mathrm{cr}} = 0$ and $\dot{\varepsilon}_{\mathrm{q}}^{\mathrm{cr}} = \dot{\varepsilon}_0 \left(\frac{\sigma_{\mathrm{q}}}{A}\right)^n$ with the material parameters A and n. A more advanced creep model is an additive combination of i Norton laws

$$\dot{\varepsilon}_{\mathrm{q}}^{\mathrm{cr}} = \dot{\varepsilon}_0 \sum_i \left\langle \frac{\sigma_{\mathrm{q}} - B_i}{A_i} \right\rangle^{n_i}, \tag{11}$$

which allows the modelling of multiple creep mechanisms like different diffusion and dislocation mechanisms. The McCauley brackets have the meaning $\langle x \rangle = (x + |x|)/2$. This ensures that a creep mechanism is only active above a respective threshold stress B_i.

In order to simulate plasticity and ductile damage the continuum damage model of Gurson [38, 39] can be used with the extensions of Tvergaard [80] and Tvergaard and Needleman [81]. This model assumes an elastic plastic continuum with a distribution of spherical voids with a void volume fraction f. The central part of the model is the yield potential

$$\Phi_{\text{pl}}^{\text{GTN}} = \left[\frac{\Sigma_{\text{q}}}{\sigma_{\text{y}}\left(\varepsilon_{\text{q}}^{\text{pl}}\right)}\right] + 2q_1 f^* \cosh\left[\frac{3}{2} q_2 \frac{\Sigma_{\text{p}}}{\sigma_{\text{y}}\left(\varepsilon_{\text{q}}^{\text{pl}}\right)}\right] - \left[1 + (q_1 f^*)^2\right] = 0 \tag{12}$$

where $\Sigma_{\text{q}} = \sqrt{\frac{3}{2} S_{ij} S_{ij}}$ denotes the macroscopic von Mises and $\Sigma_{\text{p}} = \frac{1}{3}\Sigma_{ii}$ the macroscopic hydrostatic stress, expressed by the macroscopic deviatoric stresses $S_{ij} = \Sigma_{ij} - \Sigma_{\text{p}}\delta_{ij}$. The dimensionless parameters q_1 and q_2 are used to weight the influence of the stress triaxiality. The material damage f^* depends on the void volume fraction f.

$$f^* = \begin{cases} f & \text{if} \quad f \leq f_{\text{c}} \\ f_{\text{c}} + \frac{f_{\text{f}}^* - f_{\text{c}}}{f_{\text{f}} - f_{\text{c}}}(f - f_{\text{c}}) & \text{if} \quad f_{\text{c}} < f < f_{\text{f}} \\ f_{\text{f}}^* & \text{if} \quad f \geq f_{\text{f}} \end{cases} \tag{13}$$

with $f_{\text{f}}^* = \frac{1}{q_1}$. Up to a critical void volume fraction f_{c} the damage is identical with the value of the void volume fraction. Beyond f_{c} where voids coalescence or micro crack initiation is assumed damage evolution is accelerated until a void volume fraction f_{f} is reached where the material fails. The evolution of the equivalent plastic strain of the matrix material is obtained from the plastic macroscopic strain rate $\dot{E}_{ij}^{\text{pl}}$

$$\dot{\varepsilon}_q^{\text{pl}} = \dot{\varepsilon}_q^{\text{pl}}\Big|_0 + \int_0^t \frac{\Sigma_{ij}\dot{E}_{ij}^{\text{pl}}}{(1-f)\sigma_{\text{q}}}\,\mathrm{d}t. \tag{14}$$

The evolution of the void volume fraction is composed of two terms

$$\dot{f} = \dot{f}_{\text{gr}} + \dot{f}_{\text{nucl}} \tag{15}$$

where $\dot{f}_{\text{gr}}$ describes the growth of voids based on the law of conservation of mass

$$\dot{f}_{\text{gr}} = (1-f)\dot{E}_{kk}^{\text{pl}} \tag{16}$$

and a void nucleation part, which follows a strain controlled relationship.

$$\dot{f}_{\text{nucl}} = \frac{f_{\text{n}}}{s_{\text{n}}\sqrt{2\pi}} \exp\left[-\frac{1}{2}\left(\frac{\varepsilon_{\text{q}}^{\text{pl}} - \varepsilon_{\text{n}}}{s_{\text{n}}}\right)\right]\dot{\varepsilon}_{\text{q}}^{\text{pl}} \tag{17}$$

The normal distribution of the nucleation strain has a mean value ε_{n} and a standard deviation of s_{n}. f_{n} denotes the volume fraction of void nuclei. For detailed information about the implementation into the FE-Code ABAQUS see [67, 88, 89].

The above described GTN-model belongs to the class of local damage models. It is well known that the results of these models are mesh dependent. A damage zone usually localizes within an one element thick band or plane. To avoid this one can use non-local damage models, see [51, 55, 71]. The non-local approach of Linse [55] considers the spatial vicinity for the void growth so Eq. (16) is replaced by a non-local expression

$$\dot{f}_{\text{gr}}^{\text{nl}} = (1-f)\dot{\bar{E}}_{kk}^{\text{pl}}, \tag{18}$$

where $\bar{E}_{kk}^{\text{pl}}$ is the non-local volumetric plastic strain. Thus, Eq. (15) becomes

$$\dot{f} = \dot{f}_{\text{gr}}^{\text{nl}} + \dot{f}_{\text{nucl}}. \tag{19}$$

The non-local averaging of the volumetric strain is done by implicit solving the Helmholtz-type differential equation

$$\bar{E}_{kk}^{\text{pl}} - c\bar{E}_{kk,ii}^{\text{pl}} = E_{kk}^{\text{pl}}. \tag{20}$$

The corresponding natural boundary condition for the averaged macroscopic volumetric strain is

$$\bar{E}_{kk,i}^{\text{pl}} n_i = 0, \tag{21}$$

making sure that the total averaged volumetric plastic strain remains the same as the non-averaged. Such models usually introduce a characteristic length $\sqrt{c}$ as an additional parameter which can be related to the spacing of voids or the width of localized damage bands. When using such models element sizes smaller than the characteristic length are required, which leads to larger models than those using a local damage model and solving additional field equations is necessary, which requires the use of non-standard solvers and/or elements within the finite element codes [55].

The Beremin research group [8] derived a local criterion for cleavage fracture, which is based on the assumption that cleavage fracture initiates on micro cracks, which have evolved due to inelastic deformations. The critical stress σ_{c} leading to crack propagation can be related to the crack length l_{c} and the surface energy γ

$$\sigma_{\text{c}} = \sqrt{\frac{2E\gamma}{\pi(1-\nu^2)l_{\text{c}}}}. \tag{22}$$

If the crack size distribution follows a power law $h(l) \sim l^{(-m/2)}$ the probability for brittle failure of a specimen or a component can be defined as:

$$P_{\text{f}}(L) = 1 - \exp\left[-\left(\frac{\sigma_{\text{w}}(L)}{\sigma_{\text{u}}}\right)^m\right], \tag{23}$$

where $\sigma_{\mathrm{w}}(L)$ is the Weibull stress for a specific external loading L defined by

$$\sigma_{\mathrm{w}} = \sqrt[m]{\frac{1}{V_0} \int\limits_{V_{\mathrm{pl}}} \sigma_{\mathrm{I}}^m \, \mathrm{d}V} \tag{24}$$

or in a more FEM convenient form

$$\sigma_{\mathrm{w}} = \sqrt[m]{\frac{1}{V_0} \sum_{i}^{n_{\mathrm{pl}}} \sigma_{\mathrm{I}}^m V_i}. \tag{25}$$

Within the above equations σ_{u} denotes the Weibull reference stress, m the Weibull exponent, σ_{I} the first principal stress for a material point i within an inelastic Volume V_{pl}. There exist several variations for the definition of the Weibull stress. The principal of independent action (PIA) takes all principal stresses into account, which allows also failure under large hydrostatic pressure.

$$\sigma_{\mathrm{w}} = \sqrt[m]{\frac{1}{V_0} \sum_{i}^{n_{\mathrm{pl}}} (|\sigma_{\mathrm{I}}|^m + |\sigma_{\mathrm{II}}|^m + |\sigma_{\mathrm{III}}|^m) V_i}. \tag{26}$$

4 Parameter Identification

As shown in the previous section the material model contains a set of parameters p_i, which have to be determined from experimental results. Especially for the creep parameters multiple CF-SPT experiments at different load levels are necessary. The general way to find material parameter sets is to fit the model to experimental results, which can be a single test or a set of different tests (CDR, CF and CD). Parameter identification is an optimization process, where the difference between experiments and simulations needs to be minimized by changing the model parameters. The value to be minimized is an error, which measures the quality or exactness of the simulations for each test. Each pair of experiment and corresponding simulation gets its own error value, which is multiplied by a certain weight w_i. Those weights represent the importance of an experiment or the confidence the user has for this single experiment.

For a single CDR-SPT the error e is defined as the integral of the normalized difference of the punch force between simulation and corresponding experiment in a predefined deformation interval interval $[u_0 \ldots u_1]$.

$$e^{\mathrm{CDR}} = \frac{1}{u_1 - u_0} \int_{u_0}^{u_1} \left[\frac{F_{\mathrm{sim}}(u) - F_{\mathrm{exp}}(u)}{\bar{F}_{\mathrm{exp}}} \right]^2 \mathrm{d}u \tag{27}$$

The error for a CF-SPT is the integral of the normalized difference of the punch displacement or specimen deflection between simulation and corresponding experiment in a predefined time interval $[t_0 \ldots t_1]$ plus an error which expresses the normalized difference between the times of failure for simulation and experiment. The second error term in (28) is included here, because it is much more sensitive to changes of the creep parameters than the deflection error term.

$$e^{\mathrm{CF}} = \frac{1}{t_1 - t_0} \int_{t_0}^{t_1} \left[\frac{u_{\mathrm{sim}}(t) - u_{\mathrm{exp}}(t)}{\bar{u}_{\mathrm{exp}}} \right]^2 \mathrm{d}t + \left(\frac{t_{\mathrm{fsim}} - t_{\mathrm{fexp}}}{t_{\mathrm{fexp}}} \right)^2 \tag{28}$$

The error for a CD-SPT is defined as integral of the normalized difference of the punch force for the simulations and the corresponding experiment in a predefined time interval $[t_0 \ldots t_1]$.

$$e^{\mathrm{CD}} = \frac{1}{t_1 - t_0} \int_{t_0}^{t_1} \left[\frac{F_{\mathrm{sim}}(t) - F_{\mathrm{exp}}(t)}{\bar{F}_{\mathrm{exp}}} \right]^2 \mathrm{d}t \tag{29}$$

As normalizing values the observed mean values of the punch force $\bar{F}_{\mathrm{exp}}$ or the punch displacement $\bar{u}_{\mathrm{exp}}$ are used. For a multi specimen approach the errors of different test are weighted, summed up and divided by the sum of weights for all experiments.

$$e = \frac{\sum_{i=1}^{n^{\mathrm{CDR}}} w_i^{\mathrm{CDR}} e_i^{\mathrm{CDR}} + \sum_{j=1}^{n^{\mathrm{CF}}} w_j^{\mathrm{CF}} e_j^{\mathrm{CF}} + \sum_{k=1}^{n^{\mathrm{CD}}} w_k^{\mathrm{CD}} e_k^{\mathrm{CD}}}{\sum_{i=1}^{n^{\mathrm{CDR}}} w_i^{\mathrm{CDR}} + \sum_{j=1}^{n^{\mathrm{CF}}} w_j^{\mathrm{CF}} + \sum_{k=1}^{n^{\mathrm{CD}}} w_k^{\mathrm{CD}}} \tag{30}$$

The minimization of this error is done within an optimization loop as shown in Fig. 3. One may notice that the finite element computations are not a direct part of the optimization loop. Instead finite element computations are done in advance using parameters which are varied in reasonable bounds. These computed results are used to train neural networks, which represent the approximations $F'_{\mathrm{sim}}(u, p_i)$, $u'_{\mathrm{sim}}(t, p_i)$ and $F'_{\mathrm{sim}}(t, p_i)$ of the finite element simulations. The quality of the neural network approximations can be measured by comparing predictions of the networks with simulation results that have not been part of the training. Each test type (CDR, CF or CD) requires separate neural networks. For creep tests two networks are used, one predicting the failure time and a second predicting the deflection over time.

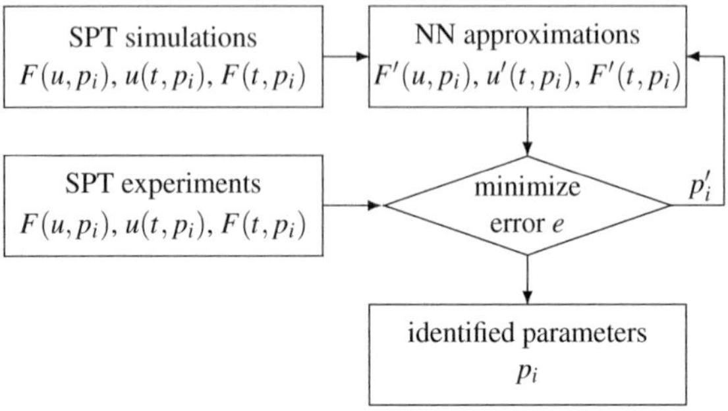

Fig. 3 Optimization loop for the identification of the damage material parameters. p_i' denotes an updated set of parameters returned by the minimization algorithm

All the networks used here have the structure of feed forward neural networks, see Fig. 4. They consist of at least three layers of neurons. All the layers are fully forward connected, which means that each neuron of one layer is connected with each neuron of the subsequent layer. The number of neurons for the first (input) layer is similar to the number of arguments of the function which the network

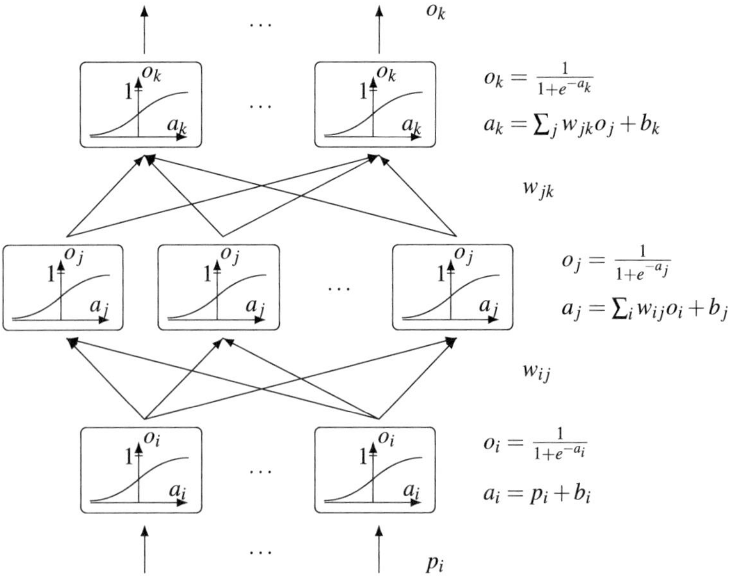

Fig. 4 Scheme of a feed forward neural network

should approximate. If we consider a classical fully connected feed forward network, having l layers of neurons, the following recursive scheme for computing the output o_j^l can be set up:

$$a_i^1 = p_i + b_i^1 \quad \text{for} \;\; i = 1\ldots n_1 \tag{31}$$

$$o_j^k = \frac{1}{1 + \exp\left(-a_j^k\right)} \quad \text{for} \;\; j = 1\ldots n_k \;\; \text{and} \;\; k = 2\ldots l \tag{32}$$

$$a_j^k = \sum_{i=1}^{n_{k-1}} w_{ij}^{k-1} o_i^{k-1} + b_j^k \quad \text{for} \;\; k = 1\ldots l \tag{33}$$

where $n_k = [n_1, \ldots, n_l]$ denotes the number of neurons in network layer k. Each neuron gets an input value a_j^k, which is the weighted and summarized output of the layer below. The first layer neurons gets a vector of input values p_i. The weights w_{ij}^k and biases b_j^k are free parameters of the network which have to be determined during a training process. Finally a trained network can be seen as a smooth mapping for an input vector p_i to an output vector o_j. Furthermore, the recursive network function is derivable at will.

The neural networks are trained with results of finite element simulations of the SPT. Each training sample is a pair of an input vector containing the punch displacement and a set of n material parameters (u, $p_1 \ldots p_n$) and an output, which is in our case the resulting punch force F. A single simulation delivers a load deflection curve which is discretized into 100 samples, where only u and F are variable. To cover a given parameter space, more simulations are done with varying parameters. It can be advantageous if the parameter variations are done randomly from an equal distribution. The total number of samples necessary depends on the complexity of the function to be approximated as well as of the wanted accuracy. For the identification of four hardening parameters we used 1000 simulations and for the identification of five damage parameters 5000 simulations. The number of necessary neurons in the middle layers also depends on the complexity of the problem. For the hardening parameters 25 neurons and for the damage parameters 50 neurons were used. For the training of the neural networks the samples are split up into two sets. One set containing 90 % of all samples is used for the training. The other 10 % are only used for evaluating the accuracy and not for training. During the training the accuracy for training and evaluation sets are monitored. The training is stopped either if the evaluation error has just passed a minimum or if training and evaluation error reach a predefined low value (0.1 %). The approximation accuracy reached was about 0.25 % of $F_{\max}$ regarding the hardening and 1 % for the damage parameter variations.

5 Applications

This section presents two applications of the SPT. One is the assessment of damage mechanical properties of a weld line of a pressure gas pipeline, with its different material zones, as there are base material (base), weld material (weld) and heat affected zone (haz) material. The second application is the assessment of brittle damage behaviour of a novel carbon bonded alumina, which is used as a refractory material or for metal melt filtration applications.

5.1 *Damage Mechanical Assessment of a Gas Pipe Weld Line*

The gas pipe under consideration is a high pressure pipe (200 bar). Segments of a pipeline are welded together when the pipeline is installed. During the installing process the pipeline can undergo rather large bending deformations which might lead to some inelastic deformations and therewith some damage especially in the vicinity of the weld lines. Due to legal issues the manufacturer wants to keep the chemical composition confidential, no further material details are presented here.

Figure 5 shows a cross section of a weld line. The wall thickness is 14 mm, the inner pipe radius is 148 mm. The weld line is made in several runs. The different

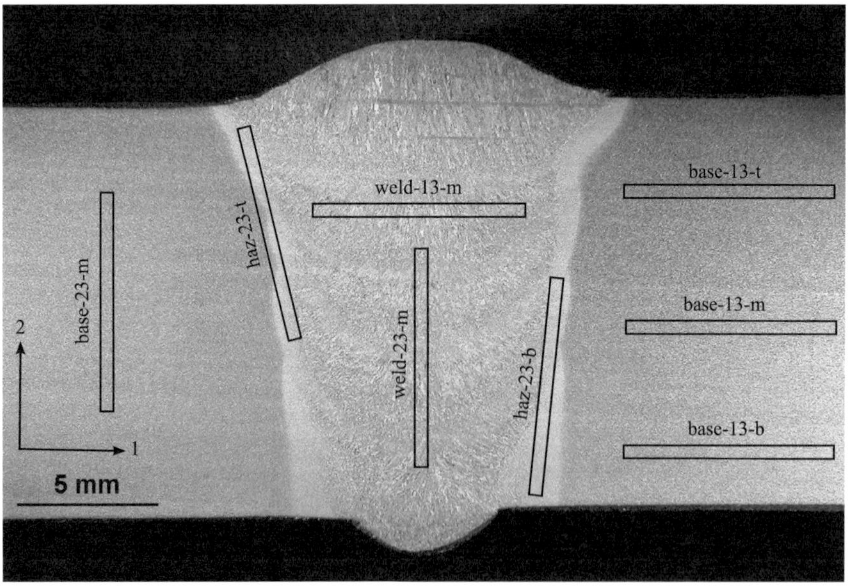

Fig. 5 Weld line width indicated specimen

layers and different material zones can be clearly distinguished. The locations of different SPT specimens are indicated, where the name is a combination of material zone (base, weld haz), orientation within the pipe (13, 23) and location (b—bottom, m—middle, t—top) within the pipe wall.

Figures 6, 7, 8 show microscopies of tested SPT specimens of each material. The left picture shows the macroscopic crack after testing. One can distinguish a primary circumferential crack and some secondary cracks evolving later perpendicular to the primary crack. The middle pictures show details of the secondary cracks. The right pictures are high resolution images of the primary crack surfaces, indicating that the failure process is purely ductile, because of the typical dimple patterns.

The results of the experimental SPTs together with the corresponding simulations, which were done using the identified material parameters for each individual test are shown in the Figs. 9, 10, 11. The parameter identification was done in two

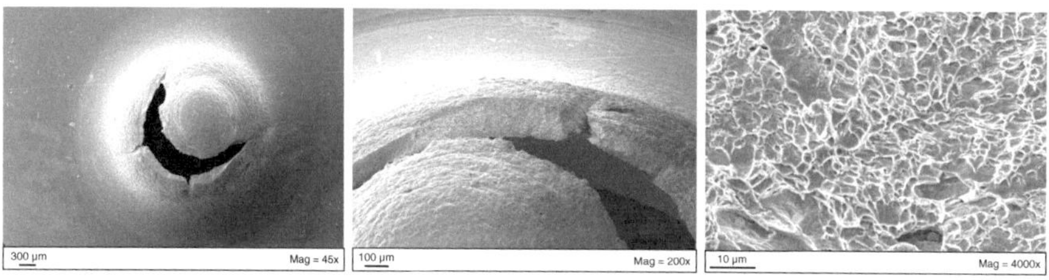

Fig. 6 Details of a tested base-13-m material CDR-SPT specimen after testing

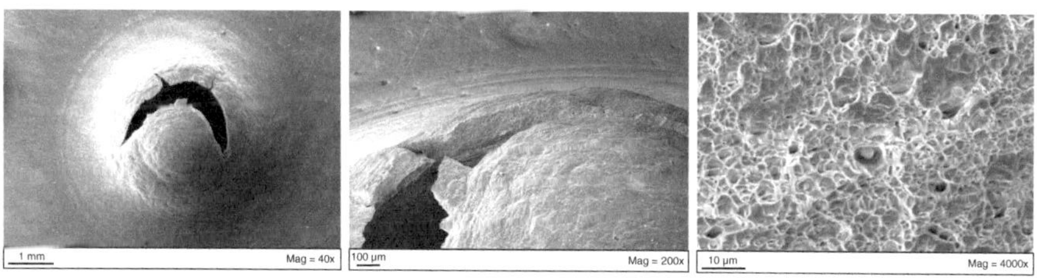

Fig. 7 Details of a tested weld-23-m material CDR-SPT specimen after testing

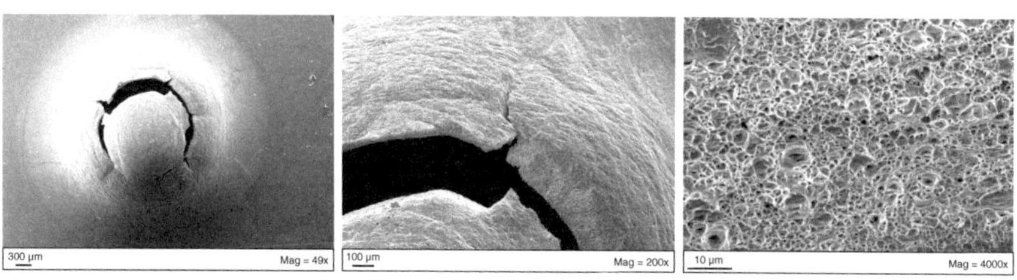

Fig. 8 Details of a tested haz-23-t material CDR-SPT specimen after testing

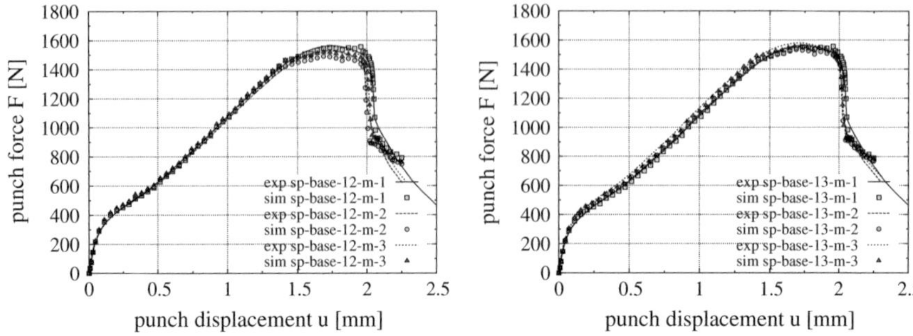

Fig. 9 Comparison of SPT LDCs obtained from experiments and corresponding simulations using the identified material parameters for the base material

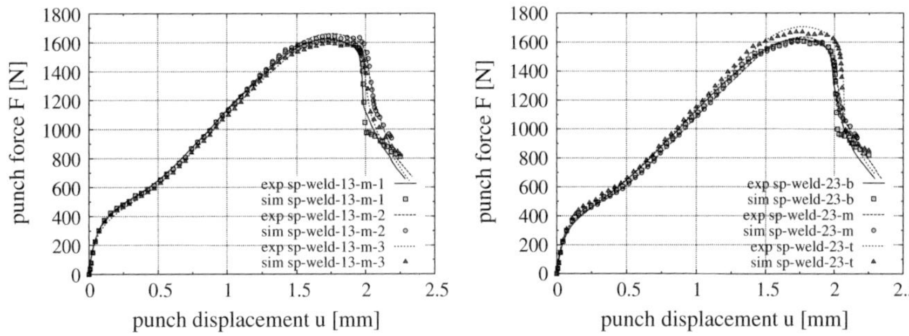

Fig. 10 Comparison of SPT LDCs obtained from experiments and corresponding simulations using the identified material parameters for the weld material

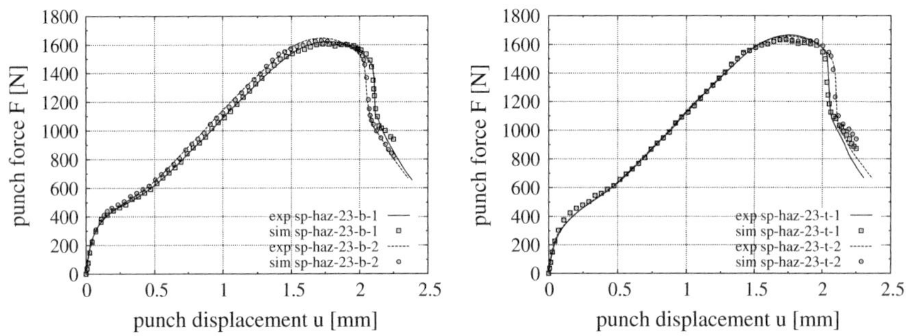

Fig. 11 Comparison of SPT LDCs obtained from experiments and corresponding simulations using the identified material parameters for the HAZ material

subsequent steps. First the hardening parameters (see Eq. 5) were identified using a network trained with 1000 simulated LDCs. Only the first three curve parts (see Fig. 1 right) were used for the identification. For the identification of the damage parameters a larger data base of 5000 simulations was used to train a network having 50 neurons within the middle layer. Here, the whole LDCs were used for the identification process. One may note that not all material parameters of the model were identified. The SPT is not very sensitive regarding the elastic properties and the Luders strain of a ductile metal. The identified hardening parameters (mean values) for each material are given in Table 1.

The identified damage parameters (mean values) are given in Table 2. Here, only the parameters f_c, f_n, ε_n and q_2 have been identified, while f_0, f_f, s_n and q_1 are fixed. This is necessary because different parameter sets may lead to similar results. Especially different combinations of f_c, f_n and f_f or combinations of q_1 and q_2 are not unique. The choice of f_0 can be related to initial impurities of the material and the void volume fraction at failure f_f can be obtained from analysis of the fracture surfaces and $q_1 = 1.5$ is a common value for ductile materials [3].

To evaluate the identified material parameters additional tensile specimen were cut from the base material as well as from the weld line (see Fig. 12). There are seven specimens according to DIN 50125-B5x25 and also notched specimens having notch radii of 0.25 and 3 mm. Some parameters of the material model (n and ε_{lu}) were directly obtained from the standard tensile tests, because the SPT isn't

Table 1 Identified hardening parameters (averaged values) for the different materials

	σ_0 (MPa)	σ_1 (MPa)	σ_2 (MPa)	n (–)	ε_{lu} (–)
Base	524.9	325.5	100.0	30.0	0.002
Weld	544.6	414.7	100.0	30.0	0.018
HAZ	553.6	380.9	89.9	30.0	0.010

Table 2 Identified damage parameters (averaged values) for the different materials

	f_0 (–)	f_c (–)	f_f (–)	f_n (–)	ε_n (–)	s_n (–)	q_1 (–)	q_2 (–)
Base	2×10^{-5}	0.16	0.20	0.03693	0.50	0.1	1.5	1.00
Weld	2×10^{-5}	0.16	0.20	0.04586	0.55	0.1	1.5	1.05
HAZ	2×10^{-5}	0.17	0.20	0.03898	0.725	0.1	1.5	1.15

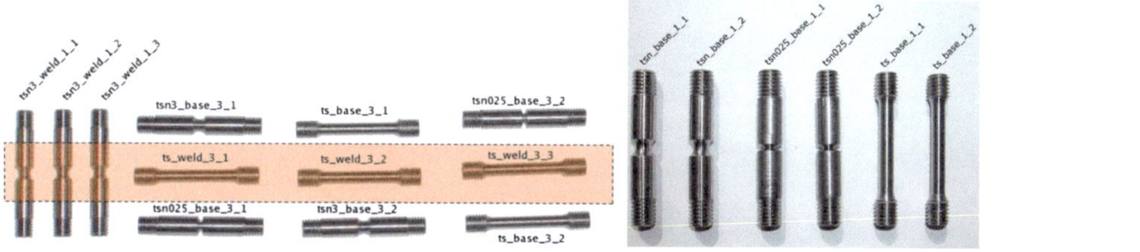

Fig. 12 Overview of all the tested tensile specimen. The *orange zone* in the left part of the figure indicates the location of the weld line

very sensitive to these parameters. The notched specimens were used to evaluate the damage parameters for different states of stress triaxiality.

In Fig. 13 the experiments for the base material tensile tests are compared with the corresponding simulations using the parameters which have been identified from SPTs. The base material shows a strength anisotropy. The material is about 20 % stronger in 1-direction (pipe axis) than in 2-direction (radial) and 3-direction (circumferential), which is caused by the pipe production process. The SPT is a biaxial test, where the specimen failure is controlled by the weaker direction. That's why the simulated tensile test curves fit better the experiments for the specimens taken in 3-direction. The failure point is slightly underestimated for the standard tensile and $r = 3$ mm tensile tests, whereas the failure points for the $r = 0.25$ mm tensile tests are in good agreement. This indicates the importance of the stress triaxiality for the failure process.

The weld material does not show an anisotropy (see Fig. 14). Here, the tensile test simulations, which were also done using parameters identified from SPTs, show a very good agreement with the corresponding experiments. The weld and the HAZ material have almost identical mechanical properties independent of the specimens orientation.

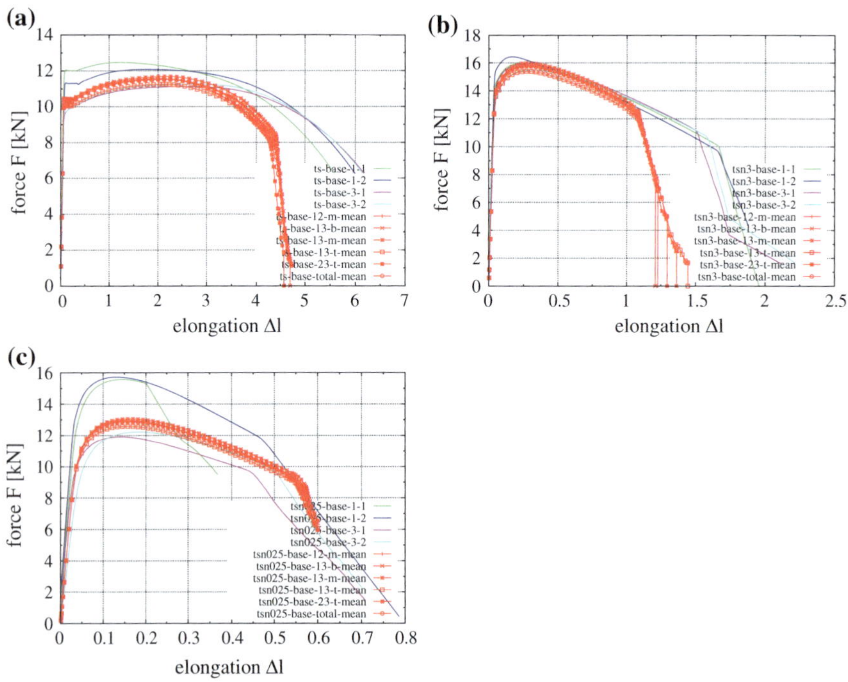

Fig. 13 Comparing tensile tests and simulations using identified material parameters from the SPT for the base material. **a** standard specimen **b** notched with r = 3 mm **c** notched with r = 0.25 mm

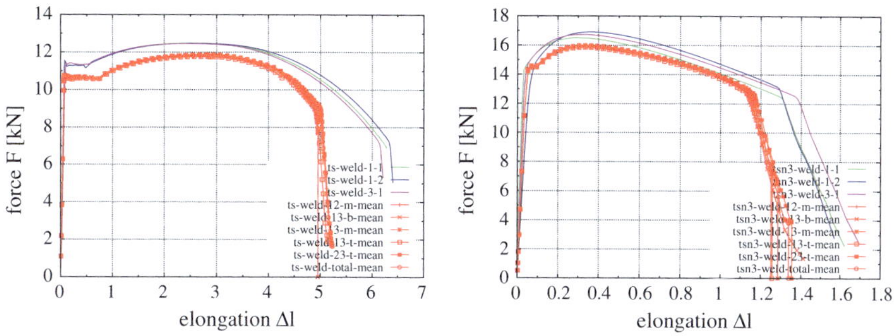

Fig. 14 Comparing tensile tests and simulations using identified material parameters from the SPT for the weld material

5.2 *Characterization of Carbon Bonded Alumina*

Carbon bonded alumina is a relatively new refractory material which is used in liquid metal casting processes as a filter to reduce impurities within the cast component. The ceramic open cell filters have to withstand a high thermal shock, which is why their mechanical integrity has to be evaluated. Testing of ceramic foams is rather complicated. The main difficulty is distributing an external load equally to the foam. Therefore, FEM simulations of the foam structure are done, using a material model for brittle failure. The material parameters can be determined using the SPT. Figure 15 left shows a filter of a carbon bonded alumina. It is an open cell foam with 10 pores per inch (ppi). Topologically it is a spatial network of struts, where theoretically always four struts met at a common node. The production of such filters is based on the Schwartzwalder process, where an open cell PU foam is coated with a slurry containing the ceramic material as powder. After drying the green body the filters are fired. During this process the PU pyrolyses and

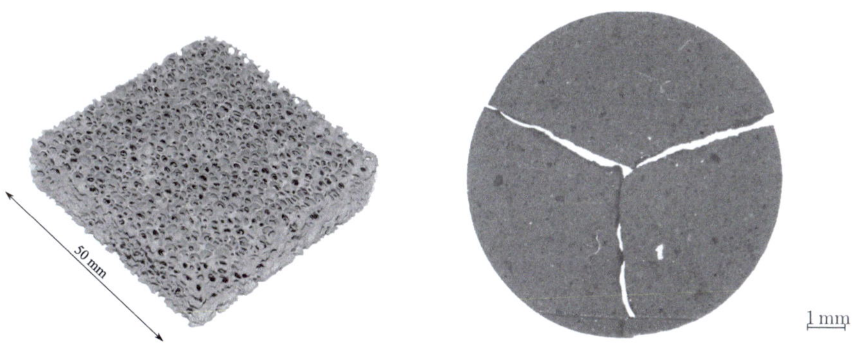

Fig. 15 C–Al_2O_3 Filter and HT-SPT specimen after testing

leaves a fine cavity structure within the strut network. The coating is mostly not perfect so there remain some closed facets around the pores.

On the right side of Fig. 15 one sees a SPT specimen after testing. It's broken into three almost equal pieces, which is a typical SPT failure pattern for brittle materials. There is an interesting fact, that the number of pieces depends on the ratio between specimen thickness and specimen radius. Thicker specimen usually break into more pieces than thinner ones, because of the larger specific elastic strain energy stored just before failure occurs. More energy can produce larger or more crack surfaces. It also shows a rather inhomogeneous micro structure of the material. Fine grained alumina particles are embedded into a matrix of amorphous graphite. The larger black particles are Carbores particles, which is tar coal having a higher strength than the pressed graphite and therefore influences the strength of the material.

The SPT was used here to compare different compositions (AC10, AC15, AC20), productions routes and thermal treatments of the ceramic material. The foam struts have a thickness which is comparable to that of the SP specimens, which excludes possible size effects of the material. Furthermore the SPT is a very efficient technique, especially at high temperatures where the heating-up period is much longer than the test period. For the results presented here more than 500 specimen have been tested.

The specimens are produced by a combined casting and cutting process. Small blocks are cast, pressed either isostatic (Pi) or uniaxial (Pu) than dried and fired by given regimes. An additional thermal treatment (T) may be used to adjust important material properties. The blocks are turned into cylindrical rods from which the SP specimens are cut using a diamond blade saw. Additional grinding at both specimen sides ensures the correct thickness and parallelism.

Figure 16 shows two typical LDCs from two CDR-SPTs, one performed at room temperature (RT) and the other at high temperature (HT) of 800 °C for the cast

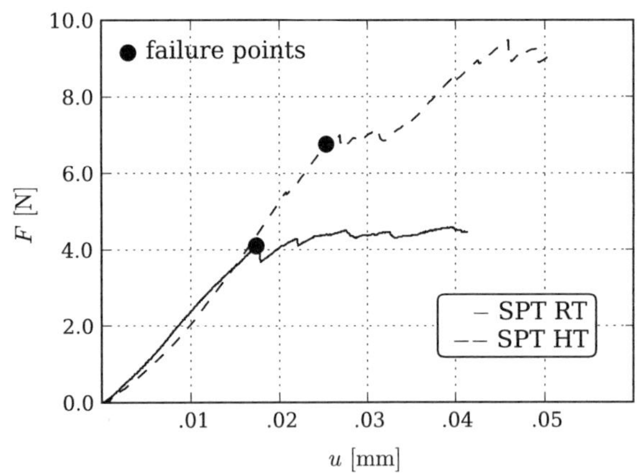

Fig. 16 SPT LDCs for the C–Al_2O_3 at RT (23 °C) and HT (800 °C)

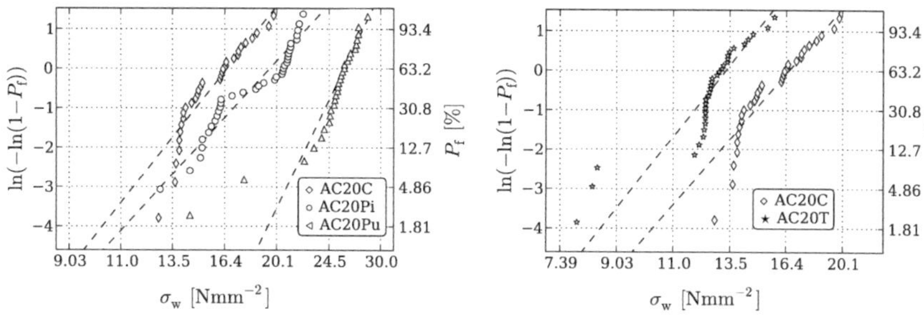

Fig. 17 Failure probability for different manufacturing routes **a** pressing **b** thermal treatment

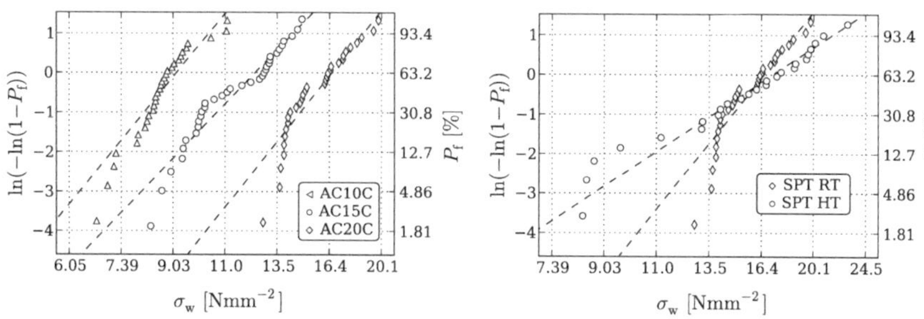

Fig. 18 Failure probability for different **a** material compositions **b** testing temperatures

carbon bonded alumina (AC20C). The two points indicate the initiation of cracking. The specimen are further deformed still bearing increasing loads but the general slope of the curve decreases, indicating that the already initiated cracks are growing until total failure. One may note that at high temperatures the material is much stronger than at low temperatures, which is a special feature of carbon bonded alumina.

Figure 17 shows the failure probability for different material states depending on the Weibull stress. On the left different manufacturing processes are compared (C—cast, Pi—isostatic pressing, Pu—uniaxial pressing). Additional pressing, especially uniaxial pressing of the cast material, increases the strength significantly. The additional thermal treatment instead does not improve the mechanical strength.

In Fig. 18 the influence of the material composition is analysed. The material is cast but contains an increasing amount of carbores (AC10C—10 %, AC15C—15 %, AC20C—20 %), which improves the strength significantly. Higher temperatures change the failure behaviour of the material. The scatter increases, indicated by a smaller Weibull exponent m, but the Weibull strength σ_u is slightly increased. Table 3 summarizes all identified Weibull parameters, where N denotes the number of tested specimens of each material state.

Table 3 Weibull parameters for all material states of the carbon bonded alumina $C–Al_2O_3$

	AC10C	AC15C	AC20C	AC20C-Pu	AC20C-Pi	AC20-CT	AC20C-HT
N (–)	30	34	31	29	37	33	25
σ_u (MPa)	9.15	12.4	16.6	26.0	19.6	13.3	17.2
m (–)	8.0	6.9	8.3	13.7	7.2	9.1	4.4

6 Conclusions and Discussion

The SPT has been presented as a suitable experimental technique to determine material properties of various materials from a small amount of material. The literature shows a wide field of applications. The crucial point of the SPT is still the problematic evaluation of the test results. The widely used empirical correlations to standard methods are material dependent. Therefore, a parameter identification procedure was developed which includes finite element simulations of the SPT considering a constitutive material model. The modelling approach presented here is able to describe ductile metals including damage behaviour but also brittle material like ceramics. It could be shown, that it is possible to fit the material law using experimental results from SPTs. For the identification of damage parameters it is important that experiments are done using specimens, which are exposed to different multi-axial stress states. This can be achieved by testing additional notched specimens. As an example the mechanical properties of a gas pressure pipe weld line were identified.

Brittle materials can be evaluated as well using statistical failure models (Weibull). Here, it was shown that it is possible to distinguish different material compositions, production routes and thermal treatments. In general the investigated carbon bonded alumina ($C–Al_2O_3$) can be well described using the Weibull model. For very low failure probabilities there is a deviation from the Weibull theory. This might be caused by the complex material composition allowing different failure mechanisms, which can't be described using a simple Weibull model.

Acknowledgments The authors gratefully acknowledge the financial support of the federal ministry of Saxony within the ADDE project as well as the support of the German Research Foundation (DFG) for the collaborative research center SFB 920.

References

1. Abendroth M, Kuna M (2003) Determination of deformation and failure properties of ductile materials by means of the small punch test and neuronal networks. Comput Mater Sci 28:633–644. doi:10.1016/j.commatsci.2003.08.031
2. Abendroth M, Kuna M (2004) Determination of ductile material properties by means of the small punch test and neural networks. Adv Eng Mater 6(7):536–540. doi:10.1002/adem.200400405

3. Abendroth M, Kuna M (2006) Identification of ductile damage and fracture parameters from the small punch test using neural networks. Eng Fract Mech 73(6):710–725. doi:10.1016/j.engfracmech.2005.10.007
4. Alegre J, Cuesta I, Lorenzo M (2014) An extension of the Monkman-Grant model for the prediction of the creep rupture time using small punch tests. Exp Mech 54:1441–1451. doi:10.1007/s11340-014-9927-6
5. Ardell AJ (1992) Mechanical behavior of ion-irradiated ordered intermetallic compounds. Mater Sci Eng A 152(1–2):212–226. doi:10.1016/0921-5093(92)90070-H
6. Baik J, Kameda J, Buck O (1983) Small punch test evaluation of intergranular embrittlement of an alloy steel. Scr Metall 17:1443–1447. doi:10.1016/0036-9748(83)90373-3
7. Baik J, Kameda J, Buck O (1986) Development of small punch tests for ductile-brittle transition temperature measurement of temper embrittled Ni-Cr steels. In: Corwin W, Lucas G (eds) The use of small-scale specimens for testing irradiated material, ASTM STP 888, Philadelphia, PA, pp 92–111
8. Beremin F, Pineau A, Mudry F, Devaux JC, D'Escatha Y, Ledermann P (1983) A local criterion for cleavage fracture of a nuclear pressure vessel steel. Metall Trans A 14(11):2277–2287. doi:10.1007/BF02663302
9. Besterci M, Dobeš F, Ballóková B, Sülleiová K, Kvačkaj T (2011) Observation of anisotropy of creep fracture using small punch test for Al-Al4C3 system produced by equal channel angular pressing. High Temp Mater Processes 30(3):205–210. doi:10.1515/htmp.2011.030
10. Besterci M, Sülleiová K, Velgosová O (2012) Kinetics of mechanical alloying, mechanical properties of micro and nanostructural Al-C systems. High Temp Mater Processes 31(4–5):359–369. doi:10.1515/htmp-2012-0068
11. Blagoeva D, Hurst R (2009) Application of the CEN (European Committee for Standardization) small punch creep testing code of practice to a representative repair welded P91 pipe. Mater Sci Eng A 510–511: (219–223). doi:10.1016/j.msea.2008.05.058
12. Budzakoska E, Carr D, Stathers P, Li H, Harrison R, Hellier A, Yeung W (2007) Predicting the J integral fracture toughness of Al 6061 using the small punch test. Fatigue Fract Eng Mater Struct 30(9):796–807. doi:10.1111/j.1460-2695.2007.01153.x
13. Bulloch J (1995) The small punch toughness test: some detailed fractographic information. Int J Press Vessels Pip 63:177–194. doi:10.1016/0308-0161(94)00050-S
14. Bulloch J (2002) A review of the ESB small punch test data on various plant components with special emphasis on fractographic details. Eng Fail Anal 9(5):511–534. doi:10.1016/S1350-6307(01)00034-6
15. Bulloch J (2004) A study concerning material fracture toughness and some small punch test data for low alloy steels. Eng Fail Anal 11(4):635–653. doi:10.1016/j.engfailanal.2003.05.020
16. CEN (2006) Workshop agreement CWA 15627:2006, small punch test method for metallic materials. Technical report, Brussels, Belgium
17. Cheon J, Kim I (2000) Evaluation of thermal aging embrittlement in CF8 duplex stainless steel by small punch test. J Nucl Mater 278(1):96–103. doi:10.1016/S0022-3115(99)00213-5
18. Chi S, Hong J, Kim I (1994) Evaluation of irradiation effects of 16 MeV proton-irradiated 12Cr-1MoV steel by small punch (SP) tests. Scr Metall Mater 30(12):1521–1525
19. Cuesta I, Alegre J (2011) Determination of the fracture toughness by applying a structural integrity approach to pre-cracked small punch test specimens. Eng Fract Mech 78:289–300. doi:10.1016/j.engfracmech.2010.09.004
20. Cuesta I, Alegre J (2012) Determination of plastic collapse load of pre-cracked small punch test specimens by means of response surface. Eng Fail Anal 23:1–9. doi:10.1016/j.engfailanal.2012.02.002
21. Cuesta I, Alegre J, Lacalle R (2010) Determination of the Gurson-Tvergaard damage model parameters for simulating small punch tests. Fatigue Fract Eng Mater Struct 33(11):703–713. doi:10.1111/j.1460-2695.2010.01481.x
22. Dobeš F, Milička K (2002) On the Monkman-Grant relation for small punch test data. Mater Sci Eng A 336:245–248. doi:10.1016/S0921-5093(01)01975-X

23. Dobeš F, Milička K (2008) Comparison of conventional and small punch creep tests of mechanically alloyed Al-C-O alloys. Mater Charact 59:961–964. doi:10.1016/j.matchar.2007.08.006
24. Dobeš F, Milička K (2009) Application of creep small punch testing in assessment of creep lifetime. Mater Sci Eng A 510–511:440–443. doi:10.1016/j.msea.2008.04.087
25. Dymáček P, Milička K (2008) Small punch testing and its numerical simulations under constant deflection force conditions. Strength Mater 40(1):24–27. doi:10.1007/s11223-008-0007-y
26. Dymáček P, Milička K (2009) Creep small-punch testing and its numerical simulations. Mater Sci Eng A 510–511:444–449. doi:10.1016/j.msea.2008.06.053
27. Dymáček P, Seitl S, Milička K, Dobeš F (2010) Influence of friction on stress and strain distributions in small punch creep test models. Key Eng Mater 417–418:561–564. doi:10.4028/www.scientific.net/KEM.417-418.561
28. Evans M, Wang D (2008) The small punch creep test: some results from a numerical model. J Mater Sci 43:1825–1835. doi:10.1007/s10853-007-2388-x
29. Evans R, Evans M (2006) Numerical modelling of small disc creep test. Mater Sci Technol 22 (10):1155–1162. doi:10.1179/174328406X118258
30. Finarelli D, Roedig M, Carsughi F (2004) Small punch tests on austenitic and martensitic steels irradiated in a spallation environment with 530 MeV protons. J Nucl Mater 328:146–150. doi:10.1016/j.jnucmat.2004.04.320
31. Fleury E, Ha J (1998) Small punch tests to estimate the mechanical properties of steels for steam power plant: I. Mechanical strength. Int J Press Vessels Pip 75(9):699–706. doi:10.1016/S0308-0161(98)00075-1
32. Fleury E, Ha J (1998) Small punch tests to estimate the mechanical properties of steels for steam power plant: II. Fracture toughness. Int J Press Vessels Pip 75(9):707–713. doi:10.1016/S0308-0161(98)00075-1
33. Foulds J, Viswanathan R (1996) Nondisruptive material sampling and mechanical testing. J Nondestr Eval 15(3–4):151–162. doi:10.1007/BF00732042
34. Foulds J, Viswanathan R (2001) Determination of the toughness of in-service steam turbine disks using small punch testing. J Mater Eng Perform 10(5):614–619. doi:10.1361/105994901770344782
35. Foulds J, Woytowitz P, Parnell T, Jewett C (1995) Fracture toughness by small punch testing. J Test Eval 23(1):3–10
36. García T, Rodríguez C, Belzunce F, Suárez C (2014) Estimation of the mechanical properties of metallic materials by means of the small punch test. J Alloy Compd 582:708–717. doi:10.1016/j.jallcom.2013.08.009
37. Guan K, Hua L, Wang Q, Zou X, Song M (2011) Assessment of toughness in long term service CrMo low alloy steel by fracture toughness and small punch test. Nucl Eng Des 241:1407–1413. doi:10.1016/j.nucengdes.2011.01.031
38. Gurson A (1977) Continuum theory of ductile rupture by void nucleation and growth: part I—yield criteria and flow rules for porous ductile materials. J Eng Mater Technol 99:2–15. doi:10.1115/1.3443401
39. Gurson A (1977) Porous rigid-plastic materials containing rigid inclusions yield function, plastic potential and void nucleation. Fracture 2:357–364
40. Hayhurst D (1972) Creep rupture under multi-axial states of stress. J Mech Phys Solids 20 (6):381–390. doi:10.1016/0022-5096(72)90015-4
41. Holmström S, Hähner P, Rippling S, Fischer B, Lapetite J, Bruchhausen M, Gupta M, Hurst R, Turba K, Gülçimen B (2014) Tensile, embrittlement and creep property determination with improved small punch testing equipment and assessment methods. In: Matocha K, Hurst R, Sun W (eds) Determination of mechanical properties of materials by small punch and other miniature testing techniques, pp 331–338
42. Hyde T, Stoyanov M, Sun W, Hyde C (2010) On the interpretation of results from small punch creep tests. J Strain Anal Eng Des 45(3):141–164. doi:10.1243/03093247JSA592

43. Ju J, Kwon D (1998) Assessment of fracture characteristics from revised small punch test using pre-cracked specimen. Met Mater 4(4):742–746. doi:10.1007/BF03026391
44. Kameda J, Mao X (1992) Small-punch and tem-disc testing techniques and their application to characterization of radiation damage. J Mater Sci 27(4):983–989. doi:10.1007/BF01197651
45. Kato T, Kohno Komazaki Y S, Tanigawa H, Kohyama A (2009) High-temperature strength analysis of welded joint of RAFs by small punch test. J Nucl Mater 386–388:520–524. doi:10.1016/j.jnucmat.2008.12.153
46. Kim J, Chung T, Lim J, Chung S (1991) The evaluation for in-service material degradation of superheat tubes of fossil boiler. Key Eng Mater 51–52:253–258
47. Kim MC, Oh Y, Lee B (2005) Evaluation of ductile–brittle transition temperature before and after neutron irradiation for RPV steels using small punch tests. Nucl Eng Des 235:1799–1805
48. Kurtz S, Foulds J, Jewett C, Srivastav S, Edidin A (1997) Validation of a small punch testing technique to characterize the mechanical behaviour of ultra-high-molecular-weight polyethylene. Biomaterials 18(24):1659–1663. doi:10.1016/S0142-9612(97)00124-5
49. Li H, Chen F, Ardell A (1991) A simple, versatile miniaturized disk-bend test apparatus for quantitative yield-stress measurements. Metall Trans A 22(9):2061–2068. doi:10.1007/BF02669873
50. Li Y, Šturm R (2006) Small Punch test for weld heat affected zones. Mater High Temp 23(3–4):225–232. doi:10.1179/mht.2006.019
51. Linse T (2013) Quantifizierung des spröd-duktilen Versagensverhaltens von Reaktorstählen mit Hilfe des Small-Punch-Tests und mikromechanischer Schädigungsmodelle. Ph.D. thesis, TU Bergakademie Freiberg, Berichte des Institutes für Mechanik und Fluiddynamik, Heft 9
52. Linse T, Kuna M, Schuhknecht J, Viehrig HW (2008) Application of the small punch test to irradiated reactor vessel steels in the brittle-ductile transition region. J ASTM Int 5(5)
53. Linse T, Kuna M, Schuhknecht J, Viehrig HW (2008) Application of the small-punch test to irradiated reactor vessel steels in the brittle-ductile transition region. J ASTM Int 5(4):1–14. doi:10.1520/JAI101008
54. Linse T, Kuna M, Schuhknecht J, Viehrig HW (2008) Usage of the smallpunch-test for the characterisation of reactor vessel steels in the brittle-ductile transition region. Eng Fract Mech 75(11):3520–3533. doi:10.1016/j.engfracmech.2007.03.047
55. Linse T, Hütter G, Kuna M (2012) Simulation of crack propagation using a gradient-enriched ductile damage model based on dilatational strain. Eng Fract Mech 95:13–28. doi:10.1016/j.engfracmech.2012.07.004
56. Linse T, Kuna M, Viehrig HW (2014) Quantification of brittle-ductile failure behavior of ferritic reactor pressure vessel steels using the small-punch-test and micromechanical damage models. Mater Sci Eng A 614:136–147
57. Manahan M (1983) A new postirradiation mechanical behavior test—the miniaturized disk bend test. Nucl Technol 63:295–315
58. Manahan M (1986) Determining mechanical behavior of solid materials using miniature specimens. US Patent Number 4,567,774
59. Manahan M, Argon A, Harling O (1981) The development of a miniaturized disk bend test for the determination of postirradiation mechanical properties. J Nucl Mater 103:1545–1550. doi:10.1016/0022-3115(82)90820-0
60. Mao X, Saito M, Takahashi H (1991) Small punch test to predict ductile fracture toughness JIC and brittle fracture toughness KIC. Scr Metall 25(11):2481–2485
61. Matsumoto Y, Morinagat M, Furui M (1997) Multiple deformation effect on the ductility of chromium. Scr Mater 38(2):321–327
62. Matsushita T, Saucedo M, Joo Y, Shoji T (1991) DBTT estimation of ferritic low alloy steels in service plant by means of small punch test. Key Eng Mater 51–52:259–264. doi:10.4028/www.scientific.net/KEM.51-52.259
63. Milička K, Dobeš F (2005) Relation between uniaxial and equi-biaxial creep and creep fracture behaviour in P91 steel. Mater Sci Forum 482:407–410
64. Milička K, Dobeš F (2006) Small punch testing of P91 steel. Int J Press Vessels Pip 83:625–634. doi:10.1016/j.ijpvp.2006.07.009

65. Misawa T, Adachi T, Saito M, Hamaguchi Y (1987) Small punch tests for evaluating ductile-brittle transition behavior of irradiated ferritic steels. J Nucl Mater 150(2):194–202. doi:10.1016/0022-3115(87)90075-4
66. Misawa T, Nagata S, Aoki N (1989) Fracture toughness evaluation of fusion reactor structural steels at low temperatures by small punch tests. J Nucl Mater 169:225–232. doi:10.1016/0022-3115(89)90538-2
67. Mühlich U, Brocks W, Siegmund T (1998) A user material subroutine of the Gurson-Tvergaard-Needleman model of porous metal plasticity for rate and temperature dependent hardening. Technical Note GKSS/WMG/98/1, GKSSForschungszentrum Geesthacht
68. Norris S, Parker J (1996) Deformation processes during disc bend loading. Mater Sci Technol 12(2):163–170. doi:10.1179/mst.1996.12.2.163
69. Rasche S (2013) Bestimmung von Materialparametern der elastischplastischen Verformung und des spröden Versagens aus Small-Punch-Kleinstproben. Ph.D. thesis, TU Bergakademie Freiberg
70. Rasche S, Bermejo R, Kuna M, Danzer R (2010) Determination of mechanical properties of brittle materials by using the small punch test and the ball on three balls test. In: Proceedings 18th European conference on fracture, Dresden
71. Reusch F, Svendsen B, Klingbeil D (2003) A non-local extension of gursonbased ductile damage modeling. Comput Mater Sci 26:219–229. doi:10.1016/S0927-0256(02)00402-0
72. Rice J, Johnson M (1970) The role of large crack tip geometry changes in plane strain fracture. In: Kanninen M (ed) Elastic-plastic behavior in solids. McGraw Hill, New York, pp 120–168
73. Saito M, Hashida T, Takahashi H (1991) Small punch test for ceramic composites at very high temperature. Key Eng Mater 51–52:477–482. doi:10.4028/www.scientific.net/KEM.51-52.477
74. Saucedo-Muñoz M, Liu S, Hashida T, Takahashi H, Nakajima H (2001) Correlationship between JIC and equivalent fracture strain determined by small punch tests in JN1, JJ1 and JK2 austenitic stainless steels. Cryogenics 41:713–719
75. Shekhter A, Croker A, Hellier A, Moss C, Ringer S (2000) Towards the correlation of fracture toughness in an ex-service power generating rotor. Int J Press Vessels Pip 77(2–3):113–116. doi:10.1016/S0308-0161(99)00091-5
76. Shekhter A, Kim S, Carr D, Croker A, Ringer S (2002) Assessment of temper embrittlement in an ex-service 1Cr–1Mo–0.25 V power generating rotor by Charpy V-Notch testing, KIc fracture toughness and small punch test. Int J Press Vessels Pip 79(8–10):611–615. doi:10.1016/S0308-0161(02)00087-X
77. Shindo Y, Yamaguchi Y, Horiguchi K (2004) Small punch testing for determining the cryogenic fracture properties of 304 and 316 austenitic stainless steels in a high magnetic field. Cryogenics 44(11):789–792. doi:10.1016/j.cryogenics.2004.04.008
78. Springmann M, Kuna M (2002) Identification of material parameters of the Rousselier model by non-linear optimization. Comput Mater Sci 26:202–209. doi:10.1016/S0927-0256(02)00400-7
79. Turba K, Gülçimen B, Li Y, Blagoeva D, Hähner P, Hurst R (2011) Introduction of a new notched specimen geometry to determine fracture properties by small punch testing. Eng Fract Mech 78(16):2826–2833. doi:10.1016/j.engfracmech.2011.08.014
80. Tvergaard V (1981) Influence of voids on shear band instabilities under plane strain conditions. Int J Fract Mech 17(4):389–407. doi:10.1007/BF00036191
81. Tvergaard V, Needleman A (1995) Effects of nonlocal damage in porous plastic solids. Int J Solids Struct 32(8–9):1063–1077. doi:10.1016/0020-7683(94)00185-Y
82. Ule B, Šuštar T, Dobeš F, Milička K, Bicego V, Maile Tettamanti K S, Schwarzkopf C, Whelan M, Kozlowski R, Klaput J (1999) Small punch test method assessment for the determination of the residual creep life of service exposed components: outcomes from an interlaboratory exercise. Nucl Eng Des 192:1–11

83. Wakai E, Ohtsuka H, Matsukawa S, Furuya K, Tanigawa H, Oka K, Ohnuki S, Yamamoto T, Takada F, Jitsukawa S (2006) Mechanical properties of small size specimens of F82H steel. Fusion Eng Des 81(8–14):1077–1084. doi:10.1016/j.fusengdes.2005.08.072
84. Wang Z, Shi H, Lu J, Shi P, Ma X (2008) Small punch testing for assessing the fracture properties of the reactor vessel steel with different thicknesses. Nucl Eng Des 238(12):3186–3193. doi:10.1016/j.nucengdes.2008.07.013
85. Yang S, Yang Z, Ling X (2014) Fracture toughness estimation of ductile materials using a modified energy method of the small punch test. J Mater Res 29(15):1675–1680. doi:10.1557/jmr.2014.205
86. Yang Z, Wang Z (2003) Relationship between strain and central deflection in small punch creep specimens. Int J Press Vessels Pip 80(6):397–404. doi:10.1016/S0308-0161(03)00069-3
87. Zhang J, Ardell AJ (1991) Measurement of the fracture toughness of CVDgrown ZnS using a miniaturized disk-bend test. J Mater Res 6(9):1950–1957. doi:10.1557/JMR.1991.1950
88. Zhang Z (1995) Explicit consistent tangent moduli with a return mapping algorithm for pressure-dependent elastoplasticity models. Comput Methods Appl Mech Eng 121(1–4):29–44. doi:10.1016/0045-7825(94)00707-T
89. Zhang Z (1995) On the accuracies of numerical integration algorithms for gurson-based pressure-dependent elastoplastic constitutive models. Comput Methods Appl Mech Eng 121(1–4):15–28. doi:10.1016/0045-7825(94)00706-S

Determination of Fracture Mechanics Parameters for Cast Iron Materials Under Static, Dynamic and Cyclic Loading

Dedicated to Professor Meinhard Kuna on the Occasion of his 65th Birthday

G. Pusch, S. Henkel, H. Biermann, P. Hübner, A. Ludwig, P. Trubitz, T. Mottitschka and L. Krüger

Abstract The results of extensive measurements of fracture mechanics parameters of cast iron materials under static, dynamic and cyclic loading at the Institute of Materials Engineering (IWT) at the TU Bergakademie Freiberg are shown. A special attention for this group of materials must be given to the microstructural influence on the crack resistance behavior and thus the component reliability. In collaboration with the Institute of Mechanics and Fluid Dynamics (IMFD) of the TU Bergakademie it was shown for castings, such as components of wind turbines made of ductile cast iron, railway wheels made of ADI or transport casks for spent nuclear fuel assemblies that the fracture mechanical strength analysis including possible casting defects extends the conventional strength analysis.

1 Introduction

Fracture mechanics concepts for assessing the strength requirements of cast materials and cast components (Fig. 1) are employed by considering casting defects or other possible casting irregularities particularly when cast materials are used due

G. Pusch · S. Henkel (✉) · H. Biermann · A. Ludwig · P. Trubitz · L. Krüger
Institute of Materials Engineering, Technische Universität Bergakademie Freiberg, Gustav-Zeuner-Str. 5, 09599 Freiberg, Germany
e-mail: henkel@ww.tu-freiberg.de

P. Hübner
Department of Mechanical Engineering, University of Applied Sciences Mittweida, Technikumplatz 17, 09648 Mittweida, Germany

T. Mottitschka
IMA GmbH Dresden (formerly Institute of Materials Engineering), Dresden, Germany

G. Hütter and L. Zybell, *Recent Trends in Fracture and Damage Mechanics*, DOI 10.1007/978-3-319-21467-2_7

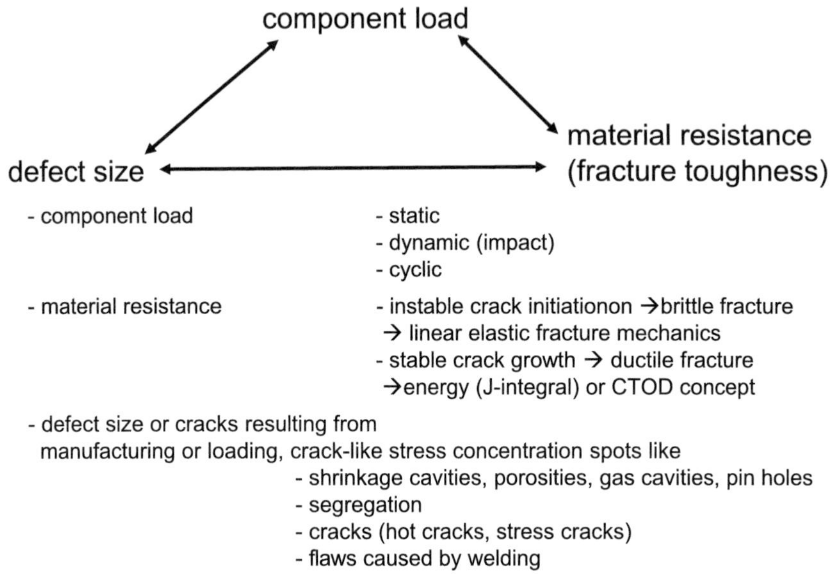

Fig. 1 Principle of fracture mechanics concepts and typical casting defects

to their specific advantages for components loaded to their ultimate strength and are potentially exposed to fracture.

The facture mechanics strength assessment, taking into account cast iron materials, is the subject matter of regulating procedures [1–5]. An essential requirement for computing the critical stresses or defect sizes is the availability of the materials fracture mechanics parameters [6].

This contribution includes the results of extensive investigations carried out at the Institute of Materials Engineering (IWT) of the Technische Universität Bergakademie Freiberg (Germany) on cast iron materials for highly loaded cast components. It includes the fracture mechanics parameters determined using different cast iron materials subject to static, dynamic and cyclic loading as a function of the microstructure, temperature and type of loading. The fracture mechanics strength assessment of particular cast components was performed in cooperation with the Institute of Mechanics and Fluid Dynamics (IMFD) of the TU Bergakademie Freiberg and is documented in the contribution's references.

2 Static Loading

The static fracture toughness values were experimentally determined according to ESIS P2-92 [7]. The K_{Ic} values are determined according to the concept of linear-elastic fracture mechanics (plane strain conditions) or the parameters of the

CTOD (crack-tip opening displacement) and J-integral concepts of elastic-plastic material behaviour (plane stress conditions). It is possible to convert J values into K values via the elastic constants E and ν using

$$K = \sqrt{\frac{EJ}{1 - \nu^2}} \tag{1}$$

To experimentally determine valid K_{Ic} values for ductile cast iron materials, the requirements for plane strain conditions lead to huge specimens. A conservative value can be determined on small specimens using the fracture mechanics parameter CTOD or J-integral concepts. The damage processes which run in ductile cast iron with ferritic matrix on the crack tip until the crack initiation are depicted in Fig. 2. For GJS-400, the critical stretch zone width (SZW_C) is in the range of about 20 μm [8]. The stable crack growth after crack inition was documented with in situ investigations in the scanning electron microscope [9] (Fig. 3).

Fracture mechanics parameters are experimentally determined by using 20 % side grooved SENB (single-edge notch bending) specimens (10 mm × 20 mm × 120 mm) and measuring static crack resistance curves (δ_R or J_R curves) according to the single specimen method (compliance method).

The physical crack initiation parameters $J_{i/BL}$ and $\delta_{i/BL}$ are determined at the intersection of the blunting line and the J_R or δ_R curve, respectively. The engineering crack initiation values $J_{0.2}$ and $\delta_{0.2}$ result for a value of stable crack

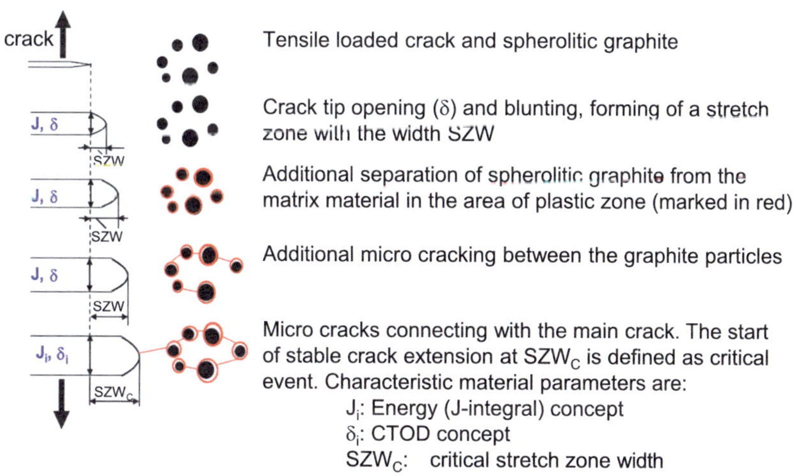

Fig. 2 Damage processes (schematic) on the crack tip of ductile cast iron with ferritic matrix under static loading

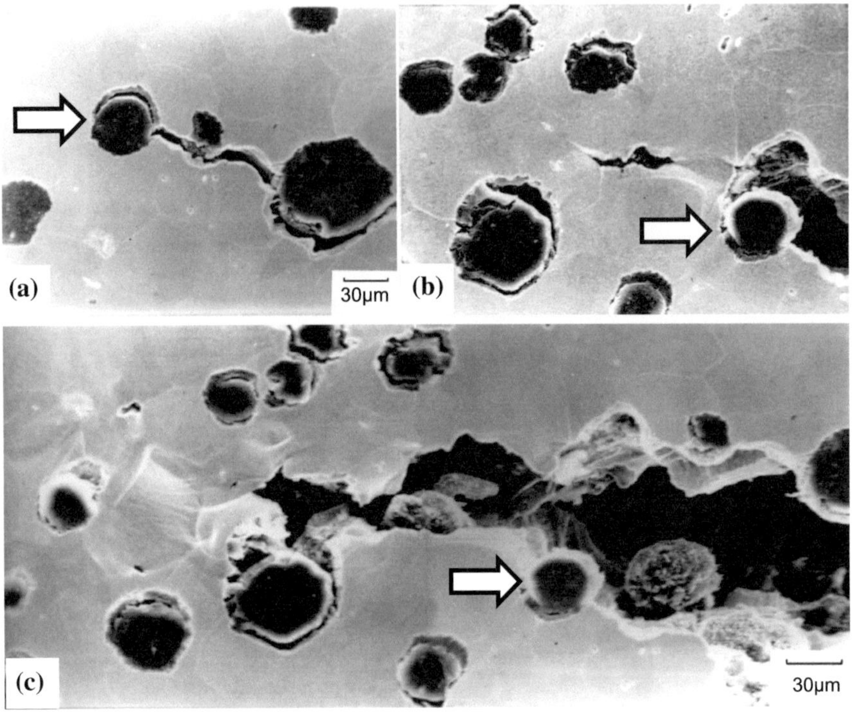

Fig. 3 In situ SEM pictures of crack extension in GJS-400, initial fatigue crack tip is marked with an *arrow*: **a** begin of crack opening, **b** additional crack opening causes debonding of graphite and matrix, **c** stable crack growth [9]

extension of $\Delta a = 0.2\,\text{mm}$. The profile of the J-Δa or δ-Δa curve is fitted via an extended power function

$$J = A(\Delta a + B)^{c} \tag{2}$$

$$\delta = A'(\Delta a + B')^{c'} \tag{3}$$

and the blunting line (BL) is computed according to

$$J = 3.75 R_{\mathrm{m}} \Delta a \tag{4}$$

$$\delta = 1.87\left(R_{\mathrm{m}}/R_{\mathrm{p0.2}}\right)\Delta a \tag{5}$$

The fracture mechanics parameters of the J-integral and CTOD concepts are compiled in Table 1.

The FKM guideline [1] refers to the potential for estimating fracture mechanics parameters via correlations with specific values from mechanical and technological testing or microstructural parameters. The available results show that it is possible

Table 1 Mechanical and fracture mechanics parameters [8–14]

Material EN-…	$R_{p0.2}$ (MPa)	R_m (MPa)	A (%)	$J_{i/BL}$ $\left(\frac{kJ}{m^2}\right)$	$J_{0.2}$ (kJ m^{-2})	$K(J_{i/BL})$ (MPa$\sqrt{m}$)	$\delta_{i/BL}$ (μm)	$\delta_{0.2}$ (μm)
GJS-400-15	264	413	26	21	51	60	37	92
GJS-800-10	637	893	13	11	35	45	31	27
GJS-1000-5	800	1062	8	8	20	37	6	15
GJV-300	240	295	5	8	19	33	21	45
GJMB-350-10	213	332	15	22	39	63	47	78
GJMB-450-6	312	453	10	28	42	72	40	57
GJMB-650-2	450	703	6	11	28	45	9	27
GJMW-360-12	166	348	21	10	27	43	24	61
GJMW-400-5	238	515	5	9	24	41	24	50
GJMW-450-7	304	536	5	8	29	38	15	52
GJMW-550-4	485	714	6	12	25	47	18	28

to estimate the physical crack initiation values of ferritic cast iron materials via microstructural and tensile test parameters [8, 9, 15, 16].

Thus, within the scope of extensive fracture mechanics measurements on ferritic GJS-400-15 and varying the graphite particle size d_G, spacing λ, shape factor f and number of particles N_A as well as the corresponding mechanical properties (Table 2), it has been identified that it is possible to approximately ascertain the $J_{i/BL}$ values via the microstructural and tensile test parameters.

Most of the structural models for ductile fracture are based on the model of pore growth according to Ritchie (Fig. 4a). This description is virtually predestined as the material model for ductile cast irons having a ferritic matrix and spherical graphite (Fig. 4b). Here, it is assumed that a critical plastic equivalent strain specific to the material φ_c must be reached over a critical length l_c ahead of the crack in order to produce crack growth (Fig. 4c).

It follows from this that the critical J-integral value $J_{i/BL}$ is

$$J_{i/BL} = cR_{p0.2}\varphi_c l_c \tag{6}$$

where c is a dimensionless constant. l_c correlates well with the average graphite particle spacing λ and φ_c is set equal to the percentage elongation after fracture A. If one plots the experimentally determined $J_{i/BL}$ values against the product $R_{p0.2}A\lambda$ (Fig. 5), then the following empirically derived correlation is valid for ferritic cast iron having spherical graphite

$$J_{i/BL} = 4.6\left(\frac{R_{p0.2}}{\text{MPa}}A\frac{\lambda}{\text{mm}}\right)\text{kJ m}^{-2} + 11.3\,\text{kJ m}^{-2} \tag{7}$$

Figure 6 and Table 3 depict the influence of the shape of the graphite particles on the crack resistance behaviour of ferritic cast iron materials.

Table 2 Microstructural parameters, mechanical and fracture mechanics parameters measured on different batches of nodular cast iron EN-GJS-400-15 [8, 9, 15, 16]

Material EN-…	d_G (μm)	λ (μm)	f	$N_A \left(\frac{1}{mm^2}\right)$	$R_{p0.2}$ (MPa)	R_m (MPa)	E (GPa)	A (%)	Z (%)	$J_{i/BL} \left(\frac{kJ}{m^2}\right)$
GJS-400-15/5	17	33	0.84	184	282	428	182	25	27	24
GJS-400-15/2	23	56	0.81	69	259	393	162	26	21	30
GJS-400-15/4	49	94	0.66	19	265	381	173	21	18	35
GJS-400-15/10	53	91	0.77	56	258	381	171	18	6	34
GJS-400-15/11	97	150	0.64	19	241	366	172	19	6	45

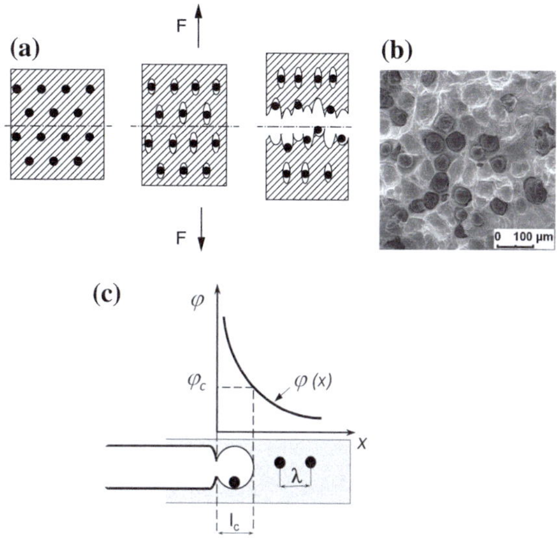

Fig. 4 Pore growth model according to Ritchie [17]. **a** Strain controlled crack formation. **b** Scanning electron microscopy image of GJS-400-15 fracture surface. **c** Strain distribution φ(x) according to Ritchie [17]

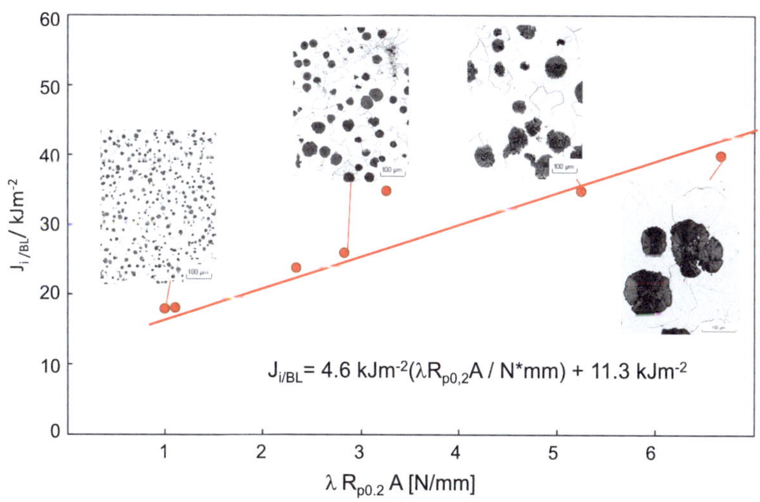

Fig. 5 $J_{i/BL}$ mean values of at least tree specimens of ferritic cast iron materials having spheritic graphite as a function of the product of fracture strain A, 0.2 % proof strength and mean graphite particle spacing λ [8, 9, 15, 16]

The different crack resistance behaviour was analysed with in situ scanning electron microscopic investigations [13, 14]. For GJS-400-18 (graphite volume fraction V_G = 12 %) damage is induced by graphite particles debonding from the

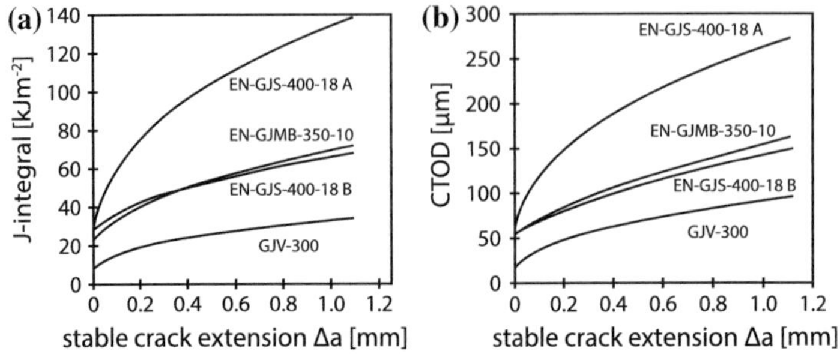

Fig. 6 Static crack resistance curves of ferritic cast iron materials (Table 3). **a** J-integral concept. **b** CTOD concept [6]

ferritic matrix. This mechanism primarily determines the crack resistance behaviour. It decreases with smaller particle spacing λ according to Eq. 7. The vermicular shape of the graphite in GJV-300 (V_G = 13 %) is expressed by the shape factor f (f = 1 for circular particles). The larger internal notch effect leads to much earlier fracture of the graphite particles. As a result the crack resistance curve's profile is significantly on a lower level. Despite the particle fractures, the superior crack resistance behaviour of the GJMB-350-10 (V_G = 10 %) is based on the matrix higher ductility due to the reduced graphite volume fraction and a lower Si content.

The project MEGAWind [18] funded by the German Bundesministerium für Umwelt investigated the development and usage of cast technologies for ductile cast iron with higher strength as well as the proof of strength under cyclic loading and the safety assessment based on the fracture mechanics analysis for the example of a wind turbine hub. With respect to offshore components a representative cast sample plate with a weight of 1.2 t was used with wall thicknesses of 60, 130 and 200 mm. Based on preliminary investigations the chemical composition as well as the optimized molten metal treatment and cast technology were fixed. Samples of the materials GJS-W1 to GJS-W7 (Table 4) were taken only from the 130 mm thick region (for details on sampling see [18]). To take welding into account which is often used on huge casting parts additional samples were taken from a multi-pass welded joint (GJS-SG). As a reference material EN-GJS-18LT was used. The static fracture toughness was determined at room temperature (RT), −20 and −40 °C using 20 % side grooved C(T)-specimens with a thickness of 25 mm according to ISO 12135 (Table 4). Depending on microstructure and temperature the linear elastic or elastic-plastic fracture mechanics concepts were used for determination of characteristic values. In the case of elastic-plastic material behavior static J-Δa crack resistance curves where measured with the single specimen technique. Table 4 shows that the ferritic reference material GJS-400-18LT has a constant high level of crack resistance at all tested temperatures. As can be taken from the designation of the fracture resistance as a $K(J_{i/BL})$ value the crack extended in a ductile, stable way. This is also the case for the welded material with the lowest Si content of all tested materials. It shows a loss in fracture resistance with lowering

Table 3 Microstructural and fracture mechanics parameters at room temperature and −40 °C of ferritic cast iron materials [6]

Material EN-…	Microstructural parameter			J-integral concept				CTOD concept			
				RT		−40 °C		RT		−40 °C	
	λ (m)	d_G (m)	f	$J_{i/BL}$ $\left(\frac{kJ}{m^2}\right)$	$J_{0.2}$ $\left(\frac{kJ}{m^2}\right)$	$J_{i/BL}$ $\left(\frac{kJ}{m^2}\right)$	$J_{0.2}$ $\left(\frac{kJ}{m^2}\right)$	δ_i (µm)	$\delta_{0.2}$ (µm)	δ_i (µm)	$\delta_{0.2}$ (µm)
GJS-400-18A	96	47	0.70	40	74	43	63	74	141	68	99
GJS-400-18B	51	22	0.35	28	42	26	42	52	75	50	72
GJMB-350-10	21	9	0.74	22	39	11	34	47	78	17	61
GJV-300	37	12	0.32	8	19	n.d.	n.d.	21	45	n.d.	n.d.

A, B internal designation
n.d. not determined

Table 4 Characteristic values from tensile tests at room temperature (RT) and static fracture mechanics testing at RT, −20 and −40 °C for different Si content. SG represents welded material

Material	Si (%)	$R_{p0.2}$ (MPa)	R_m (MPa)	$R_{p0.2}/R_m$	A (%)	Z (%)	K_{IC} (MPa$\sqrt{m}$)			$K(J_{i/BL})$ (MPa$\sqrt{m}$)		
		RT	RT	RT	RT	RT	RT	−20 °C	−40 °C	RT	−20 °C	−40 °C
GJS-400-18LT	1.85	245	403	0.61	22	17	–	n.d.	–	80	n.d.	82
GJS-W1	3.6	395	466	0.85	5	6	36	29	26	–	–	–
GJS-W2[a]	3.5	391	428	0.91	2	4	–	31	30	52	–	–
GJS-W3	2.9	326	436	0.75	11	12	–	35	36	83	–	–
GJS-W4[a]	2.8	323	410	0.79	8	9	–	–	33	73	47	–
GJS-SG	2.7	382	477	0.80	8	14	–	–	–	60	34	36
GJS-W5	3.0	355	444	0.80	8	8	–	32	29	47	–	–
GJS-W7	3.0	363	465	0.78	17	14	–	34	29	48	–	–

[a]Special inoculant treatment
n.d. not determined

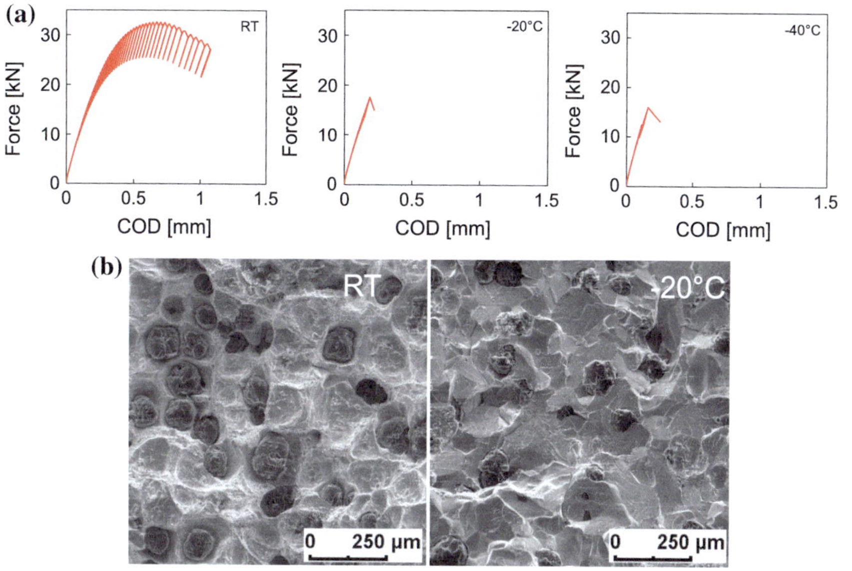

Fig. 7 Ductile behaviour at room temperature (RT) and brittle behaviour at −20 and −40 °C, **a** plot of load versus crack opening displacement, **b** SEM pictures of the fracture surface

temperature. The other variants with a Si-content of more than 2.7 % show a significant loss in crack resistance and mostly brittle crack extension with reducing testing temperature from room temperature (RT) to −20 and −40 °C. Figure 7 shows this trend for Material GJS W7. The load-COD-curves exhibit brittle pop-in behavior at −20 and −40 °C. Except variant GJS-W4 at −20 °C only cleavage fracture was observed as pop-in effect at local microstructural heterogeneities or instable crack extension.

The dependence of the measured fracture toughness from the Si content together with literature data from [19–21] is shown in Fig. 8. The correlation also contains measurements on continuous casting qualities GJS-400-18C with diameter 150 mm (Si = 2.63 %, $R_{p0.2}$ = 300 MPa, R_m = 424 MPa, A = 25.9 %) and GJS-500-14C (Si = 3.56 %, $R_{p0.2}$ = 391 MPa, R_m = 504 MPa, A = 19.7 %) with $K(J_{i/BL}) = 46.5\,\text{MPa}\sqrt{\text{m}}$ [21]. Resulting from Fig. 8 a slope of $\sim 29\,\text{MPa}\sqrt{\text{m}}$ per 1 % additional Si content can be found at room temperature. This has to be taken into account in addition to the temperature effect for Si solid solution hardened cast iron qualities.

The design and construction of a turbine's pipework made of ductile cast iron pipes can be mentioned as an example of including the fracture mechanics assessment into the design process. During the construction of a 16 MW water power plant in the Austrian state of Vorarlberg, particularly high safety requirements and difficult installation conditions in the Alpine site lead to the decision not to construct the turbine's pipework using welded steel pipes. In the range of nominal sizes up to DN 1600 for operating pressures over 30 bar ductile,

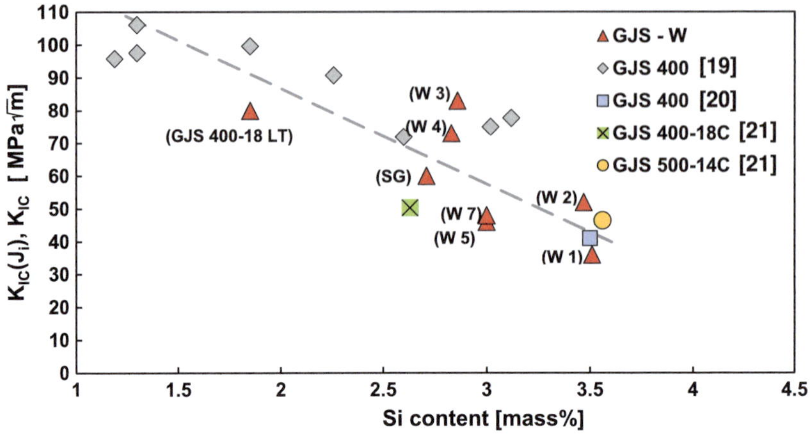

Fig. 8 Static fracture toughness values K_{IC} or $K(J_{i/BL})$ depending on Si content for ferritic cast iron at room temperature

centrifugally casted pipes made of optimised EN-GJS-400-15 material were employed. The fracture mechanics assessment was conducted based on the "leak before fracture" criterion, where the fracture mechanics parameters were determined at the IWT [8, 22]. A comparison of the experimentally determined values of $J_i = 20\,\mathrm{kJ/m^2}$ resulted in good agreement with the analytical values computed according to Eq. 7 with $J_i = 21\,\mathrm{kJ/m^2}$. Finally centrifugally cast pipes were employed as a result of the conducted fracture mechanics safety assessment, supported by burst tests carried out at the Technical Test and Research Laboratory of TU Vienna using pipe segments containing cracks.

3 Dynamic Loading

In a comprehensive assessment of the components safety, the load carrying capacity of materials in components can be characterised using fracture mechanics material properties, particularly for accident scenarios in which rapidly changing stress and strain conditions occur as a consequence of impact type loading. To experimentally determine the dynamic fracture toughness parameters, only preliminary approaches are available regarding the system of rules and standards which are based on the standards for static loading cases. The results of comprehensive investigations on different ferritic cast iron materials for determining and defining dynamic fracture toughness parameters show that attention should here be paid to the specific crack resistance behaviour of this material group [23–26]. The dynamic crack resistance curves of the J-integral concept (J_d-Δa curves) were experimentally determined at room temperature (RT) and at −40 °C by using an instrumented impact testing machine according to the "low-blow" technique with 20 % side grooved V-notched

ISO specimens (10 mm × 10 mm × 55 mm) possessing fatigue cracks. In doing this, six to eight specimens were loaded within the range $2.8 \times 10^4\,\mathrm{MPa}\sqrt{\mathrm{m}}\,\mathrm{s}^{-1} \leq \dot{K} \leq 5.7 \times 10^4\mathrm{MPa}\sqrt{\mathrm{m}}\,\mathrm{s}^{-1}$, the J_d values were calculated via the dynamic force-displacement diagram and the Δa values were determined from the specimens' fracture surfaces. The tests were performed and evaluated based on ESIS P2-92. Here however, one must ensure that no stable, ductile crack extension occurs. As a result of the scanning electron microscopy examinations of the fracture surfaces (Fig. 9), crack growth is generated for the entire range of the J_d-Δa curves as a result of a stable, cleavage crack extension.

Here, it can be assumed that a globally unstable crack extension is inhibited by means of energy dissipative crack tip blunting (crack tip radius < diameter of the graphite particles) and crack arrest (Fig. 10).

Since no stretch zone occurs for this material specific crack extension, it is not possible to define a physical crack initiation value, analogously to static loading. The practical procedure for recording and evaluating dynamic crack resistance curves of ferritic cast iron materials was published in a BDG guideline [27] and can be depicted as shown in Fig. 11.

1. Determine the data points J_d and Δa according to ESIS P2-92
2. Determine the dynamic crack initiation value $J_{\mathrm{di}/\Delta \mathrm{a}=0}$ for $\Delta a = 0\,\mathrm{mm}$. The J_d-Δa values are linearly extrapolated from the range 0.1 mm ≤ Δa ≤ 0.5 mm to $\Delta a = 0\,\mathrm{mm}$. $\Delta a_\mathrm{max} \leq 0.5\,\mathrm{mm}$ follows from the requirement according to $\Delta a_\mathrm{max} \leq 0.1(W - a_0)$, where W is the specimen width and a_0 the initial crack length.

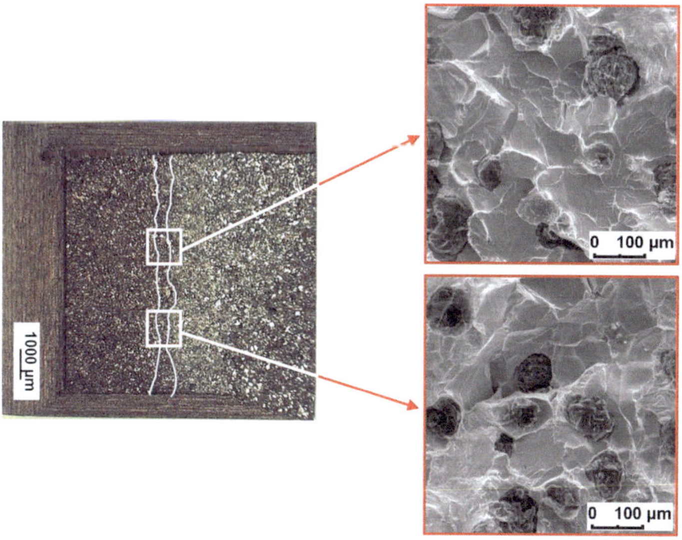

Fig. 9 Cleavage crack extension on the fracture surface at −40 °C [24]

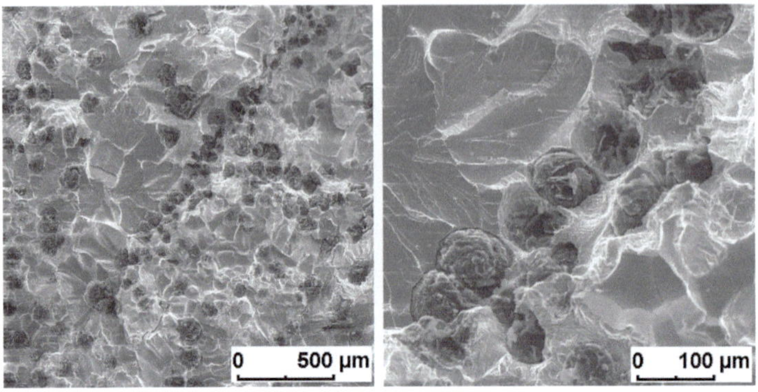

Fig. 10 Crack arrest at a graphite barrier [24]

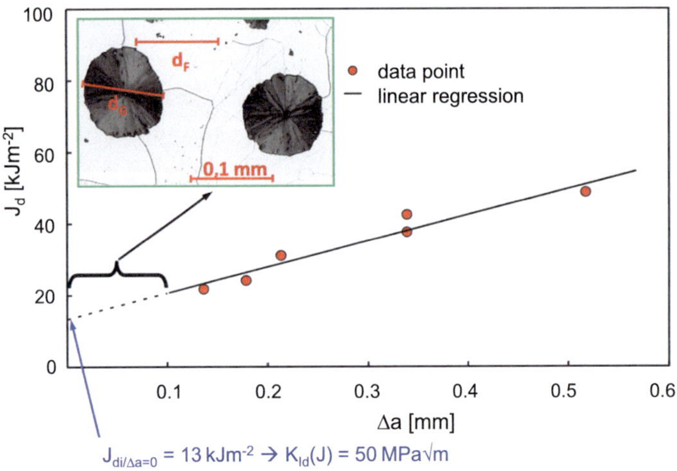

Fig. 11 J_d-Δa curve at −40 °C and the definition of the dynamic crack initiation value [24]

3. Determine an engineering, dynamic crack initiation value $J_{d0.2}$ according to ESIS P2-92 for $\Delta a = 0.2$ mm.
4. Convert the J_d values to K_{Id} (J) values via the elastic constants E and ν according to Eq. 1

The testing procedure was specifically applied to assess the crack resistance behaviour of ferritic cast materials as a function of temperature, loading rate and microstructure for applications such as transport vessels for spent nuclear fuel rods. Ferritic cast iron (pearlite fraction ≤ 5 %) possessing spherical graphite (Fig. 12) was investigated by varying the size of the graphite particles and/or ferritic grains d_F (Table 5), which are subsequently designated here as GJS (38) and GJS (62).

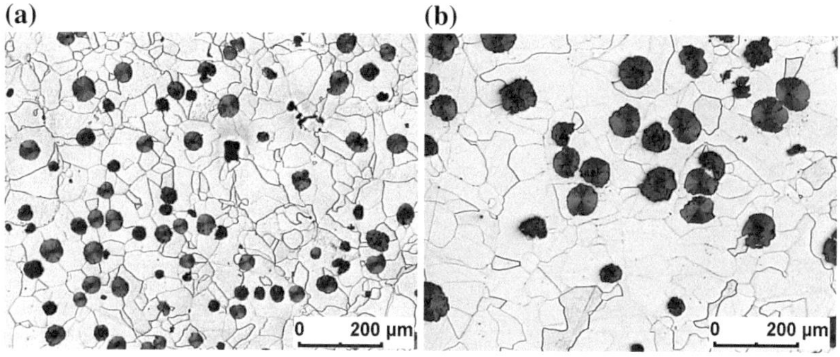

Fig. 12 Microstructure. **a** GJS (38). **b** GJS (62) [25]

Table 5 Microstructural parameters of GJS (38) and GJS (62): graphite volume fraction V_G, average number of graphite particles per Area N_A, average graphite particle diameter d_G, average ferrite grain size d_F, nodularity f, average nearest neighbor distance λ [25]

Material	V_G (%)	N_A (mm^{-2})	d_G (μm)	d_F (μm)	f	λ (μm)
GJS (38)	12	136	38	45	0.73	56
GJS (62)	15	48	62	63	0.71	86

The material was permanent mould (GJS (38)) and sand (GJS (62)) cast to specifically adjust the different microstructures [25].

The mechanical properties determined from tensile tests in microstructural and temperature dependence are given in Table 6. These values are also necessary for carrying out and evaluating the fracture mechanics tests.

The dynamic 0.2 % proof strength $R_{dp0.2}$ was determined at $\dot{\varepsilon} = 1\ \text{s}^{-1}$ based on the Arrhenius equation as a result of the tensile tests and by loading in a rotating impact machine. The result shows good agreement with the values determined in [28] for ferritic cast iron. The strain rate corresponds to the $\dot{K}$ values used for recording the dynamic crack resistance curves. The influence of the loading rate on the material's resistance to crack initiation is evident from Table 7.

A significant increase in the crack initiation resistance for dynamic loading of GJS (38) compared to GJS (62) can be attributed to the smaller ferritic grain size d_F (Table 5) which corresponds to the smaller graphite particle sizes d_G (Fig. 13).

Figures 14 and 15 depict the temperature dependence of the dynamic toughness values within the temperature range from RT to −40 °C, measured on a ferritic cast iron possessing spherical graphite ($R_{p0.2}$ = 246 MPa, R_m = 370 MPa, A = 13 %) [24]. Different from the static loading condition (Table 3), the temperature has a high influence on the dynamic toughness behaviour and is not negligible.

The fracture safety analysis of the crack initiation concept for static and dynamic loading excludes the extension of postulated and actually existing cracks. The objective of the crack-arrest concept is to increase the safety of components having

Table 6 Mechanical properties from tensile testing of GJS (38) and GJS (62) [25]

Material	T (°C)	$R_{p0.2}$ (MPa)	R_m (MPa)	A (%)	Z (%)	E (GPa)	ν	$R_{dp0.2}$ (MPa)
GJS (38)	RT	251	397	24	24	174	0.29	317
	−40	286	436	25	22	–	–	373
GJS (62)	RT	245	367	12	13	172	0.28	321
	−40	273	393	10	11	–	–	380

Table 7 Influence of loading rate at −40 °C [25]

Material	Static			Dynamic		
	J_i (kJ m^{-2})	$J_{0.2}$ (kJ m^{-2})	$K_{Ic}(J_i)$ (MPa$\sqrt{m}$)	$J_{di/\Delta a=0}$ (kJ m^{-2})	$J_{d0.2}$ (kJ m^{-2})	$K_{Id}(J_{di/\Delta a=0})$ (MPa$\sqrt{m}$)
GJS (62)	28	59	73	17	31	56

stringent requirements for fracture safety with respect to the material and /or loading. Unstable crack extension, resulting from overloading and/or material embrittlement, are arrested prior to the occurrence of catastrophic failure. The concept was developed to supply the crack initiation concept and is applied, for example, in pressure vessel technology, in pipeline and ship building as well as in offshore technology. In this way, crack-arrest represents a special case of unstable crack extension and is characterised with respect to the material by the crack arrest toughness K_{Ia}. Investigations to determine K_{Ia} values were carried out on

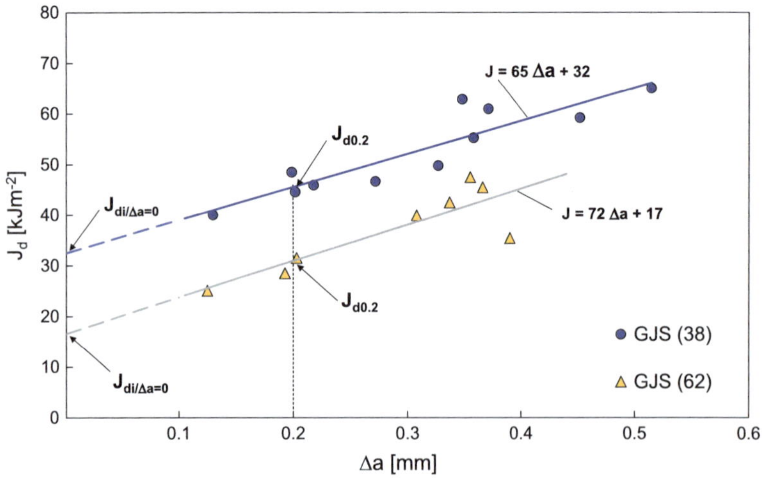

Fig. 13 Influence of the ferrite grain size and graphite nodule size on dynamic J_{dR} curve at −40 °C [25]

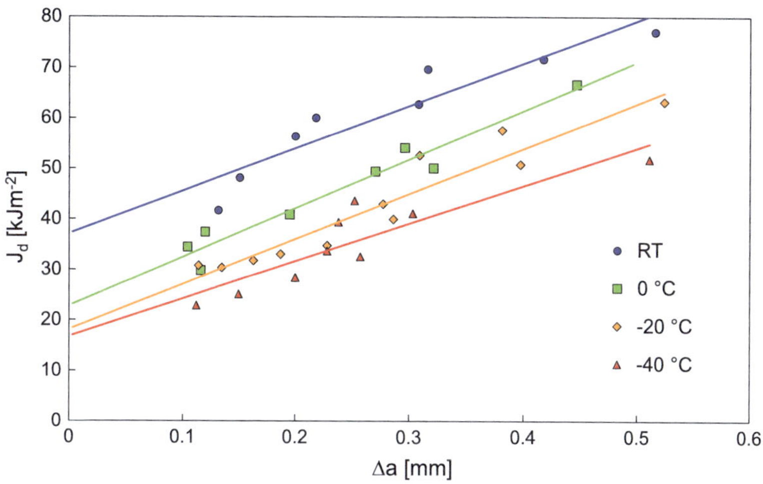

Fig. 14 Influence of temperature on the dynamic J_d-Δa curve for EN-GJS-400 measured on SENB 10 mm × 10 mm × 55 mm specimens [24]

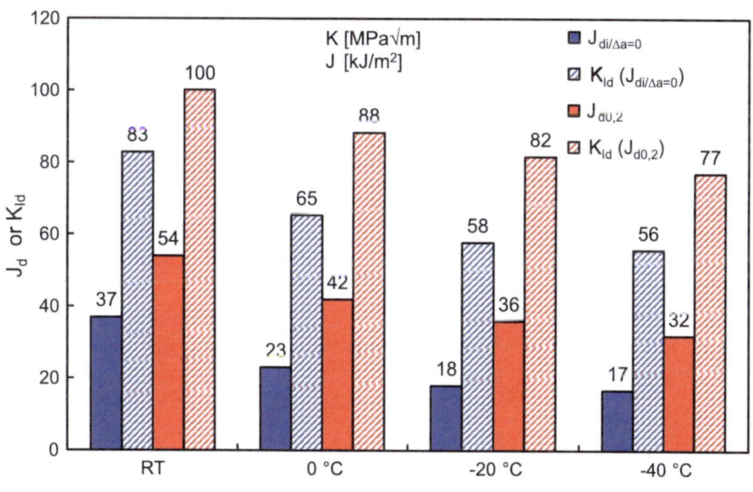

Fig. 15 Dynamic fracture toughness as a function of the temperature for EN-GJS-400 measured on SENB 10 mm × 10 mm × 55 mm specimens [24]

experimental cast iron batches (Table 8) possessing spherical graphite and different fractions of pearlite content (PC). These materials were subsequently designated as GJS (PC 4 %) and GJS (PC 29 %) [29, 30].

K_{Ia} was ascertained on 50 mm thick CCA (compact crack-arrest) specimens according to ASTM E 1221 [31]. As a crack starter, a notch was spark-eroded in a brittle weld seam across the entire specimen thickness. Details of the testing and its

Table 8 Mechanical properties at room temperature for ductile cast iron with 4 and 29 % fractions of pearlite content PC [29, 30]

Material	$R_{p0.2}$ (MPa)	R_m (MPa)	A (%)	Z (%)	E (GPa)	ν	HBW (2.5/187.5)
GJS (PC 4 %)	246	362	12	13	172	0.28	138
GJS (PC 29 %)	284	424	11	12	174	0.28	163

evaluation are extensively described in [29]. The crack-arrest tests carried out in the temperature range from −40 to −80 °C showing that at −40 to −60 °C no valid K_{Ia} values can be determined. The "minimum crack jump length" required in the standard was not reached. The high crack initiation toughness which still exists can be suggested as a cause. This is valid for both ferritic as well as for ferritic-pearlitic cast irons. Only at −80 °C the requirements were met for plane strain conditions, and valid K_{Ia} values within the range $44\,\text{MPa}\sqrt{\text{m}} \le K_{Ia} \le 61\,\text{MPa}\sqrt{\text{m}}$ were determined. The crack only extended in cleavage in the fracture surface region of crack arrest. The crack arrest can be attributed to the effect of graphite particles. Energy is dissipated when the crack runs into a graphite spherical and the crack tip radius is increased locally like pictured in Fig. 10. The K_{Ia} values determined at −80 °C are plotted together with the reference curve depending on the temperature in Fig. 16. An influence of the elevated fraction of pearlite can only be identified as a tendency.

According to [32], it is only possible to determine valid K_{Ia} values for ferritic cast iron in the temperature range from −40 to 20 °C with the aid of a duplex-arrest specimen, in which a hardened steel (AISI 4340), acting as a crack starter, is joined

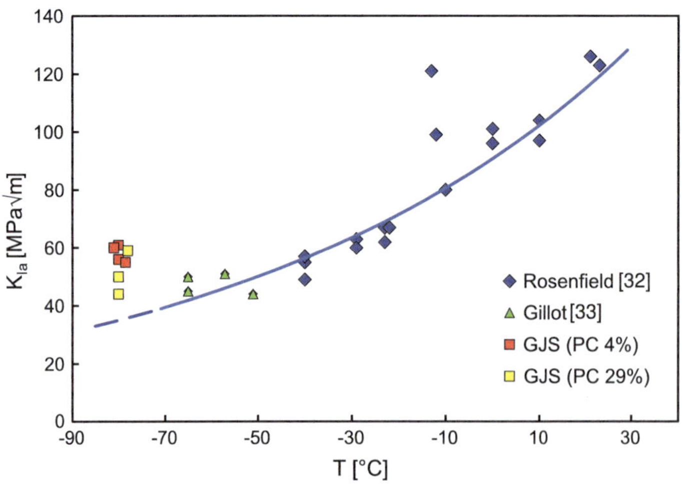

Fig. 16 K_{Ia} values for ferritic cast iron as a function of temperature; comparison with the values from the literature [32, 33]

to the GJS-400 test piece using a heat-treated electron-beam welded joint. Using this set-up, the crack starts in the hardened steel instead of in the brittle overlay weld, and runs at a high velocity into the cast iron. In this way, the K_{Ia} values are also determined in the higher temperature region above −80 °C as shown in Fig. 16.

4 Cyclic Loading Conditions

Within the scope of comprehensive structural durability analyses, fracture mechanics crack growth concepts are employed both for the microstructure dependent assessment of crack extension behaviour as well as for computing the components fatigue life. The crack resistance behaviour during cyclic loading is characterised by the experimentally determined cyclic crack growth curve (d*a*/d*N*-ΔK curve) (Fig. 17).

The crack growth rate d*a*/d*N* (d*a*—crack extension, *N*—number of load cycles) as a function of the cyclic stress intensity factor ΔK can be measured with the aid of crack gauges, by measuring the specimen's elastic compliance (compliance method) or from the period of resonant oscillations. Details of the testing and its evaluation are included in ASTM E 647 [34]. Determining the threshold value ΔK_{th} (region I) and the material specific parameters *C* and *m* of the Paris-Erdogan equation (region II)

$$\frac{da}{dN} = C[\Delta K]^m \tag{8}$$

forms the basis for computing the residual life of castings containing defects. Fracture occurs in region III at ΔK_{fc}. By introducing ΔK into a comprehensive structural durability analysis of cyclically loaded cast components, attention should be paid particularly in regions I and III to the influence of the mean stress, $R = \sigma_{min}/\sigma_{max}$ where σ_{min} and σ_{max} are the minimum and maximum stresses.

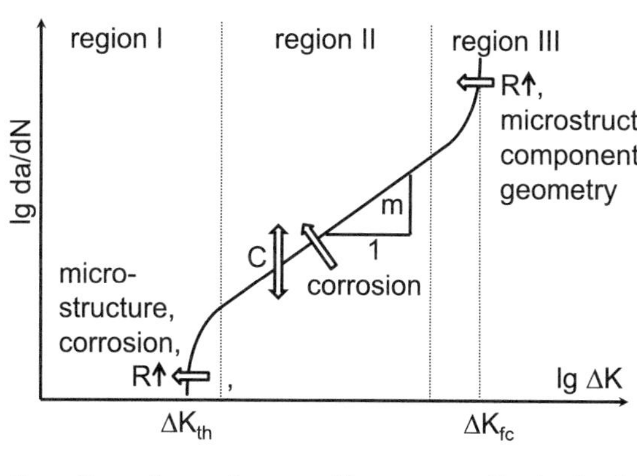

Fig. 17 Schematic cyclic crack growth curve with parameters *C* and *m* for the Paris-law, threshold value ΔK_{th}, critical fatigue stress intensity factor ΔK_{fc} and influencing factors

4.1 Constant Amplitude Loading

The fracture mechanics parameters are compiled in Table 9 as a function of the mean stress (R values) including the ΔK_{fc} values which characterise the final fracture. The parameters of the da/dN-ΔK curve are the average values each obtained from three SENB specimens having dimensions 10 × 20 × 100 mm^3. The evaluation of the field of scatter from five GJS-400 specimens resulted in a standard deviation of $\pm 0.4\,\mathrm{MPa}\sqrt{\mathrm{m}}$ for a mean value of $7.5\,\mathrm{MPa}\sqrt{\mathrm{m}}$ for ΔK_{th}. A standard deviation of this magnitude is confirmed by the totality of the tests. The relative error lies in the range of ±10 %, in agreement with measurements on steel [35–37]. By comparing the assessment of the cast iron materials listed in Table 9, a dependency of the ΔK_{th} or ΔK_{fc} values on the mean stress (R value) can be established, which can be quantified using

$$\Delta K_{\mathrm{th,R}} = (1 - R)^{\gamma} \Delta K_{\mathrm{th,R}} = 0 \tag{9}$$

where γ = 1.3 for GJV-300.

Cast iron materials possessing lamellar or vermicular graphite exhibit a microstructurally related, elevated crack growth rate [2]. If one assesses the material behaviour in region II of the da/dN-ΔK curve by using the correlation between the parameters C and m of the Paris-Erdogan equation, which was determined in [38] for brittle and ductile steels, then the crack growth rate of the investigated cast iron materials subjected to cyclic loading clearly falls into the category of brittle steels according to

$$C = \frac{2.89 \times 10^{-5}}{15.5^{m}}. \tag{10}$$

A current example for integrating fracture mechanics concepts into a comprehensive fatigue life assessment are the components for power generation in wind energy generators. With regard to the end of the projected life time of these structural components, or in cases of their damage, and the occurrence of fatigue cracks, questions concerning the residual structural life will become increasingly important for the operators and the licensing authorities. In the regulating mechanism which will then be required, reference will also be made to integrating fracture mechanics strength assessments into a comprehensive structural durability analysis [5].

The cast iron material GJS-400-18LT, which was optimised for low temperature applications, is employed in wind plant nacelle, machine bases, stratostern, blade adaptor as well as the rotor shaft. The reduction of mass of the components is necessary for the elevated power and size of construction. It requires designs for employing high-strength materials such as, for example, the Si solid solution hardened GJS materials [18] according to DIN EN 1563 [39]. Table 10 shows the parameters of the cyclic crack growth curves for such ductile iron materials GJS-W1 to GJS-W7 including the parameters of a welding (GJS-SG) in addition to the static

Table 9 Fracture mechanics parameters of the cyclic crack growth behaviour of cast iron materials according to [12, 13, 35, 37]

Material	R	ΔK_{th} (MPa$\sqrt{m}$)	m	C	ΔK_{fc} (MPa$\sqrt{m}$)
GJS-400-15	0.1	7.5	4.5	2.2×10^{-10}	40
	0.3	6.2	4.6	3.8×10^{-10}	31
	0.5	4.5	4.2	1.9×10^{-9}	22
GJS-600-3	0.1	6.9	4.1	1.2×10^{-9}	34
	0.3	6.6	3.9	1.3×10^{-9}	27
	0.5	4.6	3.7	1.9×10^{-9}	19
GJS-800-10	0.1	5.4	2.9	0.9×10^{-8}	51
	0.3	4.8	3.0	1.1×10^{-8}	40
	0.5	4.3	3.2	1.0×10^{-8}	29
GJS-1000-5	0.1	5.5	2.9	2.3×10^{-8}	45
	0.3	4.0	2.7	3.3×10^{-8}	33
	0.5	3.4	2.8	3.9×10^{-8}	24
GJV-300	0.1	7.0	6.7	1.1×10^{-11}	25
	0.3	4.6	7.2	2.0×10^{-12}	24
	0.5	3.3	5.5	1.9×10^{-10}	16
GJMB-350-10	0.1	7.6	3.4	1.2×10^{-8}	26
	0.3	6.4	3.5	9.8×10^{-9}	22
	0.5	6.0	4.0	4.0×10^{-9}	15
GJMB-450-6	0.1	6.6	3.4	6.8×10^{-9}	27
	0.3	4.9	3.2	1.5×10^{-8}	25
	0.5	5.3	3.6	5.7×10^{-9}	24
GJMB-650-2	0.1	4.4	3.1	6.7×10^{-9}	35
	0.3	4.1	3.0	1.3×10^{-8}	33
	0.5	4.4	3.1	1.3×10^{-8}	27
GJMW-360-12	0.1	5.5	3.5	7.0×10^{-9}	38
	0.3	4.6	2.9	3.0×10^{-8}	29
	0.5	3.5	2.4	1.0×10^{-7}	21
GJMW-400-15	0.1	7.5	3.6	5.0×10^{-9}	36
	0.3	6.0	4.3	1.5×10^{-9}	28
	0.5	4.4	3.1	4.0×10^{-8}	20
GJMW-450-7	0.1	5.4	3.2	8.0×10^{-9}	39
	0.3	5.1	3.1	3.0×10^{-8}	30
	0.5	4.6	3.7	8.0×10^{-9}	22
GJMW-550-4	0.1	4.1	2.2	1.5×10^{-7}	45
	0.3	3.7	2.6	7.0×10^{-8}	35
	0.5	3.2	2.6	7.0×10^{-8}	25

Table 10 Threshold values ΔK_{th} and parameters m and C for the Paris-Erdogan-equation for ferritic ductile cast iron with different Si content (ref. Table 4) [18]

Material	R	ΔK_{th} (MPa$\sqrt{m}$)	m	C
GJS-W1	0.05	9.5	5.2	6.0×10^{-11}
	−1	17.6	6.2	2.2×10^{-14}
GJS-W2	0.05	9.5	5.3	5.0×10^{-11}
	−1	18.1	7.6	1.6×10^{-16}
GJS-W3	0.05	9.6	5.5	7.3×10^{-12}
	−1	18.2	5.9	5.3×10^{-14}
GJS-W4	0.05	9.9	5.0	4.2×10^{-11}
	−1	18.8	6.3	1.4×10^{-14}
GJS-SG	0.05	8.1	4.8	1.2×10^{-10}
	−1	15.9	4.7	1.1×10^{-11}
GJS-W5	0.05	9.5	5.4	9.8×10^{-11}
	−1	17.4	5.6	2.6×10^{-12}
GJS-W7	0.05	9.8	4.8	8.3×10^{-11}
	−1	18.0	4.5	9.0×10^{-12}

toughness values given in Table 4. The tests were performed at R values which are characteristic for wind energy components. The influence of the R ratio is based on crack closure effects. From the possible mechanisms—plastic, oxide, fluid and roughness induced crack closure—the latter is supposed to be the most dominant one because of the specific microstructure. The threshold value ΔK_{th} is decreasing with increasing mean stress level. For region II of the cyclic crack growth curve the effect of mean stress can be found in the parameter C of the Paris-Erdogan-equation (Eq. 8). For symmetric tension-compression loading ($R = -1$) the C-value is considerable lower than for cyclic tensile loading with $R = 0.05$.

The mechanical properties and fracture mechanics parameters for GJS-400-18LT are listed in Tables 11 and 12. Here, WKN refers to samples directly from the rotor's hub, and WK1 and WK2 refer to specimens from the sprue which were quenched by a chilled permanent mould casting equipment and which led to a variation in the graphite particle sizes d_G from 23 to 60 m. The resulting scanning electron microscopy analyses of the fracture surface showed that rupture at RT and −40 °C is initiated by ductile, stable crack extension. It was confirmed that the J_i values increase with larger graphite particle size d_G or with corresponding graphite

Table 11 Mechanical and fracture mechanics parameters, and the microstructural parameters of EN-GJS-400-18LT [41]

Material quality[a]	$R_{p0.2}$ (MPa)	R_m (MPa)	A (%)	Z (%)	E (GPa)	d_G (μm)	λ (μm)	f	J_i (kJ m^{-2})	
									RT	−40 °C
WK1	265	396	23.0	21.0	173	23	31	0.83	19	19
WK2	256	382	18.5	23.0	170	37	53	0.74	27	26
WKN	245	403	22.5	16.2	171	60	90	0.76	38	37

[a]Internal designation

Table 12 Fracture mechanics parameters of the cyclic crack growth behaviour of EN-GJS-400-18LT as a function of the R value [41]

Material	*R*	ΔK_{th} (MPa$\sqrt{m}$)	*C*	*m*	ΔK_{fc} (MPa$\sqrt{m}$)
WK1	0.1	7.0	8.2×10^{-10}	4.2	30.4
WK2		7.7	1.8×10^{-10}	4.6	32.8
WKN		8.5	7.8×10^{-11}	4.6	37.5
WK1	0.3	6.0	6.1×10^{-10}	4.4	23.7
WK2		6.5	3.0×10^{-10}	4.4	–
WKN		7.3	1.9×10^{-10}	4.5	–
WK1	0.5	4.5	7.3×10^{-10}	4.3	17.8
WK2		4.9	3.3×10^{-10}	4.5	–
WKN		5.3	4.8×10^{-10}	4.2	–

particle spacing λ. That is, the crack initiation resistance increases. For cyclic loading, the dependency of the threshold value ΔK_{th} on the mean stress can be described according to Eq. 9, where $\gamma = 0.82$ agrees with the value determined for GJS-400-15 in [40]. By assessing the microstructural dependency of the ΔK_{th} value, it can be assumed that, when the running crack encounters a spherical graphite particle, where the crack tip opening $\delta < d_G$, crack tip blunting occurs. Generally, this relationship can be qualified by

$$\Delta K_{th} \sim \frac{\sqrt{r}}{\sqrt{\rho}} \tag{11}$$

where, $r = d_G/2$ and ρ is the radius of the notch, i.e. $\rho \approx \delta/2$. The dependency of ΔK_{th} values on the graphite particle size d_G (Fig. 18) can be described for the range of 10–70 by the empirical equation

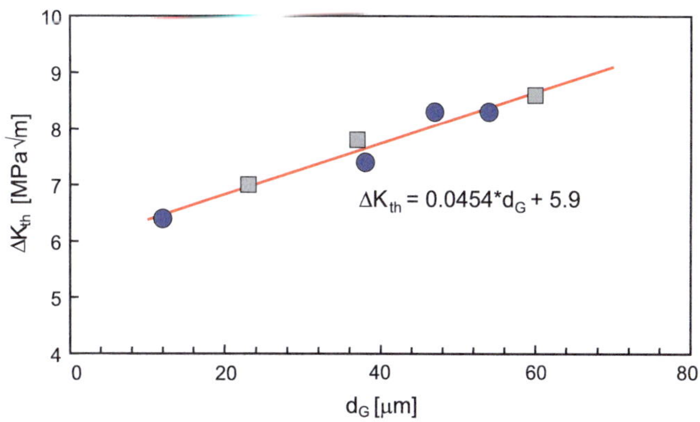

Fig. 18 Dependency of ΔK_{th} value on the graphite particle size d_G (R = 0.1): *Points* are measurements for EN-GJS-400-15, *squares* for EN-GJS-400-18-LT

$$\Delta K_{\mathrm{th}} = (0.0454 d_{\mathrm{G}}/\mu\mathrm{m} + 5.9)\,\mathrm{MPa}\sqrt{\mathrm{m}}. \tag{12}$$

The application of austempered ductile iron (ADI) materials was investigated within the scope of developing and testing alternative railway wheel materials which possess longer service lives. As generally known, these materials are characterised by a combination of high wear resistance and high fatigue strength with, at the same time, a high ductility which is atypical for cast irons in this strength class. Since a railway vehicle wheel is subjected to high static and cyclic loading, the service life assessment involving fracture mechanics concepts is extremely important. Requirements for employing ADI as a wheel material are to verify sufficient safety against fatigue and brittle fracture, and to specify suitable monitoring concepts.

A fracture mechanics assessment of an ADI wheel to rupture and fatigue crack growth was carried out using a linear-elastic fracture mechanics analysis for hypothetical crack-like defects (Fig. 19). The stress intensity factors K_{I} for these postulated cracks were ascertained using the stress distribution previously computed using FEM at the cracks location, where solutions known in the literature were assumed for comparable crack configurations. Analogously, the band width of the stress intensity factor ΔK was computed. Moreover, the fatigue crack growth rate and the number of load cycles up to fracture were calculated. The loading stresses of the solid wheel's postulated crack configuration were numerically analysed at the IMFD, the mechanical and fracture mechanics parameters of the ADI material GJS-800-8 were determined at the IWT of the TU Bergakademie Freiberg, Germany [42] (Table 13).

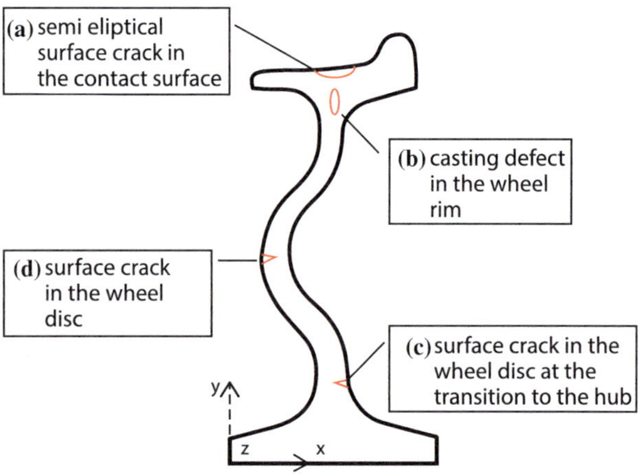

Fig. 19 Postulated crack configuration

Table 13 Mechanical and fracture mechanics parameters of ADI GJS-800-8

Properties	Parameters
Young's modulus E	170 GPa
Poisson's ratio ν	0.3
0.2 % proof strength $R_{p0.2}$	637 MPa
Tensile strength R_m	893 MPa
Yield-to-tensile ratio $R_{p0.2}/R_m$	0.71
Fracture toughness J_i or $K_{Ic}(J_i)$	11 kJ/m^2 or 45.3 MPa$\sqrt{m}$
Fatigue crack growth for R = 0.1	
Threshold value ΔK_{th}	5.4 MPa$\sqrt{m}$
C	0.94×10^{-8}
m	2.9
Fatigue crack growth for R = 0.5	
Threshold value ΔK_{th}	4.3 MPa$\sqrt{m}$
C	1.0×10^{-8}
m	3.2

In conjunction with recurring service monitoring measures and safety concepts, a permissible crack size can be derived from the computed crack length which is consistent with fracture mechanics. The critical crack size computed here can be diagnosed as safe and reliable by using non-destructive testing methods [42].

Besides the essential data about crack configuration and loading, reliable information about the crack growth behaviour is necessary for computing the

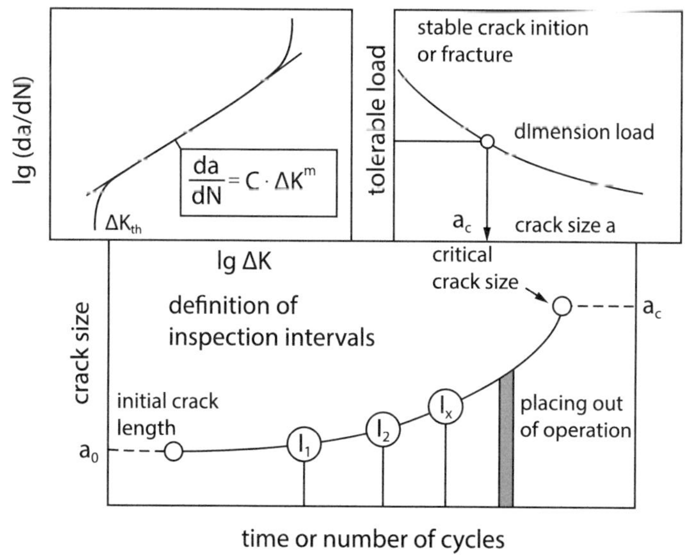

Fig. 20 Fracture mechanics safety assessment of components containing cracks subjected to cyclic loading

residual service life (Fig. 20) as an integral part of the fracture mechanics concepts. It is important that, for fatigue problems, substantially more stringent requirements must be imposed on the initial data's precision than those in cases of static loading because the fatigue life depends exponentially on the cyclic stress intensity factor.

During the design phase, conservative reference curves are frequently employed for such fracture mechanics safety analyses. These curves were derived as envelope curves via numerous measured results from material batches. Here, it is not always evident how large the safety margin proves to be for individual type of material.

It is possible to perform a statistical evaluation by deriving the quantile crack growth curves based on the NASGRO computer code, which is available as "ESACRACK 4.0" [43, 44]. The principles of this statistical evaluation and its application to cast iron materials are elucidated in [45, 46]. The crack growth model of the program "ESACRACK" is based on an extended Paris-Erdogan equation which describes all three regions of the cyclic crack growth curve (Fig. 21) as a function of the stress ratio R according to

$$\frac{da}{dN} = C_2\left[\left(\frac{1-f_1}{1-R}\right)\Delta K\right]^n \frac{\left(1-\frac{\Delta K_{\mathrm{th}}}{\Delta K}\right)^p}{\left(1-\frac{K_{\max}}{K_{\mathrm{mat}}}\right)^q}. \qquad (13)$$

The fitting constants p and q match the transition into region I (ΔK_{th} threshold value of the cyclic stress intensity factor) and into region III where, here, the critical stress intensity factor K_{mat} and the maximum stress intensity are to be determined during cyclic loading $K_{\max} = \Delta K/(1-R)$. Region II is characterised by the material specific parameters C_2 and n, which are interdependent; (C_2 and n refer to a modification of Eq. 8). The dependency of the cyclic crack growth curve on the stress ratio R is described by the crack opening function f_1 which is defined by the ratio $K_{\mathrm{op}}/K_{\max}$ (K_{op} is the crack opening stress intensity factor).

$$f_1 = \frac{K_{\mathrm{op}}}{K_{\max}} = \begin{cases} \max(R, A_0 + A_1R + A_2R^2 + A_3R^3) & R \geq 0 \\ A_0 + A_1R & -2 \leq R < 0 \\ A_0 - 2A_1 & R < -2 \end{cases} \qquad (14)$$

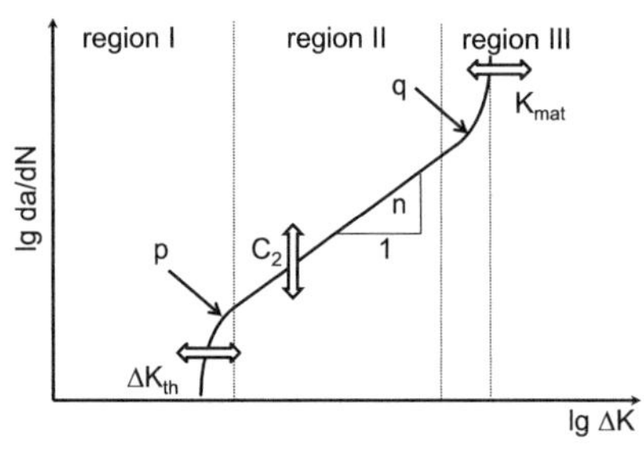

Fig. 21 Cyclic crack growth curve and definition of the statistical variables

The constants A_0 to A_3 depend on the ratio of the maximum tensile stress S_{max} to the engineering flow stress σ_0 and the constraint factor α.

$$\begin{aligned} A_0 &= \left(0.852 - 0.34\alpha + 0.05\alpha^2\right)\left[\cos\left(\frac{\pi S_{max}}{2\sigma_0}\right)\right]^{\frac{1}{\alpha}} \\ A_1 &= (0.415 - 0.071\alpha)S_{max}/\sigma_0 \\ A_2 &= 1 - A_0 - A_1 - A_3 \\ A_3 &= 2A_0 + A_1 - 1 \end{aligned} \quad (15)$$

α can take values between 1 (plane stress conditions) and 3 (plane strain conditions) and is a function of the components thickness. For ductile materials where the hypotheses according to VON MISES is valid, the quotient S_{max}/σ_0 assumes the value of 0.3. For steels, the values α = 2.5 and $S_{max}/\sigma_0 = 0.3$ lead to a good fit of experimental data. In a few cases in which no plastically induced crack closure occurs, the crack opening function f_1 can be circumvented by the so-called bypass operation because, in such cases, f_1 is set equal to R ($f_1 = R$) and thus the quotient in Eq. 13 becomes $(1 - f_1)/(1 - R) = 1$. The values $\alpha = 5.845$ and $S_{max}/\sigma_0 = 1$ have to be specified for the bypass operation (Fig. 22).

The crack opening function f_1 only describes the plastically induced crack closure. Other mechanisms, such as crack closure due to corrosion products or crack closure induced by roughness, can lead to a deviation in R depending on the da/dN-ΔK curve. This can be determined by crack growth measurements at different R values. The constants: α and S_{max}/σ_0, are then used as fitting parameters.

Fitting the crack growth curves of the investigated cast iron materials (Table 14), the following values are held constant:

$$S_{max}/\sigma_0 = 0.3;\ \alpha = 2.5;\ a = 10;\ a_0^* = 0.0381;\ C_{th} - 0.25;\ R_{cl} = 0.7;\ R_p = -1.$$

Using the example of the spherical graphite cast iron materials, it can be seen that the da/dN-ΔK curves, which were experimentally recorded for different R

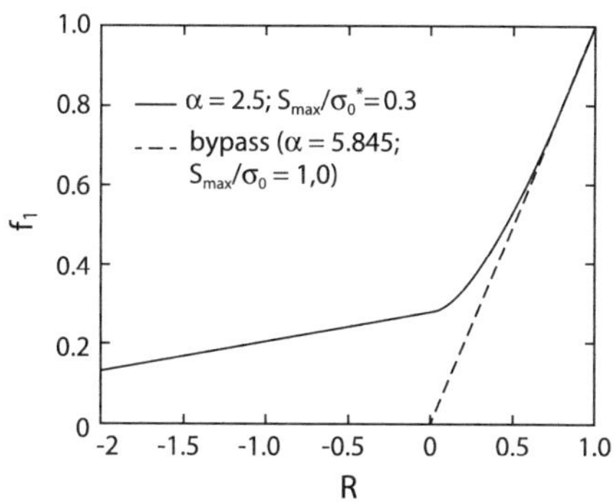

Fig. 22 Crack opening function f_1 as a function of R

Table 14 Parameters and constants of the crack growth model "ESACRACK" for cast iron materials [6]

Material EN-…	ΔK_0 (MPa$\sqrt{m}$)	ΔK_c (MPa$\sqrt{m}$)	C_2	n	p	q	C_{th+}
GJS-400-18-LT	7.8	32	3.8×10^{-9}	3.8	0.20	0.10	2.6
GJS-600-3	7.8	36	6.0×10^{-9}	3.5	0.30	0.25	1.9
GJS-800-8	6.5	58	3.5×10^{-8}	2.7	0.25	0.25	1.0
GJS-1000-5	5.0	46	5.0×10^{-8}	2.7	0.25	0.25	1.0
GJV-300	7.8	32	2.0×10^{-10}	5.5	0.25	0.25	4.0
GJMB-350-10	8.8	30	4.5×10^{-8}	3.0	0.25	0.25	1.0
GJMB-450-6	8.4	35	3.0×10^{-8}	3.0	0.25	0.25	2.0
GJMB-650-2	5.7	45	4.0×10^{-8}	2.7	0.25	0.25	1.0
GJMW-360-12	6.5	40	3.0×10^{-8}	3.3	0.40	0.50	1.3
GJMW-400-5	8.4	40	3.0×10^{-8}	3.3	0.30	0.40	1.8
GJMW-450-7	6.5	42	2.4×10^{-8}	3.3	0.50	0.30	0.1
GJMW-550-4	5.3	38	8.5×10^{-8}	2.7	0.20	0.30	0.5

Comments: The parameters are valid for the system of units: mm/cycle; MPa$\sqrt{m}$. In order to obtain the "ESACRACK" system of units mm/cycle; Nmm$^{-3/2}$, one should carry out the following transformations

ΔK_0 (N mm$^{-3/2}$) $= 31.6 \cdot \Delta K_0$ (MPa$\sqrt{m}$)

K_c (N mm$^{-3/2}$) $= 31.6 \cdot K_c$ (MPa$\sqrt{m}$)

C_2 (mm/cycle; MPa$\sqrt{m}$) $= \frac{C_2}{\sqrt{1000^n}}$

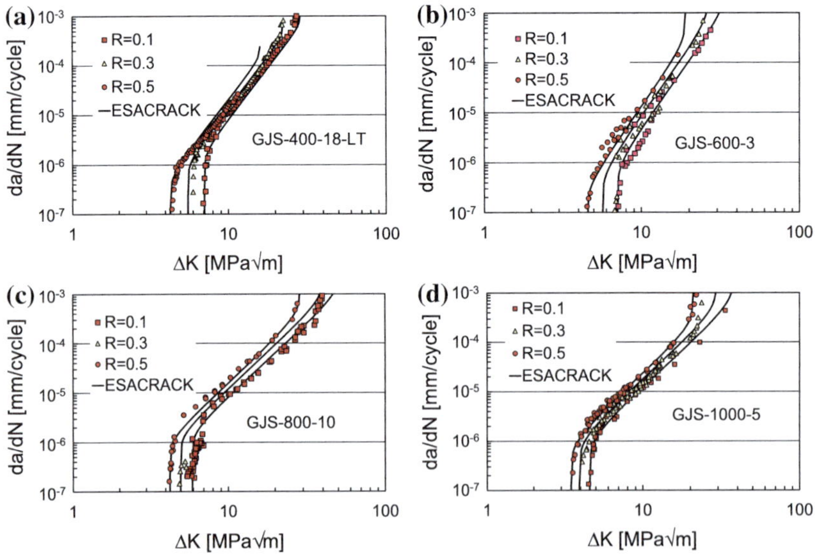

Fig. 23 Analytical description of the cyclic crack growth curves according to the ESACRACK-model for die materials. **a** EN-GJS-400-18-LT, **b** EN-GJS-600-3, **c** EN-GJS-800-8, **d** EN-GJS-1000-5 [6]

values, are described by the "ESACRACK"-model with sufficient accuracy (Fig. 23).

"ESACRACK" also permits the threshold-value region to be described for both positive and negative R values as a function of the crack depth according to

$$\Delta K_{\mathrm{th}} = \Delta K_0 \left(\frac{a}{a + a_0^*}\right)^{\frac{1}{2}} \left[\frac{1 - f_1}{(1 - A_0)(1 - R)}\right]^{-(1 + C_{\mathrm{th}} R)} \quad (16)$$

where, ΔK_0 is the threshold value for $R = 0$, a is the current crack depth, a_0^* is a microstructure constant and C_{th} is a fitting parameter which can be different for positive or negative R values ($C_{\mathrm{th}+}$ or $C_{\mathrm{th}-}$). The first part of Eq. 16 describes the influence of crack length on the threshold value and, in doing so, takes into account the lowering of the threshold value for crack depths smaller than a_0^*. The second part of Eq. 16 describes the portion of crack closure with respect to the threshold value using the crack opening function f_1, the constants A_0, and C_{th}.

For the cast iron materials listed in Table 14, a statistical evaluation was carried out which was separately performed for the three regions of the cyclic crack growth curve. From this evaluation, the resulting 5 and 95 % quantile crack-growth curves are depicted in Fig. 24 [46].

The influence which the corresponding quantiles exert on the predicted fatigue life is to be demonstrated for a component in the following computational example. A cyclic tensile load acts on an EN-GJS-400-18LT bar containing a semi-elliptical surface defect (Fig. 25). The following parameters are assumed:

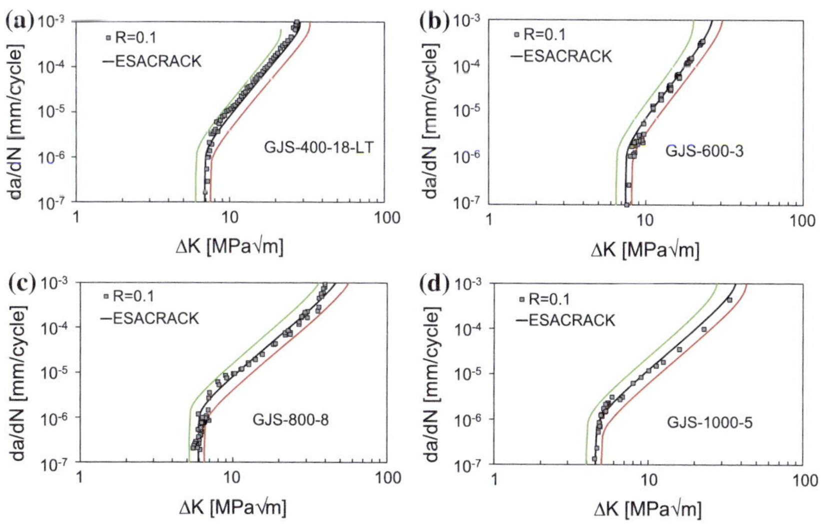

Fig. 24 Measured data points and quantile crack growth curves for 5 % (*red*) and 95 % (*green*) probability: **a** EN-GJS-400-18-LT, **b** EN-GJS-600-3, **c** EN-GJS-800-8, **d** EN-GJS-1000-5 [46]

Fig. 25 Crack geometry for the computational example

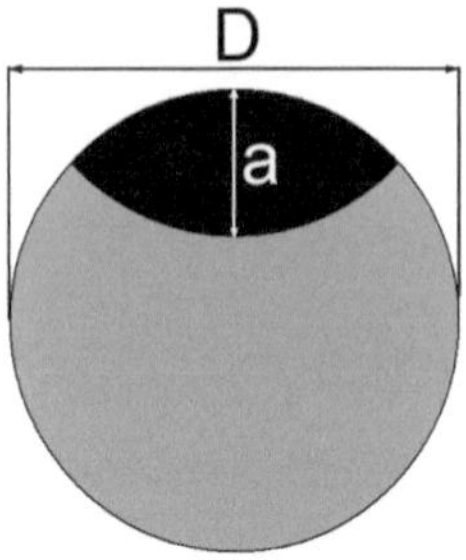

Initial crack length: $a_0 = 2.5$ mm,
Cyclic stress range: $\Delta\sigma = 150$ MPa,
Diameter: $D = 25$ mm,
Stress ratio: $R = 0$.

The computations were performed using the program "AFGROW" [47] and the crack growth curves were computed for the 5, 50 and 95 % quantiles. "AFGROW" incorporates the "ESACRACK/NASGROW" model. Table 15 lists the employed material data. The parameters ΔK_0, C_2 and K_C are defined as statistically distributed variables and varied by the factors $T_{\Delta K0}$, T_c and T_{Kmat} [46].

As a result of the computation, the crack length curves are depicted in Fig. 26 as a function of the number of cycles based on different quantile crack-growth curves.

Table 15 Parameters for the crack growth model "ESACRACK" for quantile crack-growth curves (ΔK in MPa$\sqrt{m}$, da/dN in m/cycle) [46]

Material EN-GJS-400-18LT	T_c C_2	$T_{\Delta K0}$ ΔK_0 (MPa$\sqrt{m}$)	T_{Kmat} K_C (MPa$\sqrt{m}$)	n	p	q	C_{th}^{+}
50 % quantile	3.8×10^{-12}	7.8	32	3.8	0.20	0.10	2.6
5 % quantile	1.97×10^{-12}	8.5	37	3.8	0.20	0.10	2.6
95 % quantile	7.03×10^{-12}	6.8	24	3.8	0.20	0.10	2.6

Fig. 26 Crack length versus cycles based on assuming different quantile crack-growth curves

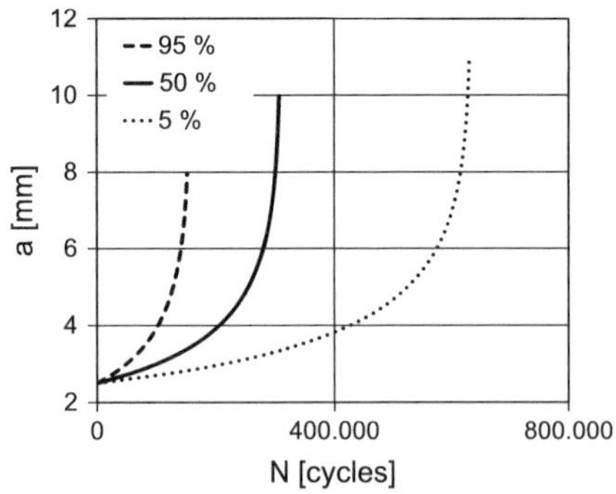

If one compares the results of the 50 % and the 95 % quantile crack-grow curves, it can be seen that the fatigue life at the 95 % probability level is half that given by the mean value curve. That shows that the frequency of inspection interval has to double at least.

4.2 Variable Amplitude Loading

The following results were determined during experimental investigations on damage mechanisms during crack propagation in ferritic cast iron materials subjected to constant and variable load amplitudes. The tests were performed within the scope of a joint research project between IMFD and IWT of the TU Bergakademie Freiberg [48–53].

The material was a GJS-400-18LT which was casted especially for the investigations in this project. Different graphite particle diameters d_G of 13 and 41 µm, subsequently designated by G10 and G50, were set by specifically influencing the cooling rate (Table 16).

The materials were characterised by measurement of da/dN-ΔK curves at constant load amplitudes for different R values, and described by means of the NASGRO-equation (Fig. 27; Table 17). The fracture mechanics parameters of the da/dN-ΔK curve, each determined from three specimens per R value, are compiled in Table 18. Both the well known dependency of the ΔK_{th} and ΔK_{fc} values on the mean stress as well as the tendency to higher ΔK_{th} and ΔK_{fc} for larger diameters of the graphite particles can be seen in Fig. 28 and Table 18.

The investigations of crack growth behaviour subject to variable loading amplitude were carried out on standard, 10 mm × 20 mm × 100 mm SENB specimens having a 3 mm starter crack. The tests were performed in a three-point-bending apparatus using load control. The load was applied using a computer controlled servo-hydraulic testing machine by MTS. Here, the crack length was recorded using crack gauges, which were applied to both specimen sides above the notch root, in combination with the crack-length measuring system FRACTOMAT made by RUMUL. In order to better quantify the sequence effects and the effectiveness of the loading reversals on the fatigue crack growth, the investigations were carried out using the following loading modes:

Table 16 Mechanical properties and graphite morphology parameters

Material	T (°C)	R_m (MPa)	$R_{p0.2}$ (MPa)	A (%)	E (GPa)	d_G (µm)	λ (µm)	f	N_a (mm^{-2})
G10	RT	388	247	23.5	176	13	21	0.75	902
	−40	420	276	26.0					
G50	RT	374	242	23.0	179	41	58	0.58	124
	−40	405	267	21.0					

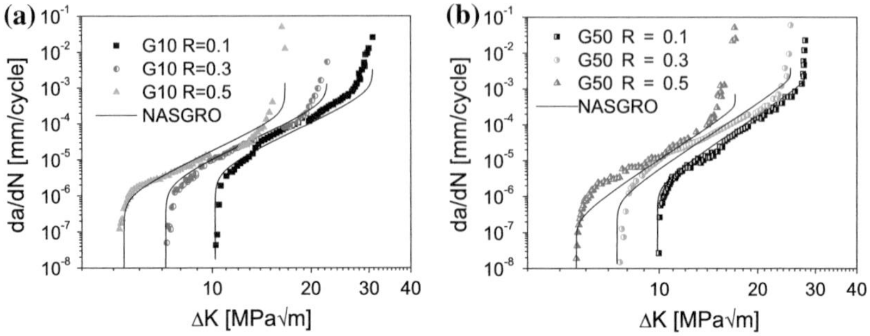

Fig. 27 Cyclic crack growth curves of G10 and G50 for different *R* values

Table 17 Parameters of the NASGRO equation

Material	ΔK_0 (MPa$\sqrt{\text{m}}$)	K_{mat} (MPa$\sqrt{\text{m}}$)	C_2	n	p	q	$C_{\text{th+}}$
G10	11.5	33.2	3.1×10^{-9}	3.8	0.4	0.4	3.5
G50	11.7	33.7	3.0×10^{-9}	3.8	0.5	0.5	3.0

Table 18 Fracture mechanics parameters (Paris parameters C and m, threshold value ΔK_{th} and critical fatigue stress intensity factor ΔK_{fc}) of the cyclic crack growth curves for EN-GJS-400-LT

Material	R value	ΔK_{th} (MPa$\sqrt{\text{m}}$)	m	C	ΔK_{fc} (MPa$\sqrt{\text{m}}$)
G10	0.1	10.1	6.0	5.1×10^{-12}	31
	0.3	8.1	5.0	1.5×10^{-10}	22
	0.5	5.1	5.0	2.8×10^{-10}	17
G50	0.1	10.2	6.1	3.3×10^{-12}	28
	0.3	8.5	5.0	5.0×10^{-10}	25
	0.5	5.5	4.6	3.1×10^{-10}	17

- Introducing of individual overloads at a base load level (R = 0.5) with variations in the overload ratios ($R_{\text{OL}} = 1.25–2.25$), which is defined by

$$R_{\text{OL}} = F_{\text{overload}} / F_{\text{max,baseload}}. \tag{17}$$

- Block load sequences low-high-low possessing different block load ratios ($R_{\text{Block}} = 1.25–2.0$), in which the block load ratio is defined by

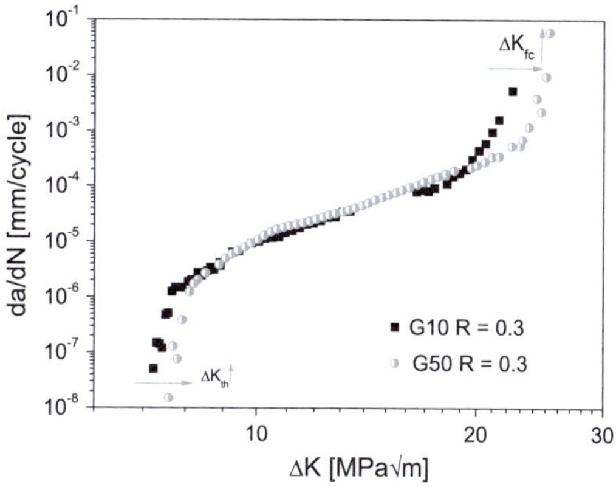

Fig. 28 Comparison of the cyclic crack growth curves of GJS-400-18LT specified as G10 and G50 for $R = 0.3$

$$R_{\mathrm{block}} = F_{\mathrm{block}} / F_{\mathrm{max,baseload}}. \tag{18}$$

The block load sequences were randomised using 5000 cycles within a base load ($R = 0.5$) of 15,000 cycles.

Using a 1.5 overload as an example, Figs. 29 and 30 depict the resulting influence on the fatigue crack growth for the test materials G10 and G50. The crack growth rate d*a*/d*N* was computed by means of the secant method. Both figures clearly show accelerations in the crack growth due to the tensile overload.

For G10, d*a*/d*N* increases from 10^{-5} up to 10^{-3} mm/cycle. Following this, d*a*/d*N* again approximates the previous level of 10^{-5} mm/cycle, which the crack requires approx. 440 cycles after the overload. For G50 (Fig. 30), d*a*/d*N* only increases to 5×10^{-4} mm/cycle and requires approx. 150 cycles in order to stabilise to its previous level. For both test materials, all the investigated overload ratios ($R_{\mathrm{OL}} = 1.25{-}2.25$) lead to initial crack acceleration phases. The block-program tests were performed for $F_{\mathrm{min}} =$ constant and 5000 block load cycles. Figure 31 shows the influence of two different block load ratios on the fatigue crack growth of G10 and depicts the corresponding d*a*/d*N*-*N* curve.

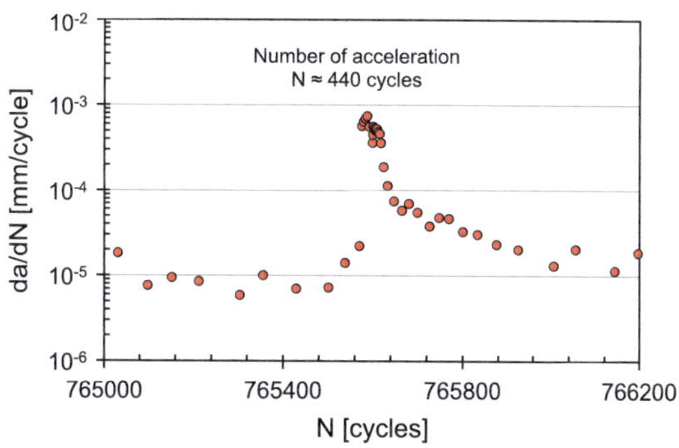

Fig. 29 Influence of an overload ($R_{\mathrm{OL}} = 1.50$) on the crack growth rate for G10

Fig. 30 Influence of an overload ($R_{OL} = 1.50$) on the crack growth rate for G50

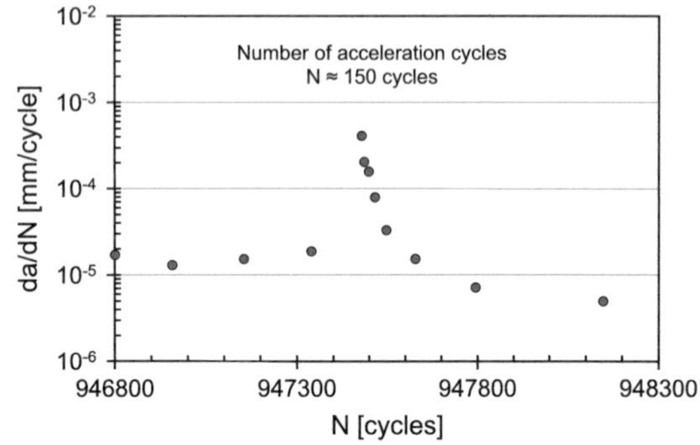

On changing from a low (low) to a high load level (high), a significant crack acceleration occurs. Within the block load, d*a*/d*N* increases over a short time to 8×10^{-5} mm/cycle for $R_{\text{block}} = 1.25$, and to 1×10^{-3} mm/cycle for $R_{\text{Block}} = 1.5$. This short time increase can be explained by the crack being opened due to the introduced block load. Subsequent to this, d*a*/d*N* stabilises to an approx. constant level during approx. 600–800 load reversals within the block load until a high-low transition. For $R_{\text{block}} = 1.25$, d*a*/d*N* stabilises over a short time to approx. 2×10^{-5} mm/cycle and, for $R_{\text{Block}} = 1.5$, to approx. 1×10^{-4} mm/cycle. The higher crack growth rate corresponds to an increase in ΔK as a consequence of the higher load level. In contrast to the well known occurrence of crack growth retardation in aluminium and steel during the high-low transition, here a short term but rather small crack acceleration phase occurs until d*a*/d*N* again stabilises to the previous base load level.

Besides this, Fig. 31 shows that d*a*/d*N* is increased with increasing R_{Block}. No crack growth retardation was measured in this case during the limited number of

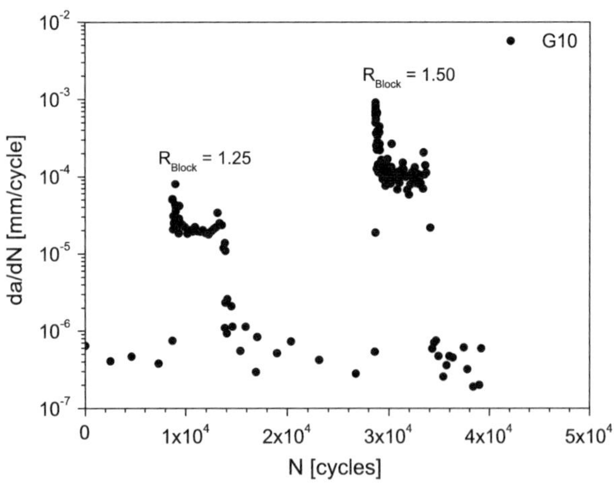

Fig. 31 Influence of two block loads on the crack growth rate for G10

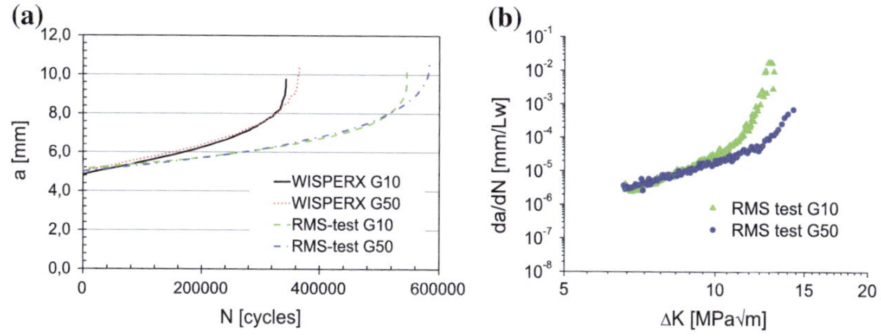

Fig. 32 **a** Comparison of the *a-N* curves for G10 and G50 from WISPERX and RMS tests; **b** Cyclic crack growth curves from the RMS tests

cycles. One reason for this acceleration effect is the predamage ahead of the crack tip in the cast iron material. In [54], it was already possible to observe the damaging process by means of scanning electron microscopy investigations. According to this, the crack acceleration effects can be clearly attributed to the spherical graphite debonding from the matrix following the overload.

Moreover, in view of the increased use of ductile cast iron in wind energy generators, the standardised load spectrum WISPERX [55] was employed in the investigations of fatigue crack growth. Using a final multiplying factor, the maximum and the minimum forces were specified within the load spectrum for WISPERX as 6 and 0.3 kN, respectively. On carrying out a global analysis based on a static description, the following effective values were specified for WISPERX: $F_{\mathrm{min,RMS}} = 3.089\,\mathrm{kN}$, $F_{\mathrm{max,RMS}} = 4.808\,\mathrm{kN}$ and $\Delta F_{\mathrm{RMS}} = 1.719\,\mathrm{kN}$. Figure 32a depicts the a-N curves for both material types loaded using WISPERX and the tests simulated from these loading sequences using the effective values (RMS test using $\Delta F_{\mathrm{RMS}} = 1.719\,\mathrm{kN}$, i.e. constant load amplitude).

As a result of these investigations, the influence of the spherical graphite size is shown. According to this, G50 exhibits a slightly longer fatigue life than G10 for the same loading. Moreover, it can be seen that WISPERX leads to a significantly shorter fatigue life than that measured during the comparable RMS test. This, in turn, is based on the occurring crack acceleration effect during the WISPERX loading sequence, which is composed of statistically distributed overloads, underloads and block loads. In addition to this, Fig. 32b shows that from $\Delta K = 10\,\mathrm{MPa}\sqrt{\mathrm{m}}$, G10 exhibits a higher crack growth rate than G50 in the RMS tests.

5 Summary

Cast iron materials show relatively low energy values in the conventional charpy impact testing compared to construction steels. Fracture mechanics based design gives the possibility to quantify the toughness needed in the construction detail. For the characteristic material properties it is important to consider the microstructure. Special attention should be given on graphite shape and size, Si content and testing temperature. The highest static fracture toughness was obtained for nodular cast iron with ferritic matrix and large graphite diameter. Also the threshold value for fatigue crack growth of such material is high. Under dynamic impact loading a smaller graphite diameter results in higher crack initiation values. Increasing the Si content leads to cleavage fracture at low temperatures and at high loading rates.

References

1. Forschungskuratorium Maschinenbau (2006) FKM-Richtlinie Bruchmechanischer Festigkeitsnachweis für Maschinenbauteile. VDMA-Verlag GmbH, Frankfurt
2. Pusch G, Hübner P, Pyttel B (2005) FKM-Richtlinie—Bruchmechanischer Festigkeitsnachweis für Maschinenbauteile. Konstruieren + Giessen 30(3):18–26
3. International Atomic Energy Agency (2002) Advisory materials for the IAEA regulations for the safe transport of radioactive material. Number safety guide TS-G-1.1 (ST-2) in safety standards series. IAEA, Vienna
4. BAM (2002) GGR-007: Leitlinie zur Verwendung von Gusseisen mit Kugelgraphit für Transport- und Lagerbehälter für radioaktive Stoffe. Bundesanstalt für Materialforschung und -prüfung, 2002 edition, Berlin
5. Lloyd Germanischer (2010) Richtlinie für die Zertifizierung von Windenergieanlagen. Germanischer Lloyd WindEnergie GmbH, Hamburg
6. Pusch G (2008) Bruchmechanische Kennwerte von Gusseisen. Konstruieren + Giessen 33 (4):2–34
7. European Structural Integrity Society (1992). ESIS P2-92: procedure for determining the fracture behaviour of materials. ESIS Office
8. Baer W (1996) Bruchmechanische Bewertung ferritischer Gusseisenwerkstoffe sowie artgleicher Schweißverbindungen bei statischer Beanspruchung. Dissertation, TU Bergakademie Freiberg
9. Rehmer B (1992) Bruchmechanische Bewertung ferritischer Gusseisenwerkstoffe mit Kugel- und Vermiculargraphit in Abhängigkeit von Beanspruchungsgeschwindigkeit und -temperatur. Dissertation, TU Bergakademie Freiberg
10. Pusch G, Liesenberg O, Hübner P, Brecht T, Krodel L (1999) Mechanische und bruchmechanische Kennwerte für Gusseisen mit Kugelgraphit. Konstruieren + Giessen 24 (2):10–19
11. Pusch G, Liesenberg O (2000) Bruchmechanische Kennwerte für Gusseisenwerkstoffe. Gießerei-Praxis 3:105–112
12. Hübner P-V (2003) Gefüge- und beanspruchungsabhängige Charakterisierung des Risswiderstandsverhaltens von entkohlend geglühtem Temperguss unter Einbeziehung der Verbundschweißung mit Stahl. Dissertation, TU Bergakademie, Freiberg
13. Brecht K, Pusch G, Liesenberg O (2002) Bruchmechanische Kennwerte von Temperguss—Teil 1: schwarzer Temperguss. Konstruieren + Giessen 27(2):21–30
14. Brecht K (2005) Bewertung des Risswiderstandsverhaltens von schwarzem Temperguss bei statischer, dynamischer und zyklischer Beanspruchung. Dissertation, TU Bergakademie, Freiberg
15. Pusch G, Ludwig A (2001–2005) Berechnung fließbruchmechanischer Kennwerte für duktile Gusseisenwerkstoffe mit unterschiedlicher Graphitausbildung aus Parametern der Mikrostruktur und der Fließkurve. Abschlussbericht DFG-Projekt Pu 104/701

16. Stroppe H, Pusch G, Ludwig A (2002) Bestimmung der Bruchzähigkeit von ferritischem Gusseisen mit Kugelgraphit aus Kennwerten des Zugversuchs und der Gefügeausbildung. Giessereiforschung 54(4):115–120
17. Ritchie RO, Server WL, Wullaert RA (1979) Critical fracture stress and fracture strain models for the prediction of lower and upper shelf toughness in nuclear pressure vessel steels. Metall Trans A 10(10):1557–1570
18. Biermann H, Pusch G, Ludwig A, Mottitschka T (2012) Werkstoffentwicklung für Windenergieanlagen im Multi-Megawatt-Bereich Offshore—Ermittlung bruchmechanischer Kenndaten, Teilbericht MEGAWind: Abschlussbericht SGK-GE 2012/06 zum F&E Vorhaben in der Zeit vom 01.08.2006 bis 31.10.2011. Technical report 0327593. Siempelkamp Giesserei GmbH, Krefeld
19. Wolfensberger S, Uggowitzer P, Speidel MO (1994) Die Bruchzähigkeit von Gusseisen. Teil II: Gusseisen mit Kugelgraphit. Giessereiforschung 39(2):71–80
20. Komatsu S, Shiota T, Matsuoka T, Nakamura K (1994) Effects of several main factors on ductile-brittle transition behaviors of fracture toughness in SG cast iron. AFS Trans 102:121–125
21. Continous Casting Section CAEF (2012) Prüfbericht: Ermittlung der Kennwerte des statischen J-Integral nach ISO 12135 an sechs unterschiedlichen Werkstoffen bei −20 °C sowie bei Raumtemperatur, 2012. Cited in DIN EN 16482:2014-09
22. Pusch G, Baer W (1995) Bruchmechanische Untersuchungen an duktilem Gusseisen GGG-40: Prüfbericht 6/95
23. Winkler H-P, Hüggenberg R, Ludwig A, Pusch G, Trubitz P (2009) Determination and definition of fracture toughness of dynamically loaded ductile cast iron. In: 12th international conference on fracture (ICF 12), Juli, Ottawa
24. Ludwig A, Pusch G, Trubitz P, Winkler HP, Hüggenberg R (2012) Ermittlung und Definition dynamischer Bruchzähigkeitswerte für ferritische Gusseisenwerkstoffe. Gießerei-Praxis 1 (2):10–13
25. Ludwig A, Pusch G, Trubitz P, Winkler HP, Hüggenberg R (2012) Gefügeabhängigkeit statischer und dynamischer Rissinitiierungswerte für ferritisches Gusseisen. Gießerei-Praxis 3:52–57
26. IWT TU Bergakademie Freiberg (2008) Bestimmung dynamischer Rissinitiierungswerte für duktiles Gusseisen (DCI). unveröffentlichter Forschungsbericht
27. BDG-Richtlinie P300 (2014) Bestimmung der dynamischen Bruchzähigkeit ferritischer Gusseisenwerkstoffe. Bundesverband der Deutschen Gießerei-Industrie
28. Baer W, Häcker R (2005) Werkstoffcharakterisierung von Gusseisenwerkstoffen mit Kugelgraphit. Materialprüfung 47(1–2):34–44
29. Trubitz P, Pusch G, Ludwig A, Winkler HP, Hüggenberg R (2012) Bestimmung der Rissauffangzähigkeit von duktilem Gusseisen. Gießerei-Praxis 7(8):294–298
30. Trubitz P, Winkler H-P, Hüggenberg R, Ludwig A, Pusch G (2013) Effect of the pearlite content of ferritic cast iron material on the crack resistance behaviour under dynamic load. In: 13th international conference on fracture (ICF 13), 16–21 Juni, Beijing
31. American Society for Testing and Materials (2006) Standard test method for determining plaine-strain crack-arrest fracture toughness, K^{Ia}, of ferritic steels. ASTM E 1221-06
32. Rosenfield AR, Ahmad J, Cialone HJ, Landow MP, Mincer PN, Papaspyropoulos V (1989) Crack arrest toughness of nodular iron. Nucl Eng Des 116(2):161–170
33. Gillot R (1988) Experimentelle und numerische Untersuchungen zum Rissstopp-Verhalten von Stählen und Gusseisenwerkstoffen. Techn.-wiss. Berichte MPA Stuttgart, 88(03):1
34. ASTM E647-08 Standard test method for measurement of fatigue crack growth rate, 2008
35. Komber B (1995) Bruchmechanische Bewertung des Rissausbreitungsverhaltens ferritischer Gusseisenwerkstoffe bei zyklischer Beanspruchung. Dissertation, TU Bergakademie, Freiberg
36. Pusch G, Komber B, Liesenberg O (1996) Bruchmechanische Kennwerte für ferritische duktile Gusseisenwerkstoffe bei zyklischer Beanspruchung. Konstruieren + Giessen 21(2):49–54

37. Pusch G, Hübner P, Liesenberg O (2002) Untersuchungen zum gefügeabhängigen Rissausbreitungsverhalten von Gusseisenwerkstoffen bei zyklischer Beanspruchung. DVM-Bericht 234 "Fortschritte der Bruch- und Schädigungsmechanik"
38. Tanaka K (1979) A correlation of Δkth-value with the exponent, m, in the equation of fatigue crack growth for various steels. Int J Fract 15(1):57–68
39. Deutsches Institut für Normung (2012) Gießereiwesen—Gusseisen mit Kugelgraphit. DIN EN 1563:2011
40. Lang K-H (1985) Das Ermüdungsverhalten von GGG-60, GGG-40, GGV-30 und GG-30 im Temperaturbereich 20°C < T < 500°C. Dissertation, Universität Karlsruhe
41. Schlosser H (2005) Bruchmechanische Bewertung des Gusseisenwerkstoffes GJS-400-18LT im Hinblick auf den Einsatz in Komponenten für Windkraftanlagen. PhD thesis, TU Bergakademie, Freiberg
42. Kuna M, Springmann M, Mädler K, Hübner P, Pusch G (2002) Anwendung bruchmechanischer Bewertungskonzepte bei der Entwicklung von Eisenbahnrädern aus bainitischem Gusseisen. Konstruieren + Giessen 27(4):27–32
43. Johnson LB (2000) Space Centre NASA. Fatigue crack growth computer program "NASGRO": JSC-22267B
44. Hübner P, Pusch G, Zerbst U (2004) Ableitung von Quantilrisswachstumskurven für Restlebensdauerberechnungen. DVM-Bericht 236 "Fortschritte der Bruch- und Schädigungsmechanik", S.121–129
45. Hübner P, Pusch G (2007) Zyklisches Risswachstumsverhalten von Gusseisenwerkstoffen—Analytische Aufbereitung für die Nutzung des Berechnungsprogramms ESACRACK. Konstruieren + Giessen 32(3):34–37
46. Henkel S, Pusch G, Hübner P (2008) Zyklisches Risswachstumsverhalten von Gusseisenwerkstoffen—Analytische und statistische Aufbereitung für die Nutzung des Berechnungsprogramms ESACRACK. Konstruieren + Giessen 41–43
47. AFGROW, Version 4.11.14.0. Stand: 01.07.2008, 2006
48. Biermann H, Kuna M, Pusch G, Mottitschka T, Zybell L (2014) DFG-Projekt—Experimentelle und theoretische Untersuchungen der Schädigungsmechanismen bei der Rissausbreitung in ferritischen Gusseisenwerkstoffen unter konstanten und variablen Lastamplituden: Geschäftszeichen: BI418/16-2 und KU929/13-2. Abschlussbericht
49. Mottitschka T, Pusch G, Biermann H, Zybell L, Kuna M, Hübner P (2010) Ermüdungsrisswachstum in Gusseisen mit Kugelgraphit unter konstanten und variablen Lastamplituden—Teil 1: Experimentelle Untersuchungen. DVM-Bericht 242:195–205
50. Zybell L, Kuna M, Chaves H, Mottitschka T, Pusch G, Biermann H (2010) Ermüdungsrisswachstum in Gusseisen mit Kugelgraphit unter konstanten und variablen Lastamplituden—Teil 2: Numerische Untersuchungen. DVM-Bericht 242:207–216
51. Mottitschka T, Pusch G, Biermann H, Zybell L, Kuna M (2012) Influence of graphite spherical size on fatigue behaviour and fracture toughness of ductile cast iron EN-GJS-400-18LT. Int J Mater Res 103(1):87–96
52. Zybell L, Chaves H, Kuna M, Mottitschka T, Pusch G, Biermann H (2012) Optical in situ investigations of overload effects during fatigue crack growth in nodular cast iron. Eng Fract Mech 95:45–56
53. Mottitschka T, Pusch G, Biermann H, Hübner P (2013) Untersuchungen zum zyklischen Risswachstum in ferritischen Gusseisen mit Kugelgraphit unter konstanten und variablen Lastamplituden. Gießerei-Praxis 10:412–417
54. Hübner P, Schlosser H, Pusch G, Biermann H (2007) Rissausbreitung in Gusseisenwerkstoffen unter variablen Lastamplituden. DVM-Bericht 239
55. ten Have AA (1992) WISPER and WISPERX: Final definition of two standardised fatigue loading sequences for wind turbine blades. Nationaal Lucht- en Ruimtevaartlaboratorium

On the Development of Experimental Methods for the Determination of Fracture Mechanical Parameters of Ceramics

Robert Danzer, Tanja Lube and Stefan Rasche

Abstract Because of their high yield strength and hardness as well as their brittle fracture characteristics the behavior of cracks in ceramics can be described within the framework of linear elastic fracture mechanics. For fracture toughness (K_{Ic}) measurements the test techniques which were developed for metallic materials are unfavorable, as an economical preparation is impossible in the case of ceramic materials. Therefore simple geometries e.g. bend bars became a preferred specimen shape for K_{Ic} measurements. A major difficulty arises when sharp and well-defined pre-cracks for crack propagation studies have to be created. Several methods to overcome this problem are introduced. Additionally, methods to investigate small amounts of material are discussed.

1 Introduction

With the increasing significance and appreciation of fracture mechanics for the reliability analysis of metal structures in the sixties and seventies of the last century also some interest in a fracture mechanical characterization of ceramics emerged. Ceramics are very brittle materials. Their yield stress is—in general—about one order of magnitude or more higher than their tensile fracture strength [1–3]. Therefore the plastic zone at the crack tip is always very small. Its diameter is in the order of just several micrometers or even less [3]. Therefore linear elastic fracture mechanics are well suited to describe the behavior of ceramic materials [3, 4]. Then the appropriate parameter to describe the propagation of cracks in ceramics is the fracture toughness, K_{Ic}.

R. Danzer · T. Lube (✉) · S. Rasche
Institut für Struktur- und Funktionskeramik, Montanuniversität Leoben, 8700 Leoben, Austria
e-mail: tanja.lube@unileoben.ac.at

R. Danzer
e-mail: isfk@unileoben.ac.at

G. Hütter and L. Zybell, *Recent Trends in Fracture and Damage Mechanics*, DOI 10.1007/978-3-319-21467-2_8

This paper starts with a short summary of the theoretical background, i.e. linear elastic fracture mechanics and toughening mechanisms of ceramics. Then the development of experimental methods for the determination of the fracture toughness of ceramics is shortly presented and the state of the art of fracture toughness measurement techniques for ceramics is described.

2 Theoretical Background

Early works on fracture mechanics of ceramics were made by Griffith [5]. For an ideal elastic solid a balance is made between the energy losses and the energy gains for a propagating crack. The losses are determined by the changes in the elastic energy and the work done by the crack propagation. The gains correspond to the energy necessary to create the new fracture surfaces, δA. Griffith mentioned that this energy is $2\gamma \cdot \delta A$, where γ is the surface energy of the solid. The factor two arises since a crack always creates two new surfaces.

Griffith's model works well for glasses, where indeed a large part of the energy necessary to create new fracture surfaces is surface energy. But for other materials (metals, polymers, ceramics) other energy dissipating processes contribute to the energy necessary to create the new fracture surfaces. This energy may become much (up to several orders of magnitude) higher than the surface energy [3].

More than 30 years later this fact has been accounted by Irwin, who proposed the concept of linear elastic fracture mechanics [6]. In his theory all involved energy changes are strictly proportional to δA and—to give "energy rates"—they can be divided by the crack surface area. The strain energy release rate, G, describes all energy losses (elastic energies and work) and the critical strain energy release rate, G_c, is equal to the energy necessary to create the new fracture surface (both divided by δA) [4, 7]. Often it is assumed that the critical strain energy release rate is a constant but in reality it may depend on the previously cracked area and the crack geometry. Then the material has a rising crack resistance curve (R-curve).

For a material without R-curve, unstable fracture occurs if the energy losses exceed the gain in fracture energy [8]:

$$G \geq G_c \tag{1}$$

In ceramics this fracture is brittle. It can be shown, that for plane stress conditions and for mode-I loading it holds [3, 7]:

$$G = K^2/E \tag{2}$$

and

$$G_c = K_{Ic}^2/E \tag{3}$$

with

$$K = \sigma Y \sqrt{\pi a}. \tag{4}$$

For plane-strain conditions and mode I loading, which are most relevant when testing ceramics, E has to be replaced by $E' = E/(1 - \nu^2)$ [8].

The symbol E refers to the Young's modulus, ν is Poisson's ratio and σ is the stress component normal to the prospective crack plane in the un-cracked body. K is the stress intensity factor, K_{Ic} is the fracture toughness and Y is a geometric factor, which takes the geometry of the crack, specimen and stress field into account. For typical loading situations such factors and their dependence on the geometric details of crack and specimen can be found in the literature [9, 10]. With this notation the well-known Griffith/Irwin criterion for brittle fracture can be defined [3, 7, 11]:

$$K \geq K_{Ic}. \tag{5}$$

Inserting Eq. (4) in Eq. (5) gives an equation for the tensile strength, σ_f, of ceramic materials:

$$\sigma_f = \frac{K_{Ic}}{Y\sqrt{\pi a_c}}. \tag{6}$$

The symbol a_c designates the length of the (critical) crack.

2.1 Strength of Ceramics

Using Eq. (6), many typical features of the fracture behavior of ceramics can be explained.

The strength of ceramic specimens is directly proportional to the fracture toughness, which describes the resistance of the material against brittle failure. As already shown by Griffith [5] and Irwin [6], the strength of ceramic specimens also depends on the inverse square root of the size of the fracture origin (critical crack): $\sigma_f \propto a_c^{-1/2}$. In smooth strength specimens many (small) cracks exist. Than the crack having the largest stress intensity factor will become fracture origin (i.e. it is the critical crack; in the case of a homogeneous uniaxial stress field and for cracks oriented perpendicular to the stress direction this is will be the largest crack in the specimen.) Since in individual specimens the value of a_c is different the strength of specimens scatters [12–14]. The strength distribution of brittle materials can be adequately described by a Weibull distribution. (This statement is exactly true if the size distribution of the flaws is homogeneous and follows a power law. Then the material is a Weibull material [15]). Since the probability of finding a large flaw in a large specimen is higher than the probability of finding a large flaw in a small specimen the strength of ceramic specimens is size dependent and the mean strength of large specimens is smaller than the mean strength of small specimens [16, 17].

The distribution of flaw sizes depends to a large extent on the processing history of the specimens [17]. Therefore the strength is not a material parameter. The strength is influenced by very specific details of the processing of the material (e.g. a small amount of coarse grains or pores may cause a low strength) and of the specimens (machining flaws [18]). Therefore the authors believe that the most significant fracture mechanical material properties of ceramic materials are the fracture toughness (the critical mode I stress intensity factor) and the critical energy release rate, which is related to the fracture toughness via Eq. (3).

2.2 *Toughness of Ceramics*

Let us come back to the fact that the critical strain energy release rate, G_c in ceramics is much higher than twice the surface energy, 2γ. This is caused by so called toughening mechanisms which can be divided into two classes: process zone and bridging mechanisms [19, 20].

It is known from Irwin's work, that in an ideal elastic body, there is a stress singularity at the crack tip, i.e. the components of the stress tensor go to infinity as the distance, r, of the analyzed volume element approaches zero: $\sigma \propto r^{-1/2}$ [6, 8]. In reality ideal elastic bodies do not exist and there is some yielding if the stresses are high enough. In metals or polymers a well-known reason for yielding is plastic deformation. This deformation only happens in a zone ahead of the crack tip, where the stresses are high enough (plastic zone). If a crack propagates the plastic zone moves with the crack tip. In the wake of the crack plastically deformed crack borders remain. The energy to deform the borders is in general much higher (up to several orders of magnitude higher) than the surface energy. This causes the high fracture toughness of many metallic materials [2]. In the case of ceramics some plastic deformation also occurs but the yield stress is very high. Therefore, the plastic zone is confined to a region very close to the crack tip and the contribution of the plastic deformation to the fracture energy is small [3]. But there exist other mechanisms causing nonlinear deformations in ceramics. Examples are phase transformations, domain switching or micro crack formation. Similar to plastic deformations they only occur at sufficiently high stresses. Therefore, they are restricted to a region ahead of the crack tip, which is called process zone. The energy to deform the wake of a propagating crack can be much larger (e.g. several thousand times larger in the case of zirconia ceramics) than the surface energy. Therefore the process zone mechanisms can cause significant contributions to the fracture energy of ceramic materials [19, 20].

The second group of mechanisms are crack bridging mechanisms caused by processes behind the crack tip. Crack bridges transmit forces across the crack faces, like nails, that fix together two boards and that have to be pulled out dissipating friction energy when the borders of the boards are pulled apart. Crack bridges may dissipate a huge amount of energy by friction or by the deformation of the bridges. Bridging is often accompanied by a crack path strongly influenced by the microstructure, e.g. if the crack follows a grain boundary phase (causing "elastic" bridges [21]), if it is deflected by particles [22, 23], or if it plastically deforms a second ductile phase (as in the case of cemented carbides [24]). The associated energy may also be much higher than the surface energy and bridging can therefore cause a significant increase in toughness.

It is obvious that some crack advance is necessary until process zones or crack bridges fully develop. Therefore the fracture energy rises at the beginning of crack propagation (R-curve behavior). To give an example the maximum of the critical stain energy rate is typically reached after a crack advance of 2–3 diameters of the process zone [19, 20]. In the case of crack bridging the R-curve can be very steep and reaches the saturation after a just a few μm of crack advance, as for example in in-situ reinforced silicon nitrides [25].

2.3 *Principle of Fracture Toughness Measurements*

Measurements of the fracture toughness or of the critical energy release rate can be done by experiments, where a specimen containing a crack with well-defined shape and size is mode-I loaded with a well-defined stress field until it becomes critically. At this moment the stress intensity factor equals the fracture toughness [see Eq. (5)]. Using Eq. (4) it holds:

$$\sigma_c Y \sqrt{\pi a_c} = K_{Ic}. \tag{7}$$

The parameter σ_c describes the stress field and its amplitude and a_c is the size of the crack. The geometric factor Y is, in general, crack length dependent. It has to be determined for each specimen and loading geometry separately [8, 11]. For a given experimental setup (specimen type), this equation can be translated into an equation for the fracture toughness in dependence of the fracture force, specimen size and crack length. Examples can be found in standards for fracture toughness testing. If Young's modulus, E, is known, the critical strain energy release rate, G_c, can be determined using Eq. (3).

Note that the stress field can have contributions caused from the externally applied load and from residual stresses which may come into existence during material processing and specimen preparation. Therefore—to make a proper analysis—residual stresses in the specimens have to be avoided or have to be considered in the data evaluation.

3 Development of Fracture Mechanics Measurements

Very soon after the first publications on the linear elastic fracture mechanics, important applications for metal structures could be found. The *leak before break*-criterion (under such conditions cracks growing in the wall of a pressure vessel only cause some leakage instead of an blow up of the vessel) was still used for the design of pressure vessels of nuclear power plants in the seventies of the last century [2].

3.1 Testing of Metals: ASTM E399

Thus it is not surprising that the measurement procedure for the fracture toughness of metals has been standardized very soon. The ASTM E399 "*Standard Test Method for Plane-Strain Fracture Toughness of Metallic Materials*" was first published in 1970. It has been revised and approved several times (the last version is from 2012, [26]). Other standards on fracture toughness testing of metals are more or less based on this ASTM E399.

ASTM E399 describes "*the determination of the plane-strain fracture toughness of metallic materials by tests using a variety of fatigue-cracked specimens ...*". It "*involves testing of notched specimens that have been pre-cracked in fatigue by loading either in tension or three-point loading*" [26]. Proposed are several specimen geometries which can easily be machined out of plates (e.g. rectangular beams or compact specimen), roads (e.g., disk-shaped compact specimen) or tubes (e.g. arc-shaped bend specimens). A very serious aspect is the introduction of the "starter crack". It has been recognized that "… *it is impractical to obtain a reproducibly sharp, narrow machined notch that will simulate a natural crack* …" [26]. It is proposed to machine a narrow notch (straight through or a Chevron notch), which is then sharpened by subsequent fatigue loading to form a short straight through fatigue crack. The crack is subsequently loaded in mode I until it starts to extend (i.e. it becomes critical; this can be recognized by a non-linearity in the force-displacement curve) and the critical force is recorded. The crack length can be measured before testing at the sides of the specimen or after testing at the fracture surface. From the critical force and geometric data describing crack length and specimen size, the fracture toughness can be determined [26].

As mentioned earlier the linear elastic fracture mechanics only applies in the case of small scale yielding and the standards demand to validate this condition. Roughly speaking this happens, if the size of the plastic zone is very small compared to crack length and the specimen dimensions. This condition may result in the requirement for a minimum specimen size for fracture toughness testing.

3.2 Early Techniques Used for Ceramics

At the beginning of scientific interest in fracture mechanics of brittle materials, especially of single crystals [27], many experiments were made on double cantilever beam specimens, which are tested in tension and which are relatively easy to machine [28]. Little later it was observed that crack growth is also possible at stress intensity factors lower than the fracture toughness (sub-critical crack growth, SCCG) [29, 30]. It was realized that the SCCG-rate depends on the stress intensity factor. Therefore specimen types and loading conditions were developed, for which the stress intensity factor is independent of the crack length, so that the experiment can easily be controlled by the applied force. This is the case for the tapered double cantilever beam, the constant moment specimen and the double torsion specimens [28]. These specimens are, on one hand, relatively large and their machining is expensive, and on the other hand some of them require complicated gripping systems for load applications. Therefore they are only used in rare cases, e.g. for some SCCG measurements.

4 Fracture Toughness Testing of Ceramics

Large efforts were made in the seventies and eighties of the last century to develop robust and easy to use fracture toughness testing procedures for ceramics that deliver reproducible results. There are some special aspects which have to be considered:

(a) Ceramics are very hard and not easy to machine. Complex shapes can hardly be manufactured or it causes huge costs. Possible (but still expensive) are cuts (abrasive cutting, disc cutting), the preparation of prismatic bars (grinding, polishing, etc.) and cylinders (cylindrical grinding). Specimens with shapes needing other machining procedures (e.g. boring, drilling) are not feasible with a standard laboratory or machine shop equipment.
(b) Ceramics are very susceptible to fatigue and pre-cracks are difficult to make by fatigue. Therefore other ideas to make “sharp” and well defined cracks have to be developed.
(c) The process zone in ceramics is very small (several 10 μm or smaller) and the conditions for small scale yielding are in general fulfilled. Larger process zones may occur in some transformation toughened zirconias [31].
(d) Very often only small amounts of material are accessible for testing. Typical sintered specimens of new materials have dimensions of several (ten) millimeters and many ceramic components are even smaller. There is a great need for testing methods for small specimens.

4.1 Tests on Bend Bars

Considering the specific aspects for the fabrication of ceramic specimens (points (a) and (d), above), bend bars soon became the favorite specimen geometry. Such specimens are the easiest to make (e.g. by surface grinding), and are most efficient in the use of material. The same geometry as used for standard strength measurements can be used. For load application three-point bending as well as four-point bending is used. Standard dimensions are 3 mm × 4 mm × 45 mm, but small specimens having cross sections of about 1 mm × 1 mm and a length of a little more than 10 mm can be tested using conventional equipment.

4.1.1 Pre-cracked Bend Bars

For ceramics it is difficult to introduce suitable pre-cracks by fatigue (point (b), above). Alternative methods were thus developed. A sharp and well defined pre-crack can be generated by letting an existing crack grow stably, until it has the expected size and geometry, and then stop it.

A problem with these methods is that the elastic energy released with crack extension can become very high, especially if the crack is loaded in pure tension and if the testing equipment is not very stiff. Then the elastic energy released in the testing machine may be much larger than the elastic energy released in the specimen. On the other hand the critical energy release rate G_c of cracks in ceramics is in general very small. To stop a crack in a typical tensile test, the loading has to be released immediately after the beginning of the crack propagation. Otherwise the specimen would fail catastrophically. But due to the inertia of the system such quick reactions are not possible with normal testing equipment. This explains the efforts to develop extremely stiff testing machines and jigs.

One way to overcome this difficulty in pre-crack generation is the "*bridging method*". A bend bar with a small notch or line of hardness indentations at mid-span is supported on a stiff anvil. The pre-crack is generated by localized loading of the notch, see Fig. 1a. The formation of the pre-crack usually occurs by unstable crack growth (pop-in) and has to be detected by acoustic emission [32–35]. Although this idea is simple the practical realization is tricky and requires sophisticated equipment [36]. In the ESIS round robin it turned out that only very few laboratories managed to produce valid cracks using this method. The conclusion was that the bridging method is not suitable for routine laboratory testing [37].

For the evaluation of K_{Ic} from the fracture force according to Eq. (7) the pre-crack length has to be measured on the fracture surface. This requires a certain amount of experience and may not be straightforward. The above described single edge pre-cracked beam (SEPB) method is standardized in ISO 15732 [38] and ASTM C1421 [39].

Alternatively, the bend bar can be notched to leave a chevron shaped area un-cut, cp. Figure 1b. When the tip of the chevron is loaded, a crack starts to grow. Due to the shape of the ligament, there is first a quadratic increase of the cracked area with

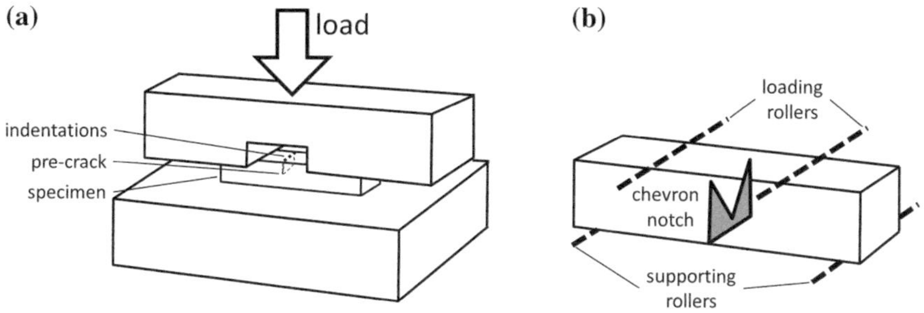

Fig. 1 Bend specimens with pre-cracks: **a** Schematic of a device for pre-cracking using the "bridging method". **b** Chevron notched bend bar

crack extension. This causes a decrease of the stress intensity factor and favors stable crack growth, which precedes subsequent instable fracture. Stable crack growth, which is mandatory for a valid test, has to be monitored on the load-displacement curve to allow for the determination of the maximum force. Together with the dimensions of the test-piece and the chevron, K_{Ic} can be determined [40]. Theoretical considerations [41] and experimental work showed that it is not easy to obtain stable crack growth [37]. In the standards EN 14425-3 [42, 43] and ASTM C1421 [39] hints for notch machining and required stiffness of the test equipment are given as to ensure valid tests.

4.1.2 Notched Bars

As an alternative to introducing sharp pre-cracks into specimens it has been tried to replace the sharp crack by a (thin) notch. Machining of the notch can be accomplished by thin grinding wheels. Best practice examples for such thin notches had a width of about 50 μm [40] but in many experiments a notch width of about 200 μm was typically. Since the stresses at the sharp crack tip ($r \rightarrow 0$) go to infinity in the ideal linear-elastic case (remember, there is a $1/\sqrt{r}$ singularity), they are always higher than at the tip of a blunt notch, were the stresses approximately follow the law: $\sigma \propto 1 + 2\sqrt{l/\rho}$, with l being the length and ρ being the half width of the notch [44]. Therefore, if notches are used instead of "sharp" cracks, the fracture toughness is always overestimated.

Systematic studies on the notch-root radius effect were made about 25 years ago [40, 45]. Experiments on bending specimens having a thin notch (single edge notched beam specimens) performed on several different classes of structural ceramics have shown that with decreasing notch width the "apparent" fracture toughness also decreases. For all investigated materials there seems to exist a lower boundary for the fracture toughness [46, 47].

Based on these observations a model was formulated that in the front of the notch a small crack is formed due to weak spots in the microstructure. If the depth of this crack is taken as a fitting parameter, the values of the "apparent" fracture toughness could be explained, if the crack depth is in the order of the mean grain size of the investigated materials [47, 48].

Based on an idea of Nishida [46] efforts were undertaken to find a procedure for making very narrow notches [49]. The basic idea is that a notch can be sharpened at its tip by a honing procedure with a razor blade that is covered with diamond paste. With this procedure the notch root radius can be reduced to a few micrometers [47, 50]. With such V-notches fracture toughness can be determined correctly (more precisely: is only slightly overestimated).

In a European round robin organized by ESIS TC 6, it could be shown that this testing procedure gives consistent results if performed by different laboratories [51]. Based on these results the single edge V-notched beam (SEVNB) test was standardized: EN 14425-5 [52] and ISO 23146 [53]. For ceramics having a grain size of several micrometers or larger, it is an easy to perform and relatively cheap procedure to determine the mode I fracture toughness.

4.1.3 Indentation Cracks in Bending

In the case of hardness testing of ceramics, e.g. by Vickers or Knoop indentation tests, an unwanted artefact is the occurrence of cracks, which emanate from the corners of the indent. These surface cracks have nearly semi-elliptical shape and a size of a few hundred micrometers. Since the indent causes non-elastic deformation of the material residual stresses are present in the vicinity of the crack. The cracks are several times deeper than the "plastic" deformation zone under the indent, so that this zone can be machined off by surface grinding and the residual stresses can be removed [54, 55]. With the same procedure sub-surface "lateral" cracks [56] that interfere with the semi-elliptical surface crack also have to be removed [57], see Fig. 2a.

Indentation cracks with ground-off plastic zone are ideal pre-cracks for fracture toughness testing. The surface-crack-in-flexure (SCF) method [58] uses this technique for generation of a pre-crack on the tensile stresses surface of a standard bend bar. Fracture toughness testing can be made by measuring the critical force for crack extension and determination of the crack size and shape on the fracture surface. The geometric factor of semi-elliptic surface cracks has been published by Newman and Raju in the seventies of the last century [59]. But the making of the indent also produces some plastic deformation under the indent, which causes residual stress. Recently a more precise solution for sections of semi elliptic surface cracks has been published by some of the authors [60]. The method is standardized in ASTM C1421 [39] and ISO 18756 [61].

An older method (indentation strength method, IS) [62] refrains from grinding off the deformed zone of the indentation placed on the bend bar prior to fracture. The effect of the residual stresses is incorporated into the theoretical analysis via an empirically calibrated factor which relates the indentation load (as a measure for the crack size) and the fracture force with the fracture toughness. However, since the

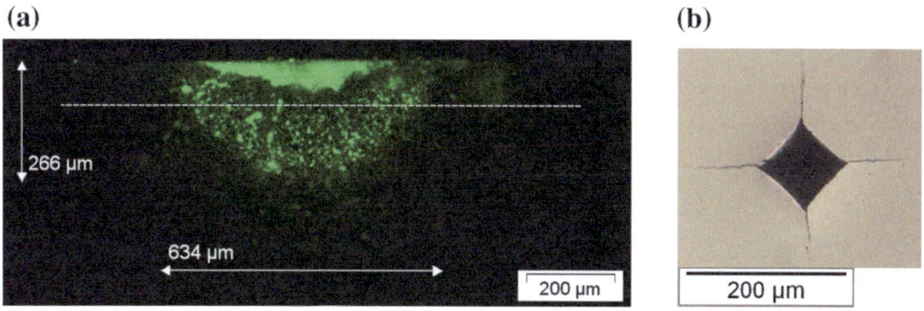

Fig. 2 Indentation cracks: **a** Knoop indentation crack on a fracture surface made visible using fluorescent dye. The thickness of the surface layer that has to be removed is indicated by a *dashed line*. **b** Vickers hardness indentation with cracks emanating from the corners

residual stress field is force and material dependent and crack shapes show a variability too, results can be dependent on indentation force [63].

4.2 Indentation Cracks—Direct Measurements

Hardness indentations can be made into very small pieces of material needing nearly no machining (apart from providing a polished section). The problems mentioned at the beginning of this chapter in aspects (a) and (d) would be solved. This observation has encouraged researchers to use these cracks for fracture toughness testing. Large efforts were made to relate the length of these cracks, which can be determined at the specimen's surface, to a value for the fracture toughness (indentation fracture method, IF) [64–66].

The driving force for the formation of indentation cracks is the multi-axial residual stress field caused by the deformation of the material by the hardness imprint. But these stresses not only depend on the yield strength (hardness) and Young's modulus of the material, they also depend on the shape of the deformed zone which depends on the microstructure and differs from material to material. Additional complications arise from a variability of the crack shape from "Palmqvist cracks" [67] to half-penny cracks and the occurrence of additional sub-surface "lateral" cracks. These facts make it impossible to set up an exact theoretical basis for the dependence of crack length on indentation force. In general an experimentally calibrated factor appears which is either only valid for a certain material or is far from precise.

Even though testing is very simple and data evaluation (measuring the length of a crack at the polished surface) can also be done easily, there is now some common opinion in the community that such measurements do not give the material property "fracture toughness" [68, 69]. IF tests are still popular if only small amounts of material are available. Based on extensive experimental work [70, 71] IF was recently incorporated into a standard ASTM F2094 for the material specification of ball bearing silicon nitrides [72]. There, the quantity determined with this method (only valid for silicon nitrides) is called "*indentation fracture resistance—IFR*" in order to be distinguished clearly form the plane-strain fracture toughness K_{Ic}.

4.3 Comparison of Methods

As pointed out by Morrell [69] methods for determination of fracture toughness that are acceptable with respect to fracture mechanical criteria (well defined crack geometries and stress field, exact calibration) and practical aspects (simple specimen and pre-crack fabrication, reliable and reproducible tests) are the SEPB, CNB, SEVNB and SCF method. They deliver equivalent results for materials having a flat R-curve. Since these methods use pre-cracks with a different length, crack opening and "crack-growth history", they may give different values if a R-curve material is tested [11].

Methods using indentation cracks with the residual stresses still present (IF, IS) are still used widely in academy and industry. Such values are highly unreliable. If at all, they should only be used as a relative measure within a given class of materials.

5 Recent Developments

For the quality control of ceramic components, fracture toughness testing techniques are required which can be applied directly to components, to (very small) specimens cut out of components or to typical specimens used in material development (small discs). Since the microstructure of ceramics sensitively depends on processing conditions, property gradients and inhomogeneity in components occur. Therefore it would be beneficial if the fracture toughness of the critical areas could be tested.

Examples are ceramic rolling elements for bearings. In the last decade the sales figures for ceramic and hybrid bearings significantly increased. Bearings with ceramic balls are used for example in offshore wind mills. They permit long service intervals combined with a high resistance to wear. The insolation of the electrically conducting parts of the mill is realized via the ceramic bearing elements. The highest loaded (critical) area of such components is the surface. Due to the sintering process bearing balls may have property gradients perpendicular to the surface. Therefore tests are required which load the surface near regions of the balls. The notched ball test (NBT) measures the tensile surface strength of the balls. A deep notch (deeper than the ball radius) is cut into the equatorial plane of the ball, which is than squeezed together at its poles. This causes an almost uniaxial stress field in the surface region of the ball opposite of the root of the notch [73, 74]. For the determination of the fracture toughness of the surface-near regions of the balls, this stress field is used to load a pre-crack, which is made by a Knoop indent, analogous to the SCF-method. Of course the plastic deformation zone under the indent has to be ground off. The theoretical details (e.g. the determination of the geometric function) and experimental details are published in [75, 76]. With this technique fracture toughness testing can be made on balls as small as a few millimeters in diameter with an accuracy of a few per cent. This corresponds to the typical scatter of fracture toughness data.

A second example is a fracture toughness test for small discs and plates. About ten years ago a new test to determine the biaxial tensile strength of thin discs and plates was developed. The specimen is positioned on three large balls and then loaded with a forth ball in the middle of the plane of the specimen opposite to the supporting balls. This causes a biaxial stress field in the center of the supported specimen surface, which is used for strength testing of the specimen. The ball on three balls (B3B) test is very precise (much more accurate than other biaxial tests or the bending test) [77–79]. This is beneficial for testing small specimens. So, very small specimens can be tested (plates with an area of about 1 mm^2 and a thickness of about 100 μm, the specimen volume can be as small as a fraction of a cubic millimeter) [80, 81].

The fracture toughness of thin discs and plates can be determined, if a modified SCF-method is applied: pre-cracks are introduced e.g. using Knoop indents, into the mid region of the supported plane of the specimen and these cracks are loaded in the B3B test until the crack becomes critical. Of course the layer containing the plastic deformed zone has to be removed, e.g. by grinding, before loading the pre-crack [82, 83].

In summary, recent trends are related to the development of fracture toughness testing procedures for specimens with special shapes and for very small dimensions.

6 Summary

The measurement of fracture toughness in ceramics confronts not only the experimenter with specific challenges, often concerning practical aspects. The designer using fracture mechanics parameters also has to be aware of specific features of the various methods in order to interpret reported fracture toughness values in a meaningful way.

- In ceramics, the crack-tip yielding zone is always small and linear elastic fracture mechanics applies. There is generally no need to choose a "correct" specimen size in order to perform valid measurements.
- Since the properties of ceramic components are defined by the processing history it is favourable to machine specimens out of components. If this is not possible, test methods have to be adapted to be applied to the components directly.

- Machining of specimens is difficult, expensive and limited to simple geometries. Pre-crack introduction is a major challenge, as is the post-mortem measurement of the pre-crack length on the fractured specimen.
- Several methods (SCF, CNB, SEPB) require a considerable amount of experience in order to be performed in a valid way. This may lead to inconsistent results between different laboratories, even if the same material is investigated.
- The standardized methods employ pre-cracks that have undergone different amounts of crack-growth. For materials exhibiting a rising crack resistance curve different methods may give different results. In some cases (CNB) it is not even possible to determine the exact amount of crack growth before instable fracture and thus a location of the measurement result on the R-curve is impossible. For a correct interpretation of a given fracture toughness value, the test method has to be known.
- Standardization is still in progress, including a part on test method selection, EN 14425-1 [84]. Even the standardized methods do not work on all materials, so that users have to be aware of limitations.

References

1. Ashby MF, Jones DRH (1986) Engineering materials 2. Pergamon Press, Oxford
2. Ashby MF (2010) Materials selection in mechanical design. Butterworth-Heinemann, Oxford
3. Danzer R, Lube T, Morrell R, Supancic P (2013) Mechanical properties of ceramics. In: Somiya S (ed) Handbook of advanced ceramics, 2nd edn. Elsevier, Amsterdam, pp 609–632
4. Danzer R (2007) Fracture mechanics of ceramics—a short introduction. Key Eng Mat 333:77–86. doi:10.4028/www.scientific.net/KEM.333.77
5. Griffith AA (1920) The phenomenon of rupture and flow in solids. Philos Trans R Soc Lond A221:163–198
6. Irwin GR (1958) Fracture. In: Flügge S (ed) Handbuch der Physik, vol 6. Springer, Berlin, pp 551–589
7. Ashby MF, Jones DRH (1980) Engineering materials 1. International series on materials science and technology. Pergamon Press, Oxford
8. Gross D, Seelig T (2006) Fracture mechanics. Mechanical engineering series. Springer, Berlin
9. Tada H, Paris P, Irwin GR (1985) The stress analysis handbook. Del Research Corporation, St. Louis
10. Murakami Y (1986) The stress intensity factor handbook. Pergamon Press, New York
11. Munz D, Fett T (1999) Ceramics. Springer series in materials science, vol 36. Springer, Berlin
12. Danzer R (1992) A general strength distribution function for brittle materials. J Eur Ceram Soc 10:461–472
13. Jayatilaka AdS, Trustrum K (1977) Statistical approach to brittle fracture. J Mater Sci 12:1426–1430
14. Evans AG (1982) Structural reliability: a process-dependent phenomenon. J Am Ceram Soc 65 (3):127–137
15. Danzer R, Lube T, Supancic P, Damani R (2008) Fracture of ceramics. Adv Eng Mater 10 (4):275–298. doi:10.1002/adem.200700347

16. Weibull W (1939) A statistical theory of the strength of materials, vol 151. Ingeniörsvetenskapsakademiens Handlingar 151. Generalstabens Litografiska Anstalts Förlag, Stockholm
17. Danzer R (2014) On the relationship between ceramic strength and the requirements for mechanical design. J Eur Ceram Soc 34:3435–3460. doi:10.1016/j.jeurceramsoc.2014.04.026
18. Pfeiffer W, Hollstein T (1997) Influrence of grinding parameters on strength-dominating near-surface characteristics of silicon nitride ceramics. J Eur Ceram Soc 17:487–494
19. Becher PF (1991) Microstructural design of toughened ceramics. J Am Ceram Soc 74(2): 222–269
20. Evans AG (1990) Perspective on the development of high-toughness ceramics. J Am Ceram Soc 73(2):187–206
21. Swanson PL, Fairbanks CJ, Lawn BR, Mai Y-M, Hockey BJ (1987) Crack-interface grain bridging as a fracture resistance mechanism in ceramics: I. Experimental study on alumina. J Am Ceram Soc 70:279–289
22. Faber KT, Evans AG (1983) Crack deflection processes—II: experiments. Acta Metall 31 (4):577–584
23. Faber KT, Evans AG (1983) Crack deflection processes—I: theory. Acta Metall 31(4): 565–576
24. Wachtman JB (1996) Mechanical properties of ceramics. Wiley-Interscience, New York
25. Fünfschilling S, Fett T, Hoffmann MJ, Oberacker R, Schwind T, Wippler J, Böhlke T, Özcoban H, Schneider GA, Becher PF, Kruzic JJ (2011) Mechanisms of toughening in silicon nitrides: the roles of crack bridging and microstructure. Acta Mater 59:3978–3989
26. ASTM E399 (2005) Standard test method for linear-elastic plane-strain fracture toughness K_{Ic} of metallic materials
27. Gilman IJ (1960) Direct measurements of surface energies of crystals. J Appl Phys 31 (12):2208–2218
28. Evans AG (1974) Fracture mechanics determination. In: Bradt RC, Hasselman DPH, Lange FF (eds) Concepts, flaws and fractography, vol 1. Fracture mechanics of ceramics. Plenum, New York, pp 17–47
29. Wiederhorn SM (1967) Influence of water vapour on crack propagation in soda-lime glass. J Am Ceram Soc 50:407–414
30. Danzer R (1994) Sub-critical crack growth in ceramics. In: Cahn RW, Brook R (eds) Encyclopedia of advanced materials, vol 4. Pergamon Press, Oxford, pp 2693–2698
31. Hannink RHJ, Kelly PM, Muddle BC (2000) Transformation toughening in zirconia-containing ceramics. J Am Ceram Soc 83(3):461–487
32. Warren R, Johannesson B (1984) Creation of stable cracks in hardmetals using 'bridge' indentation. Powder Metall 27(1):25–29
33. Nose T, Fujii T (1988) Evaluation of fracture toughness for ceramic materials by a single-edge-precracked-beam method. J Am Ceram Soc 71(5):328–333
34. Ray AK (1998) A new technique for precracking ceramic specimens in fatigue and fracture. J Eur Ceram Soc 18:1655–1662
35. Fett T, Munz D, Thun G (2001) A toughness test device with opposite roller loading. Eng Fract Mech 68(1):29–38
36. Morrell R, Parfitt M (2005) A stiff facility for controlled pre-cracking in fracture toughness tests. Measurement note DEPC (MN) 034. NPL, Teddington
37. Primas RJ, Gstrein R (1997) ESIS TC6 round robin on fracture toughness. Fatigue Fract Eng Mater Struct 20(4):513–532. doi:10.1111/j.1460-2695.1997.tb00284.x
38. ISO 15732 (2003) Fine ceramics (advanced ceramics, advanced technical ceramics)—test method for fracture toughness of monolithic ceramics at room temperature by single edge precracked beam (SEPB) method
39. ASTM C1421 (2010) Standard test methods for determination of fracture toughness of advanced ceramics at ambient temperature

40. Munz D, Bubsey RT, Shannon JI Jr (1980) Fracture toughness determination of Al_2O_3 using four-point-bend specimens with straight-through and chevron-notches. J Am Ceram Soc 63(5–6):300–305
41. Sigl LS (1991) On the stability of cracks in flexure specimens. Int J Fract 51(3):241–254
42. EN 14425-3 (2010) Advanced technical ceramics—test methods for determination of fracture toughness of monolithic ceramics—part 3: Chevron notched beam (CNB) method
43. ISO 24370 (2005) Fine ceramics (advanced ceramics, advanced technical ceramics)—test method for fracture toughness of monolithic ceramics at room temperature by chevron-notched beam (CNB) method (similar to CEN EN 14425-3)
44. Lawn BR (1993) Fracture of brittle solids. Cambridge solid state science series, 2nd edn. Cambridge University Press, Cambridge
45. Pabst RF (1974) Determination of K_{Ic}-factors with diamond saw-cuts in ceramic materials. In: Bradt RC, Hasselman DPH, Lange FF (eds) Microstructure, materials and applications, vol 2. Fracture mechanics of ceramics. Plenum, New York, pp 555–565
46. Nishida T, Hanaki Y, Pezzotti G (1994) Effect of notch-root radius on the fracture toughness of a fine-grained alumina. J Am Ceram Soc 77(2):606–608
47. Damani R, Gstrein R, Danzer R (1996) Critical notch root radius in SENB-S fracture toughness testing. J Eur Ceram Soc 16:695–702. doi:10.1016/0955-2219(95)00197-2
48. Kübler J, Danzer R, Fett T, Damani R (1999) Notch width—theory and model. In: Kübler J (ed) Fracture toughness of ceramics using SEVNB method round robin. VAMAS report no. 37, ESIS document D2-99
49. Damani R, Schuster C, Danzer R (1997) Polished notch modification of SENB-S fracture toughness testing. J Eur Ceram Soc 17(14):1685–1689. doi:10.1016/S0955-2219(97)00024-1
50. Kübler J (2002) Fracture toughness of ceramics using the SEVNB method: from a preliminary study to a standard test method. In: Salem JA, Jenkins MG, Quinn GD (eds) Fracture resistance testing of monolithic and composite brittle materials, ASTM STP 1409, vol 1409. American Society for Testing and Materials, West Conshohocken, pp 93–106
51. Kübler J (1999) Fracture toughness of ceramics using the SENVB method: round robin. VAMAS report no. 37
52. EN 14425-5 (2005) Fine ceramics (advanced ceramics, advanced technical ceramics)—determination of fracture toughness of monolithic ceramics at room temperature by the single-edge vee-notched beam (SEVNB) method
53. ISO 23146 (2005) Fine ceramics (advanced ceramics, advanced technical ceramics)—test methods for fracture toughness of monolithic ceramics—single-edge V-notch beam (SEVNB) method (similar to CEN EN 14425-5)
54. Petrovic JJ, Jacobson LA (1976) Controlled surface flaws in hot-pressed SiC. J Am Ceram Soc 59(1–2):34–37
55. Fett T, Munz D (1987) Knoop-indentations as surface flaws for subcritical crack growth measurements. Eur Appl Res Rep/Nucl Sci Technol 7:1183–1196
56. Lube T (2001) Indentation crack profiles in silicon nitride. J Eur Ceram Soc 21(2):211–218
57. Quinn GD, Salem JA (2003) Effect of lateral cracks on fracture toughness determined by the surface-crack-in-flexure-method. J Am Ceram Soc 85(4):873–880
58. Quinn GD, Gettings RJ, Kübler J (1994) Fracture toughness by the surface crack in flexure (SCF) method: results of the VAMAS round robin. Ceram Eng Sci Proc 15:846–855
59. Newman JC, Raju IS (1981) An empirical stress-intensity factor equation for the surface crack. Eng Fract Mech 15(1–2):185–192
60. Strobl S, Supancic P, Lube T, Danzer R (2012) Surface crack in tension or in bending—a reassessment of the Newman and Raju formula in respect to fracture toughness measurements in brittle materials. J Eur Ceram Soc 32:1491–1501. doi:10.1016/j.jeurceramsoc.2012.01.011
61. ISO 18756 (2005) Fine ceramics (advanced ceramics, advanced technical ceramics)—determination of fracture toughness of monolithic ceramics at room temperature by the surface crack in flexure (SCF) method (ISO 18756, 2005)

62. Chantikul P, Anstis GR, Lawn BR, Marshall DB (1981) A critical evaluation of indentation techniques for measuring fracture toughness: II. Strength method. J Am Ceram Soc 64(9): 539–543
63. Awaji H, Kon J-I, Okuda H (1990) The VAMAS fracture toughness test round robin on ceramics. VAMAS technical report 9. Japan Fine Ceramics Centre, Nagoya, Japan
64. Evans AG, Charles EA (1976) Fracture toughness determination by indentation. J Am Ceram Soc 56(7–8):371–372
65. Anstis GR, Chantikul P, Lawn BR, Marshall DB (1981) A critical evaluation of indentation techniques for measuring fracture toughness: I. Direct crack measurements. J Am Ceram Soc 64(9):533–538
66. Ponton CB, Rawlings RD (1989) Vickers indentation fracture toughness test, part 2: application and critical evaluation of standardised indentation toughness equations. Mater Sci Technol 5(10):961–976
67. Palmqvist S (1962) Rißbildungsarbeit bei Vickers-Eindrücken als Maß für die Zähigkeit von Hartmetallen. Archiv für das Eisenhüttenwesen 33(9):629–634
68. Quinn GD, Bradt RC (2007) On the Vickers indentation fracture test. J Am Ceram Soc 90 (3):673–680
69. Morrell R (2006) Fracture toughness testing for advanced technical ceramics: internationally agreed good practice. Adv Appl Ceram 105(2):88–98
70. Miyazaki H, Hyuga H, Hirao K, Ohji T (2007) Comparison of fracture resistance as measured by the indentation fracture method and fracture toughness determined by the single-edge-precracked beam technique using silicon nitride ceramics with different microstructures. J Eur Ceram Soc 27:2347–2354. doi:10.1016/j.jeurceramsoc.2006.09.002
71. Miyazaki H, Y-i Yoshizawa, Hirao K, Ohji T (2010) Indentation fracture resistance test round robin on silicon nitride ceramics. Ceram Int 36:899–907
72. ASTM F 2094-08 (2008) standard specification for silicon nitride bearing balls
73. Supancic P, Danzer R, Witschnig S, Polaczek E, Morrell R (2009) A new test to determine the tensile strength of brittle balls—the notched ball test. J Eur Ceram Soc 29:2447–2459. doi:10.1016/j.jeurceramsoc.2009.02.018
74. Lube T, Witschnig S, Supancic P, Danzer R, Schöppl O (2012) The notched ball test—charaterisation of surface defects and their influence on strength. In: Varner JR, Wightman M (eds) Fractography of glasses and ceramics VI, vol 230. Ceramic transactions. Wiley, Hoboken, pp 225–234
75. Strobl S, Supancic P, Lube T, Danzer R (2012) Toughness measurement on ball specimens, part I: theoretical analysis. J Eur Ceram Soc 32:1163–1173. doi:10.1016/j.jeurceramsoc.2011.12.003
76. Strobl S, Lube T, Schöppl O (2014) Toughness measurement on ball specimens. Part II: experimental procedure and measurement uncertainties. J Eur Ceram Soc 34:1881–1892. doi:10.1016/j.jeurceramsoc.2013.12.052
77. Börger A, Supancic P, Danzer R (2002) The ball on three balls test for strength testing of brittle discs—stress distribution in the disc. J Eur Ceram Soc 22(8):1425–1436. doi:10.1016/S0955-2219(01)00458-7
78. Börger A, Supancic P, Danzer R (2004) The ball on three balls test for strength testing of brittle discs—part II: analysis of possible errors in the strength determination. J Eur Ceram Soc 24(10–11):2917–2928. doi:10.1016/j.jeurceramsoc.2003.10.035
79. Danzer R, Supancic P, Harrer W (2009) Der 4-Kugelversuch zur Ermittlung der biaxialen Biegefestigkeit spröder Werkstoffe. In: Kriegesmann J (ed) Technische keramische Werkstoffe, 113. Ergänzungslieferung, HvB Verlag GbR, Ellerau, pp 1–48
80. Harrer W, Danzer R, Supancic P, Lube T (2009) Influence of the sample size on the results of B3B-tests. Key Eng Mat 409:176–184. doi:10.4028/www.scientific.net/KEM.409.176
81. Harrer W, Danzer R, Supancic P, Lube T (2008) The ball on three balls test: strength testing of specimens of different sizes and geometries. Proc. of 10th International Conference of the European Ceramic Society, Baden-Baden

82. Strobl S, Rasche S, Krautgasser C, Sharova E, Lube T (2014) Fracture toughness testing of small ceramic discs and plates. J Eur Ceram Soc 34(6):1637–1642. doi:10.1016/j.jeurceramsoc.2013.12.021
83. Rasche S, Strobl S, Kuna M, Bermejo R, Lube T (2014) Determination of strength and fracture toughness of small ceramic discs using the small punch test and the ball-on-three-balls test. Procedia Mater Sci 3:961–966. doi:10.1016/j.mspro.2014.06.156
84. CEN/TS 14425-1 (2003) Advanced technical ceramics—test methods for determination of fracture toughness of monolithic ceramics—part 1: guide to test method selection

Transition from Flat to Slant Fracture in Ductile Materials

William F. Hickey and Krishnaswamy Ravi-Chandar

Abstract We investigate ductile fracture in an aluminum alloy which exhibits an interesting switch in mode from flat fracture to slant fracture. While this transition is typically considered to be triggered by a change in crack tip constraint with specimen thickness, we explore this transition in rolled sheet material simply by controlling the strain hardening behavior of the material. Specifically, experiments are performed on two different heat-treatments of the same alloy, resulting in two materials that differ only in their strain-hardening behavior. Based on an examination of the microscopic and macroscopic strain evolution, we conclude that the transition arises because of strain localization that precedes fracture.

1 Introduction

Ductile fracture is a topic of enormous interest in many technological applications. The impetus for this arises from the use of materials well into their plastic response regime in the automobile, aircraft, power-generation and other industries. The theory of fracture is fairly well-established in industrial practice in the case of problems that fall in the category of small-scale yielding, where the zone of plastic deformation is limited to a small fraction of the overall structural dimensions; this is the theory of Linear Elastic Fracture Mechanics (LEFM), with the square-root singular crack tip state characterized by a single parameter, the stress intensity factor. The critical value of the stress intensity factor—*the fracture toughness*—has been established as a characteristic material parameter and is easily determined through experiments. Therefore, design and life-assessment decisions based on fracture criticality of

W.F. Hickey (✉)
Southwest Research Institute, San Antonio, USA
e-mail: wfhickey@gmail.com

K. Ravi-Chandar
University of Texas at Austin, Austin, USA
e-mail: ravi@texas.edu

G. Hütter and L. Zybell, *Recent Trends in Fracture and Damage Mechanics*, DOI 10.1007/978-3-319-21467-2_9

structures are performed appropriately in many engineering applications. The success of LEFM is related to two critical aspects: the first is the ability to determine the stress intensity factor for specific problems. Catalogs of the stress intensity factor have been published [1, 2], but given the advances in computational mechanics, the solution of specific boundary value problems and the extraction of the stress intensity factors has become a routine task. Second, and more importantly, the determination of the fracture toughness through scale model experiments has been well-established as a standard procedure by the ASTM, ESIS and other groups; with these procedures, one can determine the fracture toughness as a material parameter.

In contrast, the situation is quite different when the zone of plastic deformation increases to a significant fraction of the structural dimensions. After nearly four decades of intense research into the field, while there is much progress that has been made (see for example the review of Zerbst et al. [3]), there is still significant debate related to the possibility of characterizing failure. Numerical analysis of the stress and deformation fields in elastic-plastic problems prior to the onset of crack growth does not appear to pose significant challenges, but the primary hurdle appears to be the inability to characterize the fracture energy as a material parameter that can be used in applications where the structures are subjected to different loading conditions. We surmise that this is due primarily to the inability, in fracture experiments, to separate the intrinsic work of fracture from the work done in the inelastic deformation of the region in the vicinity of the crack tip. Different methods, such as a simple strain-to-failure criterion [4–6], phenomenological damage mechanics, and mechanistic damage models, such as the Gurson-Tvergaard-Needleman model, have been used in modeling ductile failure; while these models have been calibrated to be useful in specific applications, general predictive ability has not been clearly established as demonstrated in the recent blind-predictions challenges from the Sandia National Laboratories (see [7]). This problem is brought to significant light when considering fracture in thin sheets where the fracture occurs on a plane slanted at about ±45° with respect to the initial crack plane. Figure 1 shows the geometry of the slant facture. An initial crack was generated in a single edge-notched tension (SENT) specimen by fatigue precracking; subsequently, this specimen was loaded under monotonically increasing displacement to grow a stable crack. The resulting fracture is slanted at an angle of ±45° with respect to the flat planar crack generated by fatigue. In specimens with greater uncracked ligament, the orientation of the slant surface switches periodically between the two equivalent planes oriented at ±45° with respect to the plane defined by the fatigue precrack. A rather spectacular example of this flip-flop in the fracture plane orientation was demonstrated in the large plate tests reported by Simonsen and Törnqvist [8]. While some effort has been devoted towards examining the generation of slant fractures, it has been very difficult to generate this slant fracture in numerical simulations indicating that our models of ductile failure still face significant hurdles. Pardoen et al. [9] examined the fracture behavior of thin sheets of sixteen different materials with intermediate to high hardening capacity and showed that these exhibit flat fracture in contrast to conventional wisdom which indicates that thin sheets should exhibit slant fracture. More recently, El-Naaman and Nielsen [10] examined

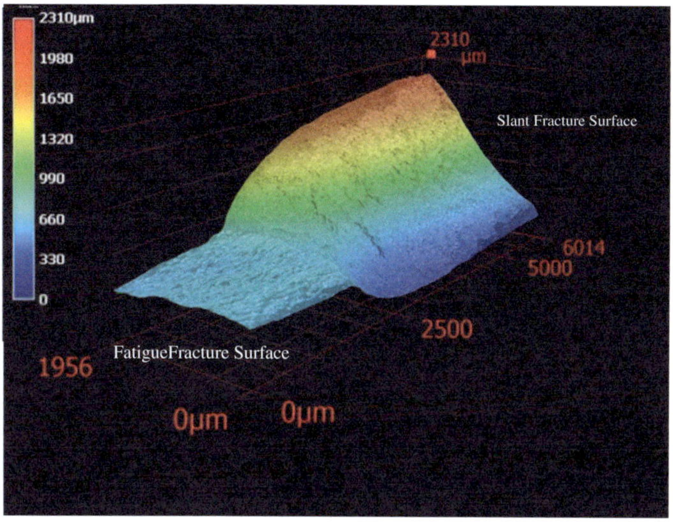

Fig. 1 Illustration of fracture in Al 6061-T6 indicating the initial fatigue fracture surface (flat) and the subsequent slant fracture under monotonic loading. A transition region can be seen near the initiation point, connecting the flat fatigue precrack and the slant growth under monotonic loading. The *color map* indicates height from the fatigue fracture plane. The height map was obtained using a Keyance microscope

fracture in a soft high purity 1050A-H14 aluminum, and a structural steel using edge cracked specimens in combined in-plane bending and tension and double-edge-notched tension (DENT) specimens; they showed that the softer material exhibited cup-cone type fracture while the material with higher hardening exhibited slant fracture, with flip-flops. Fractography indicated that failure was by ductile void growth and coalescence in the case of the steel, while it was probably through an alternating shear mechanism indicated by Orowan [11] and Ghahremaninezhad and Ravi-Chandar [12]. Rivalin et al. [13] presented an alternative point of view, based on quasi-static and dynamic tests on the same material, which suggested a possible role for material rate dependence, with slant fractures appearing at higher crack growth rates. In view of the different and sometimes contradictory hypotheses on the appearance of slant fractures, it appears that additional examination of the problem is needed. The primary goal of this manuscript is to present a set of experimental results that demonstrate that the development of a slant fracture is due to plastic strain localization that precedes fracture.

The strategy used in this work is to consider the role of strain hardening on the development of different types of fracture modes. To explore this, SENT specimens were machined from a rolled sheet stock of Al 6061-T6. The details of the grain size distribution and the uniaxial stress-strain curve for this material are described by Ghahremaninezhad and Ravi-Chandar [14]. One set of these specimens were tested in the as-received T6 condition, while another set of the specimens was annealed to the O condition through the standard heat-treatment procedure.

The specimens were heated to a temperature of 400 °C, held for 2–3 h, then cooled in the furnace at a rate of 28 °C per hour until 260 °C, and finally air cooled to room temperature. Microscopic examination of the resulting material indicated very little change in the grain morphology. However, a tension test revealed that the Al 6061-O has a lower yield strength and considerably greater ductility than the Al 6061-T6; these are quantified later in this article.

Based on LEFM considerations, the required thickness for small scale yielding to be valid is estimated to be 0.0125 m for Al 6061-T6. Since fracture toughness data are not readily available for Al 6061-O, one can make a conservative estimate of the size requirement by assuming that the fracture toughness of the annealed material remains the same as that of the T6 conditions; this would indicate that the required specimen thickness is 0.8 m; it is well known that the small-scale yielding condition is particularly difficult to satisfy in low strength materials. Based on these estimates, one would expect that in a thin sheet of 0.00244 m, neither the Al 6061-T6 nor the Al 6061-O will be able to satisfy the thickness requirements and therefore be dominated by the slant fracture. Nevertheless, the state of stress at the midplane of the specimen should be plane strain in both cases and the velocity fields must be consistent with this constraint in either case. In this paper, we explore the development of plastic deformation and fracture in these two materials in order to determine the underlying reasons for the onset of slant fracture.

2 Experimental Procedure

Uniaxial tension tests were performed on typical dog-bone specimens of Al 6061-T6 and Al 6061-O, cut from rolled sheets. Tests were performed with the specimen aligned with the rolling direction; the mild anisotropic response of the T6 material was characterized by Ghahremaninezhad and Ravi-Chandar [14], but this is neglected in the present work, noting that this must be included in future work in order to predict the correct plastic response. The experimental stress-strain curves were fit with a Ramberg-Osgood power-law model:

$$\varepsilon = \frac{\sigma}{E}\left[1 + \alpha\left(\frac{\sigma}{\sigma_Y}\right)^{N-1}\right] \tag{1}$$

where E is Young's modulus, σ_Y is the yield stress, α is a fitting constant, and N is the hardening exponent. The experimental data in terms of the true stress vs true strain are shown in Fig. 2, and the parameters of the power-law fits for the two different heat-treatments are given in Table 1. Note that in the uniaxial tensile test the deformation localizes into a diffuse neck at the Considère strain, and therefore the data presented here are shown as dotted lines beyond this point; the power-law model is usually extrapolated beyond this point as a measure of the underlying stress-strain response as indicated in the figure, and this extrapolation plays a major role in dictating localization of deformation in stress-states near the crack tip.

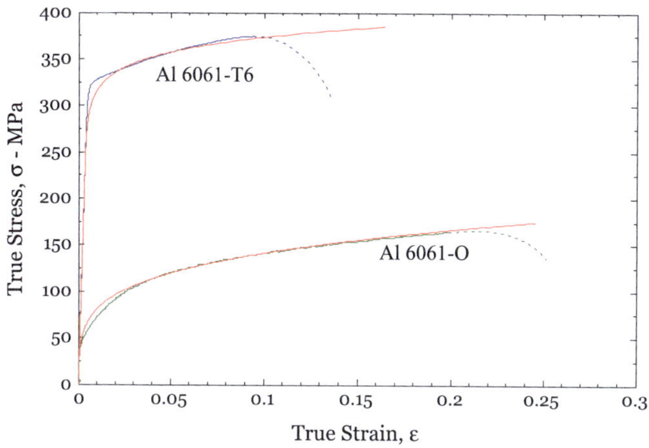

Fig. 2 True stress versus true strain in Al 6061-T6 and Al 6061-O. The *dashed lines* indicate the post-necking response and is not the true material behavior. The *red lines* indicate the Ramberg-Osgood fit and is extrapolated as indicated to higher strain levels

Table 1 Ramberg-Osgood parameters

Material	E (GPa)	σ_Y (MPa)	N	α
Al 6061-T6	70	297	17.0	0.454
Al 6061-O	70	45	4.28	1.13

Single-edge-notched-tension (SENT) specimens of Al 6061-T6 and Al 6061-O were prepared by electric discharge machining of the initial notch. The dimensions of the SENT specimens are shown in Fig. 3; the specimens were machined out of thin sheet stock, with a thickness of 2.44 mm. While the constraint in SENT specimen is different from that in the DENT or compact tension (CT) specimens commonly used, we have performed experiments in a related study with CT specimens and observed slant fracture; therefore, we do not anticipate the specimen geometry playing a critical role in the investigation of transition to slant fracture. A natural crack was generated in these specimens by fatigue cycling at load amplitudes well below 10 % of the load at which crack initiation occurs; post-test examination of the fracture surface indicated that the fatigue crack exhibited a flat surface (shown later in Fig. 9). These specimens were clamped at the top and bottom with hydraulic grips and pulled at a constant displacement rate of 0.254 mm per minute in an Instron Model 4482 testing machine until complete fracture of the specimen. Scanning electron micrographs of the fracture surface (shown later in Fig. 9) revealed a dimpled surface indicating that the failure mechanism was predominantly void growth and coalescence. The local displacement field near the crack tip region was monitored as a function of the load with digital image correlation (DIC). The surface of the fracture specimen was decorated with a stochastic pattern with black and white paint for the DIC measurement. A digital video camera

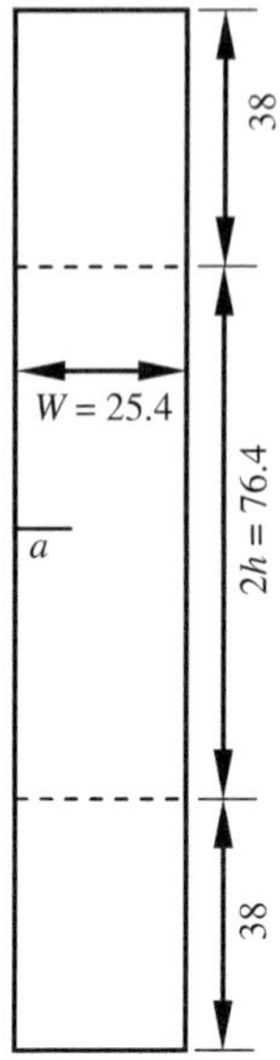

Fig. 3 Geometry of the single-edge-notched specimen used in the fracture tests. The specimens were all machined from 2.44 mm thick rolled sheets

was used to take a series of images from the beginning of the experiment until complete failure. After completion of the test, the images were analyzed with the ARAMIS™ digital image correlation software to determine the displacement fields corresponding to each image. The strain fields are then calculated using suitable smoothing procedures in order to filter the noise arising from numerical differentiation of the displacement data; the effective gage length over which the strain computation was achieved was 39 μm.

3 Experimental Results

The variation of load is plotted as a function of the crack mouth opening displacement (CMOD, measured at the specimen edge) in Fig. 4 for the Al 6061-T6 and Al 6061-O specimens; the CMOD was determined using DIC measurements. The state at the onset of crack initiation is identified in the plots by a circular marker. Stable crack growth under displacement controlled testing is evident from this response. Repeated tests in both materials indicated a remarkably consistent response not only in the load-CMOD, but also in the onset of crack initiation. Also, note that while the load is significantly greater in the Al 6061-T6, the deformation is larger in the Al 6061-O. This is illustrated further by the major strain ε_{22} contours[1] shown in Fig. 5 at four different loading stages for both the T6 and O conditions.

[1]DIC breaks down very close to the crack tip, partly due to degradation of the paint and partly from image distortion that occurs due to large curvatures near the crack tip; nevertheless, it is possible to measure strains as close as 160 μm near the crack tip.

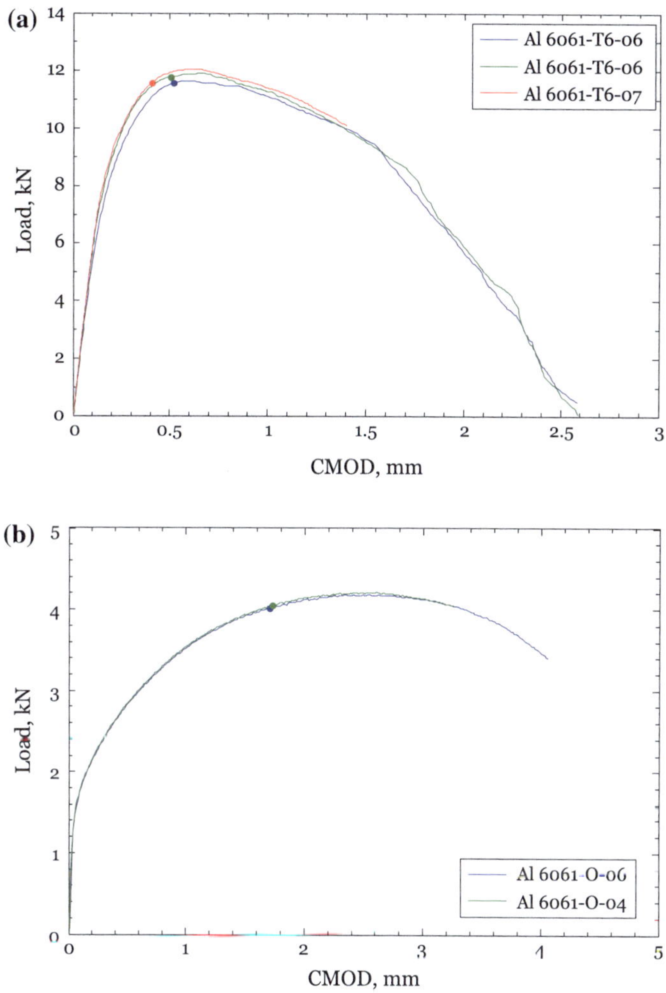

Fig. 4 The variation of the load with crack mouth opening displacement (CMOD) in the SENT specimens of **a** Al 6061-T6 (*top*) and **b** Al 6061-O (*bottom*). The state at the onset of crack growth, as identified from magnified optical images, is indicted by the *dots* in the respective plots. Repeated tests indicate nearly identical response of the specimens

The development of the strain field in the vicinity of the crack tip clearly reveals significant differences in the development of plastic strains, and is discussed further below. Noting that the microstructural morphology, the precracked state, and the specimen geometry are nearly identical in the two sets of tests in the two materials, these differences are driven primarily by the differences in the yield behavior of the two different heat treatments and their influence on the localization response.

In order to obtain fracture parameters, the DIC based displacement and strain measurements are interpreted in terms of the elastic-plastic stress field in the

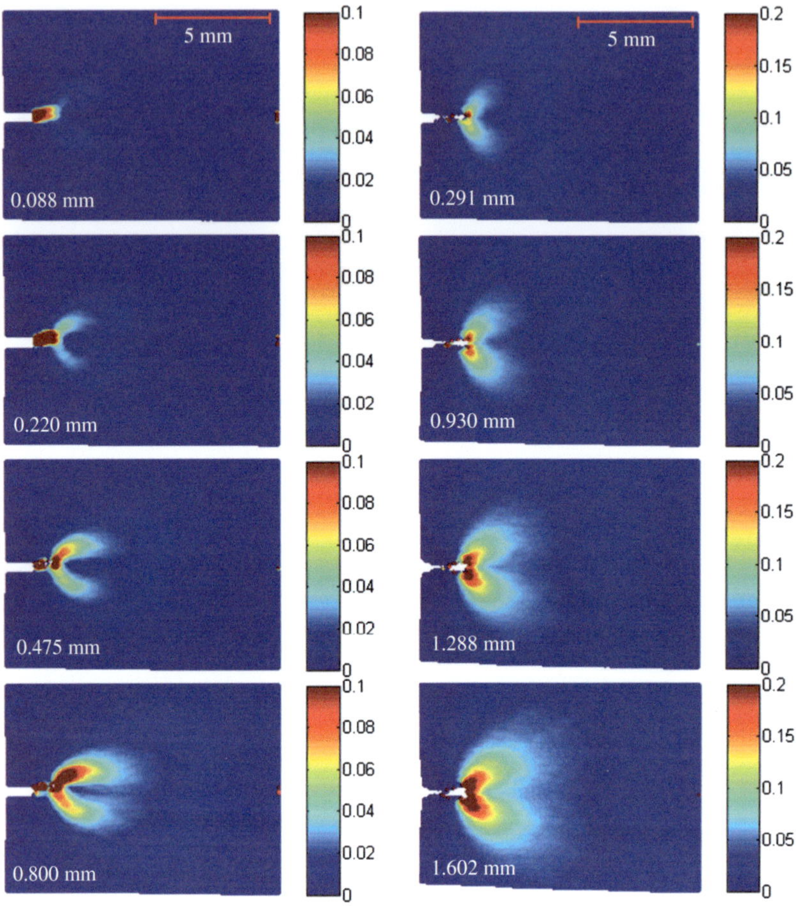

Fig. 5 The spatial variation of the major strain in the SENT specimen of Al 6061-T6 (*left column*) and Al 6061-O (*right column*) at different CMOD levels are shown

vicinity of the crack tip, and then used to extract the *J*-integral as a function of the loading and crack extension; DIC has been used in this task, but mostly with application to the elastic deformation (see for example [15–18]). The details of these calculations are given in the Appendix. Three different contours as indicated in Fig. 6 were used to determine the value of the *J*-integral for both materials. The *J*-integral for each contour shown above was calculated at each stage of the test. The variation of load versus the J-integral is shown in Fig. 7 for both materials. The top row shows the variation of J for one specimen each of the T6 and O condition, with the J value calculated along three different contours. The bottom row shows the variation of J with the load for different specimens for each material; clearly very repeatable fracture response is observed. These results provide an estimate of the fracture toughness of the two different heat-treatments of the same material; the critical values of $J_{DIC,i}$ along the three contours are shown in Table 2. The typical

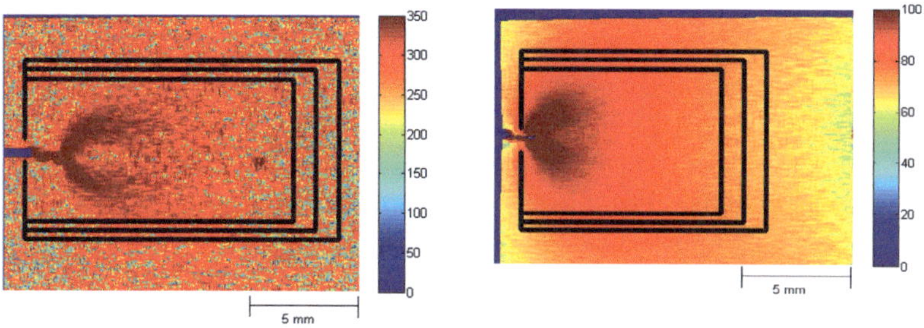

Fig. 6 Contours of the equivalent plastic stress (in unit of MPa) for the Al 6061-T6 (*left*) and Al 6061-O (*right*) are shown at the stage corresponding to crack initiation; three different contours, well away from the crack tip were selected for the calculation of the *J*-integral

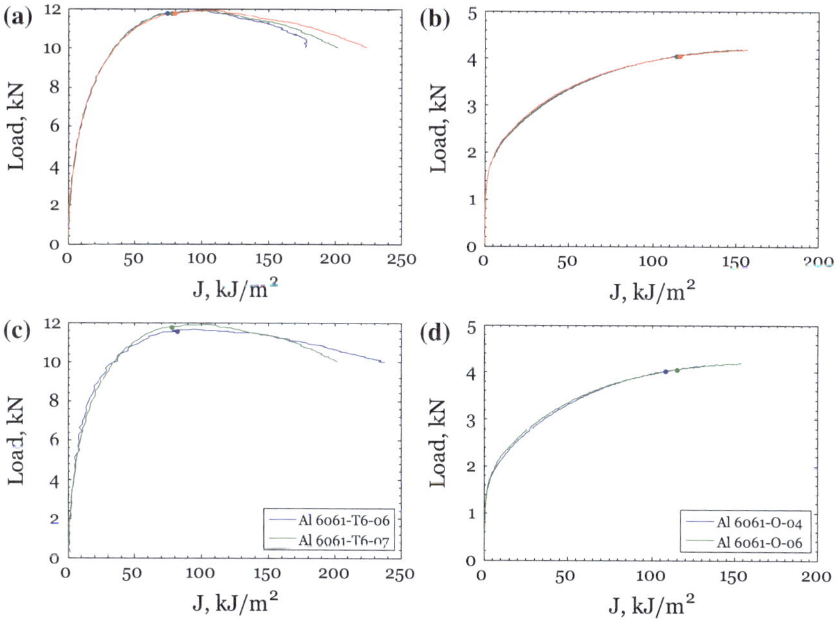

Fig. 7 Variation of the load versus the *J*-integral for different contours **a** Al 6061-T6, **b** Al 6060-O and for different specimens **c** Al 6061-T6 and **d** Al 6061-O

Table 2 Initial crack length, critical load and J-integral (along three different contours)

Test number	a (mm)	P_c (N)	$J_{DIC,1}$ (kJ/m^2)	$J_{DIC,2}$ (kJ/m^2)	$J_{DIC,3}$ (kJ/m^2)
Al6061-T6-06	8.57	11,765	74.18	77.70	79.69
Al6061-T6-07	8.61	11,561	76.42	81.53	75.69
Al6061-O-04	8.96	4046	114.1	114.7	115.9
Al6061-O-06	8.67	4014	109.3	108.1	107.3

blunting corrections applied in standard methods were not imposed since our primary focus was on the trends in fracture toughness. We note that these values of the fracture energy are significantly greater than the corresponding plane-strain value for the Al 6061-T6: $\sim 10\ \text{kJ/m}^2$ [19].

The development of initial flat fatigue crack and the local measurements of the strain field during monotonic loading clearly indicate that the mode of cracking—flat versus slant—is influenced by the stress state that develops in front of the crack tip, which in turn is influenced by the strain-hardening behavior and plastic response of the material. We resolve the issue through the following observations:

- The strain fields measured with DIC easily allow for visualization and quantitative characterization of the development of the plastic zone in the two different materials. Large differences in the development of the plastic zones for Al 6061-T6 and Al 6061-O are seen in Fig. 5. In the Al 6061-O (right column of Fig. 5), the plastic zone shape resembles that calculated using the plane-strain K-field and imposing a von Mises yield criterion. Very large strains—as large as 0.1—are generated at distances of about 5 mm from the crack tip; this results in a significant thinning of the plate as shown in the right image of Fig. 8. The high strain-hardening of the material implies that deformation instabilities in the form of localization do not occur until strain levels are extremely large. A high triaxial stress state is developed along $\theta = 0$,[2] triggering ductile failure processes along this plane and resulting in a flat fracture surface.
- In contrast to this behavior, the plastic zone shape in Al 6061-T6 bears no resemblance to estimates based on either the plane strain or the plane stress K-field! (See Fig. 5 left column). In fact, along $\theta = 0$, the strains are extremely small and remain in the elastic state, quite contrary to the expectations of plane stress. The aging heat-treatment of T6 increases the yield strength significantly in comparison to the annealed material, but the strain-hardening is decreased significantly; therefore, localization of deformation could be expected at much lower strain levels, and this is only exacerbated in the vicinity of the crack tip. Therefore, with increasing global load, the deformation in Al 6061-T6 localizes along planes that are oriented at $\theta = \pm 45°$ with respect to the normal to the crack surface (or the sheet surface); the cross-sectional micrograph shown in Fig. 8 clearly indicates a slant fracture and localized shear deformation at $\phi = \pm 45°$.[3] The visible traces of the large deformations within this band on the free surfaces can be identified in the strain fields obtained from DIC, as well as in the grain-based assessment discussed in Sect. 4. Once such localization occurs, deformation outside this plane ceases, and ductile failure processes could occur only within this plane. This localization forces a slant fracture surface, as well as very little thinning deformation of the sheet outside this plane (compare the left and right images of Fig. 8).

[2]Here theta refers to crack tip polar coordinates, and $\theta = 0$ is the line of symmetry.

[3]The angle ϕ is defined relative to the crack front along the thickness direction.

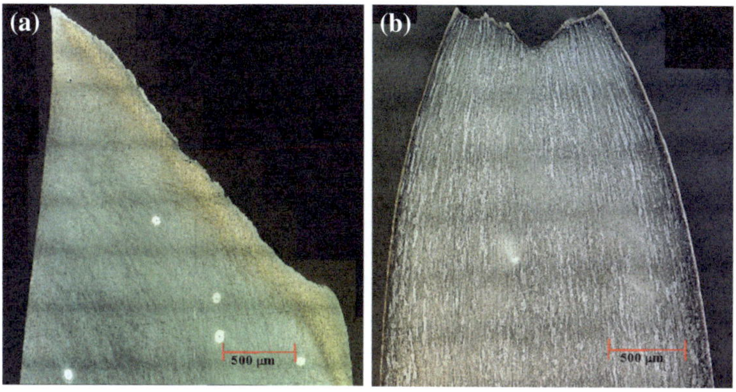

Fig. 8 Thickness section of the fractured specimens indicating slant fracture in Al 6061-T6 (a) and flat fracture Al 6061-O (b). Notice the significant thickness reduction in the right figure indicative of the large strains encountered in the Al 6061-O

- The initial fatigue precrack surface—which was created with a stress intensity level of 22 MPa $m^{1/2}$ in the T6 and 17 MPa $m^{1/2}$ in the O (equivalent $J = 0.092J_c$ for T6 and $J = 0.045J_c$ for O)—is flat. At this low loading condition, the plastic zone size can be estimated to be ~870 μm for the T6 and ~20 μm for the O. Since plasticity is confined to a small fraction of the overall specimen thickness 2.44 mm, localization of deformation as discussed above cannot appear in the Al 6061-T6, and therefore the fatigue crack grows as a flat fracture in both materials.

Combining these observations, one can generate a mechanistic understanding of flat and slant fracture in sheet metals. At low loading levels, such as those during fatigue crack growth, plane strain conditions exist near the crack tip. Under this condition, the ductile failure processes (irreversible slip etc.) are triggered along the plane $\theta = 0$, and results in a flat fracture surface. During crack growth under monotonic loading, the deformation is critically dependent on the strain-hardening, since stability of deformation depends on this parameter. For materials with high strain-hardening, localization instabilities do not appear until very large strain levels, and therefore the diffuse plastic field dictated by the crack tip stress concentration is generated; the resulting high triaxiality along $\theta = 0$ triggers ductile failure processes such as void growth, resulting in a macroscopically flat but microscopically dimpled fracture surface. On the other hand, for materials with a low strain-hardening, plastic instabilities arise at extremely low strain levels and localize deformation along two equivalent slant planes. Further straining and damage in the form of void growth occur only within these zones. Quantitative prediction/simulation of the response requires a complete characterization of the strain hardening response to very large strain levels and is deferred to future work.

4 Microscopic Observations

Both optical microscopy and scanning electron microscopy were utilized to further characterize the fracture process of Al 6061-T6 and Al 6061-O. The examination of different planes and surfaces of the post-mortem fracture specimens provides further insight into the failure mechanisms. Initial examination of the fracture surface indicates the type of fracture exhibited by the materials.

In ductile materials, failure is classically thought to occur from void growth and coalescence. Fig. 9a, b show low magnification micrographs of the fracture surfaces of Al 6061-T6 and Al 6061-O specimens, respectively. Two distinct regions are present on both fracture surfaces. The surface of the fatigue crack region is smooth with no dimples, and the monotonically fractured surface consists of many dimples as shown in the high-magnification images of both materials shown in Fig. 9c, d. Clearly, many small dimples are present on the fracture surfaces of both the Al 6061-T6 and Al 6061-O. In the Al 6061-O, the dimples appear to be of two distinct sizes. The Al 6061-T6 fracture surface dimples also appear in two sizes, but the smaller dimples in the Al 6061-T6 are smaller than in the Al 6061-O, and correspondingly the larger dimples in the Al 6061-T6 are larger than in the Al 6061-O.

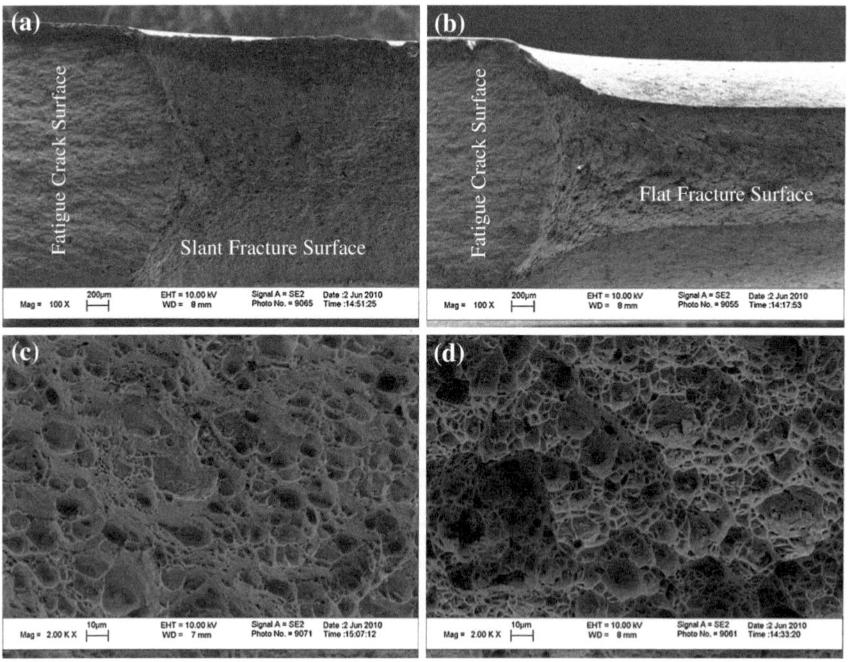

Fig. 9 Scanning electron micrographs of fracture surfaces of slant fracture Al 6061-T6 and flat fracture Al 6061-O. *Top row* contains lower magnification surface views of the fracture including the fatigue precracked region **a** 6061-T6, **b** 6061-O. *Bottom row* is a magnified view of the fracture created under monotonic loading **c** 6061-T6 and **d** 6061-O

These dimples are typically taken to be evidence of ductile failure by void nucleation, growth and coalescence with the void nucleation occurring preferentially at the sites of the second phase particles; examination of the fractographs at high resolution indicates the presence of such void nucleating particles in many of the larger dimple of the fracture surface.

4.1 Fracture Surface Observations: Cross-Sectional Planes

In order to examine the details of the progression of the void nucleation, growth and coalescence, loading of the specimens was halted before complete failure. The cracks usually tunnel deeper in the central portion of the thickness than towards the plate free surfaces. One specimen each of Al 6061-T6 and Al 6061-O was interrupted in this manner and then prepared for observation; the specimens were cut to reveal the plane perpendicular to the crack propagation direction so as to intersect the tunneling crack. These specimens were mounted in epoxy and polished for microscopic examination. Scanning electron micrographs of the tunneling cracks are shown in Fig. 10 for the Al 6061-T6 (top) and Al 6061-O (bottom), respectively. The following important observations can be made from these images.

- The SEM images indicate a number of white particles distributed over the image; these have been identified through EDAX to be iron-rich particles; they are typically elongated particles, with the small dimension being around a couple of μm, and the longer dimension ~5–10 μm. The total volume fraction of such particles is in the range of 1 %. It should be noted that the distribution of these particles is statistically the same in the T6 and O conditions since the latter was obtained from the same sheet of the T6 material through heat-treatment. Mg_2Si particles were not observed at magnifications that could observe micrometer ranged objects.
- For Al 6061-T6, the overall orientation of the fracture plane is at ±45° with respect to the loading direction; the crack opening is seen to be extremely small; it is not easy to identify the position of this slice relative to the crack front, but the overall crack opening is in the range of 1 μm. For the Al 6061-O, the overall orientation of the crack is perpendicular to the loading direction, but it appears to be made of alternating segments oriented at ±45°. The opening between the top and bottom surfaces is as large as 30 μm, but in the absence of precise location of this slice relative to the crack tip, comparison to the T6 is difficult; this requires additional exploration, perhaps through nondestructive methods such as X-ray tomography.
- Some of the second phase particles have broken into fragments in the Al 6061-T6 (see high magnification inset image in Fig. 10); there is also some debonding or growth of the void associated with the particle. However, the Al

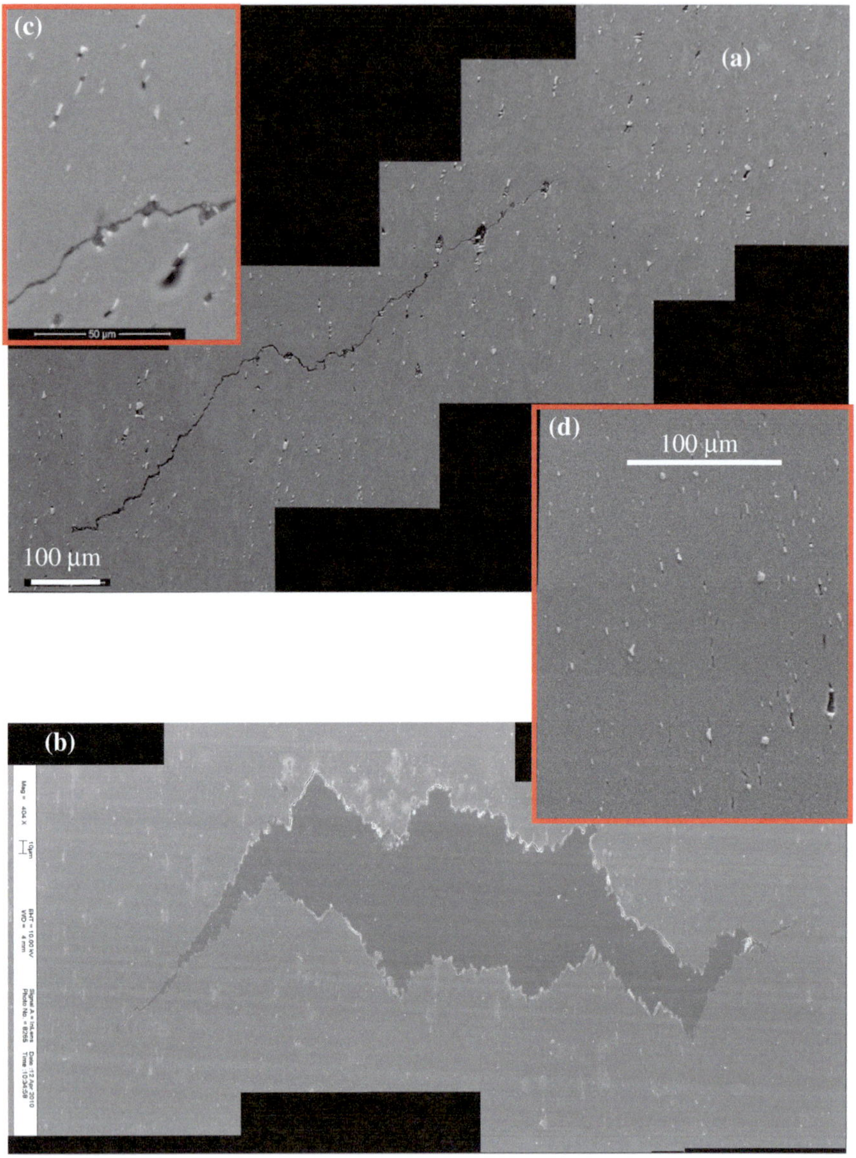

Fig. 10 Scanning electron micrographs of thickness sections of the fracture specimens of **a** Al 6061-T6 and **b** Al 6061-O. Notice the crack in the Al 6061-T6 grows at 45° while the overall crack orientation for the Al 6061-O is flat. The *inset* figures show a higher magnification of the **c** Al 6061-T6 specimen, indicating fracture of second phase particles and **d** Al 6061-O specimen indicating absence of particle fracture

6061-O shows almost no breakage of the second phase particles and no voids; since the yield strength of this material is so much lower, the material probably flows around the particles rather than develop stress in them that are high enough to break the particles.

- In both materials, fracture appears through the nucleation of damage at the second phase particles and then subsequently by the connections established between the nuclei; there appears to be very little damage outside of a small neighborhood of the fracture plane. Compared to the number of dimples that are seen in the fracture plane, damage in the form of nucleation (by debonding or fragmentation of second phase particles) and growth of cavities by plastic flow around the cavities appears to be negligible. From careful examination of the images in Figs. 10, the region in which there is any damage at all in the T6 material can be estimated to be around ~5 μm, and in the O material there exists no discernable damage outside of the fracture plane. These observations on the localized nature of damage are similar to that reported by Ghahremaninezhad and Ravi-Chandar [14] in uniaxial and flat-notched tension specimens of the same material. Recently, Morgeneyer et al. [20] have examined the development of damage using x-ray laminography in a different aluminum alloy; their examination revealed that damage appears only near the very end of the failure process, and then only in a localized band oriented at ±45°.

4.2 *Quantitative Microscopy and Grain Based Strain Measurements*

The DIC images shown in Fig. 5 indicate that the strains near the vicinity of the crack are indeed quite large. However, the DIC method decorrelates in the vicinity of the crack tip for two important reasons: first, the strains are quite large and an extreme distortion of the images arises; second, the large out-of-plane displacements near the crack cause image degradation also leading to decorrelation. On the other hand, the etched thickness section images shown in Fig. 11 permit evaluation of thickness-averaged strains that are in the range of 0.16 for the T6 and 0.84 for the O.

The local values of strain are estimated using the following procedure: first, the distribution of grain size in the initial microstructure was estimated from the images of the unstrained microstructure; this distribution was reported for this particular batch of material by Ghahremaninezhad and Ravi-Chandar [9]; it has an average grain size in the thickness direction of 15.1 μm and the standard deviation of 6.6 μm. Next, the deformed grain images were examined quantitatively: an example of a thickness section of a fracture specimen is shown in Fig. 11; a magnification view of the crack tip region is shown on the right side of this figure. This specimen corresponds to a slant fracture in Al 6061-T6. The location of each grain boundary was identified visually along the green horizontal lines marked in Fig. 11,

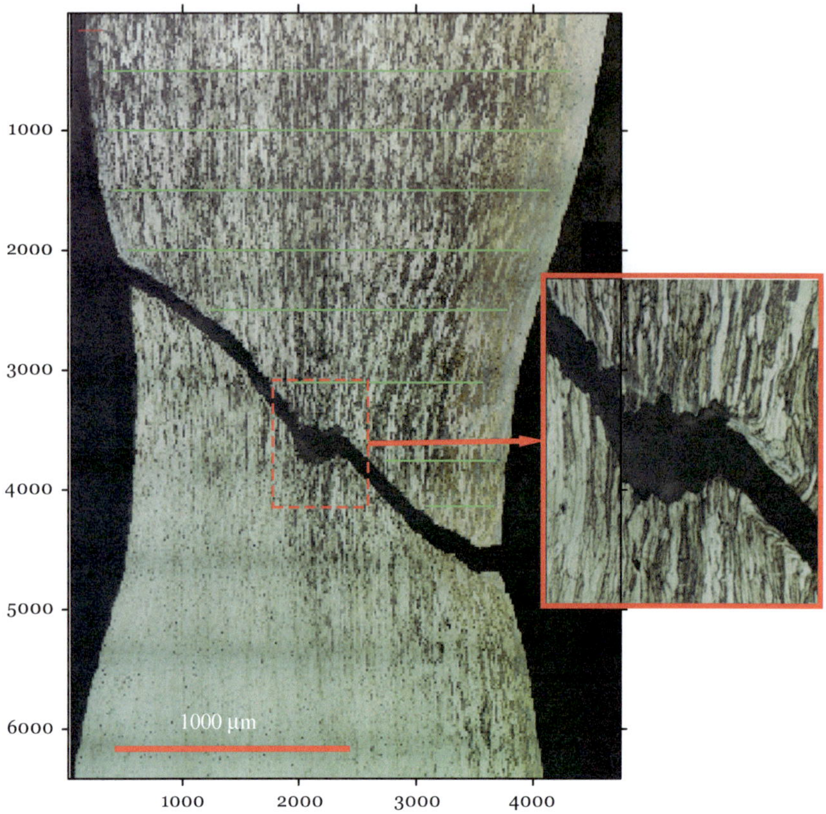

Fig. 11 Optical micrograph of thickness section of a fracture specimen of Al 6061-T6 (*left*); the specimen was etched in Weck's agent to reveal grain boundaries (see [8] for details of the etching). The *right image* shows a higher magnification view of the central portion where the shear localization is visible just below the fracture planes. The grain level deformation was measured along the *green lines* to identify the local strains

corresponding to different depths below the fracture plane and captured into a data file using a MATLAB code. The variation of the grain width t across a few lines $y = \text{const}$ was obtained and divided by the mean grain width $\hat{T}$ to obtain an estimate of the strain in the thickness direction $\varepsilon_t = \ln\left(t/\hat{T}\right)$. Then the equivalent plastic strain can be estimated assuming plastic incompressibility: $\varepsilon_{eq} = \frac{2}{\sqrt{3}}\ln\left(\hat{T}/t\right)$. Estimates for the thickness strains are shown in Fig. 12; the red-dashed lines indicate the average strain. The average grain level strain measurements appear to be similar to the macroscopic strains measured, but the local maxima are significantly greater. Very large grain rotations (macroscale shear) and grain elongations can be observed very close to the fracture plane (see magnified view in Fig. 11);

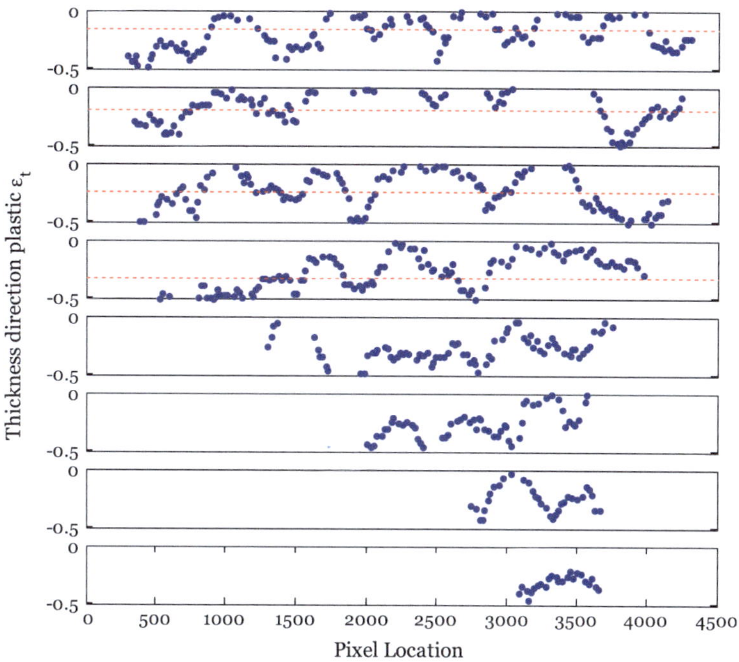

Fig. 12 Variation of the grain-level strains along each of the *green lines* shown on the fracture specimen in Fig. 11. The *red dashed lines* indicate the strain averaged along each line

grain measurements were made just below the fracture surfaces in both the T6 and O materials. For the T6 material, the grain level thickness failure strain close to the fracture plane was 0.67; this is a significant increase from the macroscopic thickness failure strain of 0.51 measured in the quasi-static tension tests. The annealed material exhibited a grain level strain of 1.06, and a macroscopic thickness average strain of 0.75 in the quasi-static tension tests at final fracture.

The large difference in the macroscopic and grain level failure strains for these materials has noteworthy implications in the modeling of failure and fracture. If estimates based on the large-scale measurements of strains are used as failure strains in simulations, the failure will be made to occur prematurely, thus greatly distorting the evolution of the stress and strain fields.

5 Conclusions

Ductile fracture in aluminum alloys has been examined experimentally, with specific interest in the transition from flat fracture to slant fracture. Specimens of the Al 6061 alloy that were nearly identical in all microstructural and geometric aspects but subjected to the T6 and O heat-treatment conditions were examined. The T6 alloy results in high yield strength and low stain hardening; this limits the extent of plastic zone but promotes localization and results in slant fracture. On the other hand, the O condition provides very low yield strength and a high strain-hardening; this allows development of an extensive plastic zone, significant thinning deformation, but inhibits strain localization and results in flat fracture. Quantitative modeling has not been considered but is the subject of a future study.

Acknowledgements This work was performed during the course of an investigation into ductile failure under two related research programs funded by the Office of Naval Research: MURI project N00014-06-1-0505-A00001 and FNC project: N00014-08-1-0189. This support is gratefully acknowledged.

Appendix: Determination of the Stress from Strain Measurements

The conversion of the strain measurements to stress requires a material model; the J_2 deformation theory of plasticity is taken to be the appropriate model. The effective stress is defined as $\sigma_e = \sqrt{3J_2} = \left(3s_{ij}s_{ij}/2\right)^{1/2}$ where s_{ij} are the components of the stress deviator, and the plastic equivalent strain is defined as $\bar{\varepsilon} = \left(2\varepsilon_{ij}\varepsilon_{ij}/3\right)^{1/2}$. The variation of the effective stress with the plastic equivalent strain is obtained from a uniaxial test and is indicated by $\sigma(\bar{\varepsilon})$. The von Mises yield criterion is then expressed as

$$\Phi(J_2, \bar{\varepsilon}) = 3J_2 - \sigma^2(\bar{\varepsilon}) = 0 \tag{2}$$

The total strain is decomposed additively into the plastic and elastic components; the plastic strain develops in the direction of the normal to the yield surface and can then be expressed as:

$$\varepsilon_{ij}^p = d\lambda s_{ij} \tag{3}$$

The stress-strain relationship is written as:

$$\varepsilon_{ij} = \frac{1+\nu}{E}\sigma_{ij} - \frac{\nu}{E}\sigma_{kk}\delta_{ij} + \frac{3}{2}\left(\frac{1}{E_s} - \frac{1}{E}\right)s_{ij} \tag{4}$$

where E_s is the secant modulus. The first two terms make up the linear elastic component of strain, and the third term is the plastic component. For conditions of plane stress imposing $\sigma_{33} = 0$, Eq. (4) can be inverted and written explicitly as

$$\begin{aligned}\sigma_{11} &= \frac{E_s}{1-v_s^2}\left[\varepsilon_{11} + v_s \varepsilon_{22}\right] \\ \sigma_{22} &= \frac{E_s}{1-v_s^2}\left[\varepsilon_{22} + v_s \varepsilon_{11}\right] \\ \sigma_{12} &= \frac{E_s}{1+v_s}\varepsilon_{12} \\ \varepsilon_{33} &= -\frac{v_s}{1-v_s}\left[\varepsilon_{11} + \varepsilon_{22}\right]\end{aligned} \tag{5}$$

where

$$v_s = \frac{1}{2} + \frac{E_s}{E}\left(v - \frac{1}{2}\right) \tag{6}$$

In order to determine the stress components, it is necessary to determine E_s and v_s with increasing strain. In the experiments, we determine the displacement vector $\mathbf{u}(\mathbf{x})$ at every point in the field of view using digital image correlation. Using the strain-displacement relations, we can determine the strain field $\boldsymbol{\varepsilon}(\mathbf{x})$. Next, we need to determine the stress field for a nonlinear material; this is accomplished by first estimating the effective stress. First, we calculate $\boldsymbol{\varepsilon} : \boldsymbol{\varepsilon}$ from Eq. (4) and simplify to get

$$\frac{2}{3}\left[\frac{1+v}{E} + \frac{3}{2}\left(\frac{E-E_s}{E_s E}\right)\right]^2 \sigma_e^2 = \left(\varepsilon_{ij}\varepsilon_{ij} - \frac{1}{3}\varepsilon_{kk}^2\right) \tag{7}$$

Since E_s in the left hand side is a function of σ_e, the above represents a nonlinear equation that can be solved for the effective stress $\sigma_e(\mathbf{x})$ from the measured strain field. This allows for conversion of the experimental measurements into contours of the Mises stress field. In particular, this can be used in a crack problem to determine the plastic zone boundary. Then E_s and v_s can be determined for the given material properties. Subsequently, the stress field $\boldsymbol{\sigma}(\mathbf{x})$ can be determined through Eq. (5). We illustrate this procedure here for the Ramberg-Osgood material model in Eq. (1). The secant modulus is a function of the stress level; this is found easily from Eq. (1):

$$E_s(\sigma_e) = E\left[1 + \alpha\left(\frac{\sigma_e}{\sigma_Y}\right)^{n-1}\right]^{-1} \tag{8}$$

One can work out a similar expression for other models of uniaxial constitutive response. Substituting for strains in the principal orientation, and utilizing the Ramberg-Osgood material model, Eq. (7) can be expanded to the following equation for the effective stress:

$$\begin{gathered}\frac{2}{3}\left[1+v+\frac{3}{2}\alpha\left(\frac{\sigma_e}{\sigma_y}\right)^{n-1}\right]^2\frac{\sigma_e^2}{E^2}-\left(\varepsilon_1^2+\varepsilon_2^2\right)\\-(\varepsilon_1+\varepsilon_2)^2\frac{\left\{\alpha\left(\frac{\sigma_e}{\sigma_y}\right)^{n-1}+2v\right\}^2-\frac{4}{3}(1-2v)^2}{\left\{\alpha\left(\frac{\sigma_e}{\sigma_y}\right)^{n-1}+2(1-v)\right\}^2}=0\end{gathered} \tag{9}$$

Equation (9) can be solved numerically for the equivalent stress at each point in the field with the measured values of the strains at each time step. This value of σ_e can be used in Eqs. (8) and (6) to calculate E_s and v_s, respectively, and then used in Eq. (5) to determine all components of stress.

Once the strain field and the stress field on the surface are determined, the J-integral can be evaluated numerically

$$J=\int_{\Gamma}\left(Wdy-T_\alpha\frac{\partial u_\alpha}{\partial x_1}ds\right)=\sum W\Delta y-\sum T_\alpha\frac{\partial u_\alpha}{\partial x_1}\Delta s \tag{10}$$

W is the strain energy, T_α are the components of the traction vector, u_α are the displacement components and ds is the length increment along the contour Γ. In addition to evaluating the J-integral, multiple contours around the crack tip can be defined, and the path independence of the J-integral can be examined.

References

1. Tada H, Paris PC, Irwin GR (2000) The stress analysis of cracks handbook. ASME Press, New York
2. Murakami Y (1987) Stress intensity factor handbook. Pergamon Press, New York
3. Zerbst U, Heinimann M, Donne CD, Steglich D (2009) Fracture and damage mechanics modelling of thin-walled structures. Eng Fract Mech 76:5–43
4. Beese AM, Luo M, Li Y, Bai Y, Wierzbicki T (2010) Partially coupled anisotropic fracture model for aluminum sheets. Eng Fract Mech 77:1128–1152
5. Barsoum I, Faleskog J (2007) Rupture in combined tension and shear: experiments. Int J Solids Struct 44:1768–1786
6. Haltom SS, Kyriakides S, Ravi-Chandar K (2013) Ductile failure under combined shear and tension. Int J Solids Struct 50:1507–1522
7. Boyce B, Gross A, Gharimaninezhad A, Ravi-Chandar K et al (2014) The sandia fracture challenge: blind predictions of ductile tearing. Int J Fract 186:5–68
8. Simonsen B, Törnqvist R (2004) Experimental and numerical modelling of ductile crack propagation in large-scale shell structures. Mar Struct 17:1–27
9. Pardoen T, Hachez F, Marchioni B, Blyth P, Atkins A (2004) Mode I fracture of sheet metal. J Mech Phy Solids 52: 423–452
10. El-Naaman SA, Nielsen KL (2013) Observations on mode I ductile tearing in sheet metals. Eur J Mech A/Solids 42:54–62
11. Orowan E (1948) Fracture and strength of solids. Report Prog Phy 12: 185–232

12. Ghahremaninezhad A and Ravi-Chandar K (2011) Ductile failure in polycrystalline OFHC copper. Int J Solids Struct 48: 3299–3311
13. Rivalin F, Pineau A, Di Fan, M, Besson J (2001) Ductile tearing of pipeline-steel wide plates I dynamic and quasi-static experiments. Eng Fract Mech 68, 329–345
14. Ghahremaninezhad A, Ravi-Chandar K (2012) Ductile failure behavior of polycrystalline Al 6061-T6. Int J Fract 174:177–202
15. Machida K, Yamada H (2006) Evaluation of mixed-mode stress intensity factor by digital image correlation and intelligent hybrid method. Proc World Acad Sci Eng Technol 12:29–34
16. Becker TH, Mostafavi M, Tait RB, Marrow TJ (2012) An approach to calculate the J-integral by digital image correlation displacement field method. Fatigue Fract Eng Mater Strcut 35:971–984
17. Vavrik D, Jandejsek I (2014) Experimental evaluation of contour J integral and energy dissipated in the fracture process zone. Eng Fract Mech 129:14–25
18. Yoneyama S, Arikawa S, Kusayanagi S, Hazumi K (2014) Evaluating J-integral from displacement fields measured by digital image correlation. Strain 50:147–160
19. Matweb. http://asm.matweb.com/search/SpecificMaterial.asp?bassnum=MA6061t6. Accessed 30 Mar 2015
20. Morgeneyer TF, Taillandier-Thomas T, Helfen L, Baumbach T, Sinclaire I, Roux S, Hild F (2014) In situ 3-D observation of early strain localization during failure of thin Al alloy (2198) sheet. Acta Mater 69:78–91

Part IV
Smart Materials

Interaction of Cracks and Domain Structures in Thin Ferroelectric Films

D. Schrade, R. Müller and D. Gross

Abstract The fracture behavior of ferroelectric materials is a complex problem that has been addressed in numerous experimental and theoretical studies. Several factors have been identified to play an important role, such as the applied electric field, the medium inside the crack, the electrical conditions on the crack faces, and polarization switching at or near the crack tip. In this investigation, a phase field model for ferroelectric domain evolution is used to calculate crack tip driving forces for mode-I cracks in barium titanate thin films. The driving forces are obtained by employing the theory of configurational forces, which is equivalent to considering the J-integral. Simulations are done for permeable, impermeable, semi-permeable, and energetically consistent crack face conditions with both air and water as crack medium. The finite element calculations are performed for films with thicknesses varying from 5 to 30 nm. The results show that the impermeable, semi-permeable and energetically consistent conditions lead to similar crack tip driving forces if air is used as crack medium. In the absence of mechanical loading, strong electric fields result in a closing crack tip driving force, while the use of water as crack medium leads to opposite driving forces. It can be confirmed that polarization switching at the crack tip has a significant effect on the driving force.

D. Schrade · R. Müller (✉)
Department of Mechanical and Process Engineering, Institute of Applied Mechanics, TU Kaiserslautern, 3049, 67653 Kaiserslautern, Germany
e-mail: ram@rhrk.uni-kl.de

D. Schrade
e-mail: schrade@rhrk.uni-kl.de

D. Gross
Division of Solid Mechanics, Department of Civil and Environmental Engineering, TU Darmstadt, Franziska-Braun-Straße 7, 64287 Darmstadt, Germany
e-mail: gross@mechanik.tu-darmstadt.de

G. Hütter and L. Zybell, *Recent Trends in Fracture and Damage Mechanics*, DOI 10.1007/978-3-319-21467-2_10

1 Introduction

Ferroelectric materials are widely used in industry e.g. for sensor- and actuator applications, transducers, and other micro-scale electromechanical components. These materials are however brittle, which makes a thorough understanding of the fracture behavior very desirable. In recent years, many experiments have been performed to study electromechanical fracture, see e.g. the review article [4]. Drawing conclusive results from the experimental research is not easy on account of a number of factors, such as the difference between single crystal or polycrystal materials, the geometry of the specimens and the type of cracks investigated, the applied loading (mechanical, electrical or combined), the medium in which the specimens are placed (air, oil), grain size, and the texture or electrical poling state. For example, the stress-induced switching of domains at the crack tip was investigated in [6, 18] for polycrystalline and single crystal barium titanate ($BaTiO_3$, BTO), respectively. In [21, 40], the influence of an applied electric field on the mode-I fracture toughness is investigated for lead zirconate titanate (PZT), and the dependency of the fracture behavior on the domain texture in PZT is addressed in [15]. The experimental findings suggest that a number of factors are crucial for an understanding of the underlying phenomena: the strength and direction of the applied electric field (relative to the macroscopic poling direction), the electrical conditions at the crack faces, domain switching at the crack tip, and the medium in the crack gap, see the discussion in [4].

Besides the experimental efforts, there is a number of theoretical studies and numerical simulations. For PZT, an analytical approach based on the Stroh formalism is presented in [40], and in [9] we can find an analytical treatment of a penny-shaped crack in an infinite piezoelectric body. Phenomenological models have been utilized to calculate fracture toughnesses and to determine switching zones at the crack tip for different electromechanical loadings, see e.g. [3, 12, 37]. Micromechanical models in conjunction with configurational forces have been used to simulate domain structures at the crack e.g. in [13, 24]. There are also a number of phase field studies which focus on the domain evolution at the crack tip in single crystals for different loading scenarios and crack face conditions. For lead titanate this has been done for permeable cracks in [38], for impermeable cracks in [31], and for both permeable and impermeable cracks in [23]. For BTO single crystals, the energy release rate at the crack tip is computed in [14] with the J-integral for the energetically consistent crack face condition. Crack propagation has been studied in [1] by a combined phase field model for crack growth and ferroelectric domain evolution.

This paper aims to study the influence of different crack face conditions on the domain evolution and the crack tip driving force in single crystal BTO. A previously published phase field model for ferroelectric domain evolution (see Sect. 2) is used to implement the permeable, impermeable, semi-permeable, and energetically consistent crack face conditions within a finite element framework. Driving forces are computed for electrical, mechanical, and combined loading.

Since the phase field model has an inner length scale, the thickness of the thin ferroelectric films is varied from 5 to 30 nm in order to investigate possible size effects. The publication [39] addresses essentially the same questions as the present paper. However, in that paper a domain wall width of about 130 nm was chosen. Here we use a more realistic value of 1.5 nm for the 180° domain wall width, which gives very different results regarding the domain structure and the crack tip driving forces.

2 Phase Field Model

There is a long history of phase field simulations for ferroelectrics. A literature review on the different model formulations and numerical implementation strategies can be found e.g. in [27]. The phase field model used for the present simulations has been published in [30]; therefore we limit ourselves to stating the relevant equations and referring to relevant past publications for further theoretical background.

Given a ferroelectric body $\mathcal{B}$, we define the linearized strain tensor $\boldsymbol{\varepsilon}$ and the electric field $\boldsymbol{E}$ by

$$\boldsymbol{\varepsilon} = \mathrm{sym}(\nabla \boldsymbol{u}), \quad \boldsymbol{E} = -\nabla \varphi, \tag{1}$$

where $\boldsymbol{u}$ is the displacement field and φ the electric potential. Furthermore, the stress $\boldsymbol{\sigma}$ and the electric displacement $\boldsymbol{D}$ satisfy mechanical equilibrium and Gauss's law:

$$\mathrm{div}\, \boldsymbol{\sigma} = \boldsymbol{0}, \quad \mathrm{div}\, \boldsymbol{D} = 0, \tag{2}$$

where volume forces and volume charge densities are not taken into account. In the present model, the spontaneous polarization field $\boldsymbol{P}(\boldsymbol{x}, t)$ is used as the (time-dependent) order parameter. A detailed discussion on the choice of the order parameter in phase field models for ferroelectrics can be found in [27]. A thermodynamical analysis based on Gurtin's notion of microforces (cf. [7, 28, 32]) yields the constitutive equations

$$\boldsymbol{\sigma} = \frac{\partial H}{\partial \boldsymbol{\varepsilon}}, \quad \boldsymbol{D} = -\frac{\partial H}{\partial \boldsymbol{E}}, \quad \boldsymbol{\Sigma} = \frac{\partial H}{\partial \nabla \boldsymbol{P}}. \tag{3}$$

Therein, $H = \tilde{H}(\boldsymbol{\varepsilon}, \boldsymbol{E}, \boldsymbol{P}, \nabla \boldsymbol{P})$ is the electric enthalpy which is obtained by a Legendre transform of the Helmholtz free energy density, i.e. $\Psi = H + \boldsymbol{E} \cdot \boldsymbol{D}$ (see e.g. [32]). The stress-like quantity $\boldsymbol{\Sigma}$ is referred to as the microstress tensor. Furthermore, the thermodynamical framework presented in [7, 28] allows for a derivation of the Ginzburg-Landau type evolution equation

$$\beta\dot{\boldsymbol{P}} = \operatorname{div}\boldsymbol{\Sigma} - \frac{\partial H}{\partial \boldsymbol{P}}, \tag{4}$$

where β is a (scalar) inverse mobility constant. The boundary conditions for the mechanical and the electrical problems are given by

$$\begin{aligned} \boldsymbol{u} - \boldsymbol{u}^* &= \boldsymbol{0} \quad \text{on } \partial\mathcal{B}_u, \quad \boldsymbol{\sigma}\boldsymbol{n} - \boldsymbol{t}^* = \boldsymbol{0} \quad \text{on } \partial\mathcal{B}_\sigma, \\ \varphi - \varphi^* &= 0 \quad \text{on } \partial\mathcal{B}_\varphi, \quad \boldsymbol{D}\cdot\boldsymbol{n} + \rho_s^* = 0 \quad \text{on } \partial\mathcal{B}_D. \end{aligned} \tag{5}$$

Therein the quantities with an asterisk are prescribed displacements, electric potentials, surface tractions, and surface charge densities. The outer unit normal to the boundary $\partial\mathcal{B}$ is denoted by $\boldsymbol{n}$. Additionally, we postulate that

$$\boldsymbol{\Sigma}\boldsymbol{n} = \boldsymbol{0} \quad \text{on } \partial\mathcal{B}, \tag{6}$$

as is commonly (and often tacitly) done in phase field models for ferroelectrics. The evolution equation (4) requires some initial conditions, which are introduced by

$$\boldsymbol{P}(\boldsymbol{x})|_{t=0} = \boldsymbol{P}^*(\boldsymbol{x}). \tag{7}$$

With reference to [28], the electric enthalpy $H = \hat{H}(\boldsymbol{\varepsilon}, \boldsymbol{E}, \boldsymbol{P}, \nabla\boldsymbol{P})$ is given by

$$\begin{aligned} H =& \frac{1}{2}\left(\boldsymbol{\varepsilon} - \boldsymbol{\varepsilon}^0\right)\cdot\left[\mathbf{C}\left(\boldsymbol{\varepsilon} - \boldsymbol{\varepsilon}^0\right)\right] - \left[\boldsymbol{e}\left(\boldsymbol{\varepsilon} - \boldsymbol{\varepsilon}^0\right) + \frac{1}{2}\boldsymbol{\epsilon}\boldsymbol{E} + \boldsymbol{P}\right]\cdot\boldsymbol{E} \\ &+ \kappa_s\frac{\gamma}{\ell}\psi(\boldsymbol{P}) + \frac{1}{2}\kappa_g\frac{\gamma\ell}{P_0^2}\|\nabla\boldsymbol{P}\|^2. \end{aligned} \tag{8}$$

The first line of this equation resembles a piezoelectric material law extended by the term $-\boldsymbol{P}\cdot\boldsymbol{E}$ and the spontaneous strain $\boldsymbol{\varepsilon}^0(\boldsymbol{P})$. The material tensors $\mathbf{C}$, $\mathbf{e}$, and $\boldsymbol{\epsilon}$ depend on the direction of polarization, i.e. on the order parameter $\boldsymbol{P}$. These dependencies have given rise to an invariant formulation of the electric enthalpy for transverse isotropy; this has been published in [28]. The resulting representations for these tensors are thus given by

$$\begin{aligned} \mathbf{C}(\boldsymbol{P}) =& \lambda\mathbf{1}\otimes\mathbf{1} + 2\mu\mathbb{I} + \omega\alpha_1\Xi + 2(\alpha_2 + \omega\alpha_4)\mathbf{m}\otimes\mathbf{m} \\ &+ \alpha_3(\mathbf{1}\otimes\mathbf{m} + \mathbf{m}\otimes\mathbf{1}), \end{aligned} \tag{9}$$

$$\boldsymbol{e}(\boldsymbol{P}) = -\omega\beta_1\boldsymbol{p}\otimes\mathbf{1} - \beta_2\boldsymbol{p}\otimes\mathbf{m} - \omega\beta_3\hat{\boldsymbol{e}}, \tag{10}$$

$$\boldsymbol{\epsilon}(\boldsymbol{P}) = -2(\gamma_1 + \omega\gamma_3)\mathbf{1} - 2\gamma_2\mathbf{m}. \tag{11}$$

Therein $\boldsymbol{p} = \boldsymbol{P}/P_0$ is the preferred direction, and $\mathbf{m} = \boldsymbol{p}\otimes\boldsymbol{p}$ is the structural tensor (the constant P_0 is the value of the spontaneous polarization of the unloaded crystal). Furthermore, $\omega = \operatorname{tr}\mathbf{m}$, $\mathbb{I}_{ijkl} = \frac{1}{2}(\delta_{ik}\delta_{jl} + \delta_{il}\delta_{jk})$, $\Xi_{ijkl} = p_i\delta_{jk}p_l + p_k\delta_{il}p_j$, and $\hat{e}_{kij} = \frac{1}{2}(\delta_{ik}p_j + \delta_{jk}p_i)$. The material parameters α_i, β_i, and γ_i are the elastic, piezoelectric, and dielectric constants, respectively. They are related to the "classic" constants by (see [28])

$$\begin{aligned}
&\lambda = c_{12}, \quad \mu = \frac{1}{2}(c_{11} - c_{12}), \quad \alpha_1 = 2c_{44} + c_{12} - c_{11}, \\
&\alpha_2 = \frac{1}{2}c_{11} - 2c_{44} - c_{13}, \quad \alpha_3 = c_{13} - c_{12}, \quad \alpha_4 = \frac{1}{2}c_{33}, \\
&\gamma_1 = -\frac{1}{2}\bar{\epsilon}_{33}, \quad \gamma_2 = \frac{1}{2}(\bar{\epsilon}_{11} - \bar{\epsilon}_{33}), \quad \gamma_3 = -\frac{1}{2}(\bar{\epsilon}_{11} - \bar{\epsilon}_{33}), \\
&\beta_1 = -\bar{e}_{31}, \quad \beta_2 = -\bar{e}_{33} + 2\bar{e}_{15} + \bar{e}_{31}, \quad \beta_3 = -2\bar{e}_{15}.
\end{aligned} \tag{12}$$

The spontaneous strain $\boldsymbol{\varepsilon}^0$ is given by

$$\boldsymbol{\varepsilon}^0(\boldsymbol{P}) = \varepsilon_a \mathbf{1}\omega + (\varepsilon_c - \varepsilon_a)\mathbf{m}, \tag{13}$$

where

$$\varepsilon_a = \frac{a_{tet} - a_{cub}}{a_{cub}}, \quad \varepsilon_c = \frac{c_{tet} - a_{tet}}{a_{cub}}, \tag{14}$$

and a_{tet} and c_{tet} are the lattice constants for tetragonal BTO, and a_{cub} is the unit cell parameter of the cubic phase.

The second line in (8) contains the non-convex polynomial $\psi(\boldsymbol{P})$ and the order parameter gradient energy density. The polynomial ψ is chosen to be of sixth degree; it has the independent constants ψ_i, $i = 1, \ldots, 4$:

$$\begin{aligned}
\psi = \kappa_s \frac{\gamma}{\ell}\Bigg[1 &+ \frac{\psi_1}{P_0^2}\left(P_1^2 + P_2^2 + P_3^2\right) + \frac{\psi_2}{P_0^4}\left(P_1^4 + P_2^4 + P_3^4\right) \\
&+ \frac{\psi_3}{P_0^4}\left(P_1^2P_2^2 + P_1^2P_3^2 + P_2^2P_3^2\right) + \frac{\psi_4}{P_0^6}\left(P_1^6 + P_2^6 + P_3^6\right)\Bigg].
\end{aligned} \tag{15}$$

This energy landscape ensures phase separation and exhibits four minima in the 2D case (see Fig. 1). For details on the construction of $\psi(\boldsymbol{P})$ the reader is referred to [28]. In conjunction with the calibration constants κ_s and κ_g, the model parameter γ is the energy density of a 180° interface, and ℓ is the width of this interface. The constants κ_s and κ_g are determined by considering a single 180° domain wall in a 1D setting while neglecting the terms in the first line in the electric enthalpy (8). As shown in [27], if

$$\kappa_s^{-1} = \frac{1}{2}\kappa_g^{-1} = \int_{-1}^{1} \sqrt{1 + \psi_1 q^2 + \psi_2 q^4 + \psi_4 q^6}\,\mathrm{d}q, \tag{16}$$

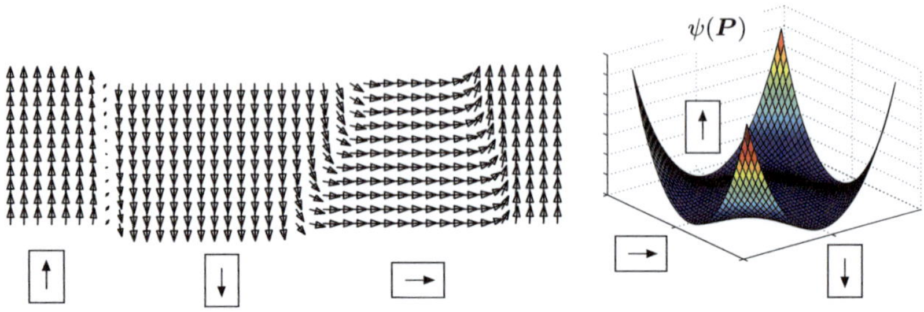

Fig. 1 Each minimum of the non-convex function $\psi(\boldsymbol{P})$ corresponds to one variant of the spontaneous polarization. This allows for the formation of domain patterns as shown on the *left hand side* of the figure

then the minimized total energy of the 1D system coincides with the parameter γ, and the width of the interface is equal to ℓ. Finally, by inserting the electric enthalpy (8) into (3), we obtain the constitutive equations

$$\begin{aligned}\boldsymbol{\sigma} &= \mathbf{C}(\boldsymbol{\varepsilon} - \boldsymbol{\varepsilon}^0) - \boldsymbol{e}^{\mathrm{T}}\boldsymbol{E},\\ \boldsymbol{D} &= \boldsymbol{e}(\boldsymbol{\varepsilon} - \boldsymbol{\varepsilon}^0) + \boldsymbol{\epsilon}\boldsymbol{E} + \boldsymbol{P},\\ \boldsymbol{\Sigma} &= \kappa_{\mathrm{g}}\frac{\gamma\ell}{P_0^2}\nabla\boldsymbol{P},\end{aligned} \tag{17}$$

and the evolution equation (4) becomes

$$\beta\dot{\boldsymbol{P}} = \kappa_{\mathrm{g}}\frac{\gamma\ell}{P_0^2}\Delta\boldsymbol{P} - \frac{\partial H}{\partial \boldsymbol{P}}, \tag{18}$$

where Δ denotes the Laplacian.

The primary aim of this investigation is to compare the crack tip driving forces resulting from different crack face conditions. The driving force at the crack tip is the negative value of the crack tip configurational force, which in turn is equivalent to the *J*-integral, see e.g. [16] and the formulation given in [14]. The configurational force can be obtained by taking the total derivative of the electric enthalpy:

$$H_{,k} = \frac{\partial H}{\partial \varepsilon_{ij}}\varepsilon_{ij,k} + \frac{\partial H}{\partial E_j}E_{j,k} + \frac{\partial H}{\partial P_j}P_{j,k} + \frac{\partial H}{\partial P_{i,j}}P_{i,jk} + \left.\frac{\partial H}{\partial x_k}\right|_{\mathrm{expl.}}. \tag{19}$$

Using Eqs. (1)–(6), one obtains in index notation

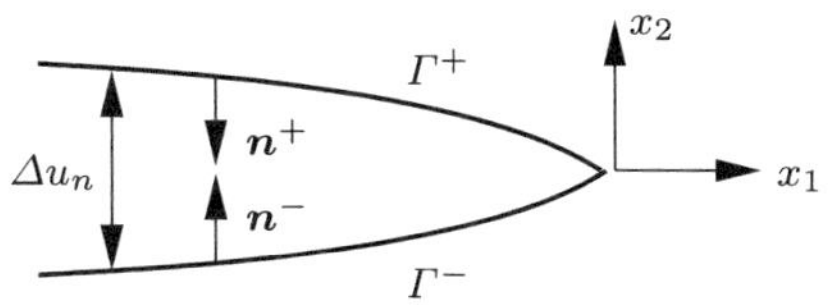

Fig. 2 Definition of the normal vectors and the crack opening displacement

$$\begin{aligned}\frac{\partial H}{\partial \varepsilon_{ij}}\varepsilon_{ij,k} &= \left(\sigma_{ij}u_{i,k}\right)_{,j}\\ \frac{\partial H}{\partial E_j}E_{j,k} &= \left(D_j\varphi_{,k}\right)_{,j}\\ \frac{\partial H}{\partial P_j}P_{j,k}+\frac{\partial H}{\partial P_{i,j}}P_{i,jk} &= \left(\Sigma_{ij}P_{i,k}\right)_{,j}-\beta\dot{P}_iP_{i,k}.\end{aligned} \tag{20}$$

With the definitions of the generalized configurational stress

$$\boldsymbol{\Xi} = H\mathbf{1} - (\nabla \boldsymbol{u})^{\mathrm{T}}\boldsymbol{\sigma} - \nabla\varphi \otimes \boldsymbol{D} - (\nabla \boldsymbol{P})^{\mathrm{T}}\boldsymbol{\Sigma} \tag{21}$$

and the configurational force vector

$$\boldsymbol{\xi} = -\left.\frac{\partial H}{\partial \boldsymbol{x}}\right|_{\text{expl.}} + \beta(\nabla \boldsymbol{P})^{\mathrm{T}}\dot{\boldsymbol{P}} \tag{22}$$

one arrives at the configurational force balance

$$\operatorname{div}\boldsymbol{\Xi} + \boldsymbol{\xi} = \mathbf{0}. \tag{23}$$

For the crack illustrated in Fig. 2 the scalar crack tip driving force τ is given by

$$\tau = -\xi_1. \tag{24}$$

The nodal crack tip driving force can be evaluated in a post-processing computation; further details are given in [39].

3 Crack Face Conditions

Several types of crack face conditions have been published in the literature; they vary in their treatment of the electrical and/or traction conditions across the crack. Here we consider four different crack face conditions: permeable, impermeable, semi-permeable, and energetically consistent. Since these types of conditions are discussed in detail e.g. in [11], we restrict ourselves to stating the different conditions without much further discussion.

In [22], a crack in a piezoelectric material is modeled as traction free and electrically permeable, i.e.

$$D_n^+ = D_n^-, \quad \varphi^+ = \varphi^-, \quad \boldsymbol{t}^\pm = \boldsymbol{0}, \tag{25}$$

where $D_n^\pm = \boldsymbol{D}^\pm \cdot \boldsymbol{n}^-$ and $\boldsymbol{t}^\pm = \boldsymbol{\sigma}^\pm \boldsymbol{n}^\pm$. The superscripts '+' and '−' indicate quantities which are evaluated at the crack faces, see Fig. 2. The permeable condition seems reasonable if the crack opening displacement is very small or vanishing. If that is not the case, the impermeable condition proposed in [5] may be applied, i.e.

$$D_n^\pm = 0, \quad \boldsymbol{t}^\pm = \boldsymbol{0}. \tag{26}$$

As a consequence, the electric field vanishes completely inside the crack. These two crack face conditions represent two extreme cases; a third possibility introduced in [8] is the semi-permeable condition, in which the crack is treated as a capacitor, i.e.

$$D_n^\pm = -\kappa_c \frac{\Delta\varphi}{\Delta u_n}, \quad \boldsymbol{t}^\pm = \boldsymbol{0}, \tag{27}$$

where $\Delta\varphi = \varphi^+ - \varphi^-$ and κ_c is the permittivity of the crack gap. As could be shown in [17], the crack tip energy release rate does not coincide with the total energy release rate given the semi-permeable condition. This inconsistency has been resolved in [11] by introducing an energetically consistent condition:

$$D_n^\pm = -\kappa_c \frac{\Delta\varphi}{\Delta u_n}, \quad \boldsymbol{t}^\pm = \frac{1}{2}\kappa_c \left(\frac{\Delta\varphi}{\Delta u_n}\right)^2 \boldsymbol{n}^\pm. \tag{28}$$

The use of these crack face conditions result in identical crack tip and total energy release rates.

4 Numerical Simulations

The presented phase field model is implemented in the finite element code *FEAP* [35]. The 2D implementation is based on quadrilateral four-noded elements with bilinear ansatz functions. The nodal degrees of freedom are the displacement field, the electric potential, and the order parameter. The occurring element integrals are computed by Gaussian integration, and the evolution equation is numerically integrated by using the first order backwards Euler method. For details on the implementation the reader is referred to [26]. The semi-permeable and energetically

Table 1 Material parameters used in the simulations

c_{11}	c_{33}	c_{12}	c_{13}	c_{44}	$\bar{e}_{31}$	$\bar{e}_{33}$	$\bar{e}_{15}$	$\bar{\epsilon}_{11}$	$\bar{\epsilon}_{33}$	a_{cub}
23.05	15.75	10.8	10.4	12.67	1.327	4.96	0	0.209	0.209	4.01
a_{tet}	c_{tet}	$P_0[\frac{\text{C}}{\text{m}^2}]$	$\beta[\frac{\text{kA}}{\text{Vm}}]$	$\gamma[\frac{\text{mJ}}{\text{m}^2}]$	ℓ[nm]	$\psi_1[-]$	$\psi_2[-]$	$\psi_3[-]$	$\psi_4[-]$	
3.992	4.032	0.26	0.3467	12	1.5	−1.147	−0.7067	5.36	0.8533	

The mechanical stiffnesses c_{ij} are stated in 10^{10}N/m^2, the piezoelectric constants $\bar{e}_{ij}$ in C/m^2, and the dielectric constants $\bar{\epsilon}_{ij}$ in 10^{-10}C/(Vm). The lattice parameters $a_{\text{cub/tet}}$ and c_{tet} are given in Å

consistent crack face conditions have been implemented by using special crack face elements, see [39] for the resulting element residuals. The numerical treatment of configurational forces in a FEM framework is discussed e.g. in [19, 20], cf. also [39].

The material parameters in (12) and (14), as well as P_0, β, γ, ℓ, and ψ_i in (15) are chosen to reflect the behavior of BTO. The values for these constants coincide with those used in [30]; for completeness they are stated in Table 1. Notably, the 180° domain wall width is chosen as 1.5 nm, and the 180° domain wall specific energy is 12 mJ/m^2. The 2D simulations are performed for plane strain conditions, see the extended explanation given in [27].

4.1 Problem Setup

We consider a cracked ferroelectric subjected to electrical and/or mechanical loading, see the sketch in Fig. 3. The electrical loading induces an electric field in the vertical direction. Additionally a mechanical stress σ_0 may be applied. All simulations were done for charge-free lateral sides and statically determinate support. The semi-permeable and energetically consistent crack face conditions require

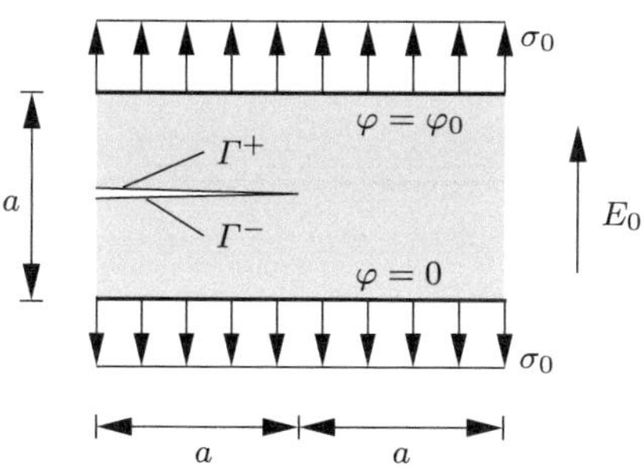

Fig. 3 Mechanical and electrical loading used in the simulations. The potential difference φ_0 induces the nominal electric field $E_0 = -\varphi_0/a$. The initial crack opening at the left lateral side is set to $a/30$

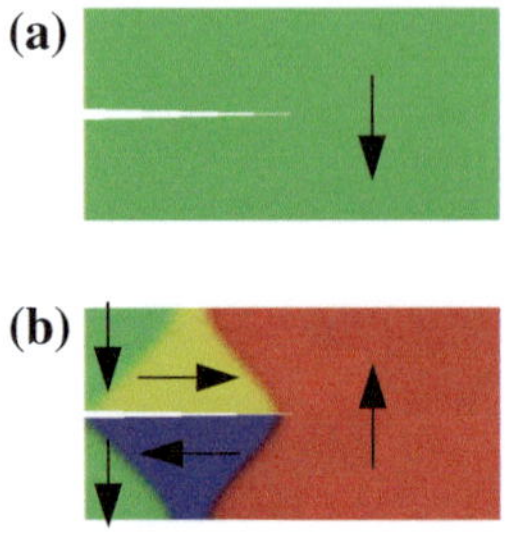

Fig. 4 Initial domain configurations are calculated starting from a randomly distributed polarization. The equilibrated states are homogeneous for the permeable crack (**a**) and inhomogeneous for all other crack face conditions (**b**). These initial states are used for the subsequent computation of the crack tip driving forces

a specification of the permittivity κ_c inside the crack. Here we consider two different media: air ($\kappa_c = 8.854 \times 10^{-12}\,\text{C/(Vm)}$) and water ($\kappa_c = 7.10 \times 10^{-10}\,\text{C/(Vm)}$).

The initial distribution of the order parameter is calculated by running a simulation for each crack face condition with random initial polarization and $\varphi_0 = \sigma_0 = 0$. The resulting polarization fields are shown in Fig. 4. The permeable condition results in a homogeneous poling state in the vertical direction (due to the charge-free lateral sides). The remaining crack face conditions lead to the domain structure depicted in Fig. 4b. These two microstructures are used as initial conditions for the subsequent calculations of the crack tip driving forces.

4.2 *Crack Tip Driving Forces*

The electrical and mechanical loadings shown in Fig. 3 are applied to ferroelectric films ranging from $a = 5$ to $a = 30$ nm in size. The first set of simulations is done for purely electric loading ($\sigma_0 = 0$) with electric fields of $E_0 = \pm 1\,\text{MV/m}$ and $E_0 = \pm 5\,\text{MV/m}$. That is to say, we apply the potential difference $\varphi_0 = -E_0 a$, where E_0 is the nominal applied electric field (cf. Fig. 3). In another set of simulations, a purely mechanical loading of 30 MPa with $\varphi_0 = 0$ is applied. Finally, simulations with combined electrical/mechanical loading are performed for electric fields between −5 and +5MV/m. Note that the loading may lead to a reorganization of the domain structure. On the other hand, if the loading is sufficiently weak, the domain structure will remain essentially unchanged. For both cases, the crack tip driving force is computed at the equilibrated state of the system.

The permeable crack face condition results in a vanishing driving force if a purely electrical loading is applied. Since the crack is permeable, the electric field in the material is essentially homogeneous (a small inhomogeneity is due to the small crack opening in the initial configuration). Furthermore, the stress, the strain, the

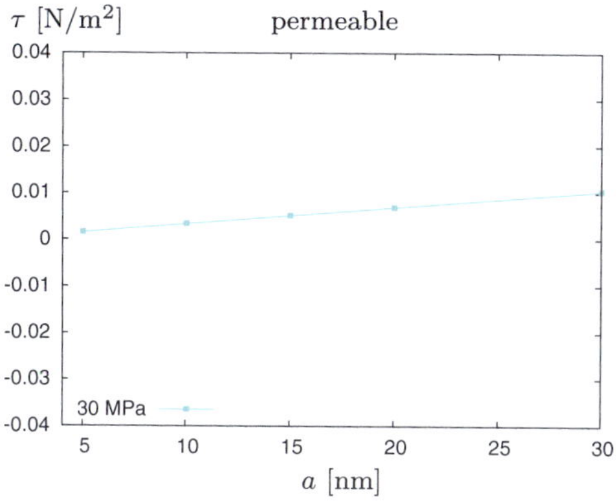

Fig. 5 For purely mechanical loading with $\sigma_0 = 30\,\text{MPa}$, the permeable crack results in a driving force which depends linearly on the crack length a. For purely electrical loading the driving forces vanish

electric displacement, the microforce tensor, and the order parameter are also homogeneous. As a result, the divergence of the configurational stress in (21) vanishes, and thus the configurational force at the crack tip also vanishes. On the other hand, the mechanical loading results in a positive driving force (see Fig. 5). Since the domain structure remains unchanged under the mechanical loading, the driving forces depend linearly on the crack length a, cf. the analytical work published in [34].

As can be seen in Figs. 6, 7 and 8, the results for the impermeable, semi-permeable, and energetically consistent crack face conditions are qualitatively very similar if air is considered as crack medium. The low permittivity of air leads to almost identical electrical conditions compared to the impermeable case. The initial

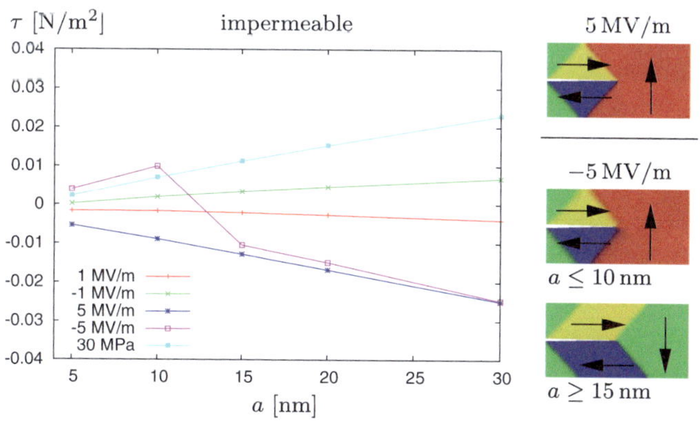

Fig. 6 Crack tip driving forces for the impermeable crack. The equilibrated domain configurations for $E_0 = \pm 5\text{MV/m}$ are shown on the *right*. The remaining final configurations coincide with the initial configurations shown in Fig. 4

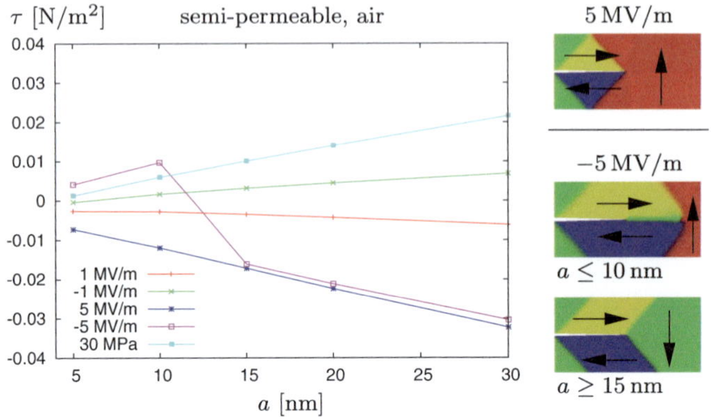

Fig. 7 The resulting driving forces for the semi-permeable crack are similar to those for the impermeable crack, although the final domain configurations for $E_0 = -5\mathrm{MV/m}$ differ if $a \leq 10\mathrm{nm}$

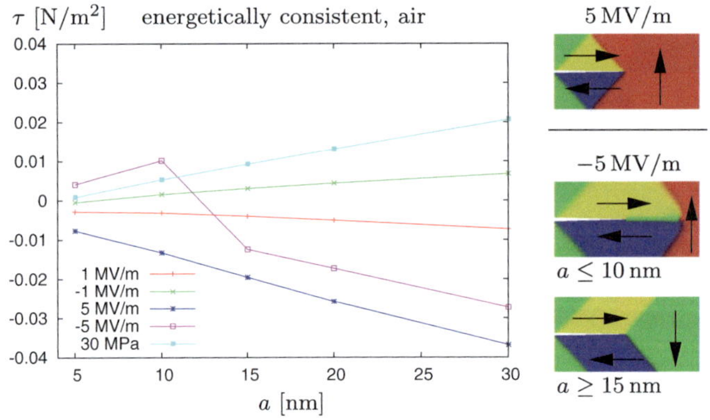

Fig. 8 The driving forces for the energetically consistent condition are similar to those shown in Figs. 6 and 7. The final domain configurations are the same as for the semi-permeable crack

domain pattern remains unchanged for all applied loadings, except for the electric loading with −5MV/m. In that case, the change in the domain structure sets in for a = 15 nm and leads to a change of sign in the driving force. Noticeably, the strong electric fields lead to negative values for τ for $a \geq 15\mathrm{nm}$, which means that the system is energetically in favor of closing the crack. The reduction of the crack tip driving force by strong electric fields is a well-established result, see e.g. [1, 4, 9].

While the use of air as crack medium for the semi-permeable and the energetically consistent condition leads to similar results compared to the impermeable case, the situation changes significantly when water is used as crack medium. As can be seen in Figs. 9 and 10, all driving forces are now close to zero or positive; this was not the case in the previous simulations. The application of the strong

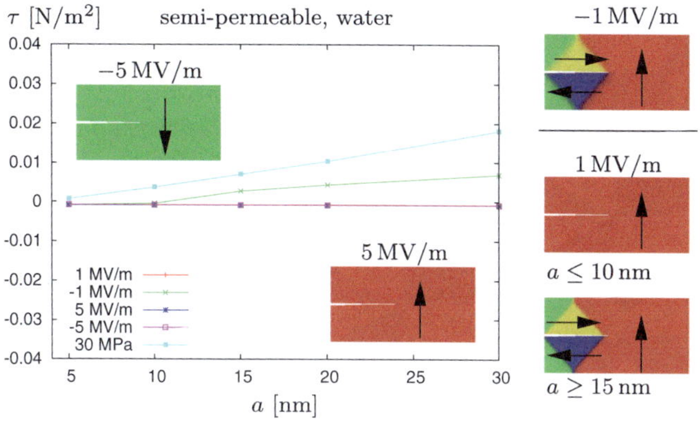

Fig. 9 For a semi-permeable gap filled with water the driving force is positive only for the mechanical loading and for $E_0 = -1\mathrm{MV/m}$. The loading with strong electric fields causes domain switching until a homogeneous poling state is reached; the driving forces vanish in this case

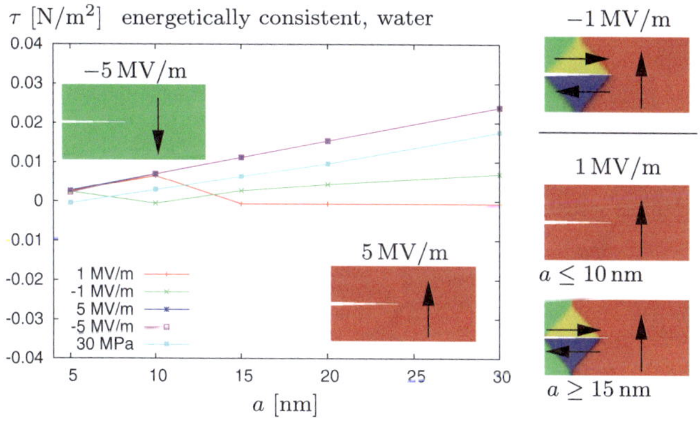

Fig. 10 Contrary to the results in Fig. 9, the driving force is maximal for strong electric fields if the energetically consistent condition is used

electric fields leads to a complete poling in the direction of the electric field. In this case the driving force vanishes as was the case for the permeable condition. The energetically consistent condition with water as medium on the other hand results in high driving forces for the strong electric fields, even though the applied fields lead to full poling as was the case for the semi-permeable condition.

The possible domain switching processes introduce a nonlinearity so that the driving forces from the purely electrical and the purely mechanical loading may not be superimposed. Therefore we performed simulations with combined electrical and mechanical loading for thin films of 60 nm × 30 nm in size with water as crack medium. The resulting driving forces and some of the equilibrated domain structures are shown in Fig. 11. The application of strong electric fields leads to a

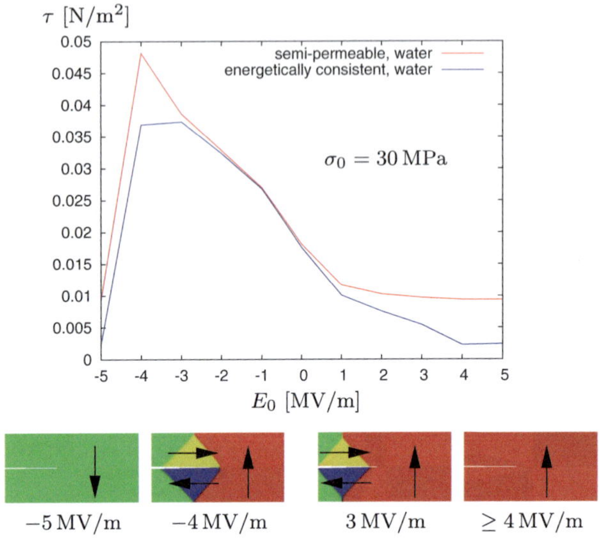

Fig. 11 Crack tip driving forces under combined mechanical and electrical loading. The final domain configurations plotted below the graphs are identical for both crack face conditions used in the simulations. Between −4 and +3MV/m the upward-poled domain grows continuously at the expense of the other domains

complete poling, whereby the driving force is drastically reduced. For applied electric fields between −4 and +3MV/m there is some switching of polarization near the crack tip, which leads to quite different results for the driving force: For both the semi-permeable and the energetically consistent condition the driving force is maximal for a negative electric field. Positive electric fields result in a reduction of the driving force compared to the case of purely mechanical loading ($E_0 = 0$).

5 Discussion

The presented simulations are based on certain assumptions. First, it was necessary to specify some boundary conditions and an initial domain configuration. Here we considered only non-periodic boundary conditions and calculated initial polarizations starting from a randomly polarized state. For different boundary conditions these initial polarizations may well be very different from the ones used here. Another relevant parameter is the permittivity inside the crack. The calculated crack tip driving forces and the microstructure evolution are quite different depending on whether the permittivity of air or that of water has been used. For other permittivities the results may again be quite different from the results obtained here. The simulations are also limited to single crystal BTO, whereby only one orientation of the crystal axes has been considered.

The results are also limited with respect to the applied loading conditions. Typically, relatively weak loading will leave the initial overall domain pattern intact; for stronger electric loading the domain structure may reorganize completely. The possible changes in the domain structures also give rise to the question of the crack tip driving force during the evolution of the system. This has not been considered here, although the theory outlined here and in [39] account for the time-dependent switching process in the configurational force.

Despite these limitations, it is possible to gain insight regarding the relevant parameters for the outcome of the crack tip driving force. First, the results for the impermeable crack and a crack filled with air are quite similar due to the almost identical electrical conditions at the crack faces. This changes when water with its two orders of magnitude higher permittivity is used as crack medium. In this case, the driving force may vanish for strong electric fields if the semi-permeable condition is used. On the other hand, the driving force has positive values if the energetically consistent condition is used. If air is used as medium, the driving force has negative values for strong electric fields. Thus the permittivity of the crack medium as well as the crack face condition play a crucial role for the prediction of the fracture behavior based on the configurational force at the crack tip.

Another important factor is the switching of polarization near the crack tip due to the applied loading, i.e. mainly the applied electric field. Even weak electric fields lead to a switching of polarization near the crack tip, which has a considerable effect on the resulting driving forces, cf. especially the results for the combined loading shown in Fig. 11. This result is in principal agreement with experimental and theoretical results published in the literature, see e.g. the review article [4].

The phase field model has an inner length scale which is given by the width of the transition region between two 180° domains. This inner length scale is known to be responsible for a variety of size effects, see e.g. the computational and experimental results published in [2, 10, 25, 29, 33, 36]. The present results show that at least for 15nm $\leq a \leq$ 30nm the driving force is approximately linear in the crack length a, but there is no noticeable size effect due to the change of the ratio of domain wall energy and bulk energy. For smaller crack lengths the domain evolution was found to be different at least for some loading scenarios. In these cases the driving force does not follow the linear relationship because of a different domain switching process, not because of a size effect in the usual sense.

Acknowledgments Financial support by the German Research Foundation (DFG) in the framework of the research group FOR 1509 "Ferroische Funktionsmaterialien—Mehrskalige Modellierung und experimentelle Charakterisierung" is gratefully acknowledged.

References

1. Abdollahi A, Arias I (2012) Phase-field modeling of crack propagation in piezoelectric and ferroelectric materials with different electromechanical crack conditions. J Mech Phys Solids 60(12):2100–2126

2. Arlt G, Sasko P (1980) Domain configuration and equilibrium size of domains in $BaTiO_3$ ceramics. J Appl Phys 51(9):4956–4960
3. Beom HG, Atluri SN (2003) Effect of electric fields on fracture behavior of ferroelectric ceramics. J Mech Phys Solids 51(6):1107–1125
4. Chen YH, Hasebe N (2005) Current understanding on fracture behaviors of ferroelectric/piezoelectric materials. J Intell Mater Syst Struct 16:673–687
5. Deeg WF (1980) The analysis of dislocation, cracks, and inclusion problems in piezoelectric solids. Ph.D. thesis, Stanford University, Stanford
6. Fang D, Jiang Y, Li S, Sun CT (2007) Interactions between domain switching and crack propagation in poled $BaTiO_3$ single crystal under mechanical loading. Acta Mater 55 (17):5758–5767
7. Gurtin ME (1996) Generalized Ginzburg-Landau and Cahn-Hilliard equations based on a microforce balance. Physica D 92(3–4):178–192
8. Hao TH, Shen ZY (1994) A new electric boundary condition of electric fracture mechanics and its applications. Eng Fract Mech 47:793–802
9. Jiang LZ, Sun CT (2001) Analysis of indentation cracking in piezoceramics. Int J Solids Struct 38(10–13):1903–1918
10. Keip MA, Schrade D, Thai HNM, Schröder J, Svendsen B, Müller R, Gross D (2015) Coordinate-invariant phase field modeling of ferroelectrics, part II: application to composites and polycrystals. GAMM-Mitteilungen 38(1):115–131
11. Landis CM (2003) On the fracture toughness of ferroelastic materials. J Mech Phys Solids 51 (8):1347–1369
12. Landis CM (2004) Energetically consistent boundary conditions for electromechanical fracture. Int J Solids Struct 41:6291–6315
13. Li Q, Kuna M (2012) Evaluation of electromechanical fracture behavior by configurational forces in cracked ferroelectric polycrystals. Comput Mater Sci 57:94–101
14. Li W, Landis CM (2011) Nucleation and growth of domains near crack tips in single crystal ferroelectrics. Eng Fract Mech 78(7):1505–1513
15. Li Y, Sun Y, Li F (2013) Domain texture dependent fracture behavior in mechanically poled/depoled ferroelectric ceramics. Ceram Int 39(8):8605–8614
16. Maugin GA (1993) Material inhomogeneities in elasticity. Chapman & Hall, London
17. McMeeking RM (2004) The energy release rate for a griffith crack in a piezoelectric material. Eng Fract Mech 71:1149–1163
18. Meschke F, Kolleck A, Schneider GA (1997) R-curve behavior of $BaTiO_3$ due to stress-induced ferroelastic domain switching. J Eur Ceram Soc 17:1143–1149
19. Mueller R, Maugin G (2002) On material forces and finite element discretizations. Comp Mech 29(1):52–60
20. Mueller R, Kolling S, Gross D (2002) On configurational forces in the context of the finite element method. Int J Numer Meth Eng 53(7):1557–1574
21. Park S, Sun CT (1995) Fracture criteria for piezoelectric ceramics. J Eur Ceramics Soc 78 (6):1475–1480
22. Parton VZ (1976) Fracture mechanics of piezoelectric materials. Acta Astronaut 3:671–683
23. Qiao H, Wang J, Chen W (2012) Phase field simulation of domain switching in ferroelctric single crystal with electrically permeable and impermeable cracks. Acta Mech Sol Sinica 25 (1):1–8
24. Ricoeur A, Kuna M (2003) Influence of electric fields on the fracture of ferroelectric ceramics. J Eur Ceram Soc 23(8):1313–1328
25. Schmitt LA, Schönau KA, Theissmann R, Fuess H, Kungl H, Hoffmann M (2007) Composition dependence of the domain configuration and size in $Pb(Zr_{1-x}Ti_x)O_3$ ceramics. J Appl Phys 101(7):074107
26. Schrade D, Mueller R, Xu BX, Gross D (2007) Domain evolution in ferroelectric materials: a continuum phase field model and finite element implementation. Comput Meth Appl Mech Eng 196(41–44):4365–4374

27. Schrade D, Müller R, Gross D (2013) On the physical interpretation of material parameters in phase field models for ferroelectrics. Arch Appl Mech 83:1393–1413
28. Schrade D, Müller R, Gross D, Keip MA, Thai H, Schröder J (2014) An invariant formulation for phase field models in ferroelectrics. Int J Solids Struct 51:2144–2156
29. Schrade D, Keip MA, Thai HNM, Schröder J, Svendsen B, Müller R, Gross D (2015a) Coordinate-invariant phase field modeling of ferroelectrics, part I: model formulation and single-crystal simulations. GAMM-Mitteilungen 38(1):102–114
30. Schrade D, Müller R, Gross D, Steinmann P (2015) Phase field simulations of the poling behavior of $BaTiO_3$ nano-scale thin films with $SrRuO_3$ and Au electrodes. Eur J Mech A/Solids 49:455–466
31. Song YC, Soh AK, Ni Y (2007) Phase field simulation of crack tip domain switching in ferroelectrics. J Phys D Appl Phys 40(4):1175–1182
32. Su Y, Landis CM (2007) Continuum thermodynamics of ferroelectric domain evolution: theory, finite element implementation, and application to domain wall pinning. J Mech Phys Solids 55(2):280–305
33. Su Y, Chen H, Li JJ, Soh AK, Weng GJ (2011) Effects of surface tension on the size-dependent ferroelectric characteristics of free-standing $BaTiO_3$ nano-thin films. J Appl Phys 110(084108):1–6
34. Suo Z, Kuo CM, Barnett DM, Willis JR (1992) Fracture mechanics for piezoelectric ceramics. J Mech Phys Solids 40(4):739–765
35. Taylor RL (2013) FEAP—a finite element analysis program. http://www.ce.berkeley.edu/projects/feap/
36. Tenne DA, Turner P, Schmidt JD, Biegalski M, Li YL, Chen LQ, Soukiassian A, Trolier-McKinstry S, Schlom DG, Xi XX, Fong DD, Fuoss PH, Eastman JA, Stephenson GB, Thompson C, Streiffer SK (2009) Ferroelectricity in ultrathin $BaTiO_3$ films: probing the size effect by ultraviolet Raman spectroscopy. Phys Rev Lett 103(177601):1–4
37. Wang J, Landis CM (2004) On the fracture toughness of ferroelectric ceramics with electric field applied parallel to the crack front. Acta Mater 52(12):3435–3446
38. Wang J, Zhang TY (2007) Phase field simulations of polarization switching-induced toughening in ferroelectric ceramics. Acta Mater 55(7):2465–2477
39. Xu BX, Schrade D, Gross D, Mueller R (2010) Phase field simulation of domain structures in cracked ferroelectrics. Int J Fract 165:163–173
40. Zhang TY, Gao CF (2004) Fracture behaviors of piezoelectric materials. Theor Appl Fract Mech 41(1–3):339–379

Modeling Approaches to Predict Damage Evolution and Life Time of Brittle Ferroelectrics

Andreas Ricoeur, Stephan Lange and Roman Gellmann

Abstract Reliability and life time of smart materials are crucial features for the development and design of actuator and sensor devices. Being widely used and exhibiting brittle failure characteristics, ceramic ferroelectrics are of particular interest in this field. Due to manifold interactions of the complex nonlinear constitutive behavior on the one hand and the damage evolution in terms of microcrack growth on the other, modeling and simulation are inevitable to investigate influence parameters on strength, reliability and life time. Two approaches are presented, both based on the same constitutive law and damage model. The one is going along with a discretisation scheme exploiting the finite element method (FEM). The so-called condensed approach, on the other hand, considers just one characteristic point in the material, nonetheless accounting for polycrystalline grain interactions. The focus of the simulations is two-fold. Life-time predictions in terms of high cycle fatigue under electromechanical loading conditions are presented based on the condensed approach. Second, the formation of macroscopic cracks at electrode tips in a stack actuator is investigated applying the FEM.

1 Introduction

Ferroelectric materials such as barium titanate (BT) or lead zirconate titanate (PZT) are used as bulk material in actuators or sensors and are constituents of micro electromechanical systems (MEMS) and composite devices. Nonlinear modeling of

A. Ricoeur (✉) · S. Lange · R. Gellmann
Institute of Mechanics, University of Kassel,
Mönchebergstr. 7, 34109 Kassel, Germany
e-mail: ricoeur@uni-kassel.de

S. Lange
e-mail: stephan.lange@uni-kassel.de

R. Gellmann
e-mail: gellmann@uni-kassel.de

G. Hütter and L. Zybell, *Recent Trends in Fracture and Damage Mechanics*, DOI 10.1007/978-3-319-21467-2_11

the constitutive behavior is crucial for ferroelectrics, if the poling process or depolarization under reverse fields are in the focus of interest. Residual stresses are initiated in a polycrystalline ferroelectric material by the incompatibility of strain in adjacent grains induced by domain wall motion. The residual fields are superimposed with the external loading, influencing switching and domain wall motion. Moreover, residual stresses give rise to microstructural damage and crack formation, thus being crucial for the development of damage models for ferroelectrics.

Summarizing the state of the art in nonlinear constitutive modeling would fill a whole chapter, so only a few shall be mentioned, e.g. Cocks and McMeeking [2], Kamlah [19], Fan et al. [6], Kamlah and Tsakmakis [20], Hwang et al. [16], Michelitsch and Kreher [27], Huo and Jiang [15], Chen et al. [1], Lu et al. [25] and Huber et al. [14]. Most of the approaches are implemented within the framework of the finite element (FE) method, enabling the solution of complex boundary value problems. On the other hand, the implementation of a discretization scheme is going along with a high effort and the solution of problems requires high computational costs. Calculating different hysteresis loops and investigating just the constitutive behavior of the material, however, does not require the application of a discretisation method. The same applies for principle investigations of damage processes. Nevertheless, it is not appropriate just considering a single grain model, since grain-grain interaction is crucial for most macroscopically observed features of a polycrystalline ferroelectric material and mechanical degradation. Comparisons of results based on a FE implementation, where grain interaction is incorporated intrinsically, and those based on the condensed approach, where kinematic constraints are introduced in terms of an effective medium theory [22], show a good agreement in many issues.

The condensed approach, however, presently meets its limits, considering problems with macroscopic field gradients, e.g. at notches or electrode edges. Current investigations deal with the exploitation of the new method with regard to this kind of problems by mapping e.g. the electric field gradient at an electrode edge to the material points. Anyhow, the application of a classical discretisation scheme is appropriate in that case and inevitable to verify a mapping technique of the condensed approach. On the other hand, the FEM could be combined with the condensed approach incorporating the latter on the element level, where each integration point represents a polycrystalline microstructure.

Concerning model-based investigations of damage in ferroelectric ceramics, some phenomenological approaches are known from literature, e.g. Wang and Han [37], Yang et al. [41] or Zheng et al. [42]. Microphysically motivated models and life-time predictions, on the other hand, are not state of the art.

2 Ferroelectric Constitutive Behavior: A Microphysical Approach

2.1 Thermodynamical Fundamentals

The local formulation of the energy balance reads:

$$\dot{u} + \rho \dot{u}_i \ddot{u}_i = \sigma_{ij} \underbrace{\left(\dot{\varepsilon}_{ij}^{\mathrm{rev}} + \dot{\varepsilon}_{ij}^{\mathrm{irr}} \right)}_{\dot{\varepsilon}_{ij}} + E_i \underbrace{\left(\dot{D}_i^{\mathrm{rev}} + \dot{P}_i^{\mathrm{irr}} \right)}_{\dot{D}_i} - q_{i,i} + \rho r, \tag{1}$$

where u and ρ are the internal energy and mass densities, respectively, u_i is the displacement field, σ_{ij} the stress tensor and E_i the electric field. The variables r and q_i characterize the heat source within the domain and the surface heat flux across the boundary. In Eq. (1), rates of strain ε_{ij} and electric displacement D_i are decomposed into reversible and irreversible parts as usual. Analytical notation is applied implying summation over double indices and $(\ldots)_{,i} = \partial / \partial x_i$ denoting the partial differentiation with respect to x_i. The terms $\dot{\varepsilon}_{ij}^{\mathrm{irr}}$ and $\dot{P}_i^{\mathrm{irr}}$ describe the irreversible strain and polarization rates as results of domain wall motion. Equation (1) neglects dissipative effects associated with linear fields (see Kessler and Balke [21]) as well as changes in elastic, piezoelectric and dielectric properties. These contributions are assumed to be small compared to those included in the equation. Furthermore, Eq. (1) implies the balance equations of momentum and charge in their simplified forms $\sigma_{ij,j} = 0$ and $D_{i,i} = 0$.

Thermodynamic consistency requires the satisfaction of the second law of thermodynamics (e.g. parton and Kudryavtsev [30])

$$\theta \dot{s} + q_{i,i} - \frac{q_i}{\theta} \frac{\partial \theta}{\partial x_i} - \rho r \geq 0 \tag{2}$$

where s is the specific entropy and θ denotes the absolute thermodynamic temperature. Inserting Eq. (1) into Eq. (2) and considering the static limit, i.e. $\dot{u}_i, \ddot{u}_i = 0$, the generalized Clausius-Duhem inequality is obtained:

$$-(\dot{u} - \theta \dot{s}) + \sigma_{ij} \dot{\varepsilon}_{ij}^{\mathrm{rev}} + \sigma_{ij} \dot{\varepsilon}_{ij}^{\mathrm{irr}} + E_i \dot{D}_i^{\mathrm{rev}} + E_i \dot{P}_i^{\mathrm{irr}} - \frac{q_i}{\theta} \frac{\partial \theta}{\partial x_i} \geq 0. \tag{3}$$

Parton and Kudryavtsev [30] discussed the inequality (3) for $\dot{\varepsilon}_{ij}^{\mathrm{irr}} = 0$ and $\dot{P}_i^{\mathrm{irr}} = 0$ and showed that it holds for any values of $\dot{\varepsilon}_{ij}^{\mathrm{rev}}$, $\dot{D}_i^{\mathrm{rev}}$ and $\dot{s}$ leaving the statement $q_i \theta_{,i} / \theta \geq 0$. Thus, considering irreversible domain processes, the inequality (3) reads

$$\sigma_{ij}\dot{\varepsilon}_{ij}^{\text{irr}} + E_i\dot{P}_i^{\text{irr}} - \frac{q_i}{\theta}\frac{\partial\theta}{\partial x_i} \geq 0. \tag{4}$$

Equation (4) has to be satisfied by any thermodynamically consistent domain evolution law.

The internal energy u is a thermodynamical potential depending on the independent variables ε_{ij}, D_i and s and found its way into Eq. (1) from the classical formulation of the first law of thermodynamics. The electric field E_i as independent variable being more easily accessible, it is useful to replace the electric displacement D_i in that function. The corresponding thermodynamical potential Ψ is obtained from a Legendre transformation [35] which is applied to Eq. (1):

$$\Psi = u - E_iD_i \quad \Rightarrow \quad \dot{\Psi}(\dot{\varepsilon}_{ij}, \dot{E}_i, \dot{s}) = \sigma_{ij}\dot{\varepsilon}_{ij} - D_i\dot{E}_i + \theta\dot{s}. \tag{5}$$

Irreversibilities are included in the total strain rate and electric displacement as well as in the dissipative part of the entropy, whereas the heat flux is covered by the exchange entropy. The latter, however, will be dropped in the following. Equation (5) reveals the relations between independent and associated variables as

$$\left.\frac{\partial\Psi}{\partial\varepsilon_{ij}}\right|_{E,s} = \sigma_{ij}, \quad \left.\frac{\partial\Psi}{\partial E_i}\right|_{\varepsilon,s} = -D_i, \quad \left.\frac{\partial\Psi}{\partial s}\right|_{E,\varepsilon} = \theta. \tag{6}$$

The nonlinear ferroelectric constitutive behavior is finally governed by the potential function which is commonly denoted as the electric enthalpy density:

$$\Psi(\varepsilon_{kl}, E_l) = \frac{1}{2}C_{ijkl}\varepsilon_{kl}\varepsilon_{ij} - e_{lij}E_l\varepsilon_{ij} - \frac{1}{2}\kappa_{il}E_lE_i - C_{ijkl}\varepsilon_{kl}^{\text{irr}}\varepsilon_{ij} + e_{ikl}\varepsilon_{kl}^{\text{irr}}E_i - P_i^{\text{irr}}E_i. \tag{7}$$

Elastic, piezoelectric and dielectric tensors C_{ijkl}, e_{ikl} and κ_{il} are assumed to be constant within an incremental range of the independent variables strain and electric field. For larger changes of state, these material parameters, themselves depending on the independent variables, have to be understood as local tangents at hysteresis loops. The nonlinear constitutive equations are finally obtained from Eqs. (6) and (7):

$$\sigma_{ij} = C_{ijkl}\left(\varepsilon_{kl} - \varepsilon_{kl}^{\text{irr}}\right) - e_{nij}E_n, \tag{8}$$

$$D_i = e_{ikl}\left(\varepsilon_{kl} - \varepsilon_{kl}^{\text{irr}}\right) + \kappa_{in}E_n + P_i^{\text{irr}}. \tag{9}$$

2.2 Evolution of Internal Variables of Domain Wall Motion

In Fig. 1 the domain structure of a single ferroelectric grain with tetragonal unit cells is depicted, exhibiting 90° and 180° domain walls. It is represented by just one

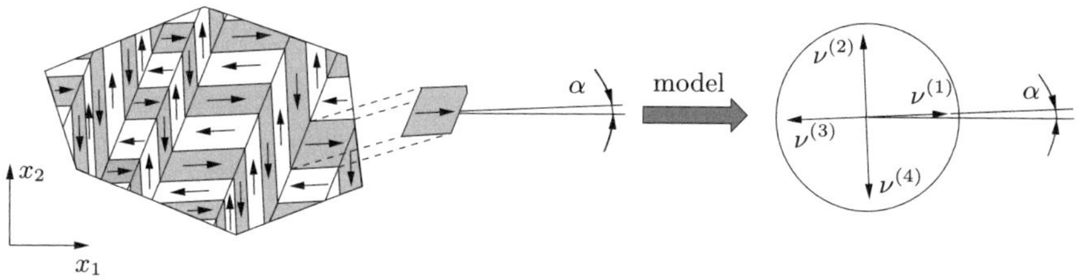

Fig. 1 Plane domain pattern of a grain and motivation of the internal variables $v^{(n)}$ [22]

local material point involving all possible polarization directions. Each one is weighted by a volume fraction or internal variable $v^{(n)}$ where $n = 1, \ldots, 4$ [14]. On the macroscopic scale, the switching process leads to a gradual change of the irreversible quantities:

$$\mathrm{d}\varepsilon_{kl}^{\mathrm{irr}} = \sum_{n=1}^{4} \varepsilon_{kl}^{\mathrm{sp}(n)} \mathrm{d}v^{(n)}, \quad \mathrm{d}P_i^{\mathrm{irr}} = \sum_{n=1}^{4} \varDelta P_i^{\mathrm{sp}(n)} \mathrm{d}v^{(n)}. \tag{10}$$

The change of polarization $\varDelta P_i^{\mathrm{sp}(n)} = P_i^{\mathrm{sp}(m)} - P_i^{\mathrm{sp}(n)}$ is defined as the vector difference of spontaneous polarizations while switching from a species (n) to a species (m). The spontaneous strain $\varepsilon_{kl}^{\mathrm{sp}(n)}$ of a species (n) switching to another orientation describes a change of state per definition, thus being unnecessary to add a Δ. The effective material constants of a grain are likewise determined as weighted averages

$$C_{ijkl} = \sum_{n=1}^{4} C_{ijkl}^{(n)} v^{(n)}, \quad e_{ijk} = \sum_{n=1}^{4} e_{ijk}^{(n)} v^{(n)}, \quad \kappa_{ij} = \sum_{n=1}^{4} \kappa_{ij}^{(n)} v^{(n)}, \tag{11}$$

where $C_{ijkl}^{(n)}$, $e_{ijk}^{(n)}$ and $\kappa_{ij}^{(n)}$ are the material properties of the polarization directions n. Keeping stresses σ_{ij} and the electric field E_i constant in a grain, strain and electric displacement are also determined by weighted averages.

The evolution equations for each volume fraction $v^{(n)}$ are given by [22]

$$\mathrm{d}v^{(n)} = -\mathrm{d}v_0 \mathscr{H}\left(\frac{w^{\mathrm{diss}(n)}}{w_{(\beta)}^{\mathrm{crit}}} - 1\right), \quad w^{\mathrm{diss}(n)} = \max\left\{w_{(\beta)}^{\mathrm{diss}(n)}\right\}, \tag{12}$$

with a model parameter $\mathrm{d}v_0$ and the Heaviside function $\mathscr{H}$. Accounting only for spontaneous terms of the dissipative work associated with the switching of a domain species n, i.e.

$$w_{(\beta)}^{\mathrm{diss}(n)} = \sigma_{ij}\, \varepsilon_{ij(\beta)}^{\mathrm{sp}(n)} + E_i\, \varDelta P_{i(\beta)}^{\mathrm{sp}(n)}, \tag{13}$$

the evolution equation satisfies the Clausius-Duhem inequality Eq. (4) for $\theta_{,i} = 0$ and bearing in mind Eq. (10). The subscript (β) distinguishes between $\pm$ 90° and 180° switching variants. Equation (12) is based on a local switching criterion originally introduced by Hwang et al. [16], requiring threshold values $w^{\mathrm{crit}}_{(\beta)}$ for the different switching variants (see e.g. [5]) depending on the macroscopic properties E_c (coercive field) and P^0 (remanent polarization).

Finally, it should be noted that both switching strain and polarization and thus the evolution of internal variables are limited, due to the fact that domain species cannot vanish totally, thus imposing the constraint $v^{(n)} > 0$. Furthermore, the presence of free charges reduces the spontaneous polarization.

3 A Condensed Approach for Polycrystalline Ferroelectrics

3.1 Theoretical Background

In Sect. 2 the constitutive behavior of a single grain has been considered. Within the framework of a FE implementation [10], the outlined equations are applied to each integration point, implying polycrystalline interactions for the discretisized structure. The condensed approach, on the other hand, considers just one representative volume element (RVE), nevertheless accounting for polycrystalline interactions. The computational and implementation efforts going along with this new approach are low compared to FE based modeling. Nevertheless, the essential features are basically included yielding smooth hysteresis loops of the bulk material under combined electromechanical loading and residual stresses in each grain. Moreover it has been proven that the condensed approach is mechanically and electrostatically consistent. Although the model has a profound thermodynamical basis, full thermodynamical consistency lacks a comprehensive formulation of the dissipative work, including minor effects like the influence of changes in material properties due to domain wall motion. The condensed approach has recently been presented elaborately by Lange and Ricoeur [22] and will therefore just be outlined shortly at this point.

Several grains with different material orientations are now considered in an RVE. In Fig. 2, three grains with different orientations α_i are exemplarily highlighted. Following Fig. 1, the four domain species in each grain are indicated by crossed arrows. Now, all grains are condensed to one material point exhibiting all possible domain orientations of an RVE. Switching, however, can only occur within the four species of each arrowed cross.

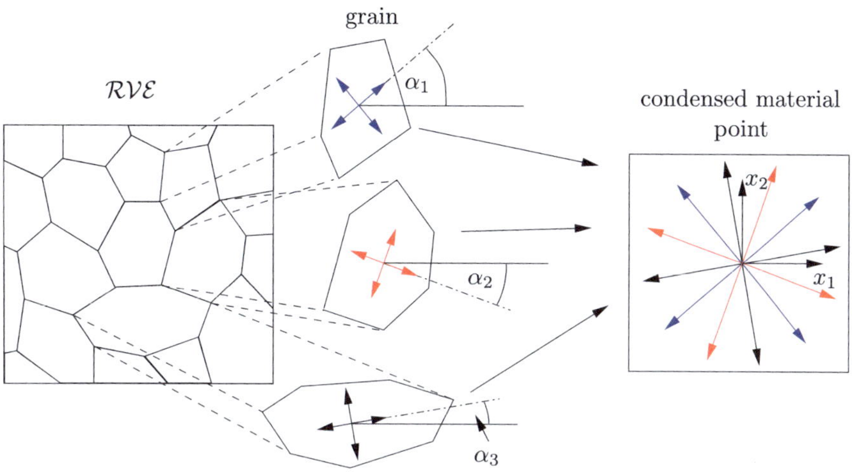

Fig. 2 Polycrystalline RVE with three grains highlighted and illustration of the idea of the condensed approach [22]

Assuming homogeneous fields in a grain $m = 1, \ldots, M$ and all grains having equal size, averaging stress and electric displacement in an RVE results in

$$\langle \sigma_{ij} \rangle = \frac{1}{M} \sum_{m=1}^{M} \sigma_{ij}^{(m)}, \quad \langle D_i \rangle = \frac{1}{M} \sum_{m=1}^{M} D_i^{(m)}, \tag{14}$$

where volume averages are specified by angled brackets. The averages of the material coefficients are determined in the same way with $C_{ijkl}^{(m)}$, $e_{ikl}^{(m)}$ and $\kappa_{ij}^{(m)}$ as material properties of a grain m calculated by Eq. (11). The constitutive Eqs. (8) and (9) are valid for both the macroscopic behavior and on the level of a grain m. The balance equations of mechanical and electrostatic equilibrium are required for averaged fields, i.e. $\langle \sigma_{ij} \rangle_{,j} = 0$ and $\langle D_i \rangle_{,i} = 0$. Assuming a generalized Voigt approximation, strain and electric field are homogeneous in an RVE and thus equal in each grain:

$$\langle E_l \rangle = \frac{1}{M} \sum_{m=1}^{M} E_l^{(m)} = E_l^{(m)} = \overline{E}_l, \langle \varepsilon_{kl} \rangle = \frac{1}{M} \sum_{m=1}^{M} \varepsilon_{kl}^{(m)} = \varepsilon_{kl}^{(m)} = \overline{\varepsilon}_{kl}, \tag{15}$$

where bars on $\overline{E}_l$ and $\overline{\varepsilon}_{kl}$ indicate the variables which are spatially kept constant. They should be distinguished from averages of non-constant quantities denoted by angled brackets. The average stress $\langle \sigma_{ij} \rangle$ is thus obtained from Eqs. (14) and (15), inserting the constitutive equations for the grain:

$$\begin{aligned}\langle\sigma_{ij}\rangle &= \frac{1}{M}\sum_{m=1}^{M}\left\{C_{ijkl}^{(m)}\left(\varepsilon_{kl}^{(m)}-\varepsilon_{kl}^{\mathrm{irr}(m)}\right)-e_{nij}^{(m)}E_n^{(m)}\right\}\\ &= \frac{1}{M}\sum_{m=1}^{M}C_{ijkl}^{(m)}\varepsilon_{kl}^{(m)}-\frac{1}{M}\sum_{m=1}^{M}C_{ijkl}^{(m)}\varepsilon_{kl}^{\mathrm{irr}(m)}-\frac{1}{M}\sum_{m=1}^{M}e_{nij}^{(m)}E_n^{(m)}\\ &= \langle C_{ijkl}\rangle\,\bar{\varepsilon}_{kl}-\langle C_{ijkl}\,\varepsilon_{kl}^{\mathrm{irr}}\rangle-\langle e_{nij}\rangle\overline{E}_n.\end{aligned} \tag{16}$$

The electric displacement $\langle D_i\rangle$ of a representative volume element is likewise derived as:

$$\begin{aligned}\langle D_i\rangle &= \frac{1}{M}\sum_{m=1}^{M}e_{ikl}^{(m)}\varepsilon_{kl}^{(m)}-\frac{1}{M}\sum_{m=1}^{M}e_{ikl}^{(m)}\varepsilon_{kl}^{\mathrm{irr}(m)}+\frac{1}{M}\sum_{m=1}^{M}\kappa_{in}^{(m)}E_n^{(m)}+\frac{1}{M}\sum_{m=1}^{M}P_i^{\mathrm{irr}(m)}\\ &= \langle e_{ikl}\rangle\bar{\varepsilon}_{kl}-\langle e_{ikl}\varepsilon_{kl}^{\mathrm{irr}}\rangle+\langle\kappa_{in}\rangle\overline{E}_n+\langle P_i^{\mathrm{irr}}\rangle.\end{aligned} \tag{17}$$

Applying a mechanical Neumann type boundary condition, e.g. to prescribe pre-stresses, the balance equation for the average stress in an RVE $\langle\sigma_{ij}\rangle_{,j}=0$ has to be taken into account to determine the unknown total strain $\bar{\varepsilon}_{kl}$. Due to the averaging procedure, both balance equations are intrinsically satisfied claiming the uniformity of all stress and electric displacement coordinates in an RVE. In case of external mechanical loads, equilibrium of a plane problem thus requires the identity

$$\begin{pmatrix}\langle\sigma_{11}\rangle\\ \langle\sigma_{22}\rangle\\ \langle\sigma_{12}\rangle\end{pmatrix}=\begin{pmatrix}\sigma_{11}^{\mathrm{ext}}\\ \sigma_{22}^{\mathrm{ext}}\\ \sigma_{12}^{\mathrm{ext}}\end{pmatrix} \tag{18}$$

with external loads $\sigma_{ij}^{\mathrm{ext}}$. With Eq. (18), the strain $\bar{\varepsilon}_{kl}=\varepsilon_{kl}^{(m)}$ is calculated from Eq. (16) as follows:

$$\bar{\varepsilon}_{kl}=\langle C_{ijkl}\rangle^{-1}(\sigma_{ij}^{\mathrm{ext}}+\langle e_{lij}\rangle\overline{E}_l+\langle C_{ijno}\varepsilon_{no}^{\mathrm{irr}}\rangle). \tag{19}$$

The residual stress in each grain is then determined inserting Eq. (19) into the constitutive equation for a grain m:

$$\sigma_{ij}^{(m)}=C_{ijkl}^{(m)}\{\langle C_{pqkl}\rangle^{-1}(\sigma_{pq}^{\mathrm{ext}}+\langle e_{rpq}\rangle\overline{E}_r+\langle C_{pqno}\,\varepsilon_{no}^{\mathrm{irr}}\rangle)-\varepsilon_{kl}^{\mathrm{irr}(m)}\}-e_{lij}^{(m)}\overline{E}_l. \tag{20}$$

Similarly, the local electric displacements $D_i^{(m)}$ are calculated and the average electric displacement results from Eqs. (17) and (19):

$$\langle D_i\rangle=\langle e_{ikl}\rangle\langle C_{pqkl}\rangle^{-1}(\sigma_{pq}^{\mathrm{ext}}+\langle e_{rpq}\rangle\overline{E}_r+\langle C_{pqno}\,\varepsilon_{no}^{\mathrm{irr}}\rangle)+\langle\kappa_{il}\rangle\overline{E}_l-\langle e_{ikl}\varepsilon_{kl}^{\mathrm{irr}}\rangle+\langle P_i^{\mathrm{irr}}\rangle. \tag{21}$$

Of course, other boundary conditions prescribing strain or electric displacement can easily be handled, see Lange and Ricoeur [22]. Also, a different kind of homogenization approach assuming electric displacements constant instead of electric fields has been investigated. Finally, it should be mentioned that since the irreversible quantities $\varepsilon_{kl}^{\mathrm{irr}}$ and P_i^{irr} as well as effective material constants $\langle C_{ijkl}\rangle$, $\langle e_{ikl}\rangle$ and $\langle \kappa_{ij}\rangle$ depend on the internal variables $\nu^{(n)}$ and thus on local mechanical and electric fields, an iteration is required for the numerical procedure.

3.2 Results for the Condensed Model: Constitutive Behavior

In all simulations, $M = 25$ grains are considered in an RVE. A larger number of grains came out to provide similar results [22]. The loading axis is the x_2-axis, with E_2 as the electrical and $\sigma_{22}^{\mathrm{ext}}$ as the mechanical load. The polarization can be arbitrary in the x_1-x_2-plane. The residual stresses $\sigma_{ij}^{(m)}$ and the electric displacement $\langle D_i\rangle$ result from Eqs. (20) and (21), while the strain is calculated from Eq. (19). In the plots, the electric load E_2 is related to the coercive field E_C, which amounts to 200 V/mm for barium titanate, the data of which have been chosen for all calculations [17]. Further, $\varepsilon_D = 1 \times 10^{-2}$ has been chosen for the switching strain and $P^0 = 0.26\ \mathrm{Cm}^{-2}$ for the spontaneous polarization [17], resulting in a remanent strain $\varepsilon^{\mathrm{r}} = 0.0002$ and remanent polarization $\mathrm{P}^{\mathrm{r}} = 0.08\ \mathrm{C/m}^{-2}$ which are found in experiments [7, 13, 24, 26].

In Fig. 3 residual stresses $\sigma_{22}^{(m)}$ and $\sigma_{12}^{(m)}$ are investigated showing results of five arbitrarily selected grains. The initial state of the material is unpoled and external mechanical tractions are zero while the electric field E_2 is increased monotonously.

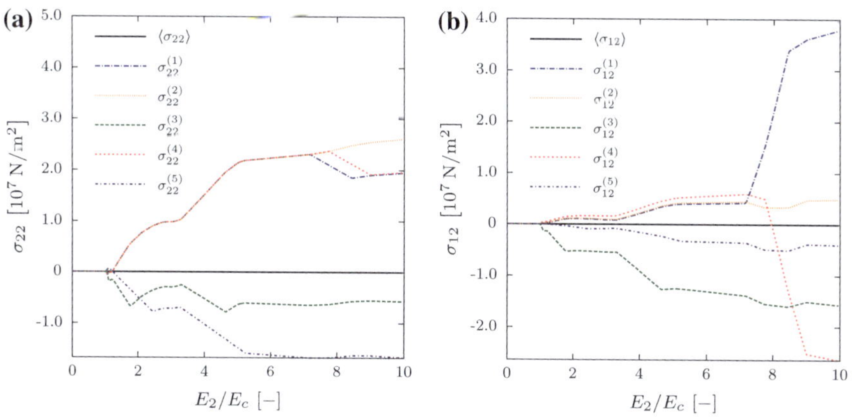

Fig. 3 Polarization process: stresses $\sigma_{22}^{(m)}$ and $\sigma_{12}^{(m)}$ in grains m and their averages $\langle\sigma_{22}\rangle$ and $\langle\sigma_{12}\rangle$ versus electric field E_2 normalized with respect to the coercive field E_C

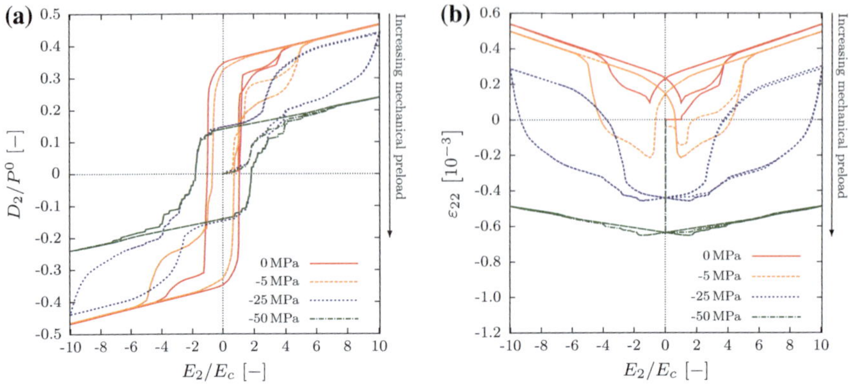

Fig. 4 Hysteresis loops for electric displacement (**a**) and strain (**b**) under cyclic electric loading for different compressive preloads

Obviously there are grains with tensile and others with compressive stress. The averages $\langle\sigma_{22}\rangle$ and $\langle\sigma_{12}\rangle$, however, are zero for all E_2 satisfying the condition of mechanical equilibrium. Looking at the values, tensile residual stresses above 20 MPa are observed at $E_2 = 5E_c$ and shear stresses attain maxima of as much as 38 MPa. The calculations give evidence that stress values are within a reasonable range being relevant for damage and mechanical degradation.

In the next example combined electromechanical loading is considered. A compressive pre-stress is raised linearly and kept constant with a value $\hat{\sigma}_{22}$ during the cyclic piecewise linear electric loading. While different values $\hat{\sigma}_{22} = 0.0, -5.0, -25.0$ and -50.0 MPa are assumed, the electric amplitude is always $\hat{E}_2 = 10E_C$. Figure 4a, b shows the results in terms of classical dielectric and butterfly hysteresis loops. Considering the butterfly loop, its area first increases with increasing compressive load but finally decreases when further raising the mechanical stress. This is due to the mobility of domains which is fostered by a moderate clamping but is suppressed to a large extent in the case of large compression. This result is confirmed by experimental investigations, e.g. those of Zhou and Kamlah [43]. Comparison with numerical results from an implementation of the constitutive model outlined in Sect. 2 into a FE code also shows good agreement concerning e.g. hysteresis loops and the prediction of residual stresses.

4 Modeling of Damage Evolution in Ferroelectrics

A continuum damage approach for brittle ferroelectrics is presented in this section. The defect phase consists of cracks, randomly and dilutely distributed in the material, i.e. neglecting crack interactions. The constitutive damage model is implemented within the context of both the condensed approach and the FEM.

4.1 Homogenisation and Effective Material Properties in Piezoelectrics

To calculate effective material properties, a linear material behavior of the cracked piezoelectric solid is considered in each load increment. Two out of four possible formulations of the constitutive behavior are (e.g. [8, 22]):

$$\begin{pmatrix} \varepsilon_{ij} \\ E_i \end{pmatrix} = \begin{pmatrix} S^*_{ijkl} & d^*_{lij} \\ -d^*_{ikl} & \beta^*_{il} \end{pmatrix} \begin{pmatrix} \sigma_{kl} \\ D_l \end{pmatrix}, \quad \begin{pmatrix} \sigma_{ij} \\ D_i \end{pmatrix} = \begin{pmatrix} C^*_{ijkl} & -e^*_{lij} \\ e^*_{ikl} & \kappa^*_{il} \end{pmatrix} \begin{pmatrix} \varepsilon_{kl} \\ E_l \end{pmatrix}. \tag{22}$$

Here, S_{ijkl}, d_{rij} and β_{ir} describe the elastic compliance, piezoelectric constants and dielectric impermeability, respectively and * denotes effective properties. C_{ijkl}, e_{ikl} and κ_{il} are the previously introduced stiffness, piezoelectric and dielectric constants. In a compressed notation Eq. (22) can be written as:

$$Z_p = F^*_{pq} \Pi_q, \quad \Pi_p = C^*_{pq} Z_q. \tag{23}$$

The variables in Eq. (23) describe generalized values for strain Z_p, stress Π_q, stiffness C_{pq} and compliance F_{pq}, where indices p, q run from 1 to 9. Following e.g. Gross and Seelig [12], Wang and Jiang [38] and Gellmann et al. [11], Z_p can be decomposed into a matrix part Z_p^{M} and a part Z_p^{C} representing the defect phase:

$$Z_p = Z_p^{\mathrm{M}} + Z_p^{\mathrm{C}}. \tag{24}$$

Based on this assumption Eq. (23) can be formulated for the matrix part:

$$Z_p^{\mathrm{M}} = F_{pq}^{\mathrm{M}} \Pi_q^{\mathrm{M}}, \quad \Pi_p^{\mathrm{M}} = C_{pq}^{\mathrm{M}} Z_q^{\mathrm{M}}. \tag{25}$$

From Eq. (24) it is obvious that $Z_p^{\mathrm{M}} = Z_p - Z_p^{\mathrm{C}}$. Inserting Z_p^{M} into Eq. (25) leads to relationships between the matrix and the cracked phase:

$$Z_p = F_{pq}^{\mathrm{M}} \Pi_q + Z_p^{\mathrm{C}}, \quad \Pi_p^{\mathrm{M}} = C_{pq}^{\mathrm{M}} \left(Z_q - Z_q^{\mathrm{C}} \right). \tag{26}$$

Two different cases will be discussed in the following: a generalized Reuss assumption, i.e. $\Pi_p = \Pi_p^{\infty}$, where generalized stresses are constant and therefore

$$Z_p^{\mathrm{C}} = F_{pq}^{\mathrm{C}} \Pi_q^{\infty} \tag{27}$$

and a generalized Voigt approximation, i.e. $Z_p = Z_p^{\infty}$, where generalized strains are constant leading to

$$Z_p^{\mathrm{C}} = D_{pq} Z_q^{\infty} \, . \tag{28}$$

F_{pq}^{C} and D_{pq} in Eqs. (27) and (28) are the influence tensors and Π_q^{∞} and Z_q^{∞} describe the loading conditions of the RVE. The determination of F_{pq}^{C} and D_{pq} for a specific defect phase consisting of microcracks will be discussed in Sect. 4.2. The effective material constants of the Reuss condition result from inserting Eqs. (23) and (27) into Eq. (26):

$$\begin{aligned} F_{pq}^{*} &= F_{pq}^{\mathrm{M}} + F_{pq}^{\mathrm{C}}, \\ C_{pq}^{*} &= \left(F_{pq}^{\mathrm{M}} + F_{pq}^{\mathrm{C}}\right)^{-1}. \end{aligned} \tag{29}$$

For the Voigt case, the effective constants follow from applying Eqs. (23) and (28) to Eq. (26):

$$\begin{aligned} C_{pq}^{*} &= C_{pr}^{\mathrm{M}} \left(\delta_{rq} - D_{rq}\right), \\ F_{pq}^{*} &= \left\{C_{pr}^{\mathrm{M}} \left(\delta_{rq} - D_{rq}\right)\right\}^{-1}. \end{aligned} \tag{30}$$

Deriving the relations (29) and (30), it has to be borne in mind that $F_{pq}^{-1} = C_{pq}$. Further, $\Pi_p = \Pi_p^M$ is assumed. Mechanically, this implies a stress-free defect phase which holds in the case of voids or cracks. From the electrostatic point of view, electric displacements must vanish inside the defects, restricting the applicability of Eqs. (29) and (30) to the limiting case of impermeable defects with charge-free boundaries.

4.2 *Modeling of the Defect Phase*

For the calculation of the effective material constants at a material point m, the microstructural RVE with an edge length of $\sqrt{A}$ shown in Fig. 5 is considered. There are several cracks with different orientations under an electromechanical loading $\sigma_I^{(m)}$ and $D_2^{(m)}$. Due to the dilute or noninteracting assumption [18], the model of the defect phase can be reduced to one single crack in an unbounded medium, where the edge length of the crack RVE is just a parameter to introduce the crack density. The coordinate system of the crack is defined by x_1^{c} and x_2^{c}. Further, only cracks perpendicular to the principle stress $\sigma_I^{(m)}$ are considered, because these are the most critical ones. Thus, the local crack coordinate system and the orientation of the defect RVE coincide with the principal axes of the stress field. Electric loads parallel to the crack faces $D_1^{(m)}$ are not considered, since they are dispensable from the fracture mechanics point of view, at least concerning classical

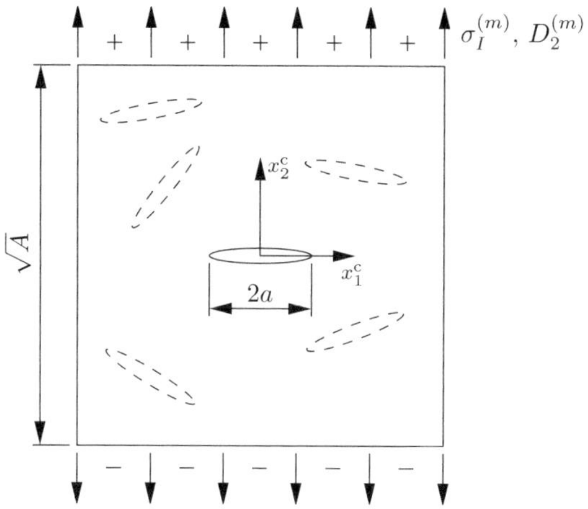

Fig. 5 Model of the defect phase in an RVE m under electromechanical loading

theories. Advanced models of piezoelectric fracture mechanics, however, reveal effects going along with electric fields parallel to the crack faces [32]. Even collinear stress loads $\bar{\sigma}_{11}$ may have an impact on the mode I SIF [9]. The principal stresses $\sigma_{I,II}^{(m)}$ of a material point m are calculated as follows:

$$\sigma_{I,II}^{(m)} = \frac{1}{2}\left(\sigma_{11}^{(m)} + \sigma_{22}^{(m)}\right) \pm \sqrt{\frac{1}{4}\left(\sigma_{11}^{(m)} + \sigma_{22}^{(m)}\right)^2 + \left(\sigma_{12}^{(m)}\right)^2}, \tag{31}$$

with $\sigma_I^{(m)} > \sigma_{II}^{(m)}$. The crack density

$$f = \frac{4a^2}{A} \tag{32}$$

is introduced as a relationship of the squares of the crack length $2a$ and the edge length $\sqrt{A}$.

Boundary conditions on the crack surfaces and coordinate systems are illustrated in Fig. 6. Here, an impermeable crack is considered, following the Neumann type boundary condition investigated e.g. by [34, 36]. Therefore, the crack faces are free of charges, i.e.

$$D_2^+(x_1^c) = D_2^-(x_1^c) = 0 \tag{33}$$

on the positive Γ^+ and negative Γ^- crack surfaces.

Besides the global coordinate system (x_1, x_2), other coordinate systems are relevant for the development of the damage model. The one related to the material point m, following the idea of the condensed approach depicted in Fig. 2, is denoted by $\left(x_1^m, x_2^m\right)$, β_1 and β_2 describe the orientation of the polarisation direction with respect to x_1^c and x_1^m. α^c and α^m denote the orientations of the local coordinate

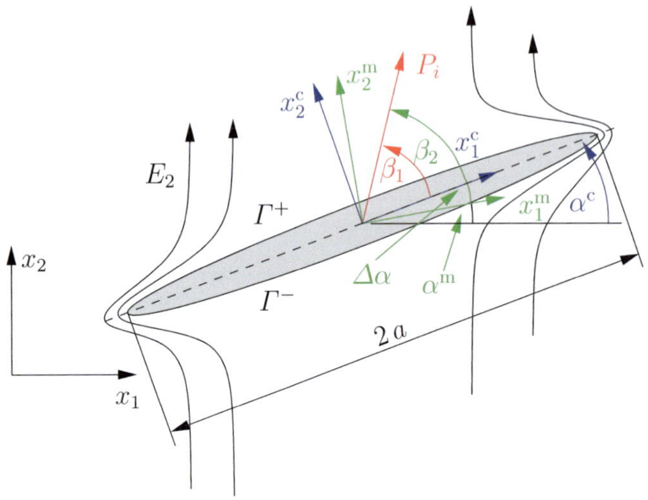

Fig. 6 Coordinate systems and parameters of an arbitrarily orientated impermeable crack

systems of the crack $\left(x_1^c, x_2^c\right)$ and the material RVE related to $\left(x_1^m, x_2^m\right)$. The angle between these two systems is defined as $\Delta\alpha = \alpha^{\rm c} - \alpha^{\rm m}$. The different angles, shown in Fig. 6, are relevant for the calculation of the Irwin matrix (see e.g. [31]) and the effective material constants.

The jump of generalized displacements Δu_M at the crack faces in a ferroelectric ceramic under an electromechanical loading is discussed e.g. by [9, 28, 31]. With the boundary conditions introduced in Eq. (33) Δu_M is given by

$$\Delta u_M = \begin{pmatrix} \Delta u_1 \\ \Delta u_2 \\ \Delta u_3 \\ \Delta\varphi \end{pmatrix} = 2 \begin{pmatrix} Y_{11} & Y_{12} & Y_{13} & Y_{14} \\ Y_{21} & Y_{22} & Y_{23} & Y_{24} \\ Y_{31} & Y_{32} & Y_{33} & Y_{34} \\ Y_{41} & Y_{42} & Y_{43} & Y_{44} \end{pmatrix} \begin{pmatrix} \sigma_{12}^\infty \\ \sigma_{22}^\infty \\ \sigma_{32}^\infty \\ D_2^\infty \end{pmatrix} \sqrt{a^2 - \left(x_1^{\rm c}\right)^2} \tag{34}$$
$$= 2Y_{MN}\Sigma_N \sqrt{a^2 - \left(x_1^{\rm c}\right)^2},$$

where Y_{MN} is the Irwin matrix depending on material constants and the orientation of the crack with respect to the poling direction. Bearing in mind that the crack surfaces are perpendicular to the direction of the first principal stress σ_I (s. Fig. 5) and only a 2D case is investigated Δu_M yields:

$$\Delta u_i = \begin{pmatrix} \Delta u_1 \\ \Delta u_2 \\ \Delta\varphi \end{pmatrix} = 2 \begin{pmatrix} Y_{11} & Y_{12} & Y_{14} \\ Y_{21} & Y_{22} & Y_{24} \\ Y_{41} & Y_{42} & Y_{44} \end{pmatrix} \begin{pmatrix} 0 \\ \sigma_I \\ D_2^\infty \end{pmatrix} \sqrt{a^2 - \left(x_1^{\rm c}\right)^2}, \tag{35}$$

where the electric potential is denoted as φ. To calculate the effective material properties, the strain of the cracked phase is given by (see e.g. [12]):

$$\langle \varepsilon_{ij} \rangle^{\mathrm{C}} = \frac{1}{2A} \int_{-a}^{a} \left(\Delta u_i n_j + \Delta u_j n_i \right) \mathrm{d}x_1^{\mathrm{c}}, \tag{36}$$

where n_i is the unit normal on the positive crack face. For the determination of the average electric field of the cracked phase $\langle E_i \rangle^{\mathrm{C}}$, it is convenient to reformulate the contribution of the matrix phase under consideration of the divergence theorem:

$$\langle E_i \rangle^{\mathrm{M}} = -\frac{1}{V_{\mathrm{M}}} \int_{V_{\mathrm{M}}} \varphi_{,i} \mathrm{d}V = -\frac{1}{V_{\mathrm{M}}} \int_{\partial V} \varphi n_i \mathrm{d}A + \frac{1}{V_{\mathrm{M}}} \int_{\partial V_{\mathrm{C}}} \varphi n_i \mathrm{d}A, \tag{37}$$

where V_M is the matrix domain and ∂V_C the boundary of the defect. Considering

$$\langle E_i \rangle = -\frac{1}{V} \int_{\partial V} \varphi n_i \mathrm{d}A \tag{38}$$

and introducing the volume fraction of the matrix $c_M = V_M / V$, the averaged electric field of the macrolevel is obtained as:

$$\langle E_i \rangle = c_M \langle E_i \rangle^M - \frac{1}{V} \int_{\partial V_{\mathrm{C}}} \varphi n_i \mathrm{d}A. \tag{39}$$

For cracks, c_M is commonly set to one due to the fact that the defect volume is small in comparison to the matrix volume. Thus, the crack density f and c_M are not related. The integral term in Eq. (39) represents the defect phase, finally relating its average electric field to the jump of the electric potential $\Delta\varphi$ by integrating along one crack face:

$$\langle E_i \rangle^C = -\frac{1}{V} \int_{-a}^{a} \Delta\varphi n_i \mathrm{d}A. \tag{40}$$

With Eqs. (36) and (40) the generalized strain of the cracked phase is defined, the jumps of displacements Δu_i and electric potential $\Delta\varphi$ resulting from Eq. (35). The influence tensor F_{pq}^{C} under Reuss conditions according to Eq. (29) is hence determined as

$$F_{pq}^{\mathrm{C}} = \frac{1}{4}\pi f \begin{pmatrix} 0 & 0 & 0 & 0 & 0 \\ 0 & Y_{22} & 0 & 0 & Y_{24} \\ 0 & Y_{12} & 0 & 0 & Y_{14} \\ 0 & 0 & 0 & 0 & 0 \\ 0 & -Y_{42} & 0 & 0 & -Y_{44} \end{pmatrix}. \tag{41}$$

Assuming a generalized Voigt approximation according to Eq. (28), the influence tensor D_{pq} results from Eq. (27) by replacing Π_q^∞ by Z_q^∞ via the constitutive Eq. (22):

$$D_{km} = \frac{1}{4}\pi f \left(\begin{matrix} 0 & 0 & 0 \\ Y_{22}C_{21}^{\mathrm{M}} + Y_{24}e_{21}^{\mathrm{M}} & Y_{22}C_{22}^{\mathrm{M}} + Y_{24}e_{22}^{\mathrm{M}} & Y_{22}C_{23}^{\mathrm{M}} + Y_{24}e_{23}^{\mathrm{M}} \\ Y_{12}C_{21}^{\mathrm{M}} + Y_{14}e_{21}^{\mathrm{M}} & Y_{12}C_{22}^{\mathrm{M}} + Y_{14}e_{22}^{\mathrm{M}} & Y_{12}C_{23}^{\mathrm{M}} + Y_{14}e_{23}^{\mathrm{M}} \\ 0 & 0 & 0 \\ -Y_{42}C_{21}^{\mathrm{M}} - Y_{44}e_{21}^{\mathrm{M}} & -Y_{42}C_{22}^{\mathrm{M}} - Y_{44}e_{22}^{\mathrm{M}} & -Y_{42}C_{23}^{\mathrm{M}} - Y_{44}e_{23}^{\mathrm{M}} \end{matrix} \right. \tag{42}$$
$$\left. \begin{matrix} 0 & 0 \\ Y_{24}\kappa_{21}^{\mathrm{M}} - Y_{22}e_{12}^{\mathrm{M}} & Y_{24}\kappa_{22}^{\mathrm{M}} - Y_{22}e_{22}^{\mathrm{M}} \\ Y_{14}\kappa_{21}^{\mathrm{M}} - Y_{12}e_{12}^{\mathrm{M}} & Y_{14}\kappa_{22}^{\mathrm{M}} - Y_{12}e_{22}^{\mathrm{M}} \\ 0 & 0 \\ Y_{42}e_{12}^{\mathrm{M}} - Y_{44}\kappa_{21}^{\mathrm{M}} & Y_{42}e_{22}^{\mathrm{M}} - Y_{44}\kappa_{22}^{\mathrm{M}} \end{matrix} \right).$$

Finally, it should be mentioned that the local polarisation directions $\beta_{1/2}$, see Fig. 6, depend on the microstructural evolution due to domain switching, see Sect. 2.2, thus being subject to a permanent change during the loading process. Therefore, the effective material properties and the Irwin matrix have to be recalculated after each switching increment dv.

4.3 Damage Evolution and Accumulation Model for High Cycle Fatigue and Life Time Prediction

Our investigations of damage in ferroelectrics are twofold. Based on FE simulations of complex boundary value problems with nonlinear ferroelectric constitutive behavior, the initiation of macroscopic cracks, e.g. at the tips of electrodes during the poling process, is investigated. This class of problems requires the consideration of maximum a few load cycles. On the microlevel, critical microcrack growth is considered with a condition $K_I = \sigma_I\sqrt{\pi a} = K_{IC}$ for local damage evolution. The fracture toughness K_{IC} is assumed constant, thus the effect of electric loads is taken into account only on the left hand side of the fracture criterion, influencing residual stresses due to domain wall motion. More sophisticated models should incorporate the dependence of K_{IC} on the electric field [8]. In the FE simulation each Gaussian point represents a crack RVE as introduced in Sect. 4.2 leading to a local calculation of effective properties according to Sect. 4.1. Locally, the material is considered as ruptured, as soon as the crack density parameter reaches a critical value f_{crit}.

The condensed approach outlined in Sect. 3 is exploited in regard to high cycle fatigue. The damage evolution is thus based on subcritical crack growth governed by the classical Paris law [29]. Most investigations about fatigue damage of ferroelectrics under electromechanical loading deal with low cycle fatigue from an experimental or modeling point of view (e.g. [39, 40, 44]). For high cycle fatigue a cycle by cycle simulation is unrealistic due to the high effort and computational costs. The idea presented in the following is based on an accumulation of many physical cycles N_P and their mapping onto much less numerical ones N.

The calculation scheme is captured in Fig. 7 and includes eight steps. In the first step, one load cycle N is simulated applying the approach outlined in Sect. 3. Next, the material RVE with the maximum peak-to-peak stress intensity factor (SIF) $\Delta\hat{K}_I$ is sought. The SIF at a material point m is calculated as

$$\Delta K_I^{(m)} = \Delta\sigma_I^{(m)}\sqrt{\pi a_0} \tag{43}$$

with $\Delta\sigma_I^{(m)} = \sigma_I^{\max(m)} - \sigma_I^{\min(m)}$. σ_I results from Eq. (31) and ΔK_I will be simply denoted as SIF in the following. In the third step it is checked if $\Delta\hat{K}_I$ is larger than or equal to the threshold value K_{Ith}. If this criterion is not satisfied, the next load cycle is simulated. Otherwise, the crack propagation increment $\Delta\hat{a}_N$ is determined as follows:

$$\Delta\hat{a}_N = \gamma\,\hat{a}_{N-1}. \tag{44}$$

Notice, that the initial crack length is a_0. Again, the superscript hat refers to the material point with the maximum stress intensity factor. γ is a numerical parameter

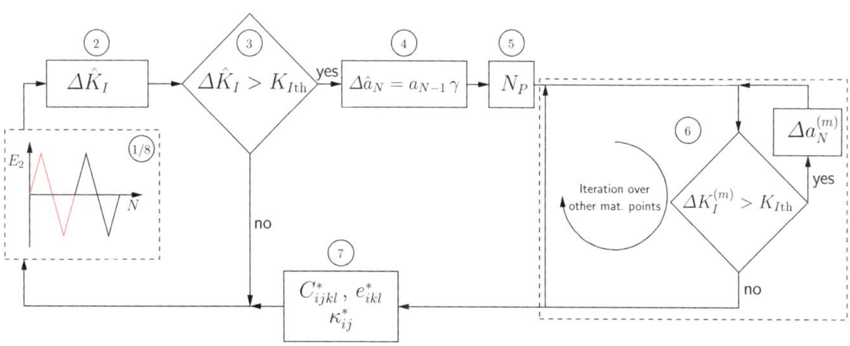

Fig. 7 Calculation scheme for high cycle fatigue damage prediction based on the condensed model

describing the ratio of the increment Δa and the crack length. Therefore, the new crack length $\hat{a}_N$ is obtained as:

$$\hat{a}_N = (1+\gamma)\,\hat{a}_{N-1}. \tag{45}$$

While the aforementioned load cycles N have a pure numerical background, the number of physical cycles N_P associated with the crack growth increment $\Delta\hat{a}_N$ is the relevant quantity. Thus, in the next step, N_P is determined from the classical Paris law [29]

$$\frac{\mathrm{d}a}{\mathrm{d}N} = C(\Delta K_I)^{\eta}. \tag{46}$$

C and η are material-dependent parameters. Separation of the independent variables under consideration of Eq. (43) leads to

$$\int\limits_0^{N_P} \mathrm{d}N = \int\limits_{\hat{a}_{N-1}}^{\hat{a}_N} \frac{\mathrm{d}a}{C(\Delta\sigma_I\sqrt{\pi a})^{\eta}}. \tag{47}$$

Integration of both sides of Eq. (47) and assuming that the peak-to-peak value $\Delta\sigma_I$ is constant during one numerical cycle N yields

$$N_P = \frac{2}{C(2-\eta)(\Delta\sigma_I\sqrt{\pi})^{\eta}}\left[(\hat{a}_{N-1}(1+\gamma))^{\frac{2-\eta}{2}} - \hat{a}_{N-1}^{\frac{2-\eta}{2}}\right]. \tag{48}$$

In step six of Fig. 7 the crack propagation $\Delta a^{(m)}$ in each of the other material points m is determined from Eq. (46) as a function of N_P, $\Delta\sigma_I^{(m)}$ and $a_{N-1}^{(m)}$:

$$\Delta a^{(m)} = C\left(\Delta\sigma_I^{(m)}\sqrt{\pi a_{N-1}^{(m)}}\right)^{\eta} N_P. \tag{49}$$

A simplifying assumption of Eq. (49) is that the crack length is constant for the integration. After having calculated all $\Delta a^{(m)}$, the effective material constants are determined according to Sect. 4.2. Now, the next numerical load cycle is simulated based on the updated crack lengths.

The scheme is repeated assembling the physical cycles to obtain their total number, finally representing the life time of the structure under electromechanical loading. The simulation is stopped, as soon as a rupture criterion is satisfied. Here, the average damage variable or crack density parameter $\langle f \rangle$, respectively, reaches a critical value, e.g. 0.5. Following the weakest link interpretation, the simulation can also be controlled by the material point $\hat{m}$, which in the first load cycle has been identified as the one with the highest stress level. Then, just the crack density $\hat{f}^{(m)}$ is inserted into the rupture criterion. The progress of damage in the other material points, however, still has an impact on the predicted life time by influencing the evolution of effective elastic, piezoelectric and dielectric constants. These material properties are in turn decisive for the domain induced residual stresses which are significantly controlling the damage process. Some other slight modifications of the simulation process are thinkable, maintaining the basic idea behind it. Those which have been implemented have all been leading to very similar life time predictions.

5 Results of Damage and Lifetime Predictions in Electromechanically Loaded Ferroelectrics

All calculations are based on the material data of bariumtitanate taken from [17]. The constants for the Paris law have been taken from [33], where experiments on fatigue crack growth in PZT under pure mechanical loading are presented. The values of C and η have been adopted here, since we are not aware of any other experimental data in this field. First, results of the life time prediction from the condensed model are presented. Then the damage evolution in a stack actuator during the poling process is investigated by FE simulation. Therefore, the constitutive model presented in Sect. 2 has been implemented as user element into the commercial FE code Abaqus [10].

5.1 *Life Time Prediction at High Cycle Fatigue*

First, the influence of a bipolar and an unipolar electrical loading is investigated for an initial damage $f_0 = 0.005$. Mechanically, the system is neither clamped nor submitted to a traction load. The loading scheme is shown in Fig. 8 in terms of the normalized electric field versus the first numerical load cycles. In Fig. 9 the crack density f normalized with respect to the initial damage f_0 and the critical crack density f_c is plotted versus the accumulated number of physical cycles.

It is obvious, that the damage evolution exhibits an asymptotic behavior, providing the life time N_P^{crit}. Although the relative damage at the end of the simulation is still below 10 %, taking $f_c = 1.0$ as a basis, the critical number of load cycles is obtained with sufficient accuracy from the dashed vertical tangent. The ratio f/f_c is depicted in Fig. 9 as an example. Obviously, the majority of load cycles is accumulated during an early stage of damage, whereas longer cracks lead to failure within a comparatively small number of cycles. The simulation further reveals that the bipolar cycling is going along with a much lower life time of $N_P^{\text{crit}} \approx 2 \times 10^5$ compared to the unipolar loading allowing for 7×10^9 cycles. The following calculations are all based on unipolar electric loading.

In Fig. 10a, b the influences of the initial crack density f_0 and the parameters of the Paris law C and η, see Eq. (46), on the critical number of load cycles are

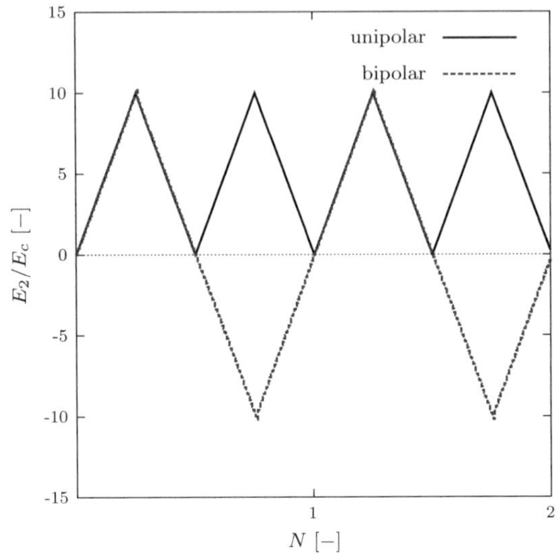

Fig. 8 Comparison of bipolar and unipolar electric loading: loading schemes

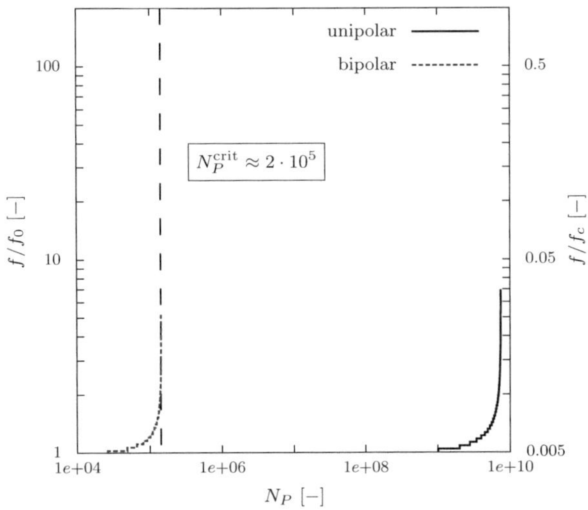

Fig. 9 Comparison of bipolar and unipolar electric loading: crack density f related to the initial and critical values f_0 and f_c to determine the number of critical cycles to failure

demonstrated. The solid lines in the graphs (a) and (b) represent the result for the unipolar loading from Fig. 9 which is taken as a reference here and in the following figures. In the logarithmic representation, there is a linear dependence of the life

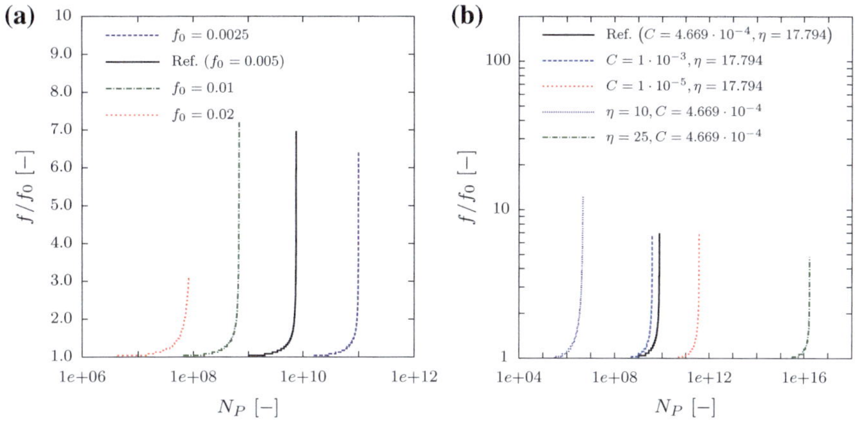

Fig. 10 Influences of the initial crack density (a) and the Paris constants C and η (b) on the critical number of load cycles

time on the initial damage. The essential finding of Fig. 10b is that the Paris constants have a crucial influence on the prediction of critical numbers of load cycles. Particularly η is a very sensitive parameter. As long as these two material constants are not known with a sufficient accuracy, the results of the predictions have to be considered as qualitative.

In Fig. 11 a combined cyclic electric and constant mechanical loading is investigated with respect to the influence of a compressive stress on both life time

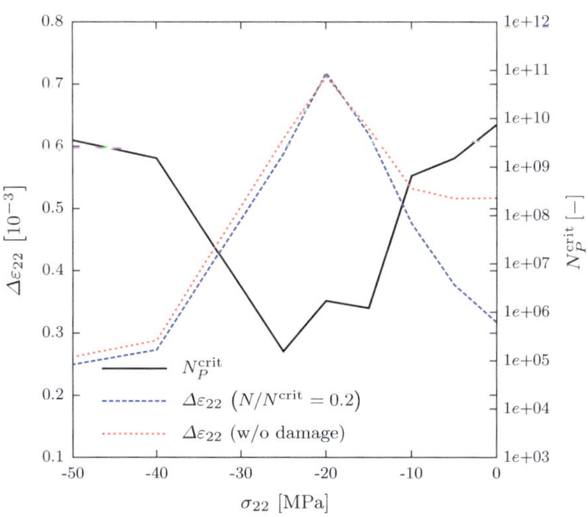

Fig. 11 Influence of a compressive preload on actuation strain and life time. The dotted line represents the strain in an undamaged structure

and actuating strain. The initial crack density f_0 and the Paris constants C, η are those of the solid lines in Fig. 10 declared as reference parameters. The unipolar electric cycles have an amplitude of $10E_C = 2$ kV/mm. Intuitively, one would maybe suppose that a compressive load basically reduces the strain and scarcely influences the life span. Experiments, however, show that a mechanical compression in the direction of electric field and thus poling fosters the efficiency of an actuator within a certain range of stress level [3, 4, 43]. Current experimental investigations also reveal that the number of bearable load cycles may be essentially reduced due to a compressive mechanical preloading [23]. In Fig. 11 both effects are visible. The solid line represents the critical number of electric load cycles N_P^{crit}, the dashed one illustrates the actuation strain $\Delta\varepsilon_{22}$. The latter changes during the service of an actuator, depending on the progressive degradation of the material. The plot shows a snapshot after 20 % of the predicted life time. The dotted line evolves from a simulation neglecting the damage process at all.

The major findings of Fig. 11 are summarized as follows. Without mechanical loading, the actuation strain is considerably reduced as the damage proceeds. Increasing the stress, the influence of damage on $\Delta\varepsilon_{22}$ is quickly reduced and almost vanishes for compressive loads exceeding 15 MPa. The actuating strain exhibits a maximum for $\sigma_{22} = -20$ MPa, not depending on the degree of damage. For large compressive loads, the strain is much smaller than for the case of zero mechanical loading, at least in an early operating stage with little damage. At approximately the same stress range where $\Delta\varepsilon_{22}$ is large, the critical number of load cycles is small. Maximum values $N_P^{\mathrm{crit}} \approx 10^{10}$ at $\sigma_{22} = 0$ are reduced to $N_P^{\mathrm{crit}} \approx 10^5$ at $\sigma_{22} = -25$ MPa. Even though the compressive stress reduces the crack tip loading, the enhancement of domain mobility finally leads to larger stress intensity factors. Increasing the stress level, the life time recovers, finally reaching the value for the stress-free specimen at $\sigma_{22} = -50$ MPa. According to the simulations, a pre-stress of -10 MPa has to be considered as convenient, providing a good compromise between actuatoric efficiency and operating life time.

In Fig. 12 different initial crack densities f_0 have been taken as a basis for simulations of unipolar electric cycling, again superimposing a constant compressive stress. The left graph illustrates the effect of stress and initial damage state on the actuation strain $\Delta\varepsilon_{22}$, the right one shows the influence on the critical number of load cycles N_P. The solid lines denoted as reference (Ref.) correspond to the parameters chosen for the other solid lines in the previous figures. The calculations reveal that the maximum strain $\Delta\varepsilon_{22}$ is scarcely affected by the initial crack density within the investigated range of $f_0 \in [0.005, 0.02]$. The life time, on the other hand, considerably depends on the quality of the material in terms of initial damage, leading to bearable load cycles of just 1000 for an initial crack density of 2 % in connection with a compressive preload of 25 MPa.

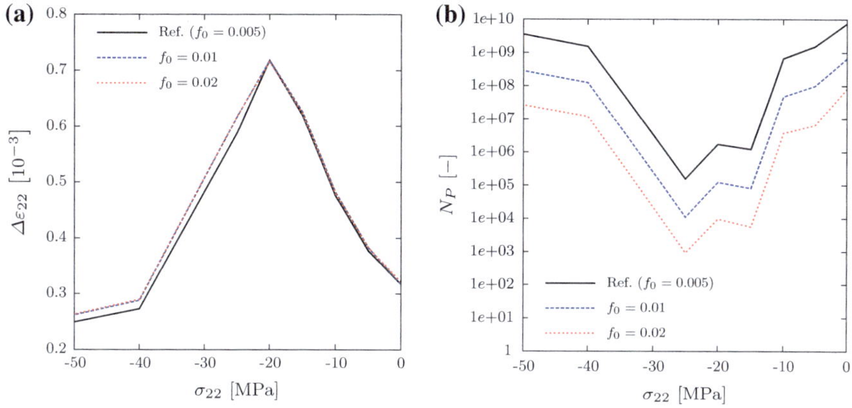

Fig. 12 Influences of compressive preloads and initial crack densities on actuation strain (**a**) and life time (**b**)

5.2 *Damage of a Stack Actuator Due to Poling*

Figure 13 shows the geometry of a model of a piezoelectric stack actuator with five electrodes being exposed to electric potentials $\varphi = 0$ and $\varphi = V_0$, respectively. A poling process is simulated with the FEM by linearly increasing V_0 to a maximum value of 5 kV, leading to a nominal electric field (sufficiently far from the electrode tips) of 5 E_C. The plot in the left column shows the maximum principal

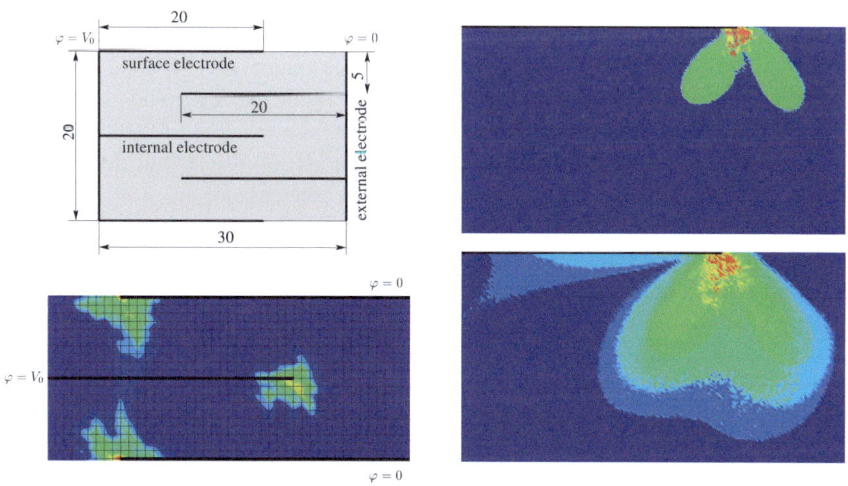

Fig. 13 Detail of a piezoelectric stack actuator with electrodes under pure electrical loading in terms of electric potentials $\varphi = 0, V_0$. The plots show the maximum principal stress neglecting microcrack growth (left) and the damage zone at an electrode tip after poling (right)

stress which is largest at the electrode tips and mainly stems from domain wall motion. In this simulation, damage has been neglected and the result corresponds to the electric load maximum. Close to the electrode tips, tensile stresses are as large as 100 MPa. At the rims of the colored zones 35 MPa are still found. The high stress level, however, will not prevail in a real piezoelectric structure but will give way to crack initiation and growth, conversely reducing the residual stresses to a much lower level.

The two plots on the right hand side of Fig. 13 result from a FE simulation taking the damage model into account. In the postprocessing, the material around just one electrode tip is considered, whereupon the lower picture shows a smaller section more detailed. The dark (red) areas close to the electrode tips represent domains, where the local microcrack density, related to Gaussian points of the FE model, is $f = 1$, thus indicating local rupture. The large bright (green) zone exhibits values $f \in [0.2, 0.9]$, the outer area of the damaged zone, which is a bit darker (pale blue), represents an interval $f \in [0.05, 0.2]$. From Fig. 13 it has to be concluded that the poling process is essential for the life time of a stack actuator under operating conditions, giving rise to cracking and thus defining the initial state of further damage evolution.

6 Conclusions

A previously presented condensed approach has been used to efficiently model the constitutive behavior of polycrystalline tetragonal ferroelectrics. It represents an alterative to a finite element simulation, which is also presented in this paper, on the level of a single grain exploiting the same microphysical constitutive model. Concerning damage predictions in ferroelectrics, it is essential that residual stresses are accurately obtained from a modeling approach, since they provide the major contribution to the loading of microcracks or other defects. The constitutive framework, which has been implemented into both the condensed approach and a FE scheme, describes an evolution of the microstructure in terms of domain wall motion and the growth of microcracks. Due to the influence of the state of damage on the effective elastic, dielectric and piezoelectric properties, there is a mutual interaction between both irreversible mechanisms. Results are discussed in terms of life time predictions under high cycle electric loading conditions, requiring an approach to efficiently handle up to 10^{10} load cycles. Further, a stack actuator model serves as an application oriented benchmark for the FE calculations, revealing a considerable material degradation at electrode tips due to the poling procedure. The model of a microcrack, however, is still very simple. To improve the predictions quantitatively, more sophisticated models have to be established accounting e.g. for crack interactions or nontrivial boundary conditions at crack faces. Also, the influence of an electric field is just taken into account on the loading side by controlling the residual stress. This might be the major contribution, however it has to be borne in mind that the intrinsic material resistance may also depend on electric fields. Finally, to get closer to applied research, the constitutive behavior of modern ferroelectric materials, including morphotropic or lead-free compositions, has to be a subject of advanced modeling.

References

1. Chen X, Fang DN, Hwang KC (1997) Micromechanics simulation of ferroelectric polarization switching. Acta Mater 45:3181–3189
2. Cocks ACT, McMeeking RM (1999) A phenomenological constitutive law for the behavior of ferroelectric ceramics. Ferroelectrics 228:219–228
3. Dittmer R, Webber KG, Aulbach E, Jo W, Tan X, Rödel J (2013) Electric-field-induced polarization and strain in 0.94$(Ba_{1/2}Na_{1/2})TiO_3$-0.06$BaTiO_3$ under uniaxial stress. Acta Mater 61:1350–1358
4. Dittmer R, Webber KG, Aulbach E, Jo W, Tan X, Rödel J (2013) Optimal working regime of lead-zirconate-titanate for actuation applications. Sensor Actuat A-Phys 189:187–194
5. Enderlein M (2007) Finite elemente Verfahren zur bruchmechanischen Analyse von Rissen in piezoelektrischen Strukturen bei transienter elektromechanischer Belastung. PhD thesis, Freiberg
6. Fan J, Stoll WA, Lynch CS (1999) Nonlinear constitutive behavior of soft and hard pzt: experiments and modeling. Acta Mater 47:4415–4425
7. Foerderreuther A (2003) Mechanische Eigenschaften von $BaTiO_3$-Keramiken unter mechanischer und elektrischer Belastung. PhD thesis, Stuttgart
8. Gellmann R, Ricoeur A (2012) Extended semi-analytical investigations of crack growth resistance behavior in ferroelectric materials. Acta Mech 223:2357–2368
9. Gellmann R, Ricoeur A (2012) Some new aspects of boundary conditions at cracks in piezoelectrics. Arch Appl Mech 82:841–852
10. Gellmann R, Ricoeur A (2014) Numerical analysis of domain switching and damage evolution in ferroelectric devices. Proc Mech Multi Mat 2:29–32
11. Gellmann R, Ricoeur A, Merkel E, Wang Z (2013) Generalized boundary conditions and effective properties in cracked piezoelectric solids. Proc Appl Math Mech 13:225–226
12. Gross D, Seelig T (2011) Fracture mechanics: with an introduction to Micromechanics. Springer, Berlin
13. Huber JE, Fleck NA (2001) Multi-axial electrical switching of a ferroelectric: theory versus experiment. J Mech Phys Solids 49:785–811
14. Huber JE, Fleck NA, Landis CM, McMeeking RM (1999) A constitutive model for ferroelectric polycrystals. J Mech Phys Solids 47:1663–1697
15. Huo Y, Jiang Q (1997) Modeling of domain switching in polycrystalline ferroelectric cermanics. Smart Mater Struct 6:441–447
16. Hwang SC, Lynch CS, McMeeking RM (1995) Ferroelectric/ferroelastic interactions and a polarization switching model. Acta Metall Mater 43:2073–2084
17. Jaffe B, Cook WR, Jaffe H (1971) Piezoelectric ceramics. Academic Press, London
18. Kachanov M (1992) Effective elastic properties of cracked solids: critical review of some basic concepts. Appl Mech Rev 45:304–335
19. Kamlah M (2001) Ferroelectric and ferroeleastic piezoceramics—modeling of electromechanical hysteresis phenomena. Continuum Mech Thermodyn 13:219–268
20. Kamlah M, Tsakmakis C (1999) Phenomenological modeling of non-linear electromechanical coupling in ferroelectrics. Int J Solids Struct 36:669–695
21. Kessler H, Balke H (2001) On the local and average energy release in polarization switching phenomena. J Mech Phys Solids 49:953–978

22. Lange S, Ricoeur A (2015) A condensed microelectromechanical approach for modeling tetragonal ferroelectrics. Int J Solids Struct 54:100–110
23. Li F (2014) Ultrahigh superelastic and actuation strains in ferroelectric crystals by reversible electromechanical domain switching. In: 5th international congress on ceramics, Beijing p 250
24. Li Q, Ricoeur A, Enderlein M, Kuna M (2010) Evaluation of electromechanical coupling effect by microstructural modeling of domain switching in ferroelectrics. Mech Res Commun 37:332–336
25. Lu W, Fang DN, Li CQ, Hwang KC (1997) Nonlinear electric-mechanical behavior and micromechanics modelling of ferroelectric domain evolution. Acta Mater 47:3181–3189
26. Lynch CS (1996) The effect of uniaxial stress on the electro-mechanical response of 8/65/35 plzt. Acta Mater 44:4137–4148
27. Michelitsch T, Kreher WS (1998) A simple model for the nonlinear material behavior of ferroelectrics. Acta Mater 46:5085–5094
28. Pak Y (1992) Linear electro-elastic fracture mechanics of piezoelectric materials. Int J Fract 54:79–100
29. Paris P, Erdogan F (1963) A critical analysis of crack propagation laws. J Basic Eng 85: 528–534
30. Parton VZ, Kudryavtsev BA (1988) Electromagnetoelasticity. Gordon and Breach Science Publishers, New York
31. Ricoeur A, Kuna M (2003) Influence of electric fields on the fracture of ferroelectric ceramics. J Eur Ceram Soc 23:1313–1328
32. Ricoeur A, Gellmann R, Wang Z (2014) Influence of inclined electric fields on the effective fracture toughness of piezoelectric ceramics. Acta Mech. doi:10.1007/s00707-014-1190-5
33. Salz CRJ, Hoffman M, Westram I, Rödel J (2005) Cyclic fatigue crack growth in pzt under mechanical loading. J Am Ceram Soc 88:1331–1333
34. Sosa H (1990) On the fracture mechanics of piezoelectric solids. Int J Solids Struct 29:1–15
35. Stephan P, Schaber K, Stephan K, Mayinger F (2010) Thermodynamik—band 2: Mehrstoffsysteme und chemische Reaktionen. Springer, Berlin
36. Suo Z (1993) Models for breakdown-resistant dielectric and ferroelectric ceramics. J Mech Phys Solids 41:1155–1176
37. Wang B, Han J (2007) An accumulation damage model for fatigue fracture of ferroelectric ceramics. Eng Fract Mech 74:1456–1467
38. Wang X, Jiang L (2003) The effective electroelastic property of piezoelectric media with parallel dielectric cracks. Int J Solids Struct 40:5287–5303
39. Westram I, Oates WS, Lupascu DC, Roedel J, Lynch C (2007) Mechanism of electric fatigue crack growth in lead zirconate titanate. Acta Mater 55:301–312
40. Westram I, Ricoeur A, Emmerich A, Roedel J, Kuna M (2007) Fatigue crack growth law for ferroelectrics under cyclic electrical and combined electromechanical loading. J Eur Ceram Soc 27:2485–2494
41. Yang X, Chen C, Hu Y (2003) A static damage constitutive model for piezoelectric materials. In: Mechanics of electromagnetic solids, Kluwer Academic Publishers, pp 259–272
42. Zheng M, Su Y, Zhou G (1999) Damage model for flexural strength variation of ferroelectric materials induced by electric field. Theoret Appl Fract Mech 32:137–145
43. Zhou D, Kamlah M (2005) Dielectric and piezoelectric performance of soft PZT piezoceramics under simultaneous alternating electromechanical loading. J Eur Ceram Soc 25:2415–2420
44. Zhu T, Yang W (1999) Fatigue crack growth in ferroelectrics driven by cyclic electric loading. J Mech Phys Solids 47:81–97

Numerical Analysis of Interface Cracks in Layered Piezoelectric Solids

Michael Wünsche, Chuanzeng Zhang, Jan Sladek and Vladimir Sladek

Abstract In this paper, transient dynamic crack analysis in two-dimensional, layered, anisotropic and linear piezoelectric solids is presented. For this purpose, a time-domain boundary element method (BEM) is developed. The layered piezoelectric solids are modelled by the multi-domain BEM formulation. The time-domain dynamic fundamental solutions for homogeneous linear piezoelectric solids are applied in the present BEM. The spatial discretization of the boundary integral equations is performed by a Galerkin-method, while a collocation method is implemented for the temporal discretization of the arising convolution integrals. An explicit time-stepping scheme is obtained to compute the discrete boundary data including the generalized crack-opening-displacements (CODs). To show the effects of the interface, the material combinations and the dynamic loading on the intensity factors, several numerical examples are presented and discussed.

1 Introduction

Piezoelectric materials offer many possibilities in advanced engineering structures due to their inherent coupling effects between mechanical and electrical fields and are widely applied in smart devices and structures like transducers, actuators and

M. Wunsche (✉)
Technical Mechanics, University of Wuppertal, 42119 Wuppertal, Germany
e-mail: wuensche@uni-wuppertal.de

M. Wünsche · C. Zhang
Structural Mechanics, University of Siegen, 57068 Siegen, Germany
e-mail: c.zhang@uni-siegen.de

J. Sladek · V. Sladek
Institute of Construction and Architecture, Slovak Academy of Sciences, 84503 Bratislava, Slovakia
e-mail: jan.sladek@savba.sk

V. Sladek
e-mail: vladimir.sladek@savba.sk

G. Hütter and L. Zybell, *Recent Trends in Fracture and Damage Mechanics*, DOI 10.1007/978-3-319-21467-2_12

sensors [7]. Important applications of piezoelectric materials are layered or laminated composites because they can be optimized to satisfy the high-performance requirements according to different in-service conditions. Beside cracks inside homogeneous domains, one of the most dominant failure mechanisms in layered or laminated piezoelectric composites is interface failure. Interface cracks or interface debonding may be induced by the mismatch of the mechanical, electrical and thermal properties of the material constituents during the manufacturing process and the in-service loading conditions. Although the static and dynamic crack problems in homogenous piezoelectric solids have been investigated by many authors (e.g., [3, 4, 6, 10, 12]) the corresponding analysis of interface cracks in layered piezoelectric solids is rather limited due to the problem complexity. This paper presents such an analysis by using a hypersingular symmetric Galerkin boundary element method (SGBEM) for crack problems in two-dimensional (2D), layered and linear piezoelectric solids.

2 Problem Formulation

Let us consider a piecewise homogeneous, layered and linear piezoelectric solid with an interface crack (Fig. 1).

In the absence of body forces, free electrical charges and applying the generalized notation, the cracked solid satisfies the generalized equations of motion

$$\sigma_{iJ,i}(\mathbf{x},t) = \rho\delta^*_{JK}\ddot{u}_K(\mathbf{x},t), \quad \delta^*_{JK} = \begin{cases} \delta_{jk}, & J=j; K=k, \\ 0, & \text{otherwise}, \end{cases} \tag{1}$$

the constitutive equations

$$\sigma_{iJ}(\mathbf{x},t) = C_{iJKl}u_{K,l}(\mathbf{x},t), \tag{2}$$

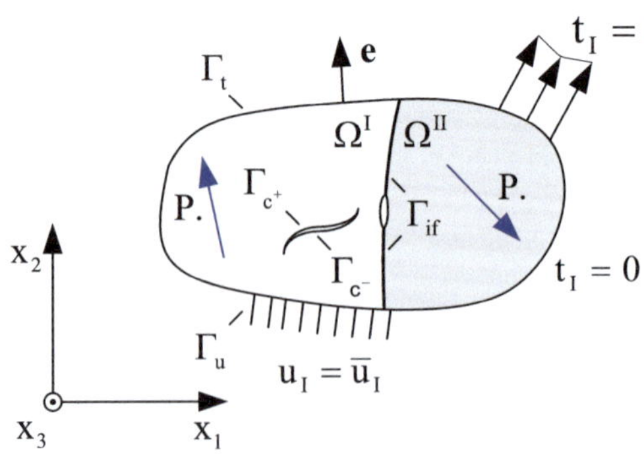

Fig. 1 A cracked piecewise homogeneous anisotropic and linear piezoelectric solid

the initial conditions

$$u_i(\mathbf{x}, t = 0) = \dot{u}_i(\mathbf{x}, t = 0) = 0, \tag{3}$$

the boundary conditions

$$u_I(\mathbf{x}, t) = \bar{u}_I(\mathbf{x}, t), \quad \mathbf{x} \in \Gamma_u, \tag{4}$$

$$t_I(\mathbf{x}, t) = \bar{t}_I(\mathbf{x}, t), \quad \mathbf{x} \in \Gamma_t, \tag{5}$$

and the continuity conditions on the interface except the crack-faces

$$u_I^I(\mathbf{x}, t) = u_I^{II}(\mathbf{x}, t) \tag{6}$$

$$t_I^I(\mathbf{x}, t) = -t_I^{II}(\mathbf{x}, t), \tag{7}$$

with the capital letter subscripts $J \in \{1, 2, 4\}$ and lower case letter subscripts $j \in \{1, 2\}$.

In Eqs. (1)–(7), the generalized displacements, stresses, elasticity matrix and tractions are defined by

$$u_I = \begin{cases} u_i, & I = i, \text{displacements} \\ \varphi, & I = 4, \text{electric potential}, \end{cases} \tag{8}$$

$$\sigma_{iJ} = \begin{cases} \sigma_{ij}, & J = j, \text{stresses} \\ D_i, & J = 4, \text{electric displacements}, \end{cases} \tag{9}$$

$$C_{iJKl} = \begin{cases} c_{ijkl}, & J = j; K = k, \text{elasticity tensor} \\ e_{lij}, & J = j; K = 4, \text{piezoelectric tensor} \\ e_{ikl}, & J = 4; K = k, \text{piezoelectric tensor} \\ -\varepsilon_{il}, & J = K = 4, \text{electric permittivity tensor}, \end{cases} \tag{10}$$

$$t_I(\mathbf{x}, t) = \sigma_{jI}(\mathbf{x}, t) e_j(\mathbf{x}), \tag{11}$$

where $u_i, \sigma_{ij}, \varphi$ and D_i represent the mechanical displacements, the stresses, the electrical potential and the electrical displacements, ρ, C_{ijkl}, e_{ijk} and ε_{ij} are the mass density, the elasticity tensor, the piezoelectric tensor and the dielectric permittivity tensor, respectively.

Further, Γ_{if} is the interface between the homogenous domains $\Omega^{\zeta}(\zeta = 1, 2, \ldots, N)$, $\Gamma_{c\pm}$ denotes the upper and the lower crack-faces, Γ_t and Γ_u stand for the external boundaries where the tractions t_i and the displacements u_i are prescribed. Three different electrical boundary conditions are considered on the crack-faces. The impermeable electrical crack-face condition

$$D_i(\mathbf{x} \in \Gamma_{c^+}, t) = D_i(\mathbf{x} \in \Gamma_{c^-}, t) = 0 \tag{12}$$

describes in a physical sense that both crack-faces are free of electrical displacements. This would be correct for a medium inside the crack with a electrical permittivity κ_c of zero. In contrast, the permeable electrical crack-face condition

$$D_i(\mathbf{x} \in \Gamma_{c^+}, t) = D_i(\mathbf{x} \in \Gamma_{c^-}, t), \quad \varphi(\mathbf{x} \in \Gamma_{c^+}, t) - \varphi(\mathbf{x} \in \Gamma_{c^-}, t) = 0 \tag{13}$$

implies identical potentials on both crack-faces. Since these two crack-face boundary conditions are physically not consistent a more realistic semi-permeable crack-face boundary condition has been introduced by taking into account the electrical permittivity of the medium inside the crack as

$$D_n(\mathbf{x} \in \Gamma_{c^+}, t) = D_n(\mathbf{x} \in \Gamma_{c^-}, t) = -\kappa_c \frac{\varphi(\mathbf{x} \in \Gamma_{c^+}, t) - \varphi(\mathbf{x} \in \Gamma_{c^-}, t)}{u_n(\mathbf{x} \in \Gamma_{c^+}, t) - u_n(\mathbf{x} \in \Gamma_{c^-}, t)}, \tag{14}$$

where D_n and u_n are the normal components of the electrical displacements and the mechanical displacements on the crack-faces, $\kappa_c = \kappa_r \kappa_0$ is defined as the product of the relative permittivity of the considered medium κ_r and the permittivity of the vacuum $\kappa_0 = 8.854 \times 10^{-12}$ C/(Vm).

The generalized crack-opening-displacements (CODs) are defined by

$$\Delta u_I(\mathbf{x}, t) = u_I(\mathbf{x} \in \Gamma_{c^+}, t) - u_I(\mathbf{x} \in \Gamma_{c^-}, t). \tag{15}$$

Throughout the paper, a comma after a quantity represents spatial derivatives while a dot over the quantity denotes time differentiation. Lower case Latin indices take the values 1 and 2 (elastic), while capital Latin indices take the values 1, 2 (elastic) and 4 (electric). Unless otherwise stated, the conventional summation rule over repeated indices is implied.

3 Time-Domain Boundary Integral Equations and Fundamental Solutions

In the sense of the weighted residual formulation, the time-domain displacement Galerkin-boundary integral equations (BIEs) for a cracked solid can be written as

$$\begin{aligned}\int_{\Gamma} \psi(\mathbf{x}) u_J(\mathbf{x}, t) d\Gamma_x = &\int_{\Gamma} \psi(\mathbf{x}) \int_{\Gamma_b} \left[u_{IJ}^G(\mathbf{x}, \mathbf{y}, t) * t_I(\mathbf{y}, t) - t_{IJ}^G(\mathbf{x}, \mathbf{y}, t) * u_I(\mathbf{y}, t)\right] d\Gamma_y d\Gamma_x \\ &+ \int_{\Gamma} \psi(\mathbf{x}) \int_{\Gamma_{c^+}} t_{IJ}^G(\mathbf{x}, \mathbf{y}, t) * \Delta u_I(\mathbf{y}, t) d\Gamma_y d\Gamma_x,\end{aligned} \tag{16}$$

where $u_{IJ}^G(x, y, t)$ and $t_{IJ}^G(x, y, t)$ are the dynamic displacement and traction fundamental solutions defined by

$$t_{IJ}^G(\mathbf{x}, \mathbf{y}, t) = C_{qIKr} e_q(\mathbf{y}) u_{KJ,r}^G(\mathbf{x}, \mathbf{y}, t). \tag{17}$$

$\psi(x)$ is the weight or test function, $\Gamma_b = \Gamma_u + \Gamma_t$ and an asterisk denotes the Riemann convolution

$$g(\mathbf{x}, t) * h(\mathbf{x}, t) = \int_0^t g(\mathbf{x}, t - \tau) h(\mathbf{x}, \tau) \mathrm{d}\tau. \tag{18}$$

The time-domain traction Galerkin-BIEs can be obtained by substituting Eq. (16) into the generalized constitutive Eq. (2), taking the limit process $x \to \Gamma$ and considering the boundary conditions as

$$\begin{aligned} \int_\Gamma \psi(\mathbf{x}) t_J(\mathbf{x}, t) \mathrm{d}\Gamma_x = & \int_\Gamma \psi(\mathbf{x}) \int_{\Gamma_b} [v_{IJ}^G(\mathbf{x}, \mathbf{y}, t) * t_I(\mathbf{y}, t) - w_{IJ}^G(\mathbf{x}, \mathbf{y}, t) * u_I(\mathbf{y}, t)] \mathrm{d}\Gamma_y \mathrm{d}\Gamma_x \\ & + \int_\Gamma \psi(\mathbf{x}) \int_{\Gamma_{c^+}} w_{IJ}^G(\mathbf{x}, \mathbf{y}, t) * \Delta u_I(\mathbf{y}, t) \mathrm{d}\Gamma_y \mathrm{d}\Gamma_x, \end{aligned} \tag{19}$$

where $v_{ij}^G(x, y, t)$ and $w_{ij}^G(x, y, t)$ are the traction and the higher-order traction fundamental solutions, which are defined by

$$v_{IJ}^G(\mathbf{x}, \mathbf{y}, t) = -C_{pIKs} e_p(\mathbf{x}) u_{KJ,s}^G(\mathbf{x}, \mathbf{y}, t), \tag{20}$$

$$w_{IJ}^G(\mathbf{x}, \mathbf{y}, t) = -C_{pIKs} e_p(\mathbf{x}) C_{qJLr} e_q(\mathbf{y}) u_{KL,sr}^G(\mathbf{x}, \mathbf{y}, t). \tag{21}$$

The fundamental solutions possess the following spatial symmetry properties

$$u_{IJ}^G(\mathbf{x}, \mathbf{y}, t) = u_{JI}^G(\mathbf{x}, \mathbf{y}, t), \tag{22}$$

$$t_{IJ}^G(\mathbf{x}, \mathbf{y}, t) = -v_{IJ}^G(\mathbf{x}, \mathbf{y}, t) = v_{JI}^G(\mathbf{y}, \mathbf{x}, t), \tag{23}$$

$$w_{IJ}^G(\mathbf{x}, \mathbf{y}, t) = w_{JI}^G(\mathbf{y}, \mathbf{x}, t). \tag{24}$$

Taking these symmetry properties (22)–(24) into account, a spatial symmetric Galerkin-method is obtained if the displacement Galerkin-BIEs (16) are applied on the external boundary Γ_u where the generalized displacements are known and the interface Γ_{if} for the generalized tractions, while the traction Galerkin-BIEs (19) are used on the external boundary Γ_t where the generalized tractions are prescribed and the interface Γ_{if} for the generalized displacements.

The dynamic time-domain fundamental solutions for homogeneous, anisotropic and linear piezoelectric solids are not available in explicit form. Here, the solutions derived by Wang and Zhang [11] using the Radon transform technique are implemented. Such solutions are expressed in the 2D case by a line integral over a unit circle as

$$u_{IJ}^{G}(\mathbf{x},\mathbf{y},t) = \frac{\mathrm{H}(t)}{4\pi^2} \int\limits_{|\mathbf{n}|=1} \sum_{m=1}^{3} \frac{P_{IJ}^{m}}{\rho c_m} \frac{1}{c_m t + \mathbf{n}\cdot(\mathbf{y}-\mathbf{x})} \mathrm{d}\mathbf{n}, \tag{25}$$

where $H(t)$, $\mathbf{n}$, c_m and P_{IJ}^{m} denote the Heaviside step function, the wave propagation vector, the phase velocities of the elastic waves and the projector defined in [11]. By integration by parts and applying the properties of the time convolution the time-domain generalized displacement fundamental solutions can be divided into a singular static and a regular dynamic part as

$$u_{IJ}^{G}(\mathbf{x},\mathbf{y},t) * f(t) = u_{IJ}^{S}(\mathbf{x},\mathbf{y}) f(t) + u_{IJ}^{D}(\mathbf{x},\mathbf{y},t) * \dot{f}(t). \tag{26}$$

Like the displacement fundamental solutions, the traction and the higher-order traction fundamental solutions can also be divided into their singular static and regular dynamic parts [12].

4 Numerical Solution Algorithm

A numerical solution procedure is presented in this section to solve the time-domain BIEs (16) and (19). The procedure uses a Galerkin-method for the spatial discretization and a collocation method for the temporal discretization. A sub-domain technique is utilized, which divides the layered piecewise homogenous solid into two or several sub-domains with homogeneous material properties and to each sub-domain the time-domain BIEs (16) and (19) are applied. In the following, some details of the numerical solution algorithm are described. For the spatial discretization, the crack-faces, the external boundary of each homogeneous sub-domain and the interfaces of the cracked piezoelectric solid are discretized by linear elements. Linear shape functions are also used for the temporal discretization in the present analysis. At the crack-tips inside a homogeneous sub-domain, special crack-tip elements are applied to describe the local behaviour of the generalized CODs near the crack-tips properly. This ensures an accurate and a direct calculation of the intensity factors from the numerically computed CODs. On the other hand, the asymptotic crack-tip field in the case of an interfacial crack between two dissimilar piezoelectric materials shows different oscillating and non-oscillating singularities in the generalized stress field [1, 8, 9], which makes an implementation of special crack-tip elements quite cumbersome. For this reason, only standard elements are applied at the crack-tips for interface cracks. The strongly singular and hypersingular boundary integrals can be computed analytically by special techniques [5]. By using linear temporal shape-functions, time integrations can also be performed analytically. Only the line integrals over the unit circle arising in the regular parts of the dynamic fundamental solutions have to be computed numerically by the standard Gaussian quadrature.

After temporal and spatial discretizations and considering the initial conditions the following systems of linear algebraic equations can be obtained for each sub-domain $\Omega^{\zeta}(\zeta = 1, 2, \ldots, N)$

$$\begin{aligned} \mathbf{C}_{\zeta}\mathbf{u}_{\zeta}^{K} &= \mathbf{U}_{\zeta}^{S}\mathbf{t}_{\zeta}^{K} - \mathbf{T}_{\zeta}^{S}\mathbf{u}_{\zeta}^{K} + \mathbf{T}_{\zeta}^{S}\Delta\mathbf{u}_{\zeta}^{K} \\ &+ \sum_{k=1}^{K}\left[\mathbf{U}_{\zeta}^{D;K-k+1}\mathbf{t}_{\zeta}^{k} - \mathbf{T}_{\zeta}^{D;K-k+1}\mathbf{u}_{\zeta}^{k} + \mathbf{T}_{\zeta}^{D;K-k+1}\Delta\mathbf{u}_{\zeta}^{k}\right], \end{aligned} \tag{27}$$

$$\begin{aligned} \mathbf{D}_{\zeta}\mathbf{t}_{\zeta}^{K} &= \mathbf{V}_{\zeta}^{S}\mathbf{t}_{\zeta}^{K} - \mathbf{W}_{\zeta}^{S}\mathbf{u}_{\zeta}^{K} + \mathbf{W}_{\zeta}^{S}\Delta\mathbf{u}_{\zeta}^{K} \\ &+ \sum_{k=1}^{K}\left[\mathbf{V}_{\zeta}^{D;K-k+1}\mathbf{t}_{\zeta}^{k} - \mathbf{W}_{\zeta}^{D;K-k+1}\mathbf{u}_{\zeta}^{k} + \mathbf{W}_{\zeta}^{D;K-k+1}\Delta\mathbf{u}_{\zeta}^{k}\right]. \end{aligned} \tag{28}$$

By invoking the continuity conditions (6) and (7) on the interface Γ_{if} as well as (12), (13) or (14) on the crack-faces Γ_{c+} and Γ_{c-}, Eqs. (27) and (28) can be summarized to the following system of linear algebraic equations

$$\Xi^{1}\mathbf{x}^{K} = \Upsilon^{1}\mathbf{y}^{K} + \sum_{k=1}^{K-1}(\Lambda^{K-k+1}\mathbf{t}^{k} - \Theta^{K-k+1}\mathbf{u}^{k}). \tag{29}$$

Here Ξ^{1} and Υ^{1} are the system matrices, $\mathbf{u}_{K}$ is the vector containing the boundary displacements and the CODs, and $\mathbf{t}_{K}$ is the traction vector on the external boundary, the interfaces and the crack-faces. By considering the boundary conditions (4) and (5), the following explicit time-stepping scheme can be obtained

$$\mathbf{x}^{K} = (\Xi^{1})^{-1}\left[\Upsilon^{1}\mathbf{y}^{K} + \sum_{k=1}^{K-1}(\Lambda^{K-k+1}\mathbf{t}^{k} - \Theta^{K-k+1}\mathbf{u}^{k})\right], \tag{30}$$

where $\mathbf{y}_{K}$ is the vector of the prescribed boundary data, while $\mathbf{x}_{K}$ represents the vector of the unknown boundary data, which can be computed time-step by time-step.

5 Intensity Factors for an Interfacial Crack

The intensity factors for an interface crack between two dissimilar anisotropic and linear piezoelectric materials can be computed from the generalized crack-opening displacements (CODs) derived by Suo et al. [9]

$$\Delta \mathbf{u}(r,t) = (\mathbf{H}+\overline{\mathbf{H}})\sqrt{\frac{r}{2\pi}}\left[\frac{K(t)r^{\mathrm{i}\varepsilon_1}\mathbf{w}}{(1+2\mathrm{i}\varepsilon_1)\cosh(\pi\varepsilon_1)}+\frac{\bar{K}(t)r^{-\mathrm{i}\varepsilon_1}\bar{\mathbf{w}}}{(1-2\mathrm{i}\varepsilon_1)\cosh(\pi\varepsilon_1)}\right.$$
$$\left.+\frac{K_4(t)r^{-\varepsilon_2}\mathbf{w}_4}{(1-2\varepsilon_2)\cos(\pi\varepsilon_2)}\right], \tag{31}$$

where $K = K_1 + \mathrm{i}K_2$ and K_4 are the complex stress intensity factor and the electrical displacement intensity factor, ε_i are the bimaterial constants, an overbar denotes the complex conjugate and i is the imaginary unit. The complex Hermitian matrix $\mathbf{H}$ is defined by

$$\mathbf{H} = \mathbf{Y}_I + \overline{\mathbf{Y}}_{II}, \quad \mathbf{Y}_I = \mathrm{i}\mathbf{A}_I\mathbf{B}_I^{-1}, \quad \mathbf{Y}_{II} = \mathrm{i}\mathbf{A}_{II}\mathbf{B}_{II}^{-1}, \tag{32}$$

where the subscripts *I* and *II* indicate the upper and the lower layers. The matrices **A** and **B** are computed from the eigenvalue problem as shown in [12]. The two bimaterial constants ε_1 and ε_2 as well as the eigenvectors $\mathbf{w}$ and $\mathbf{w}_4$ are determined by the eigenvalue problem

$$\mathbf{D}^{-1}\mathbf{E}\mathbf{w} = -\mathrm{i}\beta\mathbf{w}, \tag{33}$$

where $\mathbf{D} = \Re(\mathbf{H})$ and $\mathbf{E} = \Im(\mathbf{H})$ are the real part and the imaginary part of the complex matrix $\mathbf{H}$. Equation (33) has two complex eigenvalues with the properties $\alpha_1 = \Re(\beta) > 0$ and $\alpha_2 = \Im(\beta) > 0$ and leads therefore to the bimaterial constants

$$\varepsilon_j = \frac{1}{2\pi}\log\left(\frac{1+\alpha_j}{1-\alpha_j}\right), \quad (j = 1,2). \tag{34}$$

For interface cracks between two dissimilar piezoelectric materials, the generalized crack-opening-displacements (CODs) in the crack-tip vicinity show the well known $r^{1/2\pm\mathrm{i}\varepsilon_1}$-oscillating behaviour as for interface cracks between two dissimilar anisotropic materials [2] and additionally the non-oscillating $r^{1/2\pm\varepsilon_2}$-behavior. This makes an implementation of special spatial crack-tip shape functions rather difficult. For this reason, only standard elements near the tips of interface cracks are used in this analysis. To minimize the numerical errors, an extrapolation technique is applied to calculate the intensity factors from the numerically computed generalized crack-opening-displacements (CODs).

6 Numerical Examples

In this section, numerical test examples are presented and discussed. The following loading parameter χ is introduced to measure the intensity of the electrical loading

$$\chi = \frac{e_{22}}{\varepsilon_{22}} \frac{D_0}{\sigma_0}, \tag{35}$$

where σ_0 and D_0 are the mechanical and electrical loading amplitudes. For convenience, the mode-I, the mode-II and the mode-IV dynamic intensity factors for crack-tips inside a homogeneous layer or sub-domain are normalized by

$$K_I^*(t) = \frac{K_I(t)}{K_0}, \quad K_{II}^*(t) = \frac{K_{II}(t)}{K_0}, \quad K_{IV}^*(t) = \frac{e_{22}}{\varepsilon_{22}} \frac{K_{IV}(t)}{K_0}. \tag{36}$$

The real part K_1 and the imaginary part K_2 of the complex dynamic stress intensity factor and the electrical displacement intensity factor K_4 for interface cracks are normalized by

$$K_1^*(t) = \frac{K_1(t)}{K_0}, \quad K_2^*(t) = \frac{K_2(t)}{K_0}, \quad K_4^*(t) = \frac{e_{22}^I}{\varepsilon_{22}^I} \frac{K_4(t)}{K_0}, \tag{37}$$

with $K_0 = f\sigma_0\sqrt{\pi a}$, where $f = 1$ if a mechanical loading is applied while $f = \chi$ in the case of pure electrical loading and a is the half length of an internal crack.

For the purpose of comparison a program based on the commercial FEM program ANSYS is developed. In the FEM calculations the element type PLANE223 with quadratic shape functions is chosen and quarter-point elements (QPE) are implemented to describe the local behaviour of the CODs at crack-tips inside a homogeneous layer or sub-domain. For interface cracks only the element type PLANE223 is used. The unknown boundary values are computed numerically by an implicit time-stepping scheme in ANSYS. As in the time-domain BEM (TDBEM) the dynamic intensity factors are computed directly by the CODs in the case of crack-tips inside a sub-domain, while for interface cracks a displacement extrapolation technique is used to compute the complex dynamic intensity factors. Generalized plain strain conditions are assumed in all numerical computations.

6.1 *A Central Interface Crack in a Layered Rectangular Plate*

In the first example as shown in Fig. 2, we consider a central interface crack of length $2a$ in a rectangular, layered, anisotropic and linear piezoelectric plate. The geometry of the cracked plate is determined by $h = 20.0$ mm, $w = 10.0$ mm and $2a = 4.8$ mm.

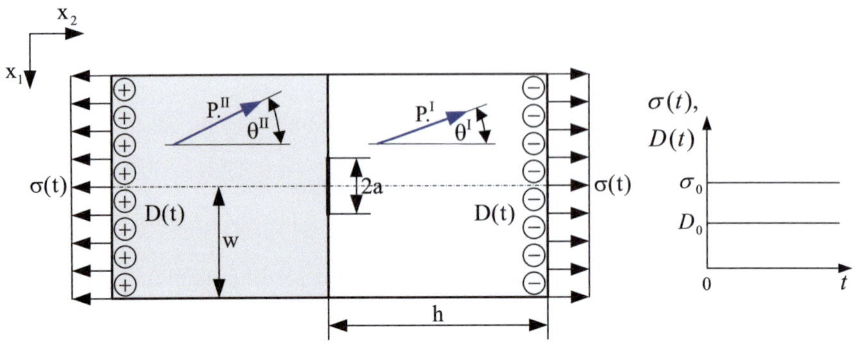

Fig. 2 An interface crack in a layered rectangular piezoelectric plate

The spatial discretization of the external boundary and the interface is performed by an element-length of 1.0 mm. Each crack-face is approximated by 12 elements and a normalized time-step of $c_L \Delta t / h = 0.05$ is chosen, where c_L is the longitudinal wave velocity. A comparison of the used BEM and FEM meshes is presented in Fig. 3. As piezoelectric material for the domain I Zirconate Titanate (PZT-5H) is chosen, which has the following material parameters

$$
\begin{aligned}
&C_{11} = 126.0\ \text{GPa}, \quad C_{12} = 84.1\ \text{GPa}, \quad C_{22} = 117.0\ \text{GPa}, \quad C_{66} = 23.0\ \text{GPa}, \\
&e_{21} = -6.5\ \text{C/m}^2, \quad e_{22} = 23.3\ \text{C/m}^2, \quad e_{16} = 17.0\ \text{C/m}^2, \\
&\varepsilon_{11} = 15.04\ \text{C/(GVm)}, \quad \varepsilon_{22} = 13.0\ \text{C/(GVm)}
\end{aligned}
\tag{38}
$$

and the mass density $\rho = 7500\ \text{kg/m}^3$, while for domain II Barium Titanate ($BaTiO_3$) with the following material constants

$$
\begin{aligned}
&C_{11} = 150.0\ \text{GPa}, \quad C_{12} = 146.0\ \text{GPa}, \quad C_{22} = 44.0\ \text{GPa}, \quad C_{66} = 66.0\ \text{GPa}, \\
&e_{21} = -4.35\ \text{C/m}^2, \quad e_{22} = 17.5\ \text{C/m}^2, \quad e_{16} = 11.4\ \text{C/m}^2, \\
&\varepsilon_{11} = 9.87\ \text{C/(GVm)}, \quad \varepsilon_{22} = 11.2\ \text{C/(GVm)}
\end{aligned}
\tag{39}
$$

and the mass density $\rho = 5800\ \text{kg/m}^3$ is applied.

The layered piezoelectric plate is subjected to a combined tensile impact loading of the form $\sigma(t) = \sigma_0 \text{H}(t)$ and an electrical impact loading $D(t) = D_0 \text{H}(t)$ on the

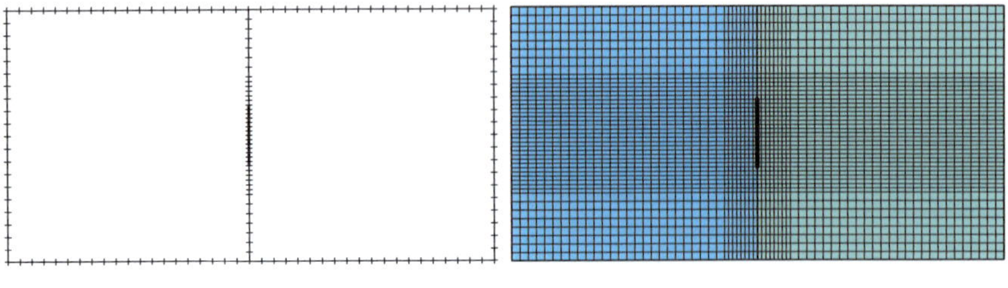

Fig. 3 Comparison of the BEM (*left*) and the FEM (*right*) meshes

left and the right boundary, where H(t) denotes the Heaviside step function. The mechanical stresses and the electrical displacements are zero on the upper and lower boundary. The crack-faces are considered as electrically impermeable described by Eq. (12). The computed results for the interface crack and the interior crack inside a homogenous plate of PTZ-5H are presented and compared in Figs. 4 and 5 with those obtained by the commercial FEM program ANSYS. For the sake of clarity only the FEM results for $\chi = -0.5$ are included.

In order to point out the influence of the orientation of the material poling direction, the normalized dynamic intensity factors for the rotation angle $\theta^I = \theta^{II} = 45°$ with respect to the axis x_2 are presented in Figs. 6 and 7.

The normalized dynamic mode-I and mode-IV intensity factors for the interior crack in a homogenous piezoelectric solid as well as the real part and the imaginary part of the complex intensity factor and the electrical displacement intensity factor for the interface crack obtained by the present TDBEM are in good agreement with those of the FEM. The peak values of the normalized dynamic intensity factors decrease with increasing electrical loading. It can be observed that, when applying an electrical impact, the normalized dynamic mode-I intensity factor and the complex intensity factor start from a non-zero value at $t = 0$. This is due to the quasi-electrostatic assumption on the electrical field, which implies that the cracked piezoelectric plate is immediately subjected to an electrical impact and therefore the crack opens at $t = 0$. For the loading configuration $\chi = -0.5$ a negative mode-I intensity factor is obtained in different time ranges. To avoid the physically meaningless crack-face intersection a non-linear crack-face contact algorithm is required. In contrast to the electrical impact, the elastic waves induced by the mechanical impact need some time to reach and excite the crack, as clearly observed for the $\chi = 0$ case. The global behaviour of the mode-I intensity factor and the real part of the complex intensity factor is similar without a significant dependence on the electrical impact loading. The mode-II intensity factors vanish, since no shear stress components are induced for all applied loadings in the case of a transversely isotropic material behaviour. Since the crack opening modes I and II are coupled each other for the interface crack the imaginary part of the complex

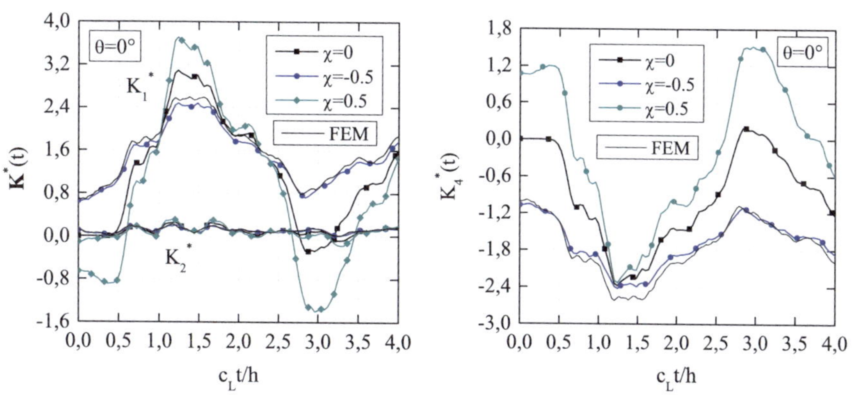

Fig. 4 Normalized dynamic intensity factors for the interface crack

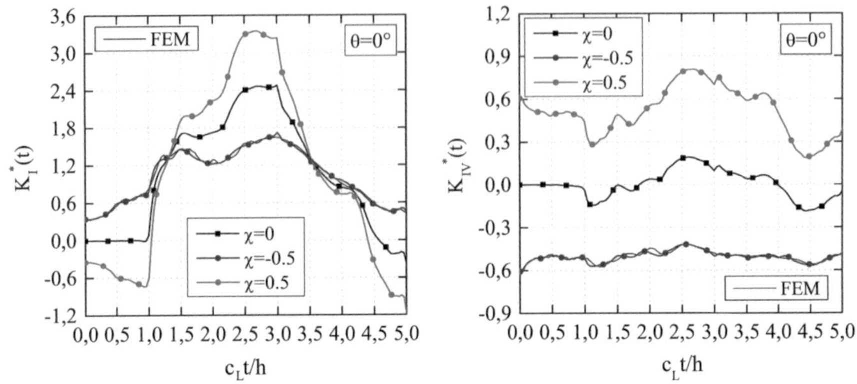

Fig. 5 Normalized dynamic intensity factors for the interior crack

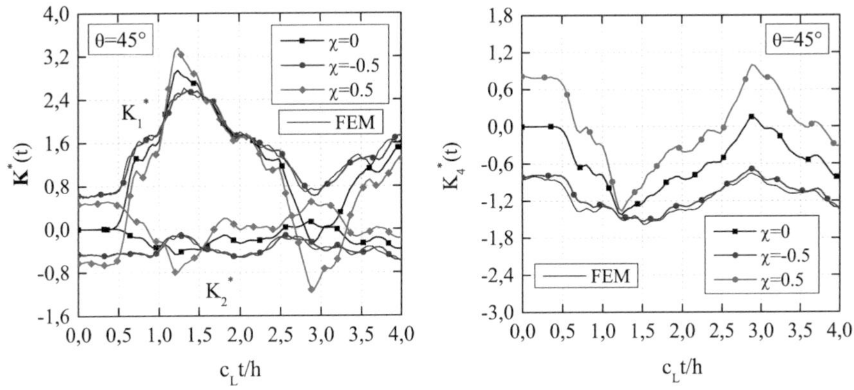

Fig. 6 Normalized dynamic intensity factors for the interface crack with $\theta^I = 45°$ and $\theta^{II} = 45°$

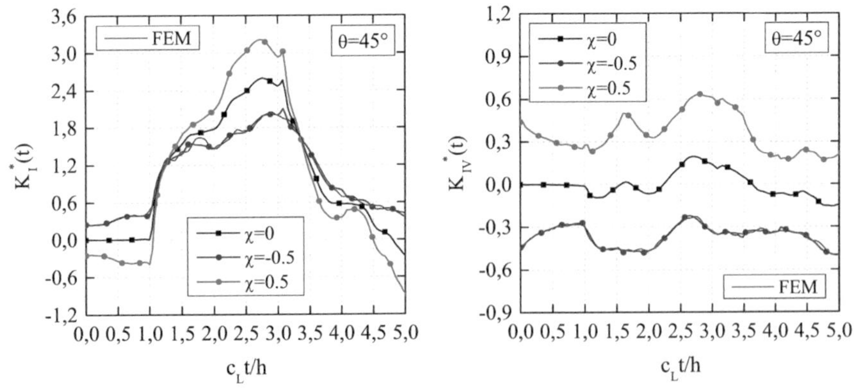

Fig. 7 Normalized dynamic intensity factors for the interior crack with $\theta = 45°$

intensity factor is unequal zero. As is expected, for both crack configurations the electrical displacement intensity factors show a strong dependence on the applied electrical loading.

In the Figs. 8 and 9, the absolute values of the scattered displacements for $\chi = 0.5$ obtained for different time-steps are shown. It can be clearly identified that the interface as well as the poling direction have significant influences on the scattered displacements.

6.2 A Rectangular Plate with Two Interface Cracks

In the next example, a rectangular plate with two interface cracks of length $2a$ is investigated.

As depicted in Fig. 10 the cracked plate is subjected to an impact tensile loading $\sigma(t) = \sigma_0 \mathrm{H}(t)$ normal to the crack-faces and an impact electrical loading $D(t) = D_0 \mathrm{H}(t)$ with $\chi = -0.5$ on the upper and lower boundary. On the left and right boundary the mechanical stresses and the electrical displacements are zero. The geometrical data are $h = 12.0$ mm, $w = 20.0$ mm, $d = 12.0$ mm and $2a = 4.0$ mm. As in the first example the material properties given in Eqs. (38) and (39) are considered. For spatial discretization the external boundary is discretized by a uniform mesh with an element-length of 1.0 mm, while the crack-faces are divided into 14 elements. A normalized time-step $c_L \Delta t / h = 0.04$ is used.

The normalized dynamic intensity factors of Tip A for the impermeable, permeable and semi-permeable crack-face boundary conditions (12)–(14) are shown in Fig. 11 for the left interface crack in a bimaterial plate and in Fig. 12 for the left interior crack in a homogenous plate. The relative permittivity $\kappa_r = 40$ is used in the computations with the semi-permeable crack-face conditions.

In comparison to the results of a single central crack, the interaction between both cracks leads to a more complicated behaviour of the normalized dynamic intensity factors. Since the geometry of the cracked plate and the external loading are symmetric with respect to the vertical midline, the dynamic intensity factors of both cracks are identical. It can be clearly seen, that the electrical permittivity of the medium inside the crack has a significant influence on the dynamic intensity factors for both investigated configurations.

Finally, it should be remarked that the present time-domain BEM has much less unknowns than the FEM, which is advantageous for the pre- and post-processing of the numerical simulation. However, the numerical implementation of the present time-domain BEM is much more complicated than the FEM, which is available as commercial software codes and widely applied in engineering sciences.

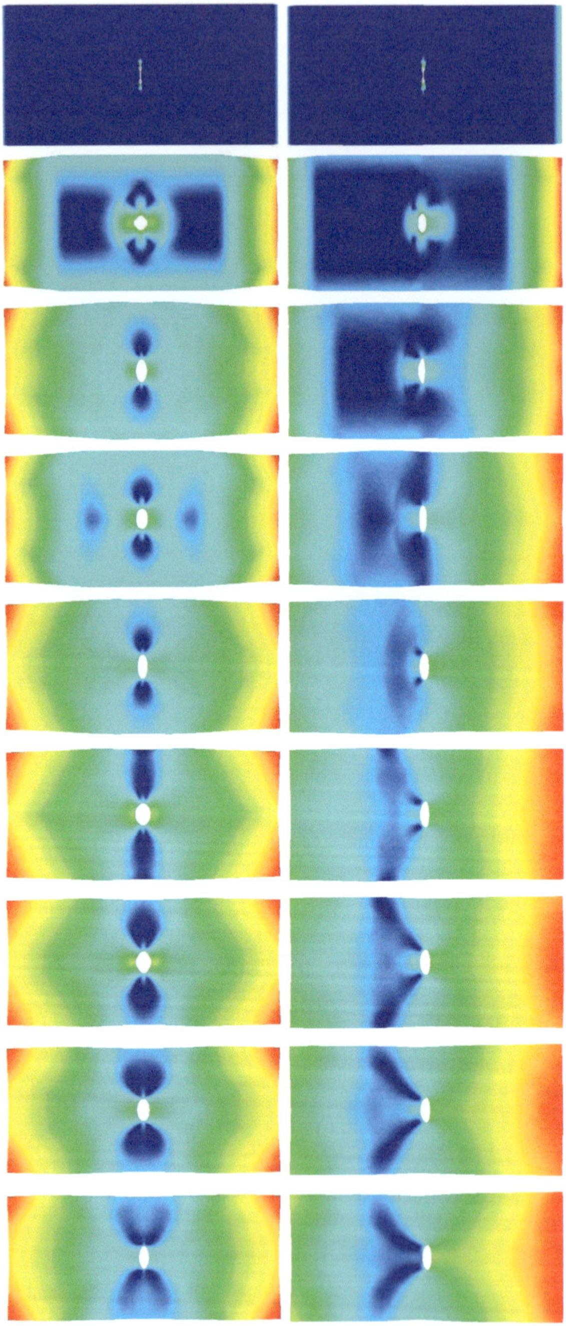

Fig. 8 Scattered wave fields at the time-steps 1, 10, 20, 30, 40, 50, 60, 70 and 80 for the interior crack with $\theta = 0°$ (*left*) and for the interface crack with $\theta^I = \theta^{II} = 0°$ (*right*)

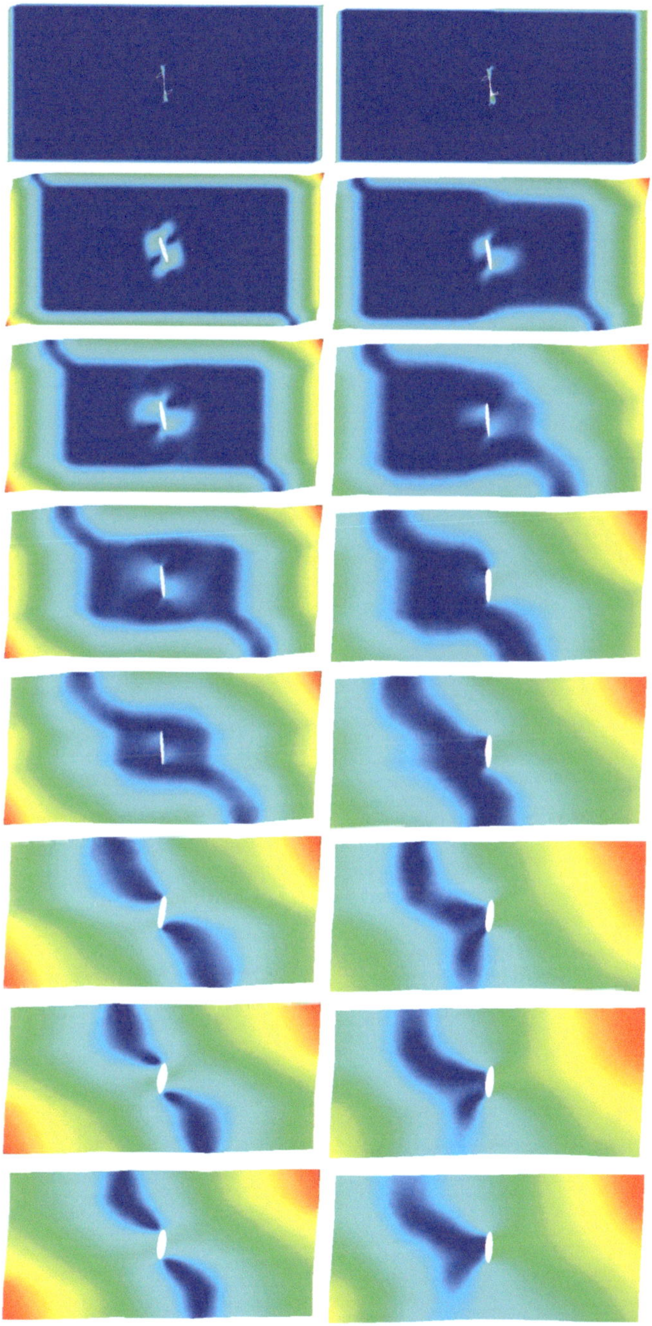

Fig. 9 Scattered wave fields at the time-steps 1, 10, 20, 30, 40, 50, 60, 70 and 80 for the interior crack with $\theta = 45^\circ$ (*left*) and for the interface crack with $\theta^I = \theta^{II} = 45^\circ$ (*right*)

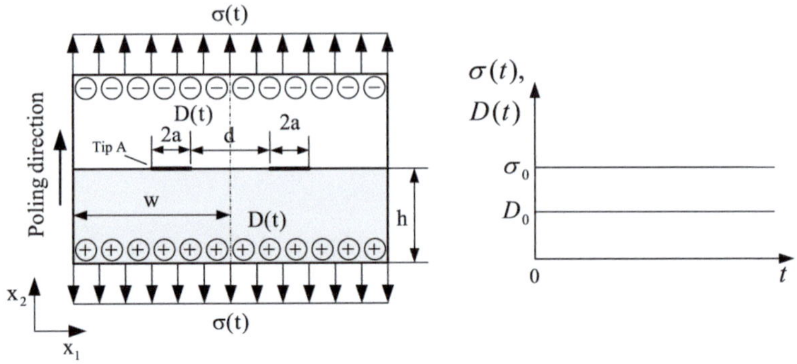

Fig. 10 A rectangular plate with two interface cracks

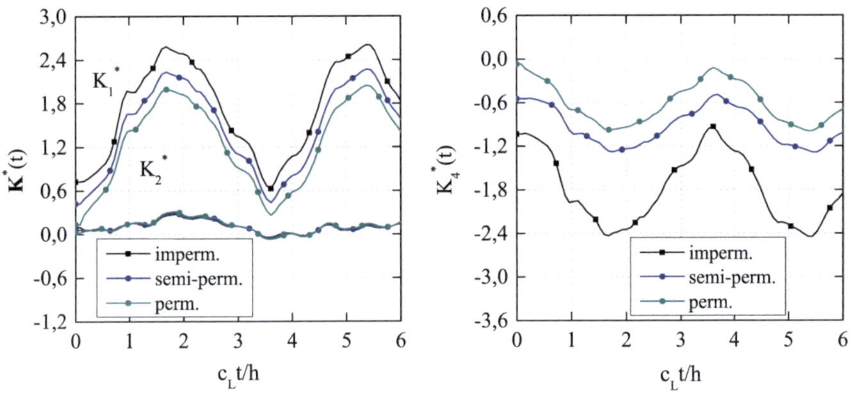

Fig. 11 Normalized dynamic intensity factors for Tip A of the left interface crack in a bimaterial plate

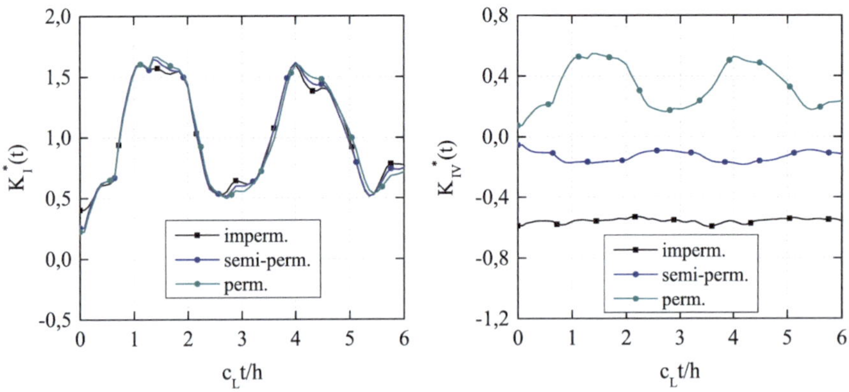

Fig. 12 Normalized dynamic intensity factors for Tip A of the left interior crack in a homogeneous plate

7 Conclusions

In this paper, transient dynamic crack analysis in layered and linear piezoelectric solids is presented. A time-domain Galerkin BEM is developed and applied for this purpose. Both temporal and spatial integrations are carried out analytically. Only the line integrals over the unit circle in the dynamic fundamental solutions are computed numerically. The developed BEM is generally applicable without limitations on the geometry and loading configurations. The obtained numerical results indicate a significant influence of the interface on the dynamic intensity factors. Further, the crack-face boundary conditions play an important role in the dynamic crack simulation of piezoelectric solids.

Acknowledgments This work was supported by the German Research Foundation (DFG) under the project number ZH 15/6-3, which is gratefully acknowledged.

References

1. Beom HG (2003) Permeable cracks between two dissimilar piezoelectric materials. Int J Solids Struct 40:6669–6679
2. Cho SB, Lee KB, Choy YS, Yuuki R (1992) Determination of stress intensity factors and boundary element analysis for interface cracks in dissimilar anisotropic materials. Eng Fract Mech 43:603–614
3. Enderlein M, Ricoeur A, Kuna M (2005) Finite element techniques for dynamic crack analysis in piezoelectrics. Int J Fract 134:191–208
4. Garca-Sánchez F, Zhang Ch, Sladek J, Sladek V (2007) 2-D transient dynamic crack analysis in piezoelectric solids by BEM. Comput Mater Sci 39:179–186
5. Gray LJ (1998) Evaluation of singular and hypersingular Galerkin boundary integrals: direct limits and symbolic computation. In: Sladek V, Sladek J (eds) Advances in boundary elements. Computational Mechanics Publishers, Berlin, pp 33–84 (Chapter 2)
6. Kögl M, Gaul L (2000) A boundary element method for transient piezoelectric analysis. Eng Anal Boundary Elem 24:591–598
7. Kuna M (2010) Fracture mechanics of piezoelectric materials—where are we right now? Eng Fract Mech 77:309–326
8. Ou ZC, Chen YH (2004) Near-tip stress fields and intensity factors for an interface crack in metal/piezoelectric bimaterials. Int J Eng Sci 42:1407–1438
9. Suo Z, Kuo CM, Barnett DM, Willis JR (1992) Fracture mechanics for piezoelectric ceramics. J Mech Phys Solids 40:739–765
10. Sladek J, Sladek V, Zhang Ch, Solek P, Pan E (2007) Evaluation of fracture parameters in continuously nonhomogeneous piezoelectric solids. Int J Fract 145:313–326
11. Wang C-Y, Zhang Ch (2005) 3-D and 2-D dynamic Green's functions and time-domain BIEs for piezoelectric solids. Eng Anal Boundary Elem 29:454–465
12. Wünsche M, Garca-Sánchez F, Sáez A, Zhang Ch (2010) A 2D time-domain collocation-Galerkin BEM for dynamic crack analysis in piezoelectric solids. Eng Anal Boundary Elem 34:377–387

Part V
Analytical Mechanics

Crack-Tip Fields of a Crack Impinging upon the Yielding/Debonding Slippage in Anisotropic Body

Qun Li, Junling Hou, Guangyan Liu and Hong Zuo

Abstract This paper presents a fundamental solution for a crack impinging normally upon the slippage in anisotropic materials under tensile loading. The slippage could occur in the form of yielding (e.g., a well-bonded ductile layer with plastic yielding) or debonding (e.g., a weak, sliding-free one). A superposition method is employed to explicitly solve the problem which combines the solution of a crack in an elastic homogeneous medium, the solution of a continuous distribution of dislocations which represent slippage, and an appendix solution which offsets the stress on the crack faces induced by the dislocations. This procedure reduces the problem to a singular integral equation which can be numerically solved by using Chebyshev polynomials. Numerical implementations are performed to analyze the influence of slippage on cracking and stress redistribution near the crack tip in anisotropic bodies. It is found that yielding or debonding slippage redistributes the stress ahead of the crack tip. The presence of yielding or debonding lowers the high stress concentration in the tensile stresses ahead of the crack tip. It is also concluded that debonding appears to be more effective in lowering the stress concentration than yielding.

1 Introduction

The fracture behavior with a crack impinging normally upon the slippage in materials under tensile loading has received more attentions by many scholars [1–12]. The slippage can originate from the interface between two laminated same

Q. Li (✉) · J. Hou · H. Zuo
State Key Laboratory for Strength and Vibration of Mechanical Structures,
School of Aerospace, Xi'an Jiaotong University, Xi'an 710049, China
e-mail: qunli@mail.xjtu.edu.cn

G. Liu
Department of Mechanics, School of Aerospace Engineering,
Beijing Institute of Technology, Beijing 100081, China

G. Hütter and L. Zybell, *Recent Trends in Fracture and Damage Mechanics*, DOI 10.1007/978-3-319-21467-2_13

bodies e.g., shale gas reservoirs [1], or from the interface between fiber and matrix in composites where the thickness of fiber is negligible compared that of matrix [2–6], or from the interface of two dissimilar bodies [7–12].

Many investigations about the crack propagation in anisotropic body for the engineering designs have been performed. The fracture problem in anisotropic body can be explicitly analyzed according to the Lekhnitskii's theory [13]. In most cases, however, it is believed that the slippage behaviors between the constituents or between the plies in composites have significant contributions to the materials' toughness and resistance. It has been reported that anisotropic material will exhibit slippage emanating from the crack tips, such as intra-ply splitting along the fiber direction or inter-ply delamination [12, 14]. To enable good strength and toughness, the stress redistribution caused by the presence of a crack impinging upon a slippage should be considered for accurate prediction. Depending on the nature of material, the slippage could occur in the form of yielding (e.g., a well-bonded ductile layer with plastic yielding) or debonding (e.g., a weak, sliding-free one). The presence of yielding/debonding at a crack tip can alleviate the stress concentration effect of the crack, reducing the probability of material failure. For instance, Cook and Erdogan [15] considered the elastostatic problem of a crack running into and terminating at the interface of two bonded half planes, especially, they emphasized the important case of the crack perpendicular to the interface. Erdogan and Biricikoglu [16] formulated a system of singular integral equations to analyze the problem of two bonded elastic half planes containing a finite crack perpendicular to and going through the interface. He and Hutchinson [17] considered the role of the slippage between dissimilar elastic materials when approached by a crack. Seyoung [18] considered the asymptotic stress field of composite with transverse cracks by the method of the eigenfunction expansion and Leknitskii's complex-variable potentials for generalized plane strain deformations.

In this paper, we consider a general issue of a crack impinging upon the slippage in one anisotropic material which is not reported in the open literatures. To do this we pose and solve an asymptotic model problem and the role of slippage in terms of yielding and debonding on the crack propagation is analyzed. The analysis concerns the stress redistribution ahead of the crack tip induced by the yielding and debonding slippage.

2 Problem Statements

The problem concerns a crack impinging upon a slippage and the corresponding asymptotic problems to be studied are schematically shown in Fig. 1a. A slippage having zero thickness lies at $x = 0$ where yielding or debonding happens extending a distance d above and below the crack tip, i.e. $-d \leq y \leq d$, $x = 0$. The shear flow stress τ, equivalently, the yielding strength is exerted along the yielding slippage while the shear stress $\tau = 0$ is applied along the debonding under condition such that

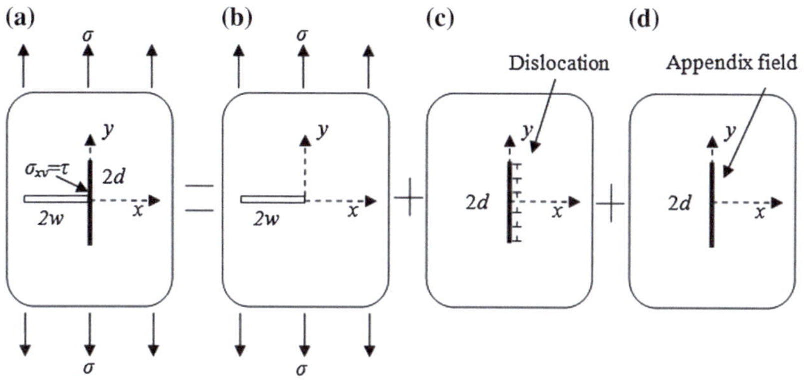

Fig. 1 An asymptotic model by superposition method of a crack impinging upon a slippage in anisotropic body. **a** The asymptotic model; **b** a crack in an elastic medium; **c** a continuous distribution of dislocations which represent slippage; **d** the appendix field for relieving the tractions on the crack surface induced by dislocations

the materials remain in contact. A crack of width $2w$ is impinging upon the slippage. A remote tension stress $\sigma_{yy} = \sigma$ is acting normal to the crack.

To obtain the fundamental solutions of the problem and to analyze cracking and stress redistribution ahead of the crack tip accurately, the superposition method is proposed to handle this problem. The aimed asymptotic problem (Fig. 1a) can be superposed by three subproblems. The first one is an analytical solution of the elastic field associated with a crack in a single infinite anisotropic homogeneous medium subjected to remote tension (Fig. 1b). The second one is a continuous distribution of dislocations which represents slippage (Fig. 1c). The third one is an appendix field which eliminates the tractions on the crack faces induced by the dislocations (Fig. 1d). The present superposition of three fields can really achieve

Fig. 2 Stress redistribution ahead of the crack tip for yielding slippage

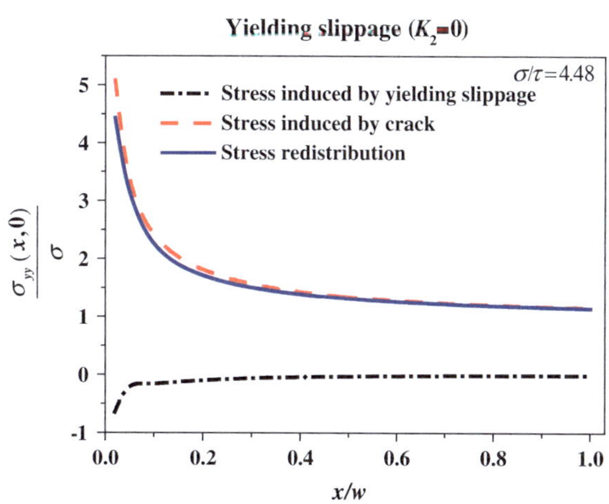

the traction-free condition on the crack faces. We assume that the yielding or debonding slippage can be viewed as an unknown continuous distribution of sliding dislocation singularities, so that an integral equation could be established to solve the dislocation distribution for meeting the shear stress equilibrium along the slippage. In the following, the fundamental solutions associated with the three proposed subproblems are formulated in explicit forms and then numerical implementation is performed to solve the integral equation and determine the dislocation distribution.

3 Fundamental Solutions

3.1 *Stress Fields of a Crack in an Anisotropic Homogeneous Body*

The crack problem of Fig. 1b in a two-dimensional infinite anisotropy can be solved according to the Lekhnitskii's theory [13]. For a homogeneous anisotropic material, the stress fields (σ_{xx}, σ_{yy}, σ_{xy}) which readily expressed in terms of the complex functions $\phi_1(z_1)$ and $\phi_2(z_2)$ as,

$$\begin{aligned} \sigma_{xx} &= 2\,\mathrm{Re}[\mu_1^2\phi_1'(z_1) + \mu_2^2\phi_2'(z_2)] \\ \sigma_{yy} &= 2\,\mathrm{Re}[\phi_1'(z_1) + \phi_2'(z_2)] \\ \sigma_{xy} &= -2\,\mathrm{Re}[\mu_1\phi_1'(z_1) + \mu_2\phi_2'(z_2)] \end{aligned} \tag{1}$$

where the superscript comma denotes a partial derivative with respect to the complex variable $z = x + \mu y$; Re denotes the real part of a complex variable; The parameters μ_1 and μ_2 are the complex roots of the characteristic equation in anisotropic material determined by

$$s_{11}\mu^4 - 2s_{16}\mu^3 + (2s_{12} + s_{66})\mu^2 - 2s_{26}\mu + s_{22} = 0 \tag{2}$$

where s_{11}, s_{22}, s_{12}, s_{16}, s_{26}, s_{66} are the elastic compliance constants of the material.

Additionally, the displacements (u_x, u_y) can be given by

$$\begin{aligned} u_x &= 2\,\mathrm{Re}[p_1\phi_1(z_1) + p_2\phi_2(z_2)] \\ u_y &= 2\,\mathrm{Re}[q_1\phi_1(z_1) + q_2\phi_2(z_2)] \end{aligned} \tag{3}$$

Here we have used the designations

$$\begin{aligned} p_1 &= s_{11}\mu_1^2 + s_{12} - s_{16}\mu_1; \quad p_2 = s_{11}\mu_2^2 + s_{12} - s_{16}\mu_2 \\ q_1 &= s_{12}\mu_1 + s_{22}/\mu_1 - s_{26}; \quad q_2 = s_{12}\mu_2 + s_{22}/\mu_2 - s_{26} \end{aligned} \tag{4}$$

For the problem of a crack in an infinite anisotropic body bearing a uniform tensile stress at infinity, the complex functions are well-known and solved earlier [19, 20]

$$\begin{aligned}\Psi_1(z_1) = \phi_1'(z_1) &= \frac{\sigma\mu_2}{2(\mu_2-\mu_1)\sqrt{z_1(z_1+2w)}}\left[z_1 + w - \sqrt{z_1(z_1+2w)}\right] \\ \Psi_2(z_2) = \phi_2'(z_2) &= \frac{\sigma\mu_1}{2(\mu_1-\mu_2)\sqrt{z_2(z_2+2w)}}\left[z_2 + w - \sqrt{z_2(z_2+2w)}\right]\end{aligned} \tag{5}$$

Substituting Eq. (5) into Eq. (1), the stress fields near the crack tip are written in the form of polar coordinates as

$$\begin{cases} \sigma_{xx}^0 = \frac{\sigma\sqrt{w}}{\sqrt{2r}}\operatorname{Re}\left\{\frac{\mu_1\mu_2}{\mu_1-\mu_2}\left[\frac{\mu_2}{\sqrt{\cos\theta+\mu_2\sin\theta}} - \frac{\mu_1}{\sqrt{\cos\theta+\mu_1\sin\theta}}\right]\right\} + O(r^0) \\ \sigma_{yy}^0 = \frac{\sigma\sqrt{w}}{\sqrt{2r}}\operatorname{Re}\left\{\frac{1}{\mu_1-\mu_2}\left[\frac{\mu_1}{\sqrt{\cos\theta+\mu_2\sin\theta}} - \frac{\mu_2}{\sqrt{\cos\theta+\mu_1\sin\theta}}\right]\right\} + O(r^0) \\ \sigma_{xy}^0 = \frac{\sigma\sqrt{w}}{\sqrt{2r}}\operatorname{Re}\left\{\frac{\mu_1\mu_2}{\mu_1-\mu_2}\left[\frac{1}{\sqrt{\cos\theta+\mu_1\sin\theta}} - \frac{1}{\sqrt{\cos\theta+\mu_2\sin\theta}}\right]\right\} + O(r^0) \end{cases} \tag{6}$$

3.2 *Fundamental Solution of a Sliding Dislocation in an Anisotropic Body*

In order to consider the problem of Fig. 1c in a two-dimensional infinite anisotropic medium, the slippage may be viewed as a continuous distribution of dislocation singularities. Thus it is necessary to study firstly the fundamental solution induced by the dislocation singularity in an anisotropic body.

Assume the Burger's vector of a dislocation at the point (x_0, y_0) in an anisotropic body given by

$$\mathrm{b} = b_x + ib_y \tag{7}$$

In the context of Lekhnitskii's complex-potential formulation, the dislocation singularity appears as a branch point in the complex plane. The corresponding complex solution $\phi_1^D(z_1)$, $\phi_2^D(z_2)$ for a single dislocation at the point (x_0, y_0) in an infinite medium can be written as

$$\begin{aligned}\phi_1^D(z_1) &= A\ln(z_1 - s_1) \\ \phi_2^D(z_2) &= B\ln(z_2 - s_2)\end{aligned} \tag{8}$$

where $z_1 = x + \mu_1 y$; $z_2 = x + \mu_2 y$; $s_1 = x_0 + \mu_1 y_0$; $s_2 = x_0 + \mu_2 y_0$; The coefficients A and B will be determined by relying on the monodrome condition of displacements and stress equilibrium of dislocation. They are explicitly formulated as follows.

According to the definition of Burger's vector, it is corresponding to the jump of displacements along the clockwise contour Γ^+ enclosing the dislocation singularity. That is,

$$\begin{aligned} b_x &= [u_x]_{\Gamma^+} = \left[p_1\phi_1(z_1) + p_2\phi_2(z_2) + \overline{p_1\phi_1(z_1)} + \overline{p_2\phi_2(z_2)}\right]_{\Gamma^+} \\ b_y &= [u_y]_{\Gamma^+} = \left[q_1\phi_1(z_1) + q_2\phi_2(z_2) + \overline{q_1\phi_1(z_1)} + \overline{q_2\phi_2(z_2)}\right]_{\Gamma^+} \end{aligned} \tag{9}$$

Substituting Eq. (8) into Eq. (9) reduces to

$$\begin{aligned} b_x &= 2\pi i[p_1(A - \bar{A}) + p_2(B - \bar{B})] \\ b_y &= 2\pi i[q_1(A + \bar{A}) + q_2(B + \bar{B})] \end{aligned} \tag{10}$$

In addition, considering that the medium is in an equilibrium state, the value of the principle stress vector $\mathbf{P} = -Y + iX$ acting on the outside contour should be zero where the complex potential expression of **P** is given by

$$\mathrm{P} = (1 + i\mu_1)\phi_1(z_1) + (1 + i\mu_2)\phi_2(z_2) + (1 + i\bar{\mu}_1)\overline{\phi_1(z_1)} + (1 + i\bar{\mu}_2)\overline{\phi_2(z_2)} \tag{11}$$

The equilibrium equation of principle stress vector can be written as

$$\oint \mathrm{P}d\Gamma = 0 \tag{12}$$

Substituting Eqs. (8) and (11) into Eq. (12), one obtains

$$(1 + i\mu_1)2\pi iA + (1 + i\mu_2)2\pi iB - (1 + i\bar{\mu}_1)2\pi iA - (1 + i\bar{\mu}_2)2\pi iB = 0 \tag{13}$$

Thus, the coefficients A and B in Eq. (8) could be determined by the linear equations composed by Eqs. (10) and (13). And then the complex potentials $\phi_1^D(z_1)$ and $\phi_2^D(z_2)$ induced by a dislocation are explicitly derived.

For the problem we are interested in here, the Burger's vector along the slippage can be specialized as the case of a sliding dislocation (0, b_y). Herein, we will treat the slippage as a group of sliding dislocations, and also assume that the contact is maintained with regions of slippage. Therefore, the explicit expressions of coefficients A and B corresponding to a sliding dislocation can be obtained as

$$A = \frac{-\mu_2 b_y}{4\pi i(q_2\mu_1 - q_1\mu_2)}, \quad B = \frac{\mu_1 b_y}{4\pi i(q_2\mu_1 - q_1\mu_2)} \tag{14}$$

3.3 Appendix Fields for Removing the Stress on the Crack Surface Induced by the Dislocations

The final fundamental solution of a crack impinging upon the slippage in anisotropic body must leave the crack free of traction. It is known that the stress fields for a crack in homogeneous anisotropy given in the Sect. 3.1 satisfy this traction-free condition naturally. Therefore, the distributed dislocations must also leave the crack faces free of traction. Thus, it is essential to impose an appendix field (Fig. 1d) for eliminating the stress on the crack faces which is induced by the presence of dislocations.

In order to keep the crack free of traction, the solutions for dislocations are no longer given by $\Psi_1^D(z_1) = \phi_1^{D'}(z_1), \Psi_2^D(z_2) = \phi_2^{D'}(z_2)$ alone. An appendix field has to be proposed to satisfy the traction-free boundary conditions for a single dislocation at the point (x_0, y_0). The corresponding complex solution can be written in the form [21, 22]

$$\begin{aligned}\Psi_1(z_1) &= \Psi_1^D(z_1) + \Psi_1^R(z_1)\\ \Psi_2(z_2) &= \Psi_2^D(z_2) + \Psi_2^R(z_2)\end{aligned} \tag{15}$$

where Ψ_1^R, Ψ_2^R represent the appendix solution associated with eliminating the tractions induced by Ψ_1^D, Ψ_2^D on the crack faces, and can be determined by enforcing the stresses on the crack due to Ψ_1, Ψ_2 to be zero. Equivalently, the problem may be reduced to finding Ψ_1^R, Ψ_2^R with prescribed stresses on the crack equal to the negative of those induced by Ψ_1^D, Ψ_2^D.

Considering that the traction-free boundary conditions on the crack faces requires that

$$\sigma_{yy}^+ = \sigma_{yy}^- = 0 \quad -2w \le x \le 0 \tag{16}$$

$$\sigma_{xy}^+ = \sigma_{xy}^- = 0 \quad -2w \le x < 0 \tag{17}$$

where the superscripts + and − refer to the upper and lower crack surface, respectively.

Combining Eqs. (1), (8) and (15) with Eq. (16), the normal traction-free boundary condition on the crack surface leads to

$$\Psi_1^{R+}(x) + \Psi_2^{R+}(x) + \bar{\Psi}_1^{R-}(x) + \bar{\Psi}_2^{R-}(x) = -2\mathrm{Re}\left(\frac{A}{x - s_1} + \frac{B}{x - s_2}\right) \tag{18}$$

$$\Psi_1^{R-}(x) + \Psi_2^{R-}(x) + \bar{\Psi}_1^{R+}(x) + \bar{\Psi}_2^{R+}(x) = -2\,\mathrm{Re}\left(\frac{A}{x - s_1} + \frac{B}{x - s_2}\right) \tag{19}$$

And the shear traction-free boundary condition on the crack surface leads to

$$\mu_1 \Psi_1^{R+}(x) + \mu_2 \Psi_2^{R+}(x) + \mu_1 \bar{\Psi}_1^{R-}(x) + \mu_2 \bar{\Psi}_2^{R-}(x) = -2\,\mathrm{Re}\left(\frac{A\mu_1}{x - s_1} + \frac{B\mu_2}{x - s_2}\right) \tag{20}$$

$$\mu_1 \Psi_1^{R-}(x) + \mu_2 \Psi_2^{R-}(x) + \mu_1 \bar{\Psi}_1^{R+}(x) + \mu_2 \bar{\Psi}_2^{R+}(x) = -2\,\mathrm{Re}\left(\frac{A\mu_1}{x - s_1} + \frac{B\mu_2}{x - s_2}\right) \tag{21}$$

Subtracting Eq. (18) to Eq. (19), one obtains

$$\left[\Psi_1^R(x) + \Psi_2^R(x) - \bar{\Psi}_1^R(x) - \bar{\Psi}_2^R(x)\right]^+ - \left[\Psi_1^R(x) + \Psi_2^R(x) - \bar{\Psi}_1^R(x) - \bar{\Psi}_2^R(x)\right]^- = 0 \tag{22}$$

Since Eq. (22) yields a homogenous Riemann-Hilbert problem, the solution of formulation can be given by specifying the remote boundary condition at infinity. One obtains

$$\Psi_1^R(z_1) + \Psi_2^R(z_2) = \bar{\Psi}_1^R(z_1) + \bar{\Psi}_2^R(z_2) \tag{23}$$

The same treatment can be used to deal with subtracting Eq. (20) to Eq. (21), and finally the solution can be written in the form

$$\mu_1 \Psi_1^R(z_1) + \mu_2 \Psi_2^R(z_2) = \bar{\mu}_1 \bar{\Psi}_1^R(z_1) + \bar{\mu}_2 \bar{\Psi}_2^R(z_2) \tag{24}$$

Combining Eq. (23) with Eq. (24), the relationships between $\Psi_1^R, \bar{\Psi}_1^R$ and $\Psi_2^R, \bar{\Psi}_2^R$ can be achieved by

$$\begin{cases} \Psi_1^R(z_1) = \dfrac{(\bar{\mu}_1 - \mu_2)\Psi_2^R(z_2) - (\bar{\mu}_1 - \bar{\mu}_2)\bar{\Psi}_2^R(z_2)}{\mu_1 - \bar{\mu}_1} \\ \bar{\Psi}_1^R(z_1) = \dfrac{(\mu_1 - \mu_2)\Psi_2^R(z_2) - (\mu_1 - \bar{\mu}_2)\bar{\Psi}_2^R(z_2)}{\mu_1 - \bar{\mu}_1} \end{cases} \tag{25}$$

Alternatively,

$$\begin{cases} \Psi_2^R(z_2) = \dfrac{(\bar{\mu}_2 - \mu_1)\Psi_1^R(z_1) - (\bar{\mu}_2 - \bar{\mu}_1)\bar{\Psi}_1^R(z_1)}{\mu_2 - \bar{\mu}_2} \\ \bar{\Psi}_2^R(z_2) = \dfrac{(\mu_2 - \mu_1)\Psi_1^R(z_1) - (\mu_2 - \bar{\mu}_1)\bar{\Psi}_1^R(z_1)}{\mu_2 - \bar{\mu}_2} \end{cases} \tag{26}$$

Using Eq. (25), Eqs. (18) and (19) can be rearranged as

$$\left[\frac{(\mu_2-\mu_1)\Psi_1^R(x)-(\bar{\mu}_2-\bar{\mu}_1)\bar{\Psi}_1^R(x)}{\mu_2-\bar{\mu}_2}\right]^+ + \left[\frac{(\mu_2-\mu_1)\Psi_1^R(x)-(\bar{\mu}_2-\bar{\mu}_1)\bar{\Psi}_1^R(x)}{\mu_2-\bar{\mu}_2}\right]^- = -2\,\mathrm{Re}\left(\frac{A}{x-s_1}+\frac{B}{x-s_2}\right) \tag{27}$$

$$\left[\frac{(\mu_1-\mu_2)\Psi_2^R(x)-(\bar{\mu}_1-\bar{\mu}_2)\bar{\Psi}_2^R(x)}{\mu_1-\bar{\mu}_1}\right]^+ + \left[\frac{(\mu_1-\mu_2)\Psi_2^R(x)-(\bar{\mu}_1-\bar{\mu}_2)\bar{\Psi}_2^R(x)}{\mu_1-\bar{\mu}_1}\right]^- = -2\,\mathrm{Re}\left(\frac{A}{x-s_1}+\frac{B}{x-s_2}\right) \tag{28}$$

In the same way, Eqs. (20) and (21) can be rearranged as

$$\left[\frac{\bar{\mu}_2(\mu_2-\mu_1)\Psi_1^R(x)-\mu_2(\bar{\mu}_2-\bar{\mu}_1)\bar{\Psi}_1^R(x)}{\mu_2-\bar{\mu}_2}\right]^+ + \left[\frac{\bar{\mu}_2(\mu_2-\mu_1)\Psi_1^R(x)-\mu_2(\bar{\mu}_2-\bar{\mu}_1)\bar{\Psi}_1^R(x)}{\mu_2-\bar{\mu}_2}\right]^- = -2\,\mathrm{Re}\left(\frac{A\mu_1}{x-s_1}+\frac{B\mu_2}{x-s_2}\right) \tag{29}$$

$$\left[\frac{\bar{\mu}_1(\mu_1-\mu_2)\Psi_2^R(x)-\mu_1(\bar{\mu}_1-\bar{\mu}_2)\bar{\Psi}_2^R(x)}{\mu_1-\bar{\mu}_1}\right]^+ + \left[\frac{\bar{\mu}_1(\mu_1-\mu_2)\Psi_2^R(x)-\mu_1(\bar{\mu}_1-\bar{\mu}_2)\bar{\Psi}_2^R(x)}{\mu_1-\bar{\mu}_1}\right]^- = -2\,\mathrm{Re}\left[\frac{A\mu_1}{x-s_1}+\frac{B\mu_2}{x-s_2}\right] \tag{30}$$

The solving procedure follows the complex potential theory in elasticity [23, 24], and two new complex potentials $G(z)$and $H(z)$ are introduced by principle of analytic continuation. They are presented through $\Psi_1^R(z_1), \bar{\Psi}_1^R(z_1)$ or $\Psi_2^R(z_2), \bar{\Psi}_2^R(z_2)$ as following

$$\begin{cases} G(z_1) = \frac{(\mu_2-\mu_1)\Psi_1^R(z_1)-(\bar{\mu}_2-\bar{\mu}_1)\bar{\Psi}_1^R(z_1)}{\mu_2-\bar{\mu}_2} \\ H(z_1) = \frac{\bar{\mu}_2(\mu_2-\mu_1)\Psi_1^R(z_1)-\mu_2(\bar{\mu}_2-\bar{\mu}_1)\bar{\Psi}_1^R(z_1)}{\mu_2-\bar{\mu}_2} \end{cases} \tag{31}$$

$$\begin{cases} G(z_2) = \frac{(\mu_1-\mu_2)\Psi_2^R(z_2)-(\bar{\mu}_1-\bar{\mu}_2)\bar{\Psi}_2^R(z_2)}{\mu_1-\bar{\mu}_1} \\ H(z_2) = \frac{\bar{\mu}_1(\mu_1-\mu_2)\Psi_2^R(z_2)-\mu_1(\bar{\mu}_1-\bar{\mu}_2)\bar{\Psi}_2^R(z_2)}{\mu_1-\bar{\mu}_1} \end{cases} \tag{32}$$

According to Eqs. (31) and (32), the Riemann-Hilbert boundary problems of Eqs. (27)–(30) can be rearranged in the following forms

$$G^{+}(z_1) + G^{-}(z_1) = -2\,\mathrm{Re}\left(\frac{A}{z_1 - s_1} + \frac{B}{z_1 - s_2}\right) \tag{33}$$

$$H^{+}(z_1) + H^{-}(z_1) = -2\,\mathrm{Re}\left(\frac{A\mu_1}{z_1 - s_1} + \frac{B\mu_2}{z_1 - s_2}\right) \tag{34}$$

$$G^{+}(z_2) + G^{-}(z_2) = -2\,\mathrm{Re}\left(\frac{A}{z_2 - s_1} + \frac{B}{z_2 - s_2}\right) \tag{35}$$

$$H^{+}(z_2) + H^{-}(z_2) = -2\,\mathrm{Re}\left(\frac{A\mu_1}{z_2 - s_1} + \frac{B\mu_2}{z_2 - s_2}\right) \tag{36}$$

The representations are the non-homogeneous Riemann-Hilbert problems with square root singularity. The analytical solutions of complex potentials can be obtained by referring to the solving procedure in references [23, 24]. Thus, the explicit expressions of $G(z_1)$, $G(z_2)$, $H(z_1)$, and $H(z_2)$ are given by

$$\begin{aligned}
G(z_1) &= -\left[AF(z_1, s_1) + \overline{AF(z_1, s_1)} + BF(z_1, s_2) + \overline{BF(z_1, s_2)}\right] \\
G(z_2) &= -\left[AF(z_2, s_1) + \overline{AF(z_2, s_1)} + BF(z_2, s_2) + \overline{BF(z_2, s_2)}\right] \\
H(z_1) &= -\left[A\mu_1 F(z_1, s_1) + \overline{A\mu_1 F(z_1, s_1)}\right] - \left[B\mu_2 F(z_1, s_2) + \overline{B\mu_2 F(z_1, s_2)}\right] \\
H(z_2) &= -\left[A\mu_1 F(z_2, s_1) + \overline{A\mu_1 F(z_2, s_1)}\right] - \left[B\mu_2 F(z_2, s_2) + \overline{B\mu_2 F(z_2, s_2)}\right]
\end{aligned} \tag{37}$$

where

$$\begin{aligned}
&F(z_1, s_1) = \frac{1}{2}\frac{1}{z_1 - s_1}\left[1 - \frac{X_{ref}(s_1)}{X_{ref}(z_1)}\right], \quad F(z_1, s_2) = \frac{1}{2}\frac{1}{z_1 - s_2}\left[1 - \frac{X_{ref}(s_2)}{X_{ref}(z_1)}\right] \\
&F(z_2, s_1) = \frac{1}{2}\frac{1}{z_2 - s_1}\left[1 - \frac{X_{ref}(s_1)}{X_{ref}(z_2)}\right], \quad F(z_2, s_2) = \frac{1}{2}\frac{1}{z_2 - s_2}\left[1 - \frac{X_{ref}(s_2)}{X_{ref}(z_2)}\right] \\
&X_{ref}(s_1) = \sqrt{s_1(s_1 + 2w)}, \quad X_{ref}(s_2) = \sqrt{s_2(s_2 + 2w)}, \\
&X_{ref}(z_1) = \sqrt{z_1(z_1 + 2w)}, \quad X_{ref}(z_2) = \sqrt{z_2(z_2 + 2w)}.
\end{aligned}$$

Finally, the complex functions $\Psi_1^R(z_1)$ and $\Psi_2^R(z_2)$ of appendix fields for eliminating the stress on the crack faces are formulated by

$$\Psi_1^R(z_1) = -AF(z_1, s_1) + \frac{[A(\bar{\mu}_1 - \mu_2)F(z_1, \bar{s}_1) + B(\bar{\mu}_2 - \mu_2)F(z_1, \bar{s}_2)]}{\mu_2 - \mu_1}$$
$$\Psi_2^R(z_2) = -BF(z_2, s_2) + \frac{A(\bar{\mu}_1 - \mu_1)F(z_2, \bar{s}_1) + B(\bar{\mu}_2 - \mu_1)F(z_2, \bar{s}_2)}{\mu_1 - \mu_2} \tag{38}$$

3.4 Singular Integral Equation to Determine the Dislocation Distribution and Its Numerical Implementation

In this study, the slippage ahead of the crack tip is assumed to be with shear flow stress τ. With reference to Fig. 1b, let $b_y(y_0) = -d\delta_y/dy_0$ denote the amplitude of the sliding dislocations extending from $-d$ to d along $x = 0$. A singular integral equation can be formed as follows.

An integral equation governing $b_y(y_0)$ is derived by enforcing the condition of $\sigma_{xy} = \pm\tau$ along $x = 0$ between $-d$ and d. This leads to the following singular integral equation

$$\sigma_{xy}^0(y) + \int_0^d H(y, y_0)b_y(y_0)dy_0 = \tau \quad 0 \le y \le d \tag{39}$$

$$-\sigma_{xy}^0(y) + \int_{-d}^0 H(y, y_0)b_y(y_0)dy_0 = -\tau \quad -d \le y \le 0 \tag{40}$$

where $\sigma_{xy}^0(y)$ is the shear stress along $x = 0$ due to the remote tensile loading in the absence of the slippage; $H(y, y_0)$ denotes the shear stress $\sigma_{xy}(y)$ at the point y due to the dislocation $b_y(y_0)$ at the point y_0 along $x = 0$. The expression of $H(y, y_0)$ can be derived by substituting Eqs. (8) and (38) into the formulas of shear stress in Eq. (1). One obtains

$$\begin{aligned} H(y, y_0) = -2\,\mathrm{Re}\Bigg[& \frac{A\mu_1}{z_1 - s_1} - A\mu_1 F(z_1, s_1) + \frac{A\mu_1(\bar{\mu}_1 - \mu_2)F(z_1, \bar{s}_1) + B\mu_1(\bar{\mu}_2 - \mu_2)F(z_1, \bar{s}_2)}{\mu_2 - \mu_1} \\ & + \frac{B\mu_2}{z_2 - s_2} - B\mu_2 F(z_2, s_2) + \frac{A\mu_2(\bar{\mu}_1 - \mu_1)F(z_2, \bar{s}_1) + B\mu_2(\bar{\mu}_2 - \mu_1)F(z_2, \bar{s}_2)}{\mu_1 - \mu_2}\Bigg] \end{aligned} \tag{41}$$

where $z_1 = \mu_1 y, z_2 = \mu_2 y, s_1 = \mu_1 y_0, s_2 = \mu_2 y_0$; The coefficients of A and B are given in Eq. (14) corresponding to the dislocation distribution b_y.

The singular integral equation can be numerically solved by using Chebyshev polynomials [25, 26]. After doing so, the dislocation distributions are available through the above numerical implementation and then full fields for the crack impinging upon the slippage can be finally reached.

4 Numerical Results and Discussions

As an example, the material is taken to be one orthotropic material with the main axes aligned with the coordinate axes (i.e., $s_{16} = s_{26} = s_{36} = 0$ in Eq. 2). The material constants are taken as $E_x = 172.4$ GPa, $E_y = 10.3$ GPa, $\nu_{xy} = 0.32$, $G_{xy} = 5.52$ GPa. The corresponding characteristic roots are numerically calculated as $\mu_1 = 0.7465i$, $\mu_2 = 5.4804i$ according to the Lekhnitskii's complex theory. The present solutions are used to analyze the cracking and stress redistribution in anisotropic body containing a crack impinging on the slippage and the geometry of the problem to be studied is shown in Fig. 1a. A central cracked anisotropic body is subjected to remote tension $\sigma_{yy} = \sigma$ acting normal to the crack. The crack length is $2w$ while the slippage has a length of $2d$ perpendicular to the crack surface.

Two distinct mechanisms of slippage including yielding and debonding are under considerations. In first case, the slippage is taken to be a very thin ductile layer of an elastic-perfectly plastic material with yielding strength τ. To ensure well-behaved shear stresses at the end of the yielding zone and establish the zone length d, the condition that the Mode II stress intensity factor equals zero i.e., $K_{\mathrm{II}} = 0$ should be enforced at the end of the yielded slippage. The relationship between σ/τ and d/w can be established by the criterion of $K_{\mathrm{II}} = 0$, that is, the shear stress at the slippage tip induced by the tensile loading in the anisotropic body in the absence of slippage is equal to the yielding strength τ. The yielding length can be determined by $\sigma_{xy}^{0}(d, \pi/2) = \tau$, that is

$$\frac{\sigma\sqrt{w}}{\sqrt{2d}}\,\mathrm{Re}\left\{\frac{\mu_1\mu_2}{\mu_1-\mu_2}\left[\frac{1}{\sqrt{\mu_1}}-\frac{1}{\sqrt{\mu_2}}\right]\right\} = \tau \tag{42}$$

Actually, the criterion of $K_{\mathrm{II}} = 0$ to determine the yielding length in the present analysis is same as the procedure in Dugdale's mode [27] in fracture mechanics.

Figure 2 shows the distribution of normal stress $\sigma_{yy}(x, 0)$ normalized by the applied far field stress σ ahead of the crack tip where the tensile loading $\sigma = 4.48\tau$ is applied at infinity. In order to allow relative slipping of the layers joined by plastic yielding, the preferred condition of $K_{\mathrm{II}} = 0$ yields a slippage length $d = 2.0w$. It can be seen from Fig. 2 that the normal stress at the crack tip induced by the crack in the absence of yielding slippage presents an evident singularity which is determined by the analytical solution Eq. (6) of a crack in anisotropic body. Furthermore, a compressive normal stress is obtained from the stress field induced by yielding slippage only. The negative values of the stresses attributed from yielding ensure that when yielding occurs, the singularity of crack tip could be slacking down, and the crack propagation could also be effectively prevented. A drastic reduction of the corresponding stress concentration factor can be found near the crack tip. For instance, the concentration factor is 5.07 at $x = 0.02w$ for the crack without yielding while the contribution of the yielding behavior reduces the concentration factor to 4.42. Although the superposed stress fields from solutions of Hilbert problem have square root singularity, the present numerical results show that the degree of singularity at the crack tips due to the yielding slippage is in the weak singular form.

Next, we turn our attention to the stress redistribution induced by debonding slippage. In this case, the slippage is assumed to be a weak plane that debonds and slips. In contrast, the condition $\tau = 0$ is enforced within the deboned zones of the slippage. Herein, no frictional resistance is exerted across the debonded slippage. The role of debonding slippage on stress redistribution ahead of the crack tip is seen in Fig. 3, where the distribution of normal stress $\sigma_{yy}(x, 0)$ is normalized by the applied far field stress σ and the length of debonded slippage is $d = 2.0w$. It can be seen that debonding clearly has a significant effect on lowering the power of the stress concentration near the crack tip. The values of normal stress induced by a debonding slippage are always negative, which means that the debonding could reduce the stress concentration ahead of the crack tip, thus hinder the crack propagation in anisotropic bodies, and reinforce the strength and toughness of materials. The feature of debonding slippage to redistribute the normal stress near the crack tip is similar as that for yielding slippage as shown in Fig. 2. It should be mentioned that the present results are consistent with those by the present author's FEM simulation (see Figs. 10–13 in the paper by Liu and Tang [2]).

Although slippage in terms of yielding and debonding results in the same conclusion that the stress redistribution is available and the concentration at the near crack tip will be extremely decreasing and in turn enhances the loading carrying capacity of the material, the magnitudes of the reduced stress concentrations are quite different for the yielding and debonding slippage. Figure 4 displays a

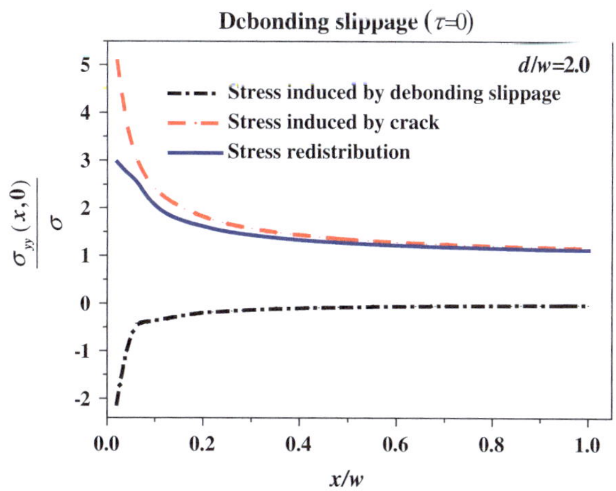

Fig. 3 Stress redistribution ahead of the crack tip for debonding slippage

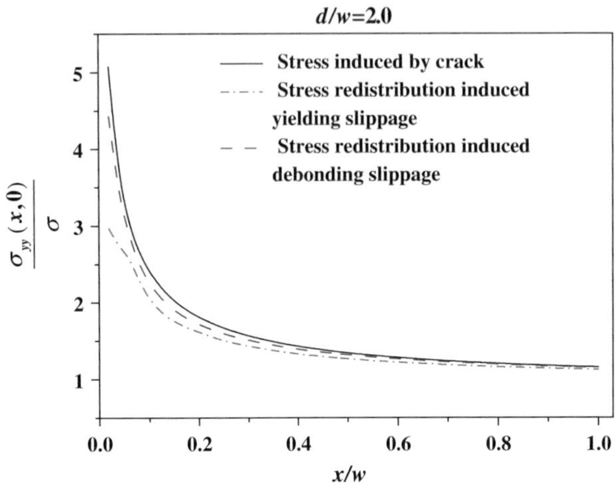

Fig. 4 Comparison of stress redistribution ahead of the crack tip between yielding slippage and debonding slippage

comparison of stress redistribution between yielding and debonding slippage at the same length of slippage, that is $d/w = 2.0$. Wherein, $\tau = 0$ is exerted for the debonding slippage while $\sigma/\tau = 4.48$ is enforced according to the condition $K_{\text{II}} = 0$ for the yielding slippage. It can be seen that the stress reduction due to debonding is more dramatic compared with that of yielding. For instance, the concentration stress factor is 4.42 at $x/w = 0.02$ for the yielding while the factor is reduced to 2.96 for the debonding although both values are smaller than the values of 5.07 for the crack in the absence of slippage. It can be concluded that debonding appears to be more effective in decreasing the stress concentration than yielding. Additionally, another intriguing conclusion can be made is that the diminish effect of yielding or debonding on the normal stress is obvious within a small range near the crack tip $x/w < 0.25$ and otherwise the effect is slight. The normal stress ahead of the crack tip is drastically reduced only when approaching the crack tip, as shown in Fig. 4.

5 Conclusions

An effective approach is provided for investigating the stress distribution in the presence of a crack impinging upon the slippage in anisotropic body. The slippage is assumed to occur in terms of yielding or debonding. Fundamental solutions are explicitly derived according to the superposition method. The present work obtains an analytical solution of the appendix fields with eliminating the tractions on the crack faces due to any dislocation by solving the Riemann-Hilbert boundary problem. Assuming a continuous distribution of dislocations which represent slippage, a tension crack impinging upon the slippage in anisotropic body is studied and the influence of slippage including yielding and debonding on defeating cracks is concluded. It can be found that yielding or debonding will blunt the crack and alleviate the stress concentration at the crack tip. Numerical implementation also shows that debonding has a more significant effect on lowering the stress concentration factor at the crack tip than yielding. Particularly, the present fundamental solutions can be useful for analyzing the influence of characterization parameters (e.g., friction or residual stress, slippage length, angle of main axes orientation in anisotropy and so on) concerning the slippage on the fracture toughening in materials. They will be discussed in the sequent work.

Acknowledge This work was supported by the National Natural Science Foundation of China with grant No. 11472205, No.11202156, No. 11321062, No.11242015 and No.11172228 and the Fundamental Research Funds for the Central Universities in China.

References

1. Lam KY, Cleary MP (1984) Slippage and re-initiation of (hydraulic) fractures at frictional interfaces. Int J Numer Anal Met 8(6):589–600
2. Liu GY, Tang KL (2015) Study on stress concentration in notched cross-ply laminates under tensile loading. J Compos Mater. doi:10.1177/0021998315573802
3. Folsom CA, Zok FW, Lange FF, Marshall DB (1992) Mechanical behavior of a laminar ceramic/fiber-reinforced epoxy composite. J Am Ceram Soc 75(11):2969–2975
4. Aveston J, Cooper GA, Kelly A (1971) Single and multiple fracture, the properties of fiber composites. In: Conference proceedings on properties of fiber composites. IPC Science and Technology Press Ltd, National Physical Laboratory, Guildford, UK, pp 15–26
5. Budiansky B, Hutchinson JW (1986) Matrix fracture in fiber-reinforced ceramics. J Mech Phys Solids 34(2):167–189
6. Chiang YC (2001) The influence of possion contraction on matrix cracking stress in fiber reinforced ceramics. J Mater Sci 36(13):3239–3246
7. Dalgleish BJ, Trumble KP, Evans AG (1989) The strength and fracture of alumina bonded with aluminum alloys. Acta Metall Mater 37(7):1923–1931
8. Chartier T, Merle D, Besson JL (1995) Laminar ceramic composites. J Eur Ceram Soc 15 (2):101–107
9. Gotsis PK, Chamis CC, Minnetyan L (2002) Application of progressive fracture analysis for predicting failure envelopes and stress–strain behaviors of composite laminates: a comparison with experimental results. Compos Sci Technol 62(12–13):1545–1559
10. Hutchinson JW, Suo Z (1991) Mixed mode cracking in layered materials. Adv Appl Mech 29:63–191
11. Beuth JL Jr (1992) Cracking of thin bonded films in residual tension. Int J Solids Struct 29 (13):1657–1675
12. Chan KS, He MY, Hutchinson JW (1993) Cracking and stress redistribution in ceramic layered composites. Mater Sci Eng 167(1–2):57–64
13. Lekhnitskii SG, Tsai SW, Cheron T (1968) Anisotropic Plates. Gordon and Breach Science Publishers, Philadelphia
14. Kortschot MT, Beaumont PWR (1990) Damage mechanics of composite materials: I—measurements of damage and strength. Compos Sci Technol 39(4):289–301
15. Cook TS, Erdogan F (1972) Stresses in bonded materials with a crack perpendicular to the interface. Int J Eng Sci 10(8):677–697

16. Erdogan F, Biricikoglu V (1973) Two bonded half planes with a crack going through the interface. Int J Eng Sci 11:745–766
17. He MY, Hutchinson JW (1989) Crack deflection at an interface between dissimilar elastic materials. Int J Solids Struct 25(9):1053–1067
18. Seyoung I (1990) Asymptotic stress field around a crack normal to the ply-interface of an anisotropic composite laminate. Int J Solids Struct 26(1):111–127
19. Hoenig A (1982) Near tip behavior of a crack in a plane anisotropic elastic body. Eng Fract Mech 16:393–403
20. Ting TCT (1996) Anisotropic elasticity, theory and applications. Oxford University Press, Oxford
21. Dollar A, Steif PS (1989) A tension crack impinging upon frictional interfaces. J Appl Mech 56(2):291–298
22. Lo KK (1978) Analysis of branched cracks. J Appl Mech 45(4):797
23. Muskhelishvili NL (1958) Some basic problems of mathematical theory of elasticity. Noordhoff International Publishing, Leyden
24. England AM (1971) Complex variable methods in elasticity. Wiley Interscience, New York
25. Erdogan F, Gupta GD (1972) On the numerical solution of singular integral equations. Q Appl Math 28(6):525–534
26. Erdelyi A, Bateman H (1953) Higher transcendental functions, vol 2. McGraw-Hill, New York
27. Dugdale DS (1960) Yielding of steel sheets containing slits. J Mech Phys Solids 8:100–104

On Conservation Laws and Reciprocity in Configurational Mechanics

R. Kienzler and S. Boettcher

Abstract Material conservation laws and associated path-independent integrals play a prominent role in the assessment of defects in structures. Especially Rice's *J*-integral is widely used in fracture mechanics. For systems governed by a Lagrangian, the usual tool for the derivation of material conservation laws is the application of Noether's first theorem in combination with Bessel-Hagen's extension. The so-called Neutral-Action (NA) method is a different approach. Its advantage in comparison with the classical Noether's approach lies in the fact that it is applicable to field equations that are not necessarily the Euler-Lagrange equations of a variational principle, i.e., for systems not governed by a Lagrangian. After a short review of the NA method, a complete set of characteristics and the associated conserved currents are derived and interpreted in physical terms. As an example, path-independent integrals are evaluated around a crack tip and a defect-interaction problem is treated in terms of reciprocity. Finally, the application of conservation laws in defect mechanics and its potential are discussed.

1 Introduction

The usual way to derive conservation laws in field theories is via Noether's first theorem [15], as long as the governing equations are the Euler-Lagrange equations of a variational problem. This method is based on Lie's theory of continuous groups and invariance under group action in jet spaces [14]. Roughly speaking, the action integral A under consideration, which is the integral of the Lagrangian L over the

R. Kienzler (✉) · S. Boettcher
Bremen Institute for Mechanical Engineering (bime), University of Bremen,
Am Biologischen Garten 2, 28359 Bremen, Germany
e-mail: rkienzler@uni-bremen.de

S. Boettcher
e-mail: boettcher@mechanik.uni-bremen.de

G. Hütter and L. Zybell, *Recent Trends in Fracture and Damage Mechanics*, DOI 10.1007/978-3-319-21467-2_14

domain B, is subjected to an infinitesimal transformation of **both** the independent variables (e.g., x_i) and the dependent variables (e.g., $u_i = u_i(x_j)$)

$$\begin{aligned} x_i \mapsto x_i^* &= x_i + \varepsilon\xi_i(x_j, u_k), \\ u_i \mapsto u_i^* &= u_i + \varepsilon\phi_i(x_j, u_k). \end{aligned} \tag{1}$$

The single constant parameter ε is supposed to be small in the sense that the ensuing terms without ε (which usually cancel each other) and linear in ε be retained, while terms with ε^2 and with higher powers will be omitted. Working through the analysis (cf., e.g., [10]), it turns out that the change of the action integral ΔA is given by

$$\Delta A = A^* - A = \varepsilon \int_B \left(P_{i,i} + Q_j \boldsymbol{E}_j(L)\right) dV. \tag{2}$$

An index followed by a comma denotes partial differentiation with respect to the indicated independent variable and summation is implied over repeated indices; here, the range of indices (3D-elasticity) is 1, 2, 3.

In [14], the quantities P_i and Q_j are called currents and characteristics, respectively, and are given for the setting we will discuss below by

$$\begin{aligned} P_i &= \phi_j \frac{\partial L}{\partial u_{j,i}} + \xi_j \left(L\delta_{ij} - u_{k,j} \frac{\partial L}{\partial u_{k,i}} \right), \\ Q_i &= \phi_i - \xi_j u_{i,j}, \end{aligned} \tag{3}$$

and $\boldsymbol{E}_j(L)$ are the Euler-Lagrange equations of the underlying variational problem.

Now, one seeks for transformations, which leave the action integral invariant, i.e., $\Delta A = 0$. Then, along solutions u_i of the Euler-Lagrange equations $\boldsymbol{E}_j(L) = 0$, the current P_i is conserved, and with the divergence theorem, we arrive at a conservation law

$$\int_B P_{i,i} dV = \int_{\partial B} P_i n_i dA = 0 \tag{4}$$

giving rise to path-independent integrals (unit outward-normal vector n_i).

If a Lagrangian function is not available, and the system is given only by some set of partial differential equations

$$\Delta_i(x_j, u_k, u_{k,\ell}) = 0, \tag{5}$$

the Neutral-Action (NA) method [7] might be used to advantage. Firstly, we need the notion of a "null Lagrangian". If a Lagrange function $\tilde{L}$ is expressible as a divergence of a vector-valued function $g_i(x_j, u_k, u_{k,\ell})$ then it follows [18]

$$\tilde{L} = g_{i,i} \Leftrightarrow \boldsymbol{E}_j(\tilde{L}) \equiv 0 \Leftrightarrow \delta\tilde{A} = 0, \tag{6}$$

i.e., the action integral $\tilde{A} = \int \tilde{L}dV$ is insensitive (or behaves neutral with respect) to a (classical) variation δ of only the dependent variables u_i, and we arrive at a so-called trivial variational principle, which is valid independent of whether u_k are solutions of the governing differential equations or not.

Now, instead of the characteristics Q_i being specified by the transformation functions ξ_j and ϕ_i (cf. (3b)), we determine Q_i, employing the symbol $-f_i$ instead (in order to avoid confusion), such that the product $f_i\Delta_i$ forms a null Lagrangian

$$f_i\Delta_i = P_{i,i}. \tag{7}$$

The functions f_i, therefore, have to be determined from

$$\boldsymbol{E}_j(f_i\Delta_i) = 0. \tag{8}$$

As soon as suitable characteristics f_i are found from (8), the conserved currents P_i follow from (7), and due to (5), conservation laws in the form (4) are established.

It may be mentioned that the NA method to construct conservation laws might be applied also to systems possessing a Lagrangian. In that case, it leads to the same result as long as an unrestricted version of Noether's theorem [16] is employed together with the Bessel-Hagen extension [2].

2 Conservation Laws of Linear Elasticity

We adopt the Navier-Lamé equations for a three-dimensional body made of a homogeneous isotropic material (Lamé constants λ and μ) in the absence of body forces

$$\Delta_i = \mu u_{i,jj} + (\lambda + \mu)u_{j,ji} = 0 \tag{9}$$

(cf., e.g., [23]). We restrict the characteristics f_i to depend on the independent variables x_i, the dependent variables, i.e., the displacements u_k and the displacement gradients $u_{\ell,m}$

$$f_i = f_i(x_j, u_k, u_{\ell,m}). \tag{10}$$

Application of the Neutral-Action method leads to equations to determine f_k as

$$E_k(f_i\Delta_i) \equiv 0 \Rightarrow \mu f_{k,jj} + (\lambda + \mu)f_{j,jk}. \tag{11}$$

It turns out that the f_k are governed by the same differential equations as the displacements u_i (9) are, which is not surprising since the Navier-Lamé operator (9) is self-adjoint. If we consider two boundary-value problems {1} and {2} for the same body B with the solutions

$$u_i = \overset{\{1\}}{u_i}, \quad f_i = \overset{\{2\}}{u_i}, \tag{12}$$

then from (7), Betti-Maxwell's reciprocity relations in physical space are recovered as

$$\int_{\partial B} \overset{\{1\}}{\sigma_{ji}}\, n_j\, \overset{\{2\}}{u_i}\, dA = \int_{\partial B} \overset{\{2\}}{\sigma_{ji}}\, n_j\, \overset{\{1\}}{u_i}\, dA \tag{13}$$

(cf. [10]) with the Cauchy stress tensors $\overset{\{1\}}{\sigma_{ji}}$ and $\overset{\{2\}}{\sigma_{ji}}$ of problem {1} and {2}, respectively. A corresponding reciprocity relation in material space will be discussed in Sect. 4.

In order to reach further conclusions, the characteristics (10) are inserted into (11) and the differentiations have to be carried out in detail. We arrive at equations involving second- and third-order derivatives of the displacement fields. Since the characteristics f_i depend on derivatives up to the first order only, the coefficients of higher derivatives have to vanish. The results indicate (cf. [12, 13]) that f_i are linear in the displacements and the displacement gradients

$$f_i = f^1_{imn}(x_j)u_{m,n} + f^2_{im}(x_j)u_m + f^3_i(x_j). \tag{14}$$

Preceding further along this line of reasoning, the functional dependence of f_i can further be restricted to

$$f^1_{ijk} = a(x_\ell)\varepsilon_{ijk} + b_k(x_\ell)\delta_{ij} + c_m(x_\ell)\left[(\lambda + 2\mu)\delta_{jk}\delta_{im} + \mu\delta_{ik}\delta_{jm}\right] \tag{15}$$

The scalar- and vector-valued quantities a and b_k, c_m, respectively, are functions of the independent variable x_ℓ and will be further restricted by comparing equal coefficients of terms involving different orders of derivatives of u_i (δ_{ij} and ε_{ijk} are the Kronecker symbol and the permutation tensor, respectively).

The conservation laws resulting from $a(x_\ell)$ will be dealt with elsewhere and will not be considered further in what follows. We also discharge the quantities $f^3_i(x_j)$. They lead to physical conservation laws, which have been thoroughly discussed in [12]. Considering, for the moment, only the b_k-term, Eq. (14) reads as follows

$$f_i = b_k(x_\ell)u_{i,k} + f^2_{ik}u_k. \tag{16}$$

Comparing (16) with (3b) ($f_i = -Q_i$) we can identify

$$\begin{aligned} \xi_k(x_\ell, u_m) &= b_k(x_\ell), \\ \phi_i(x_\ell, u_m) &= -f^2_{ik}(x_\ell)u_k, \end{aligned} \tag{17}$$

i.e., b_k describe material transformations (see (1)). From the one-dimensional theory of elasticity, i.e., tension and compression of bars, we know (cf. [10]) that these functions involve constant linear and quadratic terms in x_ℓ. Guided by this knowledge, we investigate the functional dependence of $b_k = b_k(x_\ell)$ further, and it can be shown (cf. [12]) that the characteristic (14) has the following appearance

$$\begin{aligned} f^1_{ijk} &= \delta_{ij} b_k \\ &= \delta_{ij}\left(\beta^{(0)}_k + \varepsilon_{nmk} x_m \beta^{(1)}_n + x_k\beta + (2x_k x_m - \delta_{km} x_n x_n)\beta^{(2)}_m\right) \\ f^2_{ij} &= \varepsilon_{ijk}\beta^{(1)}_k + \frac{n-2}{2}\delta_{ij}\beta + (n-2)\delta_{ij} x_m \beta^{(2)}_m . \end{aligned} \tag{18}$$

The integer n designates the dimensionality of the problem, whether we treat a three-dimensional ($n = 3$), a two-dimensional (plane strain, $n = 2$) or a one-dimensional (tension and compression of a bar, $n = 1$) body.

The terms $\beta^{(0)}_k, \beta^{(1)}_n, \beta^{(2)}_m$ and β are vector-valued and scalar-valued constants, respectively.

Before we precede to the corresponding conservation laws, let us interpret the material translations in geometrical terms. Obviously, $\beta^{(0)}_k$ describe material translations (see Fig. 1a)

$$\begin{aligned} x_k &\mapsto x^*_k = x_k + \varepsilon\beta^{(0)}_k, \\ u_k &\mapsto u^*_k = u_k. \end{aligned} \tag{19}$$

For reasons of clarity we sketch the transformations in the (x_1, x_2)-plane only ($\beta^{(0)}_3 = 0$).

Obviously again, $\beta^{(1)}_n$ describe material rotations

$$\begin{aligned} x_k &\mapsto x^*_k = x_k + \varepsilon\, \varepsilon_{nmk} x_m \beta^{(1)}_n, \\ u_k &\mapsto u^*_k = u_k + \varepsilon\, \varepsilon_{nmk} u_m \beta^{(1)}_n \end{aligned} \tag{20}$$

whilst the displacement field is co-rotated. In the (x_1, x_2)-plane, $\beta^{(1)}_n$ has only one possible component $\beta^{(1)}_3 = \omega$, see Fig. 1b.

The constant β describes scaling (see Fig. 1c)

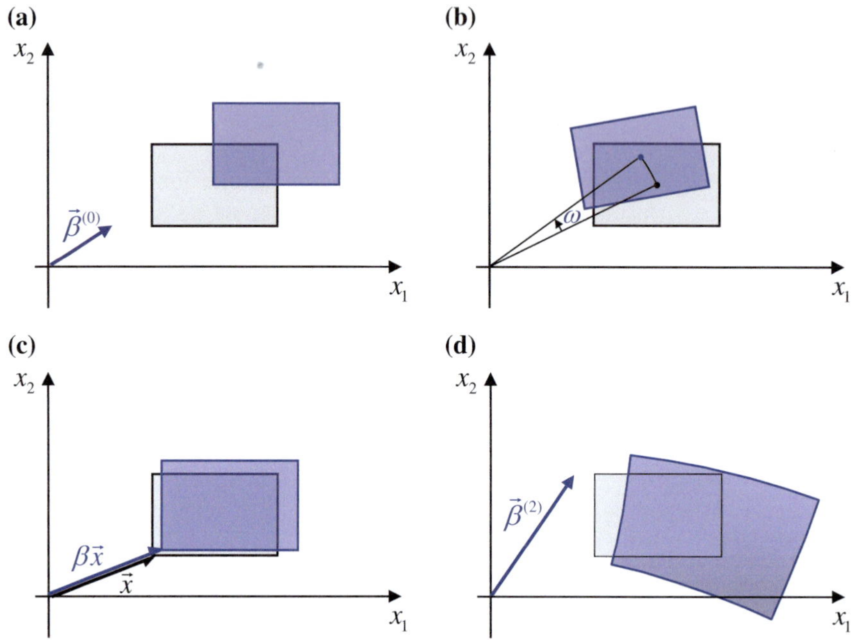

Fig. 1 Material transformations: **a** translation, **b** rotation, **c** scaling, **d** inversion

$$\begin{aligned} x_k \mapsto x_k^* &= x_k + \varepsilon\beta x_k, \\ u_k \mapsto u_k^* &= u_k + \varepsilon\frac{n-2}{2}\beta u_k. \end{aligned} \tag{21}$$

Due to this transformation the body under consideration is expanded (or shrunk) self-similarly and translated. The corresponding displacement transformation depends on the dimension of the problem. For plane strain ($n = 2$), u_k is not changed.

The transformation described by $\beta_m^{(2)}$ is rather strange. If we introduce the length of the vector $\vec{x} = x_i\vec{e}_i$ by $|\vec{x}| = (x_n x_n)^{1/2}$ and introduce unit vectors by $n_k = x_k/|\vec{x}|$, the transformation reads as

$$\begin{aligned} x_k \mapsto x_k^* &= x_k + \varepsilon|\vec{x}|^2(2n_k n_m - \delta_{km})\beta_m^{(2)}, \\ u_k \mapsto u_k^* &= u_k + \varepsilon(n-2)u_k x_m \beta_m^{(2)}. \end{aligned} \tag{22}$$

The matrix $W_{km} = 2n_k n_m - \delta_{km}$ is proper orthogonal

$$\boldsymbol{W}^{-1} = \boldsymbol{W}^T, \quad \det \boldsymbol{W} = +1, \tag{23}$$

and rotates the vector $\vec{\beta}^{(2)} = \beta_m^{(2)} \vec{e}_m$ around the position vector $\vec{x}$ by an angle of π (cf. [19, 20, 22]). In addition, the vector is scaled by $|\vec{x}|^2$. This transformation is called "inversion" (cf. [17]). Figure 1d shows a qualitative sketch for $\beta_1^{(2)} = 1$ and $\beta_2^{(2)} = 3$ in the (x_1, x_2) plane.

The four transformations $\beta_j^{(0)}$, $\beta_n^{(1)}$, β and $\beta_m^{(2)}$ lead to the four conservation and balance laws [12], respectively

$$
\begin{array}{ll}
\text{Translation} & \beta_j^{(0)} \neq 0 : b_{ij,i} = \left[W\delta_{ij} - \sigma_{ik}u_{k,j}\right]_{,i} = 0, \\
\text{Rotation} & \beta_n^{(1)} \neq 0 : \varepsilon_{nkj}\left[x_k b_{ij} + u_k \sigma_{ij}\right]_{,i} = 0, \\
\text{Scaling} & \beta \neq 0 : \left[x_j b_{ij} + \frac{2-n}{2} u_j \sigma_{ij}\right]_{,i} = 0, \\
 & \beta_m^{(2)} \neq 0 : [(2x_m x_k - x_\ell x_\ell \delta_{mk}) b_{ik} \\
\text{Inversion} & \quad +(2x_k u_m + (2-n)x_m u_k - 2x_\ell u_\ell \delta_{mk})\sigma_{ik} \\
 & \quad +n\mu\left(u_m u_i + \frac{1}{2}u_\ell u_\ell \delta_{mi}\right)\Big]_{,i} = (n\lambda + (4+n)\mu)u_{\ell,\ell}u_m.
\end{array}
\tag{24}
$$

The term b_{ij} is the well-known Eshelby tensor involving the strain-energy density W. On integration over the volume V of a body B and application of the divergence theorem, Eq. (24a) gives rise to Rice's J-integral [21], which describes the energy-release rate due to the translation of a material inhomogeneity within the body. In a similar way, (24b, c) resemble the L- and M-integrals introduced in [3], but discussed much earlier in [5]. The integrals L and M indicate the energy-release rates due to a rotation and self-similar expansion of the inhomogeneity, respectively.

Inversion does not give rise to a conservation law but rather a (more or less) trivial balance law. The right hand side of (24d) vanishes either under the unphysical condition (n = 3): $3\lambda + 7\mu = 0$, i.e., Poisson's ratio $\nu = 7/8$ or for the special case of an isochoric deformation, i.e., $u_{k,k} = 0$. In the one-dimensional case we have

$$
\begin{aligned}
& n = 1 \\
& x_k \to x_1 = x, \\
& u_i \to u_1 = u, \\
& u_{i,j} \to u_{1,1} = u', \\
& \mu \to \frac{E}{2}, \quad \lambda \to 0
\end{aligned}
\tag{25}
$$

(Young's modulus E), and the right-hand side of (24d) becomes

$$
(n\lambda + (4+n)\mu)u_{\ell,\ell}u_m \to \frac{5}{2}Eu'u = \frac{5}{4}E(u^2)' \tag{26}
$$

which is a "one-dimensional divergence" and as such can be combined with the divergence on the left-hand side, leading to a conservation law which coincides with that given in [10].

We turn back now to Eq. (15) and realize that, first of all, the transformations involving $c_m(x_\ell)$ follow from Noether's theorem only, if we admit an unrestricted or extended form of the transformation (1) as [18]

$$\begin{aligned} x_i \mapsto x_i^* &= x_i + \varepsilon a_{ij}\xi_j(x_k, u_\ell), \\ u_i^* \mapsto u_i^* &= u_i + \varepsilon b_{ij}\phi_j(x_k, u_\ell) \end{aligned} \tag{27}$$

(a_{ij}, b_{ij} = const.). Secondly, the transformation coefficients c_m are scaled with the material constants λ and μ of the elastic body under consideration. Finally, the transformations c_k leading to conservation or balance laws have a similar form as the transformations b_k have, cf. (18), and so have the governing conservation laws (cf. [12])

$$\begin{aligned} &\text{Translation} \quad \gamma_i^{(0)} \neq 0 : c_{ji,j} = 0, \\ &\text{Rotation} \quad \gamma_i^{(1)} \neq 0 : \varepsilon_{ik\ell}\big[(\lambda+\mu)x_k c_{j\ell} + \mu(\lambda+3\mu)u_k\sigma_{j\ell} \\ &\qquad\qquad + 2\mu^3 u_k\big(u_{m,m}\delta_{j\ell} - u_{m,\ell}\delta_{jm}\big)\big]_{,j} = 0, \\ &\text{Scaling} \quad \gamma \neq 0 : \big[x_i c_{ji} + \mu u_i\big(\sigma_{ij} + \mu(u_{k,k}\delta_{ij} - u_{j,i})\big) \\ &\qquad\qquad + \mu^2\big(u_j u_{i,i} - u_i u_{j,i}\big)\big]_{,j} = \tfrac{1}{2}(n\lambda + (n+4)\mu)(\lambda+2\mu)u_{i,i}u_{j,j} \end{aligned} \tag{28}$$

$$\begin{aligned} \text{Inversion} \quad \gamma_\ell^{(2)} = 0 : &\Bigg[\left(x_k x_\ell - \frac{1}{2}x_n x_n \delta_{k\ell}\right)c_{mk} \\ &+ \mu(x_k u_\ell - x_\ell u_k - x_n u_n \delta_{k\ell})\sigma_{mk} \\ &+ 2\mu^2(x_m u_{k,k}u_\ell + x_\ell u_{k,m}u_k - x_n u_{k,k}u_n\delta_{m\ell}) + 2\mu(\lambda+\mu)x_\ell u_{k,k}u_m \\ &+ \frac{2\mu^2(\lambda+2\mu)}{\lambda+\mu}u_m u_\ell + \frac{2\mu^3}{\lambda+\mu}x_k\big(u_{k,m}u_\ell - u_{\ell,m}u_k\big)\Bigg]_{,m} \\ &= (n\lambda + (n+4)\mu)(\lambda+2\mu)u_{k,k}\left[\frac{1}{2}x_\ell u_{m,m} + \frac{\mu}{\lambda+\mu}u_\ell\right]. \end{aligned}$$

with f^1_{ijk} and f^2_{ij} given in this case as

$$\begin{aligned} f^1_{ijk} &= c_m\big[(\lambda+2\mu)\delta_{jk}\delta_{im} + \mu\delta_{ik}\delta_{jm}\big] \\ c_m &= \gamma^0_m + \varepsilon_{knm}x_n\gamma^1_k + x_m\gamma + \left(x_m x_\ell - \frac{1}{2}x_n x_n \delta_{m\ell}\right)\gamma^2_\ell \\ f^2_{ij} &= \frac{\mu(\lambda+3\mu)}{(\lambda+\mu)}\varepsilon_{ijk}\gamma^1_k + \mu\frac{n-2}{2}\delta_{ij}\gamma + \mu\frac{n-2}{2}\delta_{ij}x_m\gamma^2_m \end{aligned} \tag{29}$$

(Note that some minor flaws have been corrected and some terms have been specified in comparison to [13]).

The tensor c_{ij} is given in displacement gradients as

$$\begin{aligned} c_{ij} = &\frac{1}{2}(\lambda + 2\mu)(\lambda + \mu)u_{k,k}u_{\ell,\ell}\delta_{ij} + \mu^2 u_{j,k}(u_{k,i} - u_{i,k}) \\ &+ \mu(\lambda + 2\mu)u_{k,k}u_{j,i} \end{aligned} \tag{30}$$

and coincides with Q_{ij} in Olver's paper [17]. It will be applied to a crack in the next paragraph. Rotation leads again to a conservation law (28b), whilst scaling (28c) and inversion (28d) yield rather balance laws, the right-hand side being proportional to the same factor $3\lambda + 7\mu$ discussed above.

3 Application in Fracture Mechanics

For later use, we modify the tensor c_{ij} (30) by a linear combination with b_{ij} (24a) and a trivial conservation law t_{ij}, i.e., a conservation law which is satisfied independently of whether or not the displacement field u_i satisfies the Navier-Lamé equations (9)

$$\begin{aligned} t_{ij} &= \varepsilon_{i\ell n}\varepsilon_{jkm}u_{k,\ell}u_{m,n}, \\ t_{ij,i} &\equiv 0. \end{aligned} \tag{31}$$

The resulting tensor is called d_{ij} and is defined as

$$d_{ij} = \frac{\lambda + \mu}{2(\lambda + 2\mu)}\left(b_{ij} + \frac{1}{\mu}c_{ij} + \mu t_{ij}\right). \tag{32}$$

This tensor has been derived in a different way in [8]. Replacing the Lamé constants λ and μ by Young's modulus E and Poisson's ratio v via

$$\lambda = \frac{Ev}{(1 + v)(1 - 2v)}, \quad \mu = \frac{E}{2(1 + v)} \tag{33}$$

and writing b_{ij} and d_{ij} in terms of displacement gradients

$$
\begin{aligned}
b_{ij} &= \frac{E}{4(1+\nu)(1-2\nu)}\left\{\delta_{ij}\left[2\nu u_{\ell,\ell}u_{m,m} + (1-2\nu)u_{\ell,m}(u_{\ell,m}+u_{m,\ell})\right]\right.\\
&\quad \left.-2\left[2\nu u_{\ell,\ell}u_{i,j} + (1-2\nu)u_{\ell,j}(u_{\ell,i}+u_{i,\ell})\right]\right\},\\
d_{ij} &= \frac{E}{4(1+\nu)(1-2\nu)}\left\{\frac{1}{2}\delta_{ij}\left[\frac{2(1-\nu)}{1-2\nu}u_{\ell,\ell}u_{m,m} + \frac{1-2\nu}{2(1-\nu)}u_{\ell,m}(u_{\ell,m}-u_{m,\ell})\right]\right.\\
&\quad \left.-\left[u_{\ell,\ell}(u_{i,j}-u_{j,i}) + \frac{1-2\nu}{2(1-\nu)}(u_{i,\ell}-u_{\ell,i})(u_{j,\ell}-u_{\ell,j})\right]\right\},
\end{aligned}
\tag{34}
$$

it turns out that b_{ij} and d_{ij} have quite a similar appearance.

The Eshelby tensor b_{ij} serves as integrand of Rice's J-integral as

$$
J_i = \int_{\partial B} b_{ji} n_j dA\,. \tag{35}
$$

Accordingly, we introduce an N-integral in which d_{ij} serves as integrand

$$
N_i = \int_{\partial B} d_{ji} n_j dA\,. \tag{36}
$$

In plane fracture mechanics, the J-integral is used to calculate stress-intensity factors K_I and K_{II} (cf., e.g., [4]). On evaluating both integrals along a path Γ within the near-crack-tip field around a crack tip under mixed-mode-loading conditions in plane elasticity (see Fig. 2) it turns out that the following relations hold

$$
\begin{aligned}
J_1 &= \frac{K_I^2+K_{II}^2}{E^*}, \quad & N_1 &= \frac{K_I^2-K_{II}^2}{E^*},\\
J_2 &= -\frac{2K_IK_{II}}{E^*}, \quad & N_2 &= -\frac{2K_IK_{II}}{E^*},
\end{aligned}
\tag{37}
$$

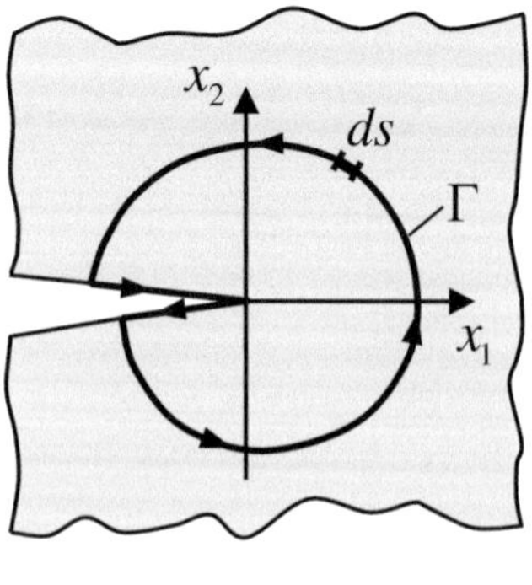

Fig. 2 Integration path Γ in the vicinity of a crack tip under mixed-mode-loading conditions in plane elasticity

with

$$E^* = \begin{cases} E & \text{for plane stress,} \\ \frac{E}{1-\nu^2} & \text{for plane strain.} \end{cases} \tag{38}$$

As discussed in [13], linear combinations of J_1 and N_1 provide favorable tools to calculate K_I and K_{II} separately

$$K_I = \sqrt{\frac{E^*}{2}(J_1 + N_1)},$$
$$K_{II} = \sqrt{\frac{E^*}{2}(J_1 - N_1)}.$$

Also, advantages in the numerical implementation and the obtained accuracy are reported.

4 Material Reciprocity Relations

Reciprocity relations play an important role in various branches of applied mechanics. Let us return to the physical Betti-Maxwell-reciprocity relation, i.e., reciprocity in Newtonian mechanics, of the second paragraph (see (13)) and let us consider a specific situation depicted in Fig. 3a.

Let $\vec{u}_{12}$ be the change of the displacement of point ① due to the application of a point force $\vec{F}_2$ at point ② and $\vec{u}_{21}$ the change of the displacement of point ② due to the application of a force $\vec{F}_1$ at point ①. Betti-Maxwell's reciprocity relation states (cf. e.g., [1])

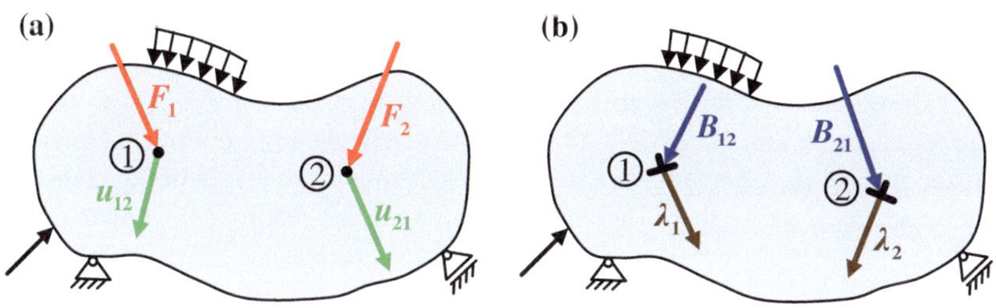

Fig. 3 **a** Two forces $\vec{F}_1$ and $\vec{F}_2$ acting on an elastic body, **b** two defects displaced within the elastic body by $\vec{\lambda}_1$ and $\vec{\lambda}_2$

$$\vec{F}_1 \cdot \vec{u}_{12} = \vec{F}_2 \cdot \vec{u}_{21}. \tag{39}$$

Eshelbian mechanics deals with motions (material translations, or more general, configurational changes) of defects or inhomogeneities within the material. Physical forces have their counterpart as material forces (configurational forces), which are defined as negative gradients of the total energy with respect to the position of a defect. In a recent paper [6], material reciprocity relations, i.e., reciprocity relations in Eshelbian mechanics have been established. It turns out that in material space, the material displacement of a defect plays the role of physical force application. In physical space, work is done by the applied physical force (cause) in a physical displacement (effect), whereas in material space work is done by the material force (effect) in the applied material displacement (cause).

Let $\vec{B}_{12}$ be the change of the material force acting at defect ① due to the material translation $\vec{\lambda}_2$ of defect ② and $\vec{B}_{21}$ the change of the material force acting at defect ② due to the material translation $\vec{\lambda}_1$ of defect ① (cf. Fig. 3b), the material reciprocity theorem states

$$\vec{\lambda}_1 \cdot \vec{B}_{12} = \vec{\lambda}_2 \cdot \vec{B}_{21}. \tag{40}$$

Equations (39) and (40) imply linear dependencies of relevant quantities. They are based on the result that the energy stored in an elastic body after the application of two physical forces (in Newtonian mechanics) or two material displacements (Eshelbian mechanics) is independent of their sequel of application, and equals the external work exerted on the body.

As an example of application let us consider a defect configuration in plane elastostatics consisting of an edge dislocation at the origin of a Cartesian coordinate system (x_1, x_2) with component b of the Burgers vector in x_2-direction, and a circular hole of radius r_0 positioned at (ξ_1, ξ_2), i.e., at distance $d (d > r_0, \delta = d/r_0 > 1)$ from the origin under the angle φ measured form the x_1-axis, as depicted in Fig. 4a.

The change of the energy due to a material translation λ_1 in x_1-direction, λ_2 in x_2-direction, a material rotation ω with respect to the x_3-axis and a self-similar expansion of the hole ($r_0 \rightarrow \alpha r_0$) are calculated by means of J_1-, J_2-, L- and M-integrals, respectively, using (24a–c) as integrands. The configurational changes are sketched in Fig. 4b. Choosing for L as point of reference the origin of the coordinate system and for M the center of the hole, the integrals can readily be evaluated as (cf. [10, 11])

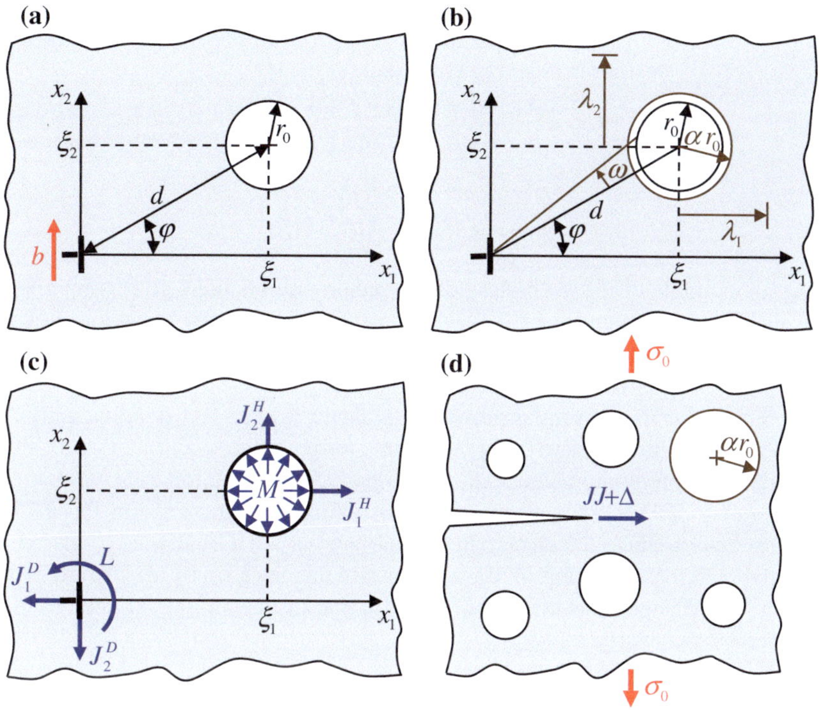

Fig. 4 **a** Circular hole and edge dislocation, **b** configurational changes, **c** free-body diagram, **d** crack in damaged configuration

$$
\begin{aligned}
J_1 &= -\frac{E^* b^2 \cos\varphi}{4\pi r_0 \delta^3}\left(\frac{1}{\delta^2 - 1} + 1 + 2\sin^2\varphi\right),\\
J_2 &= -\frac{E^* b^2 \sin\varphi}{4\pi r_0 \delta^3}\left(\frac{1}{\delta^2 - 1} + 2\sin^2\varphi\right),\\
L &= -\frac{E^* b^2 \sin\varphi\cos\varphi}{4\pi\delta^2},\\
M &= +\frac{E^* b^2}{4\pi\delta^2}\left(\frac{1}{\delta^2 - 1} + 1 + \sin^2\varphi\right)
\end{aligned}
\tag{41}
$$

with E^* given in (38). As in physical space, free body diagrams can be sketched in material space as shown in Fig. 4c. The change of the integrals due to changes of the material transformations $\Delta_{\lambda 1}, \Delta_{\lambda 2}, \Delta_{\omega}, \Delta_{\alpha}$ can also be evaluated in closed form [9] and the following material reciprocity relations can be established

$$
\begin{aligned}
\lambda_2\Delta_{\lambda 1}(J_2) &= \lambda_1\Delta_{\lambda 2}(J_1),\\
-\omega\Delta_{\lambda 1}(L) &= \lambda_1(\Delta_\omega(J_1)+\omega J_2),\\
-\omega\Delta_{\lambda 2}(L) &= \lambda_2(\Delta_\omega(J_2)+\omega J_1),\\
\lambda_1\Delta_\alpha(J_1) &= (\alpha-1)\Delta\lambda_1(M),\\
\lambda_2\Delta_\alpha(J_2) &= (\alpha-1)\Delta_{\lambda 2}(M),\\
-\omega\Delta_\alpha(L) &= (\alpha-1)\Delta_\omega(M).
\end{aligned}
\tag{42}
$$

As an example of the usefulness of the reciprocity relations given above let us consider a crack surrounded by damaged material characterized by various holes of different radii (n-holes) as depicted in Fig. 4d.

Assume that we would be interested in the change of the J_1-integral at the crack tip (defect 1) due to self-similar growth α_i of each void i (defect i) individually. We would thus have to calculate $\Delta_{\alpha i}(J_1)$. For this purpose, we would have to evaluate the original configuration first and, additionally, construct for each void a new finite-element (FE) mesh with an extended radius $r_i \mapsto \alpha_i r_i$, perform the FE calculations and calculate the change in J_1 each time (n + 1 calculations). Instead, by using the reciprocity relation (42d), we translate the crack tip by an amount λ_1 and calculate the change in M of void i, $\Delta_{\lambda 1}(M_i)$, due to this translation, and we have

$$
\Delta_{\alpha i}(J_1) = \frac{\alpha_i - 1}{\lambda_1}\Delta_{\lambda 1}(M_i). \tag{43}
$$

Note that the change of J_1 due to the growth of any void is obtainable from only one remesh (overall, 2 calculations). In this way, it is straight forward to construct influence surfaces for the J-integral to assess the risk of voids in the neighborhood of a crack. Of course, the voids may have different forms, they could also be cracks.

5 Conclusion

In continuum defect mechanics, instead of localized defects, we have to deal with defect densities giving rise to material body forces. The conservation laws thus change to balance laws and new failure criteria have to be established, presumably based on components of the Eshelby tensor or equivalent material stresses. As it has been shown, the material balance and conservation laws are by far not completely investigated and the reciprocity relations may be successfully applied to continuously damaged materials. The generalization to continuously distributed defects, i.e., material forces per unit of volume, and material translations fields $\lambda_i = \lambda_i(x_j)$ would result in reciprocity relations involving surface-and volume-integral expressions with probably useful and far reaching applications. Along this line of reasoning, a material boundary-element method could be established.

References

1. Barber JR (2002) Elasticity, 2nd edn. Kluwer, Dordrecht
2. Bessel-Hagen E (1921) Über die Erhaltungssätze der Elektrodynamik. Math Ann 84:258–276
3. Budiansky B, Rice JR (1973) Conservation laws and energy-release rates. J Appl Mech 40:201–203
4. Gross D, Seelig T (2006) Fracture mechanics-with an introduction to micromechanics. Springer, Berlin
5. Günther W (1962) Über einige Randintegrale der Elastomechanik. Abh Braunschw Wiss Ges 14:53–72
6. Herrmann G, Kienzler R (2007) Reciprocity relations in Eshelbian mechanics. Mech Res Commun 34:338–343
7. Honein T, Chien N, Herrmann G (1991) On conservation laws for dissipative systems. Phys Lett A 155:223–224
8. Kienzler R (1993) Konzepte der Bruchmechanik. Vieweg, Braunschweig
9. Kienzler R (2007) Reciprocity in fracture and defect mechanics. Int J Fract 147:3–11
10. Kienzler R, Herrmann G (2000) Mechanics in material space. Springer, Berlin
11. Kienzler R, Kordisch H (1990) Calculation of J_1 and J_2 using L and M integrals. Int J Fract 43:213–225
12. Kienzler R, Rohde L, Schröder R (2010) On path-independent integrals within the linear theory of elasticity. Int J Fract 166:53–60
13. Kienzler R, Rohde L, Schröder R, Kutz K (2010) Treating mixed-mode problems with path-independent integrals. Eng Fract Mech 77:3604–3610
14. Lie S (1891) Vorlesungen über Differentialgleichungen mit bekannten infinitesimalen Transformationen. Teubner, Leipzig
15. Noether E (1918) Invariant variations problem. Nachr. Ges. Wiss. Göttingen, Math Phys Kl 2:235–257
16. Olver PJ (1984) Conservation laws in elasticity, I. general results. Arch Rat Mech Anal 85:111–129
17. Olver PJ (1984) Conservation laws in elasticity II. Linear homogeneous isotropic elastostatics. Arch Rat Mech Anal 85:111–129
18. Olver PJ (1993) Application of lie groups to differential equations, 2nd edn. Graduate texts in mathematics, vol 107. Springer, New York
19. Podio-Guidugli P (2000) A Primer in elasticity. Kluwer, Dordrecht
20. Podio-Guidugli P (2010) Private communications
21. Rice JR (1968) A path independent integral and the approximate analysis of strain concentrations by notches and cracks. J Appl Mech 35:379–386
22. Schade H, Neemann K (2009) Tensoranalysis, 3rd edn. de Gruyter, Berlin
23. Timoshenko SP, Goodier JN (1970) Theory of elasticity, 3rd edn. McGraw-Hill, New York

Part VI
Local Approach to Fracture

A Model for Predicting Fracture Toughness and Scatter in Thermally Embrittled Steels

A. Pineau and A. Andrieu

Abstract Temper embrittlement in a thick plate of a Mn-Ni-Mo pressure vessel steel was investigated using fracture toughness tests on Charpy and CT type specimens. A shift of the ductile-to-brittle transition temperature ($\sim$30–40 °C) was measured when the material was aged at 450 °C for 5000 h. Moreover an unusual scatter in fracture toughness tests was determined on aged material, such as $K_{IC}(P_R = 90\,\%)/K_{IC}(P_R = 10\,\%) \sim 5$, where P_R is the probability of failure. Scanning electron micrographs (SEM) indicated that the fracture surface was partly intergranular along micro-segregated zones (MSZ). This observation was made both on the initial and the aged conditions. Intergranular facets were largely covered by phosphorus segregation. A fully predictive model involving a combination of a local approach to fracture based on Beremin theory and accounting for MSZ distribution, and on the modelling of segregation kinetics in ternary (Fe-C-P) systems is developed to analyze these results. This model predicts the scatter in fracture toughness measurements and the shift in DBT. Moreover the statistical distribution of MSZ leads to a size effect in fracture toughness measurements which is different from the K^4B law inferred from the Beremin model applied to a homogeneous material.

Keywords Bimodal fracture · Cleavage · Intergranular fracture · Statistical approach · Impurity interaction

A. Pineau (✉)
Centre des Matériaux Pierre Marie Fourt, Mines ParisTech, 10 rue Henri Desbruères, Evry Cedex 91003, France
e-mail: andre.pineau@ensmp.fr

A. Andrieu
AREVA, 1 Place Jean Millier, Courbevoie 92400, France
e-mail: antoine.andrieu@areva.com

G. Hütter and L. Zybell, *Recent Trends in Fracture and Damage Mechanics*, DOI 10.1007/978-3-319-21467-2_15

1 Introduction

Heavy components of Pressurized Water Reactors (PWR) are made of low alloy Mn-Ni-Mo steels (A508, A533) used in the quenched and tempered conditions. Most of the French shells used to fabricate pressurizers and steam generators in nuclear industry were manufactured before 1993 using rolled and welded plates. This manufacturing process generates characteristic microstructures composed of thin strips of micro-segregated material (MSZ) surrounded by base material (BM), i.e. having a chemical composition, close to that of the selected steel. Once these components are welded together a Post Weld Heat Treatment ($\sim$615 °C few hours cooling rate ≤50 °C/h) is applied both to relax residual stresses introduced during the welding operations and to temper the fresh martensite or bainite microstructures located in the heat affected zone of the welds (HAZ). These Mn-Ni-Mo steels are relatively sensitive to thermal aging which produces a shift of the ductile-to-brittle transition temperature (DBTT) after long term exposure at high temperature ($\sim$300–550 °C). This shift is due to temper embrittlement and is linked to the local material chemical composition and in particular to the content in residual impurities such as phosphorus (see e.g. [1, 3, 5]). Druce and his co-workers [8, 9] have modelled intergranular segregation of phosphorus using McLean's approach to equilibrium segregation [18]. These authors have also proposed a simple equation relating the shift of the DBTT measured at an energy of 80 J to the amount of phosphorus segregated at grain boundaries. The first expression proposed in 1986 [8] was:

$$\Delta TK_{80\,\mathrm{J}} = 1300 \times (C^{P}_{gb} - 0.065) \tag{1}$$

where C^{P}_{gb} is the P/Fe peak height ratio (PHR) measured on Auger spectra. In the following, the same definition for C^{P}_{gb} was adopted. This relation (Eq. 1) was revised later in 1988 by Druce et al. [9] from further measurements on real HAZ made of A533B steel.

$$\Delta TK_{80\,\mathrm{J}} = 1080 \times (C^{P}_{gb} - 0.065) \tag{2}$$

These expressions in which 0.065 represents the initial concentration in P at grain boundaries, are drawn in Fig. 1. Joly et al. [15] showed that the application of these empirical relations to the AREVA aging program which included HAZ, base metal (BM) of 18MND5 (French equivalent to A533B) and the welded metal leads to a large over-estimate of the embrittlement compared to the shift measured experimentally. The results are illustrated in Fig. 1 where the observed DBTT shift corresponding to 68 J is compared to the shift at 80 J predicted from Eqs. 1 and 2. This observation can be attributed to the fact that another chemical element such as carbon has to be considered when calculating the kinetics of segregation.

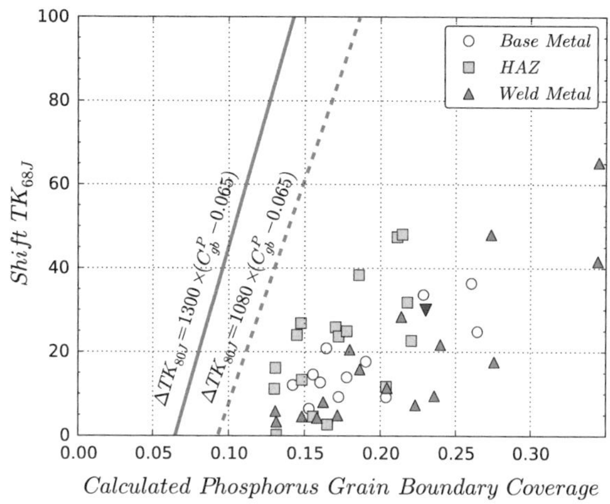

Fig. 1 DBTT shift induced by thermal aging on base metals, heat affected zones and welds of 18MND5 grade steel. Comparison of experimental results with embrittlement models. The symbol *black down pointing triangle* represents the DBTT shift measured on the 18MND5 base metal used in this study and aged during 5000 h at 450 °C

The aims of the present study were the following:

1. Determine experimentally the shift of the DBTT after aging a material representative of the fabrication of pressurizers. It is well to remember that these components work at relatively high temperature (320–340 °C), as compared to the PWR core shell.
2. Investigate to what extent the shift in DBTT is correlated with a modification in the fracture modes, from cleavage to intergranular.
3. Improve the modelling of the existing kinetics of impurity segregation at grain boundaries, in particular the models introduced by Druce [8, 9, 12] and McLean [18].
4. Model the variation of fracture toughness with temperature and aging conditions.
5. Investigate the scatter in fracture toughness in relation with the MSZ distribution.

The present article is organized as follows. In Sect. 2 the experimental procedures are briefly introduced. Experimental data are presented in Sect. 3 which includes results of mechanical testing, fractography and Auger spectrometry. Section 4 is devoted to the modelling of impurity segregation kinetics and its link with fracture toughness evolution.

2 Material and Experimental Procedures

The composition of the investigated steel is (in weight %): C = 0.192, Mn = 1.44, Ni = 0.083, Mo = 0.473, Cr = 0.194, Si = 0.249, S = 0.0053, P = 0.010. This material was received as a thick plate (100 mm) produced for simulating welding operations. This composition is very close to the plates used in the past fabrication of pressurized vessels. The material was given the conventional heat-treatment (8 h at 615 °C followed by air cooling at 30 °C/h). The prior austenite grain size was between 0 and 30 μm and the bainitic packet size of 10–15 μm. The yield strength determined at room temperature was 490 MPa. The macrostructure of the material is shown in Fig. 2. Segregated zones aligned along the rolling direction (L) and spaced of about 2 mm in the short-transverse (ST) direction are observed. These inhomogeneities result from the micro-segregations of alloying elements and impurities during solidification. Micro-segregated zones (MSZ) are deformed and aligned by the rolling process. These zones are much harder than the matrix (~350 HV instead of ~210 HV) (Fig. 2). Micro-segregated zones are enriched in all alloying elements (C, Mn, Mo, Si) and in P as revealed by microprobe analysis (Fig. 2). The surface fraction covered by these MSZ is about 10 %. Quantitative image analysis was used to determine the dimension and the shape of these segregated zones [2]: mean size in a plane TL was 350 × 750 μm^2.

Conventional Charpy specimens (TL orientation) were tested to determine the DBTT curve. Standard CT specimens (thickness, B = 20 mm, orientation TL) were also used to investigate the effect of thermal aging (450 °C, 5000 h in the present study) on the fracture toughness. SEM and Auger spectrometry were used to examine the fracture surfaces and to measure the amount of phosphorus and carbon

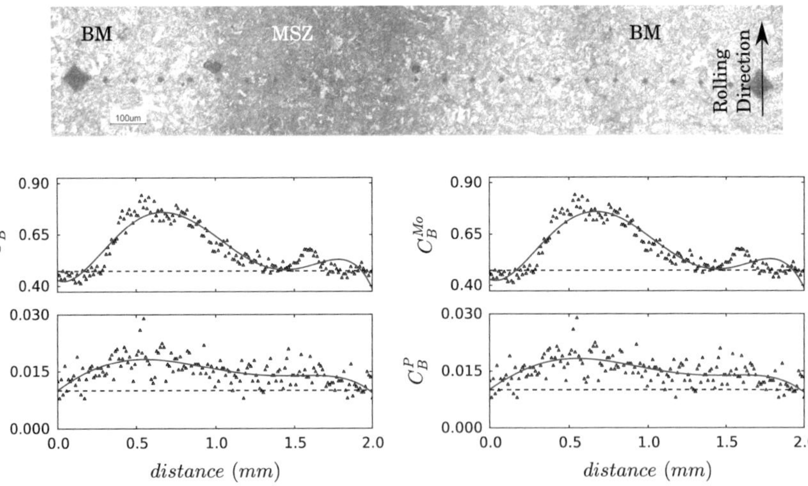

Fig. 2 Measurement of the bulk content in C, P and Mo while crossing a micro-segregated zone (MSZ) and microhardness measurements

segregated along the grain boundaries. An Auger apparatus (JEOL JAMP 9500F) equipped with a cooling and in situ fracture device working at liquid nitrogen temperature was used to break specimens before performing measurements of intergranular segregation on fresh fracture surfaces.

Most of the observations were made with the following electron gun conditions: 20 keV/10 nA or 5 keV/1–3 nA. It was confirmed that the fractured grain boundaries contained a coverage in phosphorus of at least 10 %. This means that such level of segregation is necessary to be reached for triggering intergranular fracture, as already published by Naudin [21] and confirmed by Ding et al. [7].

SEM observations of fracture surfaces were made using either a LEO microscope equipped with a EDX analysis or an Hitachi apparatus equipped with an EBSD camera. Quantitative stereography was used to measure the tilt (ϕ) and the twist (ψ) angles between two adjacent grains (For further details, see [2, 22]).

3 Results

3.1 Mechanical Tests

The variation of the yield strength ($\sigma_{0.2}(T)$) measured at a strain rate of about 0.0005 s^{-1} is shown in Fig. 3. These results were fitted with the Lean's law [17].

$$\frac{1}{\sigma_{0.2}(T)} = \frac{1}{\sigma(0K)} + BT \quad (3)$$

where T (in K) is the temperature and $\sigma(0K) = 1043\,\text{MPa}$, $B = 3.57 \times 10^{-6}\ \text{MPa}^{-1}\ \text{K}^{-1}$.

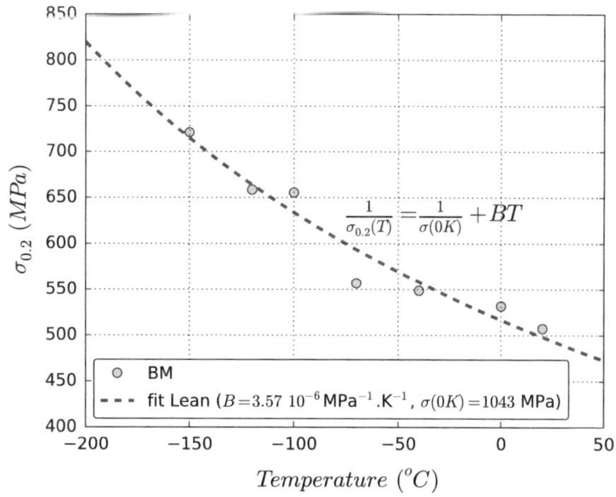

Fig. 3 Evolution of the yield strength as a function of temperature for the base metal —experimental measurements

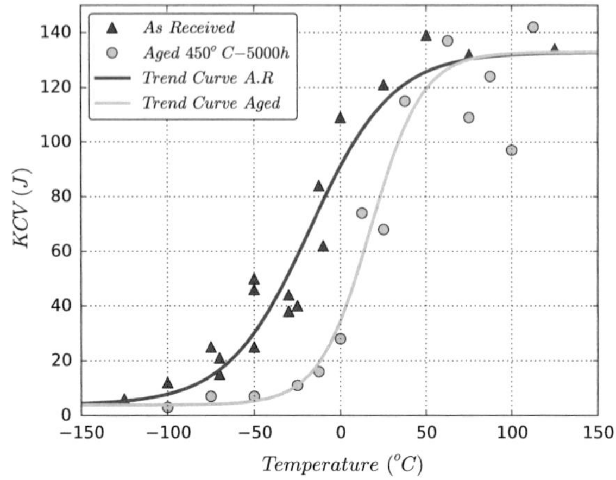

Fig. 4 Charpy impact tests results obtained on 18MND5 grade steel in the as-received condition and after aging at 450 °C for 5000 h

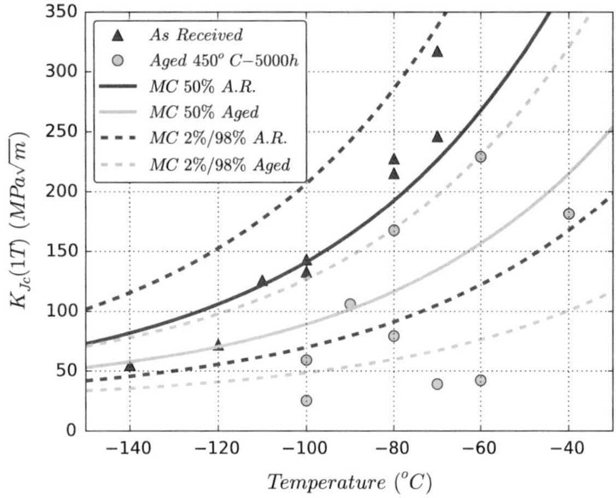

Fig. 5 Fracture toughness results obtained on 18MND5 grade steel in the as-received condition and after ageing at 450 °C for 5000 h. For both conditions ASTM E1921 master curve [4] is plotted

The results obtained on Charpy specimens tested in the initial condition after stress relieving heat-treatment (TTDS) and after aging (450–5000 h) are reported in Fig. 4. A shift of the DBTT by about 40 °C measured at an energy of 41 J appears in this figure when comparing the as-received and the aged material. Similarly the fracture toughness determined on CT_{20} is lowered by aging (Fig. 5). The temperature T_0, estimated according to ASTM E1921 [4], is shifted from −124 °C in the as-received condition to −91 °C in the aged condition. Figure 5 shows that the ASTM E 1921 master curve [4] is unable to account for the scatter observed in the aged condition.

Fig. 6 Characteristic fracture surface observed on a compact tensile specimens, made of 18MND5 grade steel, tested at low temperature and in the as received condition. Arrows indicate intergranular, cleavage and MnS particles

3.2 *Fractography*

The fracture surfaces of the material tested in the as-received condition are mainly covered by cleavage facets (Fig. 6). However inclusions identified as MnS particles aligned along the rolling direction are surrounded by intergranular fracture. Figure 6 was taken on a as-received condition specimen tested at −80 °C ($K_{JC} = 204$ MPa $\sqrt{m}$). Inclusions are located within the MSZ evidenced in Fig. 2. This observation is similar to previous results obtained on another PWR material (see Fig. 1 in [1]).

Examples of the fracture surfaces of the aged material are given in Fig. 7. Figure 7a, b were taken on an aged specimen tested at −80 °C ($K_{JC} = 159\,\text{MPa}\sqrt{m}$). In the embrittled condition the fracture surfaces remain mixed: partly cleavage in the 'matrix' and partly intergranular along the initial segregated zones. It is noted that the cavities initiated from MnS inclusions are much smaller in the aged condition compared to the as-received condition. This suggests that the grain boundaries in the MSZ are much more brittle in the embrittled condition, as expected, due to P enrichment in the segregated zones. It should be emphasized that in this specific case the shift of the DBTT cannot be attributed to a transition in fracture modes between cleavage and intergranular since the ratio of intergranular/cleavage fracture is very similar in the as-received and in the aged material. Moreover these observations suggest that the shift in DBTT might be related to the segregation of impurities in MSZ during the aging treatment at 450 °C.

3.3 *Auger Spectroscopy*

Figure 8 shows an example of an Auger spectrum obtained on a specimen. This spectrum clearly shows the presence of phosphorus and carbon. The peak height ratio of these elements compared to that of Fe703 was used to calculate the amount

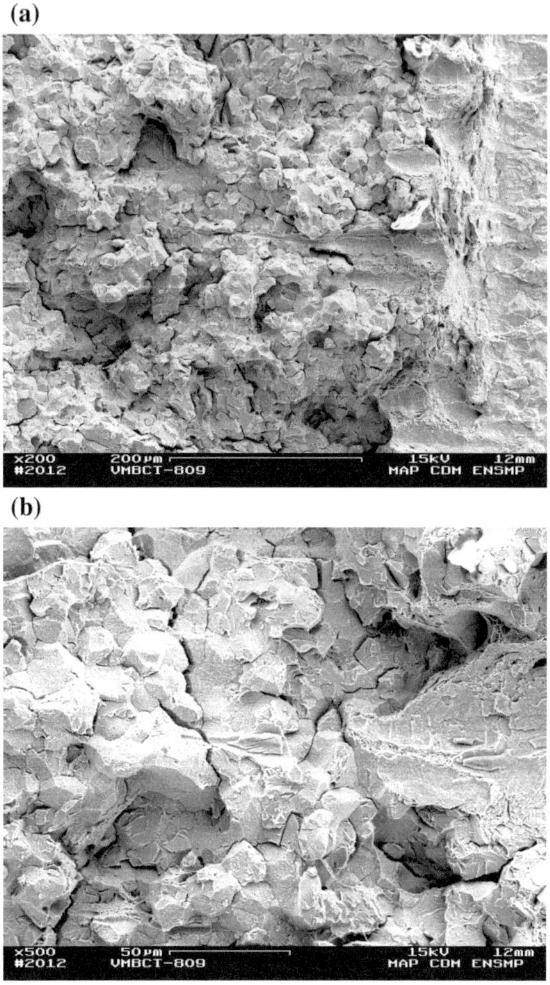

Fig. 7 **a** Characteristic fracture surfaces observed on a compact tensile specimens, made of 18MND5 grade steel, tested at low temperature and after aging 5000 h at 450 °C. **b** Enlargement of **a**, arrows indicate intergranular, cleavage and MnS particles

of impurities segregated on the grain boundaries. Guttmann et al. [11] and Erhart and Grabke [10], showed that C can modify the kinetics of the segregation of P on grain boundaries because these elements segregate on the same sites. Moreover, following the classification established by Hondros and Seah [13], carbon and phosphorus have an antagonistic effect: phosphorus decreases the cohesion of grain boundaries while carbon reinforces them. See also Rice and Wang [25] for the embrittlement of grain boundaries by solute segregation.

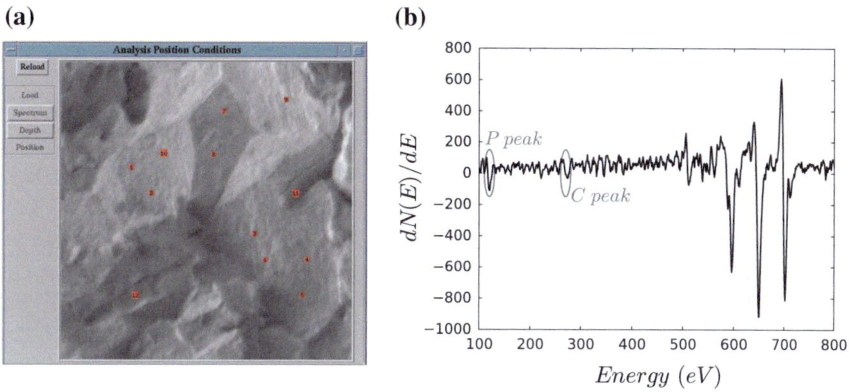

Fig. 8 **a** Points analysed by probe tracking to determine grain boundary segregation. **b** Example of derivative spectrum used to calculate the P/Fe and C/Fe peak height ratios

4 Modelling

This part begins with the prediction of phosphorus (and carbon) intergranular segregation before dealing with a model for the prediction of the fracture toughness in inhomogeneous tempered embrittled materials. These predictions are compared to experimental results. Then the influence of grain boundary segregation on fracture toughness is discussed.

4.1 *Intergranular Segregation*

The Mc Lean theory [18] was developed to predict the equilibrium segregation of impurities along grain boundaries. According to this theory the concentration of phosphorus segregated on grain boundaries, C_{gb}^{P} depends on: the concentration of phosphorus in the bulk, C_1, the time, t, and the temperature of ageing. The kinetics of enrichment of grain boundaries in phosphorus is expressed as:

$$C_{gb}^{P}(t) = C_{gb}^{P}(t \rightarrow \infty) - C_1(\alpha_2 - \alpha_1)\exp\left(\frac{4Dt}{\alpha_2^2\delta^2}\right) erfc\left(\frac{2\sqrt{Dt}}{\alpha_2\delta}\right) \qquad (4)$$

with $C_{gb}^{P}(t \rightarrow \infty)$ given by the Langmuir[16]–McLean [18] equilibrium model:

$$C_{gb}^{P}(t \rightarrow \infty) = \frac{C_1 \exp\left(\frac{\Delta G_P^0}{RT}\right)}{1 + C_1 \exp\left(\frac{\Delta G_P^0}{RT}\right)} \qquad (5)$$

The free energy of segregation, ΔG_0, and the diffusion coefficients, D, were determined from measurements of P segregated at grain boundaries by Druce et al. [8]. In Eqs. 4 and 5 R is the gas constant, δ the 'thickness' of the grain boundaries arbitrarily chosen as equal to 3×10^{-8} cm by McLean [18], α_1 is the ratio between the initial concentration in P at grain boundaries (i.e. prior to aging and resulting from the stress relieving heat-treatment) divided by the P bulk content in the MSZ and α_2 is the ratio between the concentration of P at grain boundaries at any time during the aging treatment divided by the concentration of P in the bulk of the MSZ. After having performed many aging heat treatments on a simulated HAZ in A533 steel, Druce et al. [8] have proposed to relate the concentration in Phosphorus with the shift in DBTT using Eqs. 1 and 2. This theory does not take into account the interaction between P and C. Guttmann and al. [11] and more recently Nakata et al. [20], among others, have highlighted the competition between these two elements. To describe the kinetics of segregation in the ternary {Fe, C, P} system the theory developed by Guttmann et al. [11], was used in the present study. Preliminary results were presented elsewhere [3]. In this theory, the {Fe, C, P} system is no longer considered as ideal and the free energy of segregation of P and C is described by the two following coupled equations:

$$\begin{aligned} \Delta G_C &= \Delta G_C^0 - 2\alpha_{CC} C_C^{gb} + \alpha_{CP} C_P^{gb} \\ \Delta G_P &= \Delta G_P^0 - 2\alpha_{PP} C_P^{gb} + \alpha_{CP} C_C^{gb} \end{aligned} \tag{6}$$

where the coefficients α_{ij} account for the modification of the free energy created by the interaction between the element i and the element j at grain boundaries. This interaction can be either attractive (leading therefore to an increase of the segregation kinetics) or repulsive (leading to a deceleration of the segregation kinetics). These interaction coefficients were determined using the study of Erhart and Grabke [10] while the calibration of the free energy was made using Nakata et al. [20] work. Militzer and Wieting [19] demonstrated that, in the case of ternary systems with site competitions, the kinetics of segregation can be written as:

$$\begin{aligned} C_i^{gb}(t) &= C_i^{gb,0} + 2\frac{C_i^{B,0}}{\sqrt{\pi}\delta/2}\sqrt{\int_0^t D_i(t')dt'} - \frac{1}{\sqrt{\pi}\delta/2}\int_0^t \frac{q_i(t')D_i(t')}{\sqrt{\int_{t'}^{t} D_i(t'')dt''}}dt' \\ D_i(t) &= D_i^0 \exp(-Q_i/RT(t)) \\ q_i(t) &= \frac{C_i^B \exp(-\Delta G_i(t)/RT(t))}{1 + \sum_{j=C,P} C_j^B \exp(-\Delta G_j(t)/RT(t))} \end{aligned} \tag{7}$$

All terms in Eqs. 6 and 7 have been determined and identified from the literature and are presented in Table 1. These equations can also be used under

Table 1 Thermodynamic parameters identified from literature and used to describe 18MND5 grade steel

Guttmann model
$\Delta G_C^{seg} = 9030 + 65.12T - 2\alpha_{CC} * C_C^{gb}/c^0 + \alpha_{CP} * C_P^{gb}/c^0 (in\,\mathrm{J})$
$\Delta G_P^{seg} = 65273 - 23.69T - 2\alpha_{PP} * C_P^{gb}/c^0 + \alpha_{CP} * C_C^{gb}/c^0\ (in\,\mathrm{J})$
$\alpha_{PP} = 139\ (in\,\mathrm{J\,mol^{-1}})$
$\alpha_{CC} = 4000\ (in\,\mathrm{J\,mol^{-1}})$
$\alpha_{CP} = -4500\ (in\,\mathrm{J\,mol^{-1}})$
Diffusion coefficients, $(in\,\mathrm{cm^2\,s^{-1}})$
$D_P(T)$ = 0.25 exp (−200,000/(RT)) [27]
$D_C(T)$ = 0.003 exp (−76,000/(RT)) [14]

non-isothermal conditions and thus are able to account for P and C segregation taking place during the cooling (30 °C/h) subsequent to post weld heat treatment.

This model for ternary systems was applied and validated using Nakata et al. [20] results. Segregations kinetics with or without taking into account P and C competition are presented in Fig. 9a. It can be observed that the kinetics obtained by neglecting P and C competition leads to a large overestimate of the amount of P segregated at grain boundaries.

Application of Eqs. 6 and 7 to our results (limited to as received and aged conditions) is presented in Fig. 9b. A good agreement between calculated and experimental segregation kinetics is observed. Calculations showed that in the material representative of the MSZ (P ~ 150 ppm), segregation taking place during cooling, from 615 °C to room temperature, leads to $C_{gb}^P \sim 13\ \%$ before aging [2]. It was shown by Naudin et al. [21] that this concentration is sufficient to trigger intergranular fracture, as indicated earlier. This is consistent with observations highlighting intergranular fracture surfaces within the MSZ in the as-received condition.

4.2 *Effect of Grain Boundary Segregation on Fracture Toughness*

The prediction of the shift in DBTT measured on Charpy impact specimens requires time-consuming numerical calculations (see e.g. Tanguy et al. [26]). This explains why in the present study this shift was measured and modelled using the results of fracture toughness tests on CT specimens shown in Fig. 5. Numerical modelling of these specimens tested in static conditions is much easier than that of Charpy specimens. It has been noted that the fracture mode of our material, when tested in the DBTT regime, was bimodal: transgranular cleavage in the base metal and intergranular in the MSZ. A recent model accounting for the contribution of each material to the overall fracture toughness of a given specimen has been proposed

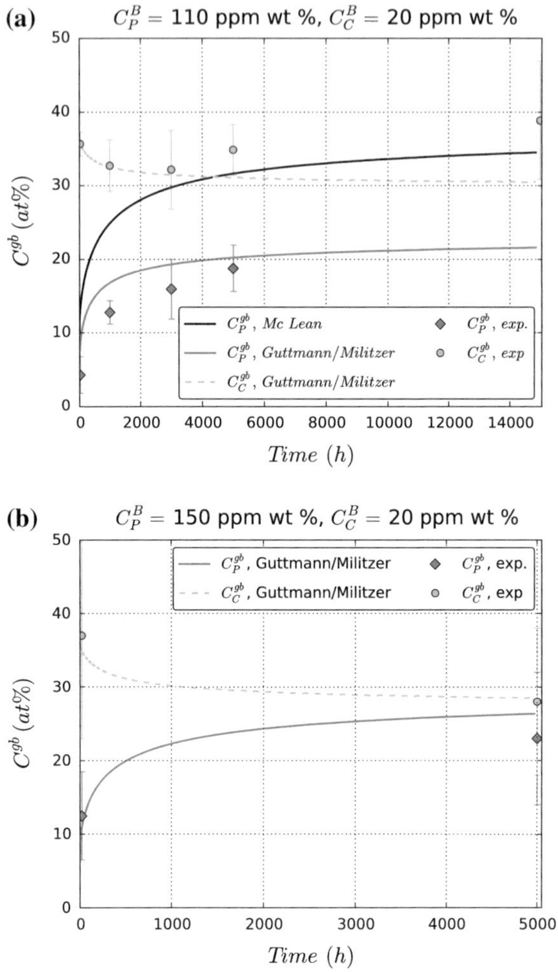

Fig. 9 Application of the Guttmann-Militzer segregation model to: **a** results obtained by [20] on a A533B steel containing 110 ppm of phosphorus and aged at 450 °C, **b** results obtained by [2] on a 18MND5 grade steel containing 90 ppm of phosphorus and aged at 450 °C

recently [1, 3].[1] This model is based on the weakest link theory and the introduction of the 'Weibull stress' which is calculated both in the matrix and in the segregated zones. Values of the shape parameter, m, and of the characteristic stress σ_u depend on the material: in the base metal the characteristic stress represents the cleavage fracture stress while in the MSZ it represents the intergranular fracture stress.

[1]It should be noted that our model is somewhat different from the bimodal master curve approach [28] in which inhomogeneity between specimens is introduced but not within an individual specimen.

The cleavage stress in the matrix depends on the aging conditions since it has been shown that even in this mode of failure some intergranular fracture takes place at the crossing of grain boundaries by transgranular cleavage microcracks [2, 22–24]. The matrix cleavage stress decreases therefore with aging time. It can be assumed that the intergranular fracture stress in the MSZ decreases linearly with the grain boundary P content, C_{gb}^{P}, as shown by [7, 21]. The values of C_{gb}^{P} can be calculated using the model developed in the previous section.

In the present situation another source of scatter is related to the statistical distribution of the segregated zones within a given specimen. Andrieu et al. [1] have shown that the predicted probability of failure can thus be written as:

$$P_r = \sum_{i=0}^{B/e} P_i P_r(i) \tag{8}$$

where B is the thickness of the CT specimen, e, a distance (for example 50 μm), P_i, the probability that $i \times e$ mm of MSZ intercepts the crack front and $P_r(i)$ the probability to fracture, given by:

$$P_r(i) = 1 - \exp\left(- \frac{(B - ie)K^4 (\sigma_0^{BM})^{m_{BM}-4} C_{m_{BM}}}{V_0 (\sigma_u^{BM})^{m_{BM}}} \right) \times \exp\left(- \frac{ie(E_{MSZ}\sigma_0^{MSZ}/E_{BM}\sigma_0^{BM})^2 K^4 (\sigma_0^{MSZ})^{m_{MSZ}-4} C_{m_{MSZ}}}{V_0 (\sigma_u^{MSZ})^{m_{MSZ}}} \right) \tag{9}$$

In this expression m the Weibull exponent, V_0 an elementary volume, σ_u the characteristic stress and C_m a numerical parameter which depends on the value of the shape parameter, m and of the work hardening exponent, n. The values of $C_m(n)$ have been tabulated elsewhere (Andrieu et al. [1]).

The initial Weibull parameters have been calibrated, using the least mean square likelyhood, in order to describe the results obtained in the as-received conditions: $m_{BM} = 30, m_{MSZ} = 28, \sigma_u^{BM} = 3364$ Mpa, $\sigma_u^{MSZ} = 3200$ MPa. Figure 10a shows that these values lead to a good representation of the measurements in the initial condition. The scatter measured by the ratio $K_{JC}(P_R = 95\,\%)/KJC(P_R = 5\,\%)$ is close to 3.5 which is a little larger than the value predicted from Beremin theory [6] when there is a unique mode of failure in an homogeneous material. The above model is also able to represent the shift of the DBTT and the increase in the scatter in test results, as observed in Fig. 10b for the aged condition. The scatter measured by $K_{JC}(P_R = 95\,\%)/K_{JC}(P_R = 5\,\%)$ is close to 5, as observed experimentally. This constitutes the second salient feature of the present study.

As noted in Fig. 1, Eqs. 1 and 2 applied to our material largely overestimate the DBTT shift due to thermal aging at 450 °C. Our experimental results ($\Delta T \approx 30\,°C$, $C_{gb}^{P} = 0.23$) lead to a proportionality factor (the slope of the lines plotted in Fig. 1)

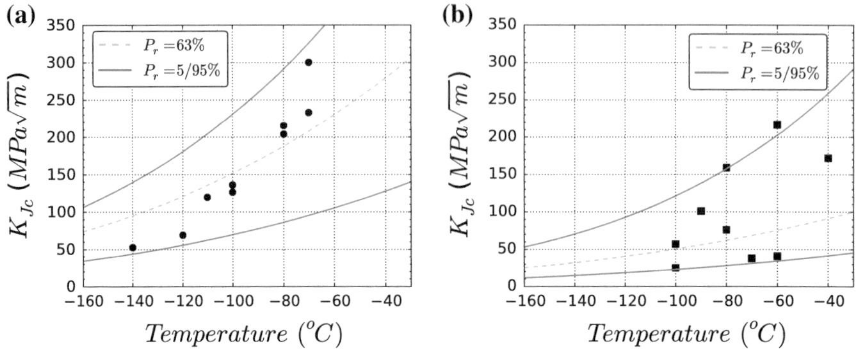

Fig. 10 Application of the Beremin bimodal model to fracture toughness results obtained **a** in the as-received condition **b** in the aged condition: 450 °C for 5000 h

much smaller, of the order of 175 instead of 1080. This clearly indicates that very much work remains to be done before a fully predictive formulation of the temper embrittlement is accepted.

5 Conclusions

1. It has been demonstrated that in a Mn-Ni-Mo steel used in the fabrication of heavy components in nuclear power industry and containing micro-segregated zones (MSZ) enriched in impurities (P, C), the slow cooling rate applied after stress relieving heat-treatment produces a mixed mode of failure (cleavage in the matrix and intergranular in the MSZ at low temperature).
2. Impurity segregation at grain boundaries produces a shift of the ductile-to-brittle transition temperature (DBTT) of the order of 30–40 °C when the material is tested at temperatures such as $K_{JC} \sim 100$ MPa$\sqrt{m}$.
3. The amount of P and C segregated at grain boundaries corresponding to the slow cooling rate after post weld heat treatment and subsequent aging at 450 °C for 5000 h has been calculated using a modified McLean-Militzer model accounting for C-P interactions. This amount has been compared with experimental data obtained from Auger spectrometry measurements. A good agreement between experimental and calculated results is observed.
4. The shift of the DBTT observed after aging at 450 °C is satisfactorily predicted using a statistical Beremin type model in which the statistical distribution of MSZ within a specimen is introduced.

Acknowledgments This study has been made in the frame of the AREVA-Ecole des Mines industrial professorship. Support from AREVA is acknowledged. The authors would like to thank also D. Loison, S. Cremel and O. Bouaziz from Arcelor-Mittal Research for spectrometry measurements. They acknowledge also P. Joly and F. Roch from AREVA and D. Ryckelynck from Ecole des Mines for many stimulating discussions.

References

1. Andrieu A, Pineau A, Besson J, Ryckelynck D, Bouaziz O (2012) Bimodal Beremin-type model for brittle fracture of inhomogeneous ferritic steels: theory and applications. Eng Fract Mech 95:84–101
2. Andrieu A (2013) Mécanismes et modélisation multi-échelle de la rupture fragile trans- et inter-granulaire des aciers REP en lien avec le vieillissement thermique. PhD Ecole des Mines de Paris
3. Andrieu A, Pineau A, Joly P, Roch F, Ryckelynck D (2014) Influence of P and C intergranular segregation during manufacturing and aging on the fracture toughness of nuclear pressure vessel steels. Procedia Mater Sci 3:655–660
4. ASTM E 1921-12 (2012) Standard test method for determination of reference temperature, T0, for ferritic steels in the transition range. American Society of Testing and Materials, West Conshohocken
5. Bhadeshia KHDH, Honeycombe RWK (2006) Steels-microstructure and properties, 3rd edn. Butterworth-Hennemann, Oxford
6. Beremin FM (1983) A local criterion for cleavage fracture of a nuclear pressure vessel steel. Metall Trans 14A:2277–2287
7. Ding R, Islam A, Wu S, Knott J (2005) Effect of phosphorus segregation on fracture properties of two quenched and tempered structural steels. Mater Sci Technol 21:467–475
8. Druce SG, Gage G, Jordan G (1986) Effect of ageing on properties of pressure vessel steels. Acta Metall 34:641–652
9. Druce SG, Gage G, Popkiss EW (1988) Thermal ageing embrittlement of the heat affected zone in a PWR RPV steel weldment. ANS Topical Meet Nucl Power Plant Life Extension 2:86–97
10. Erhart H, Grabke HJ (1981) Equilibrium segregation of phosphorus at grain boundaries of Fe–P, Fe–C–P, Fe–Cr–P, and Fe–Cr–C–P alloys. Met Sci 15:401–408
11. Guttmann M, Dumoulin P, Wayman M (1982) The thermodynamics of interactive co-segregation of phosphorus and alloying elements in iron and temper-brittle steels. Metall Trans A 13:1693–1711
12. Hippsley CA, Druce SG (1983) The influence of phosphorus segregation to particle/matrix interfaces on ductile fracture in a high strength steel. Acta Metall 31:1861–1872
13. Hondros ED, Seah MP (1977) The theory of grain boundary segregation in terms of surface adsorption analogues. Metall Trans A 19:1363–1371
14. Honeycombe RWK (1981) Steels–microstructure and properties. Edward Arnold Ltd., London
15. Joly P, Roch F, Primault C (2013) Effect of thermal ageing on properties of pressure vessel low alloy steel. In: Proceedings of the ASME 2013 pressure vessels and piping conference
16. Langmuir I (1916) The constitution and fundamental properties of solids and liquids. Part I Solids J Am Chem Soc 38:2221–2295
17. Lean JB, Plateau J, Crussard J (1958) Etude des propriétés mécaniques et de la rupture fragile de l'acier doux. Comptes Rendus de l'Acadḿie des Sciences 247:1458–1461
18. McLean D (1957) Grain boundaries in metals. Oxford Press, Oxford
19. Militzer M, Wieting J (1986) Theory of segregation kinetics in ternary systems. Acta Metall 34:1229–1236
20. Nakata H, Fujii K, Fukuya K, Kasada R, Kimura A (2006) The thermodynamics of interactive co-segregation of phosphorus and alloying elements in iron and temper-brittle steels. J Nucl Sci Technol 43:785–793
21. Naudin C, Frund JM, Pineau A (1999) Intergranular fracture stress and phosphorus grain boundary segregation of a Mn–Ni–Mo steel. Scripta Mater 40:1013–1019
22. Pineau A (2015) Crossing grain boundaries in metals by slip bands, cleavage and fatigue cracks. Phil Trans A. To appear
23. Qiao Y, Argon AS (2003) Cleavage crack-growth-resistance of grain boundaries in polycrystalline Fe–2 % Si alloy: Experiments and modeling. Mech Mater 35:129–154

24. Qiao Y, Argon AS (2003) Cleavage cracking resistance of high angle grain boundaries in Fe–3 % Si alloy. Mech Mater 35:313–331
25. Rice JR, Wang JS (1988) Embrittlement of interfaces by solute segregation. Mater Sci Eng A 107:23–40
26. Tanguy B, Besson J, Piques R, Pineau A (2005) Ductile to bittle transition of an A508 steel characterized by Charpy impact test. Part II: modeling of the Charpy transition curve. Eng Fract Mech 72:413–434
27. Vatter IA, Hippsley CA, Druce SG (1993) Review of thermal ageing data and its application to operating reactor pressure vessels. Int J Press Vessels Pip 54:31–48
28. Wallin K, Nevasmaa P, Laukkanen A, Planman T (2004) Master curve analysis of inhomogeneous ferritic steels. Eng Fract Mech 71:2329–2346

Micromechanical-Based Models for Describing Damage of Ferritic Steels

Michael Seidenfuss and Thomas Linse

Abstract Usually the safety margin against failure for precracked components is calculated with fracture mechanics approaches. Due to several severe limitations of these approaches, it was searched for alternative calculation models. Starting with McClintock and Berg in the sixties, so-called damage models have been developed for describing ductile fracture on the basis of micromechanical processes. The development of such kind of models is in progress now for nearly 50 years, but until today no model is generally accepted and incorporated into the international standards. In an extended introduction, the micromechanical phases of ductile rupture of metal and alloys are presented. Against this background, a summary of the evolution and the different kinds of micromechanical-based model approaches is given. The theoretical background, the advantages/ disadvantages and the limitations of the models are discussed critically. Finally non-local formulations of damage models are presented. Combinations of ductile damage models and models for cleavage to describe fracture in the brittle-ductile transition region are discussed.

1 Introduction

To guarantee safe operation of technical components and systems, the safety margins against failure must be quantified. One possible approach to predict these safety margins is the use of numerical simulation methods with advanced material models. For the prediction of crack initiation, crack growth and fracture of ductile metals so-called micromechanical-based damage models based on the early work of

M. Seidenfuss (✉)
Institut für Materialprüfung, Werkstoffkunde und Festigkeitslehre, Universität Stuttgart, Pfaffenwaldring 32, 70569 Stuttgart, Germany
e-mail: michael.seidenfuss@imwf.uni-stuttgart.de

T. Linse
Institute of Solid Mechanics, Technische Universität Dresden, 01062 Dresden, Germany
e-mail: thomas.linse@tu-dresden.de

G. Hütter and L. Zybell, *Recent Trends in Fracture and Damage Mechanics*, DOI 10.1007/978-3-319-21467-2_16

McClintock [169] and Berg [33] have been established. In this context micromechanical-based means that the models try to simulate the processes on the microscopic level with continuum mechanical approaches.

For the development and application of damage models it is fundamental to understand the micromechanical processes in the material leading to fracture. In ductile fracture, these micromechanical processes can be divided into three phases: void formation, void growth and void coalescence [74, 207, 261, 276]. A detailed summary of this failure development is given in Sect. 2.

The classical micromechanical-based damage models known from literature try to describe the three fracture phases with continuum mechanical approaches. Generally for each phase, a separate model is needed: void initiation [6, 34, 72, 115], void growth [115, 154, 169, 211, 224, 277] and void coalescence [26, 61, 272].

The classic micromechanical-based damage models are derived for high stress multiaxialities and a pronounced void growth. Such models are described in Sect. 3. Mechanisms observed in pure shear mode are not or insufficiently described by these models. Micromechanical-based models to describe the failure at such low stress multiaxialities are not in the focus of interest here. Models which describe both high and low stress multiaxialities are usually empirical in nature.

Nearly all models discussed so far are of local nature. This means that the material behaviour depends only on the local state variables. Neighboring points have no influence on the local material behaviour. If material softening occurs, this can lead to so-called bifurcation problems. This means that a homogeneous strain or damage field will get unstable against a strongly localized one [209]. In finite element calculations, this means that strains and damage locate in one element layer [244]. The so-called pathological mesh dependence of results is observed.

In practice, this problem can be overcome by keeping the mesh size constant. Often the mesh size is directly coupled to the microstructure [45, 62, 86, 182, 216, 240, 241, 262]. To eliminate the pathological mesh dependence different concepts have been published, for example [21, 55, 92, 205]. Together, all these concepts and the derived models introduce a material-specific characteristic length. A summary of the most common approaches is given in Sect. 4.

Following the concept of the Local Approach [201], fracture toughness values can be predicted by numerical calculations only. Different possibilities for the description of competing brittle and ductile damage in the entire toughness region are discussed in Sect. 5.

2 Failure by Void Initiation, Growth and Coalescence [244]

Materials and components can be deformed up to a characteristic extent. Fracture limits the deformation and often leads to a catastrophic failure of vehicles, machines and plants with consequences for safe operation. Therefore it is essential to investigate the causes and the mechanism of fracture to find adequate simulation approaches for being able to predict the failure behaviour.

There is no uniform classification of fractures in literature. Classifications which are based on load type or macroscopic phenomena can be helpful, however, they are not suitable to be used in a damage mechanical calculation model [244] which describes the local material behaviour.

The microscopic description of the failure behaviour by means of micromechanical processes only requires the local state and the local kinematic laws. Therefore, this description is more suitable to be used in a damage model. With such damage models the macroscopic definition of fracture which is necessary for practical application can be calculated by means of the finite-element-method. Damage models belong to the group of the so-called advanced material models.

To understand the local failure process, it is indispensable to know the micromechanical processes which occur on the micro level. They determine the microscopical and macroscopical processes in the material as well as the future appearance of the fracture (cleavage fracture or dimple/shear fracture). Hereinafter it is referred only to the dimple fracture.

The dimple fracture [240] is characterized by locally very high plastic deformations on the fracture surface. The dissipated energy is much higher than in cleavage fracture. This reflects the rough dimpled fracture surface [83, 113] which is characteristic for all technical metals and metal alloys, see Figs. 1 and 2.

It depends on the surface energy, on the stress state as well as the elastic energy released during the crack growth whether dimple fracture is stable or not.

Investigations done in the 1940s and 1950s [74, 207, 276] have already shown that for almost all metal alloys, the micro-mechancial processes which lead to dimple fracture can be divided into 3 phases, see Fig. 3:

I. void initiation,
II. void growth,
III. void coalescence.

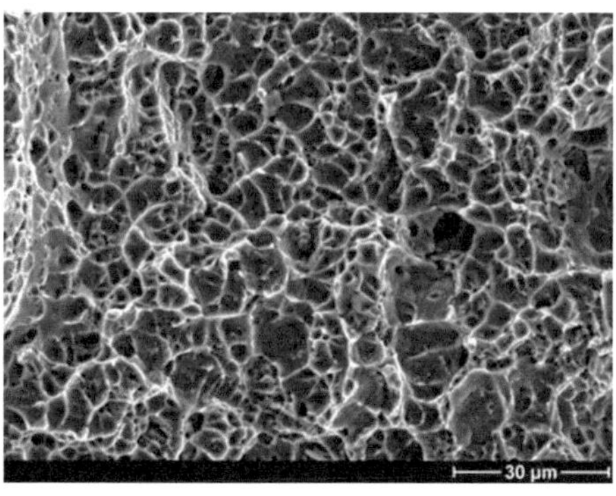

Fig. 1 Dimple fracture, copper, REM image

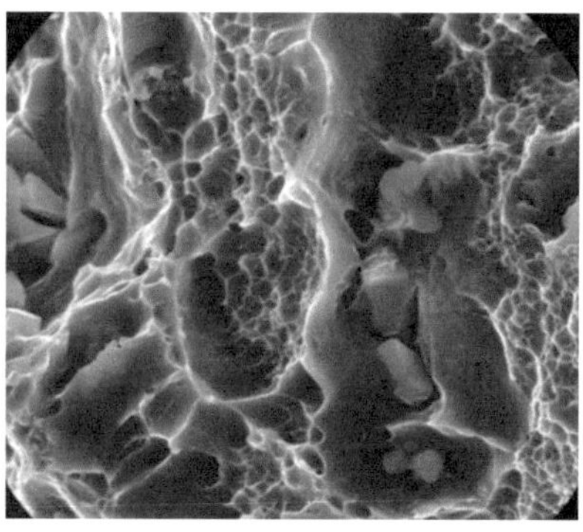

Fig. 2 Dimple fracture, Al-Cu-alloy, REM image

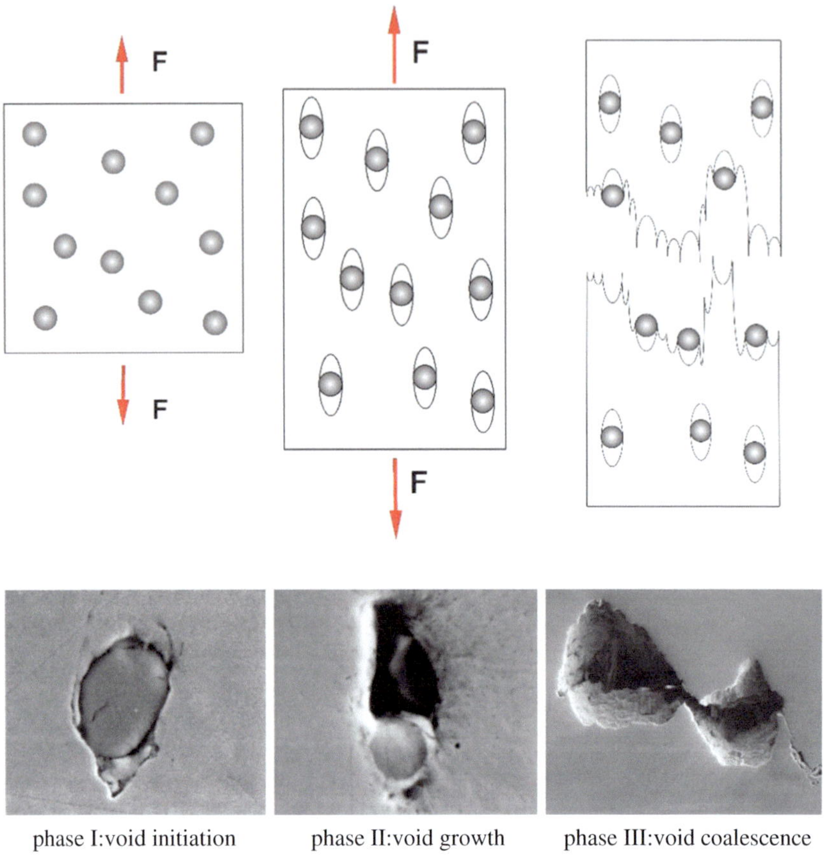

Fig. 3 Phases of dimple fracture [240, 261]

Not only ferritic [113, 240] and austenitic [17, 242] steels show this behaviour but also for example aluminium [60, 241], nickel [11], magnesium [68], cobalt [119] and titanium [166] alloys. This kind of failure behaviour affects many technically pure metals [113, 207] too.

In the following sections, the three phases of failure are described in detail.

2.1 Formation of Voids

The first phase of dimple fracture, i.e. the formation of voids, can occur [240] at:

- particles of a secondary phase [113, 119, 146, 207],
- grain boundaries [120, 166],
- perlite cracks [118, 202, 218] and
- dislocation cell boundaries [108, 290].

In the majority of technical metals, inclusions and precipitations are relevant for the primary formation of voids. A distinction is often made between primary and secondary voids [267]. Primary voids occur at the beginning of the deformation process at relatively small plastic strains. Generally, they show a significant void growth [65] and are relevant for initiating the failure process [268]. Secondary voids often initiate very late within the load history [44]. Compared to primary voids, they are small and play an important role [60, 228] during the coalescence of the voids.

The procedures during the formation of voids at particles are determined by a variety of possible factors, see for example as well [93, 94, 251, 274]:

- atomic structure, micro- und macroscopic defect and homogeneity of the particles
- size and form of the particles
- arrangement of the particles, clustering
- distance of the particles to each other
- different populations of particles
- position of the particles in the microstructure (i.e. on grain boundaries)
- the orientation of the slip or cleavage planes in the matrix and in the particles
- plastic deformation at and in the particle
- cohesive strength between particle and matrix
- deformation behaviour (elastic/ plastic) of particles and matrix
- stresses and stress multiaxiality in particle and matrix
- grain size of the matrix
- hardening behaviour of the matrix
- free surface energy
- manufacturing process and damage of particles and/ or matrix which may be caused

In principle voids can be initiated by the following two mechanisms:

- debonding between matrix (see Figs. 4 and 5) and particle and
- particle fracture (see Figs. 6 and 7).

It depends on various factors whether voids are initiated by particle fracture or by debonding. An essential factor is the particle shape.

In loading direction, elongated particles often fail by particle fracture [14, 93, 111, 119, 146, 149, 213]. It seems as it is not so important whether the particles behave ductile, like i. e. certain manganese sulphides [119, 123] or more brittle, like i. e. carbides [93, 146, 149]. The extent to which the fragments of an inclusion remain attached to the matrix [29, 146] or whether they debond with increasing plastic deformation from the matrix [27, 122] mainly depends on the cohesive strength between the matrix and particle as well as the multiaxiality of the stress state [29].

At more spherical particles debonding is often observed between particle and matrix [14, 111, 119, 240]. But also at elongated or sheet-like particles, which are arranged perpendicular to the major principal stress, voids can be caused by

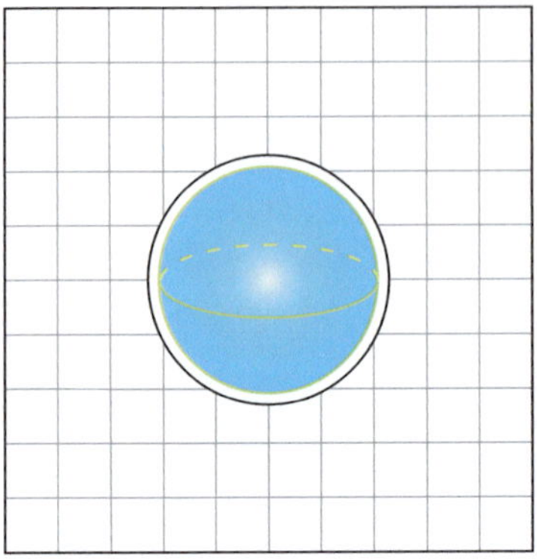

Fig. 4 Void initiation by debonding between matrix and particle

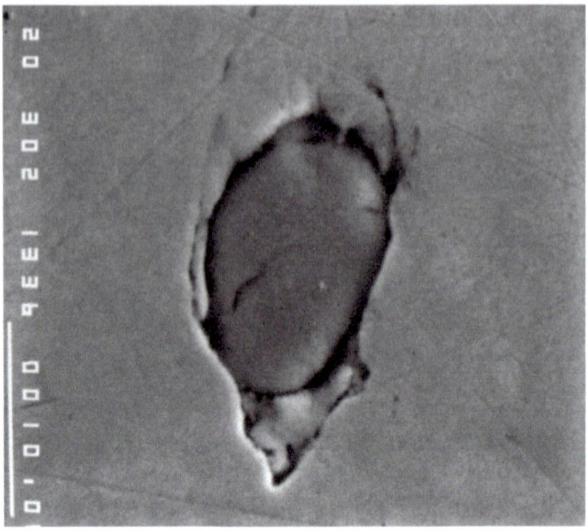

Fig. 5 Manganese sulfid with total debonding between matrix and particle, 20 MnMoNi5-5, REM image

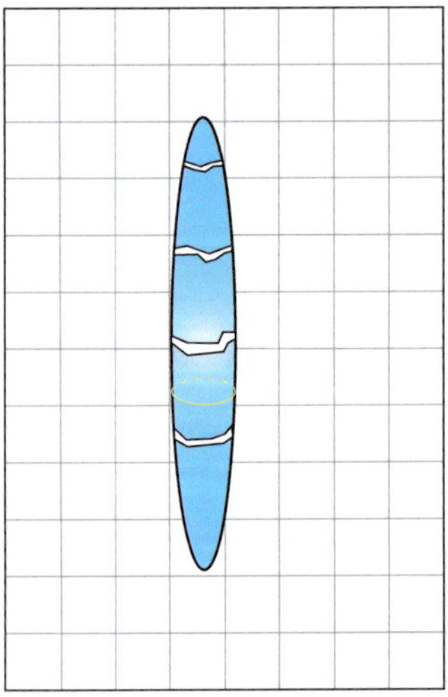

Fig. 6 Void initiation by particle fracture

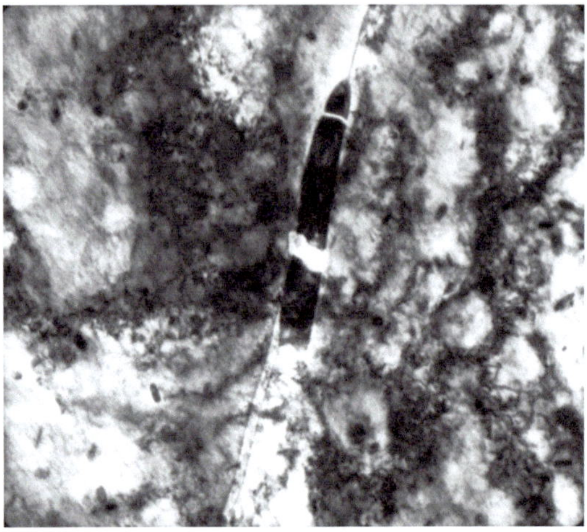

Fig. 7 Cracked iron carbide, steel, TEM foil

decohesion beween particle and matrix [77, 123]. Whether the debonding is only partially [93] or completely depends among other things on the applied stress condition. At high multiaxiality a complete debonding is observed frequently, while low stress multiaxiality ($\sigma^{m}/\sigma^{vM} < 2/3$) only leads to partial separation [29].

However, the link between particle shape and failure mechanism described before is not mandatory. Initiation by particle cracking can be found as well at perfectly spherical particles [122, 125] and debonding at elongated ones.

Experimental studies and simulations on the microstructure level show that not only the cohesive strength between particle and matrix, toughness and shape of the particles have an influence on the initiation mechanism (fracture or debonding). Soppa et al. [250, 251] show that both the arrangement and the volume fraction of the particles as well as the hardening behaviour of the matrix influence the mechanism leading to void initiation.

2.1.1 Which Deformations Lead to Voids?

The presence of plastic deformations [213] is considered as a prerequisite for void initiation. During plastic deformation, dislocations accumulate at particles which can be deformed worse than the matrix [120, 171, 261] and slip bands are blocked [113]. These processes lead to stress peaks at and in the particles. Void initiation will take place if these stress peaks are higher than the cohesive strength between particle and matrix or the tensile strength of the particle. If the yield strength of the particle is lower than the one of the surrounding matrix, slip bands in the inclusion are blocked at the interface between particles and matrix, thus leading to a stress peak [75, 291].

Void formation by particle fracture or debonding from the matrix can either be observed soon after exceeding the yield strength [147, 190] or only after large plastic deformation [149, 274]. At which deformation void initiation at particles will be observed depends primarily on

- cohesive strength between particle and matrix,
- deformation behaviour of the particles
- deformation behaviour of the matrix and
- the degree of stress multiaxiality.

Many materials already contain voids initiated during the production process [8, 27, 63, 70, 236, 291].

A very early void initiation is often observed at particles that can deform plastically. As examples, manganese sulfides in steels, [8, 27, 63, 70, 236, 291], or spherical graphite cast iron [258] have to be mentioned. Numerous authors observe void initiation at zero or very low plastic deformations in steels containing manganese sulphides [8, 30, 77, 146, 213, 293]. Likewise sometimes, a very early void initiation can be observed at brittle particles [27, 93, 113, 250].

In other particles with a high cohesive strength between particle and matrix, very large deformations are needed to initiate voids. For iron carbides in steel, strains of over 50 % were measured until the void initiation started [7, 93, 149]. Even at very small manganese sulfides ($\phi < 0.22$ μm)[1] in a high-strength steel, voids are initiated at strains of about 50 %.

[1]Manganese sulfides at which voids initiate at low plastic deformations can have a size larger than several μm.

The onset of void formation is also affected by the yield strength of the matrix material. For example, Huber et al. [125] observed at a near eutectic Al-Si casting alloy that for a low-strength version voids initiation occurs by particle fracture at much higher strains than for a high-strength version of the alloy.

In addition the multiaxial stress state affects the amount of plastic deformation which is necessary for void initiation.

2.1.2 At Which Particles Void Initiation Takes Place?

At which particles in a given alloy void initiation is observed is mainly determined by the chemical composition, the origin and the size distribution of the particles. The strength of the interface between particles and matrix depends not only on the material characteristics of the particle, but also on the chemical composition and the micro-structure of the matrix. In [93] for example, it is observed in a steel with globular cementite that void formation by debonding primarily occurs at particles on grain boundaries.

Depending on the material void initiation can often be observed simultaneously at very different precipitations and inclusions. For steels, these are often impurity-related inclusions, such as manganese sulfides and oxides as well as precipitations in combination with carbon and nitrogen. For example in the inclusions of a 22MnMoNi3 7 steel aluminum, calcium, magnesium, titanium and zirconium [44, 212, 215, 267,] are often detected, see Fig. 8.

Most metallographic studies show that void initiation takes place first at above-average-sized particles [60, 65, 75, 93, 113, 114, 228, 268, 289]. "Above-average" does not necessarily reflect the absolute size, but the size in relation to the present distribution. After voids occurred at the above-average-sized particles, voids at smaller particles initiate as well with increasing deformation. With decreasing particle size, larger plastic strains are required for the void initiation [75]. From these observations it can be deduced directly that there is a more or less large initiation interval depending on the size distribution of the particles. This statement contradicts with experimental studies on a copper-chromium alloy in which no size-dependent initiation time point was found [7].

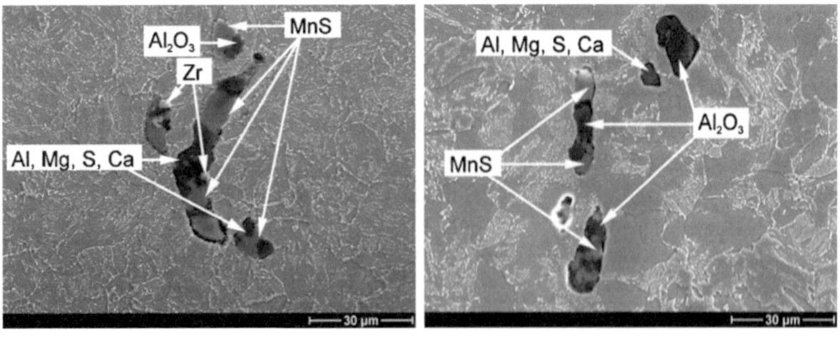

Fig. 8 Void initiation at different particles, material 22NiMoCr3-7, REM image, [215]

It is also observed that particles below a certain size neither have voids [27, 28, 119, 217] nor existing voids grow any more [27]. It is also reported of niobium carbides (> 1μm) in a steel (X52) that no damage occurs because of the strong binding with the matrix [27].

2.2 *Void Growth*

A more or less pronounced void growth follows void initiation, see Figs. 9, 10 and 11. The void volume can grow by a multiple compared to the initial volume [14]. For example, Benzerga et al. [27] observed in the low alloy steel X52 a void growth up to a factor of 50.

2.2.1 Dependence of Void Growth on Stress Multiaxiality

Whether and how strongly voids grow largely depends on the multiaxiality of the stress state. With the help of tomographic studies, this can be directly observed [165]. Many authors state that the void growth increases with increasing plastic deformation and increasing stress multiaxiality [27, 46, 75, 165, 168, 282].

Under uniaxial loading a void, initiated at a particle, deforms in the direction of the external force. A growth perpendicular to the main direction of loading is hardly observed [75]. Thus the volume growth is low. This behaviour can be demonstrated very well with Finite Element calculations. In Fig. 12, one-eighth of a spherical void is shown. While under uniaxial loading the void is only streched in the loading direction, see Fig. 13, an increased volume void growth can be observed [75] under multiaxial loading, see Fig. 14.

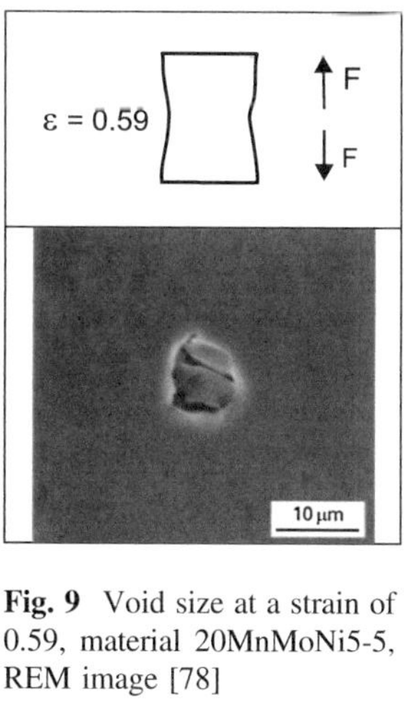

Fig. 9 Void size at a strain of 0.59, material 20MnMoNi5-5, REM image [78]

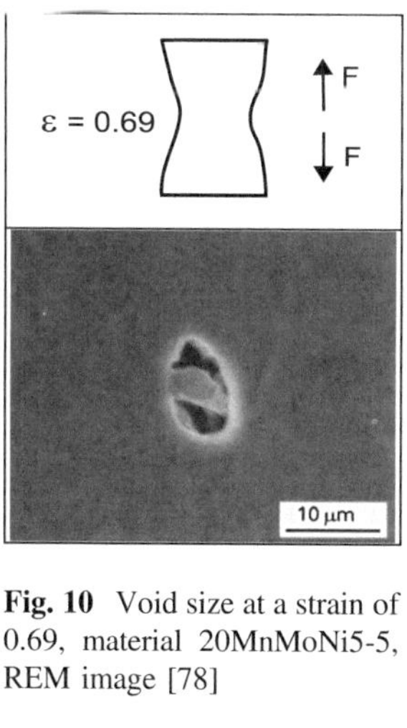

Fig. 10 Void size at a strain of 0.69, material 20MnMoNi5-5, REM image [78]

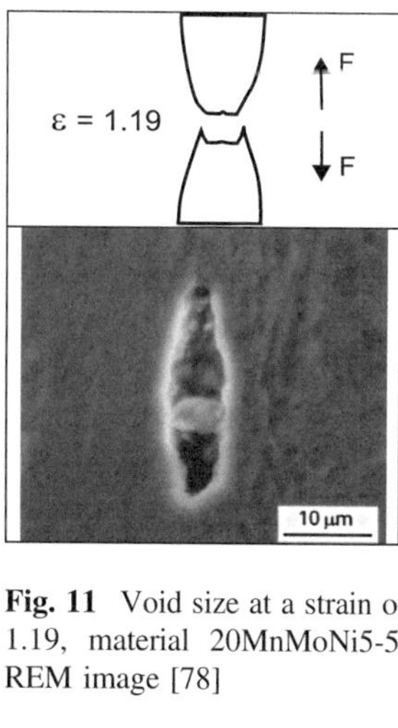

Fig. 11 Void size at a strain of 1.19, material 20MnMoNi5-5, REM image [78]

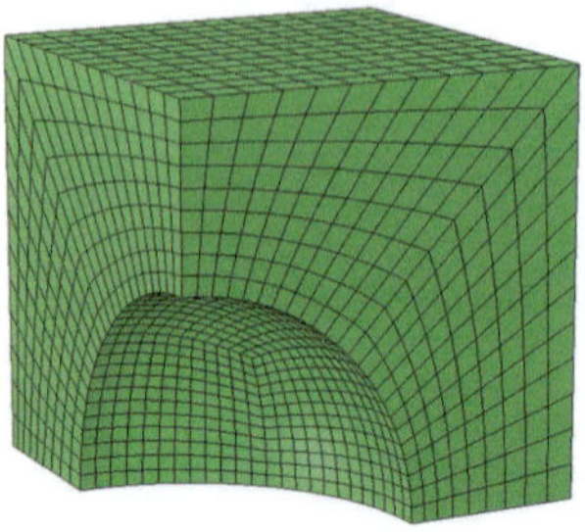

Fig. 12 Initial void shape, FE model, quarter model

Fig. 13 Elongated void caused by uniaxial loading, FE model, quarter model

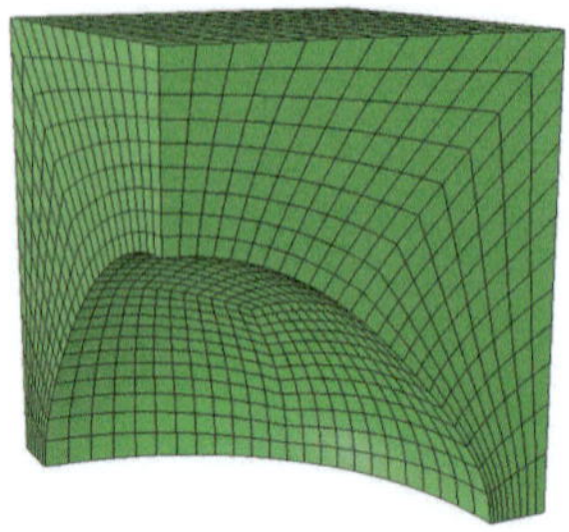

Fig. 14 Spherical void caused by multiaxial loading, FE model, quarter model

It is also assumed that voids loaded with negative stress multiaxiality, i. e. within the pressure range, can become smaller again. Experimental studies showing such a decrease in void volume, however, cannot be found in the relevant literature.

2.2.2 Dependence of Void Growth on Particle Form and Size

The void growth is highly dependent on the absolute size of the void [75]. Large voids grow much faster than small ones. For flat particles with a loading perpendicular to the major particle axes, bigger voids can be formed [27, 213] by planar delamination of inclusions and matrix.

2.2.3 Void Locking

At low stress multiaxiality, elongated voids can be formed as described above. In pure shear (e.g. torsion) even a decrease of the void diameter, perpendicular to the main loading direction, i.e. a closure, is predicted by cell model calculations. The particles leading to void initiation can hinder such a void closure. Due to their finite dimensions, they block the transverse contraction of a void [29]. Benzerga [29] indicates that this type of 'void locking' especially occurs in a multiaxiality range of $\sigma^{m}/\sigma^{vM} \leq 2/3$.

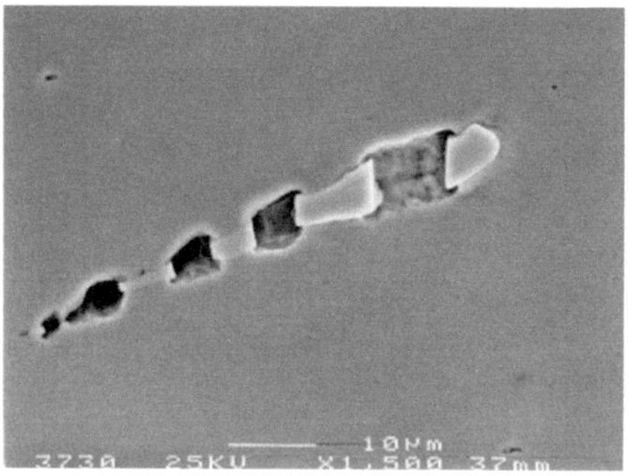

Fig. 15 Void growth at the fracture sites of a niobium carbide, material X10CrNiNb18-10

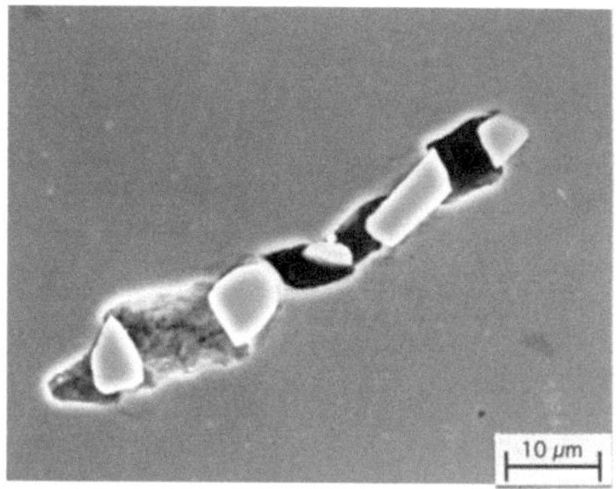

Fig. 16 Void growth at the fracture sites of a niobium carbide, material X10CrNiNb18-10

Even at higher stress multiaxialities where a strong growth in void volume can be observed, the remaining particles can influence the growth behaviour. For example the void volume growth can be hindered by particle fragments with a good adhesion between particle and matrix [149], see Figs. 15 and 16.

2.2.4 Dependence of Void Growth on the Yield Strength and the Hardening of the Matrix Material

The void growth is of course also affected by the yield behaviour of the matrix material. Van Stone et al. [282] show in their literature review that the void growth is more pronounced in high strength materials with low hardening than in comparable low strength materials with high hardening.

2.3 *Coalescence of Voids*

Void growth is limited. Depending on the material and the stress multiaxiality, the materials bridges between the voids are teared apart. This merging of voids is called void coalescence.

By breaking of the materials bridges between the voids, a dimpled structure is being formed on the fracture surface. Within the individual dimples the complete or broken particles which led to void initiation can often still be found, see Figs. 17 and 18.

Dimpled fracture surfaces can be observed in almost all technical metals and alloys, see Figs. 19, 20, 21, 22, 23 and 24. Size and shape of dimples vary strongly depending on the materials and load conditions.

In most cases coalescence of voids is initiated by a strain localization between the large primary voids. Two fundamentally different mechanisms of void coalescence can be observed in experiments and are predicted in simulations:

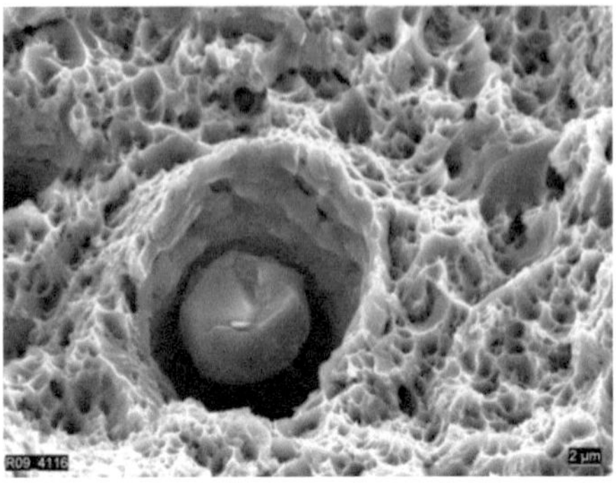

Fig. 17 Dimple with inclusion, material 20MnMoNi5-5, [MPA-archive]

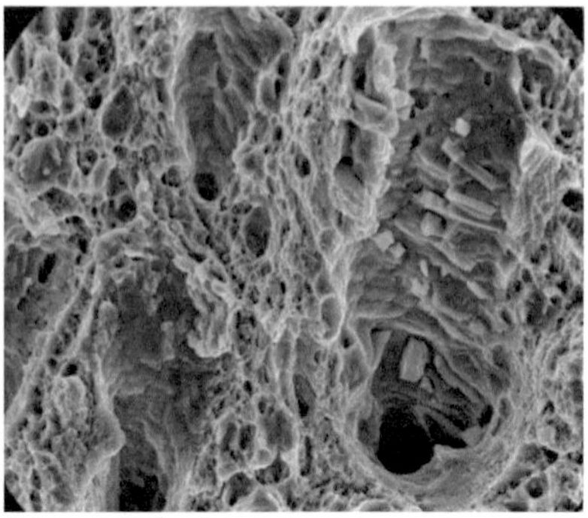

Fig. 18 Fracture surface with voids containing fractured niobium carbids, material X10CrNiNb18-10

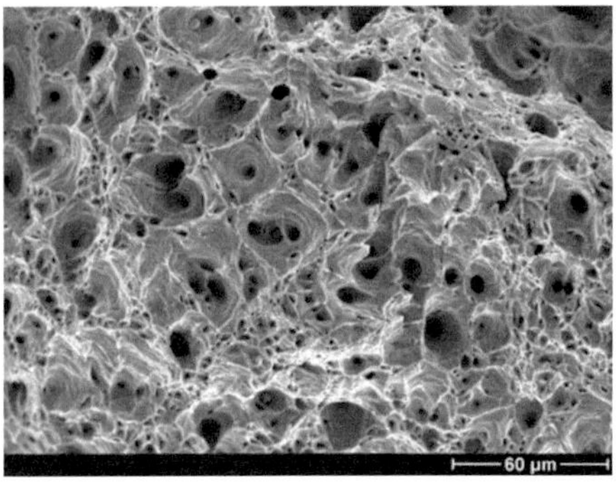

Fig. 19 Fracture surface copper

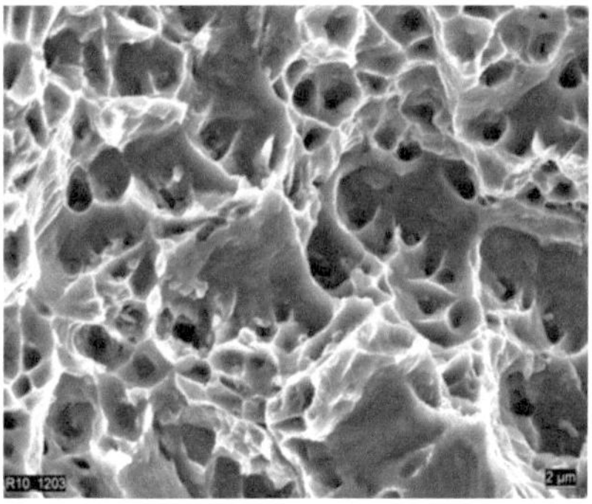

Fig. 20 Fracture surface NiCr70Nb

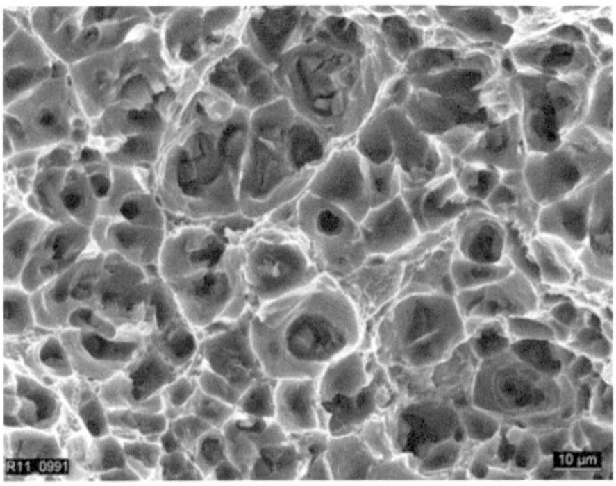

Fig. 21 Fracture surface aluminium

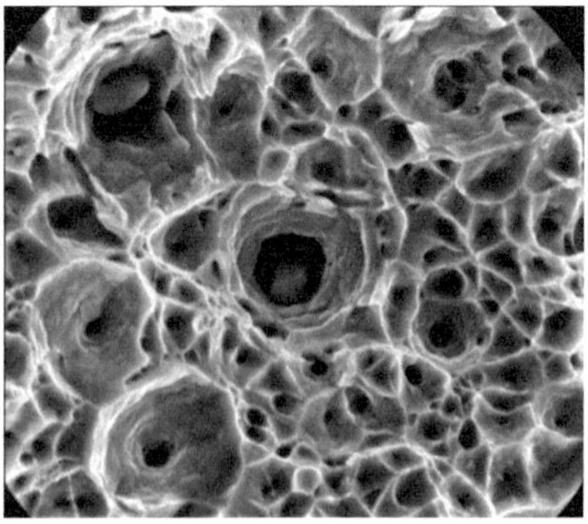

Fig. 22 Fracture surface austenite

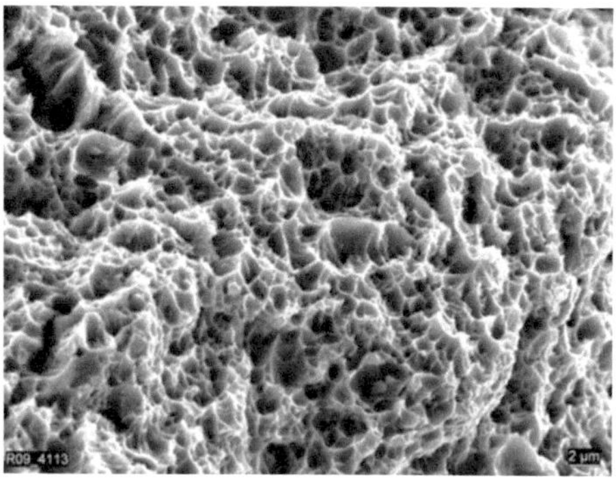

Fig. 23 Fracture surface 20MnMoNi5 5

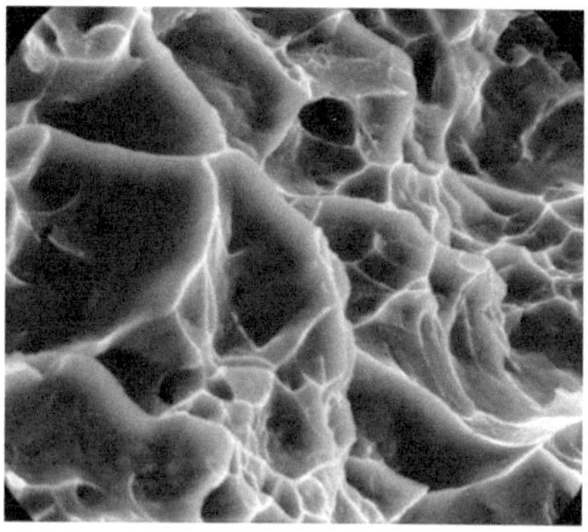

Fig. 24 Fracture surface Ti–Al alloy

- Formation of shear bands between neighbouring primary voids [36, 148, 190], see Fig. 25.
- Plastic collapse of the material bridge between two neighboring primary voids [27, 76, 119, 213, 274], see Fig. 26.

Often secondary voids initiated at much smaller particles are involved in the void coalescence process [117, 171, 228]. The model for this is that due to the strain localisation very large local plastic strains occur, which lead to the initiation of the small secondary voids [75, 274]. For example these secondary voids can be seen on shear bands between the larger voids [28, 75, 76, 148, 268], see Fig. 27. Small secondary voids can play a role, too, when the materials bridges fail by plastic collapse, see Fig. 28. In this failure mode secondary voids have the effect that the large voids do not fully grow together and the residual ligament is not stretched to a tip, but being connected via the secondary voids [119, 274]. Usually not only one of the described mechanisms leads to void coalescence, but several mechanisms are observed simultaneously.

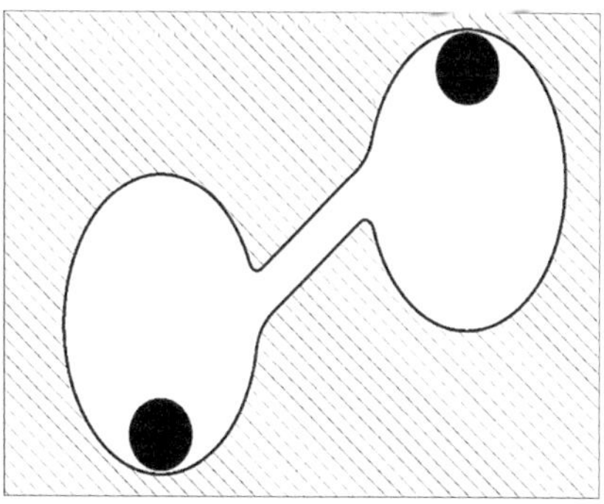

Fig. 25 Shear band between two primary voids

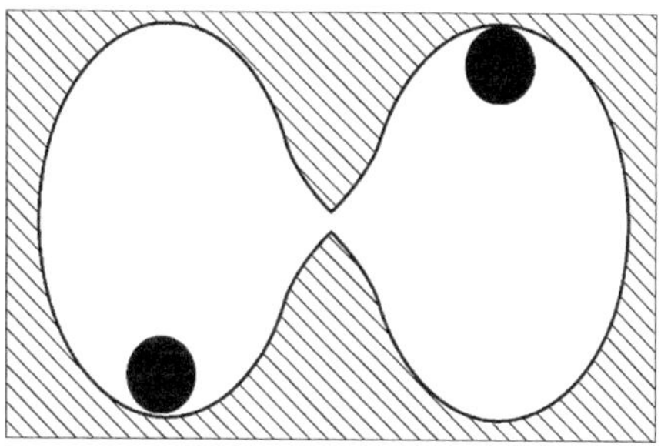

Fig. 26 Plastic collapse of the material bridge between two primary voids

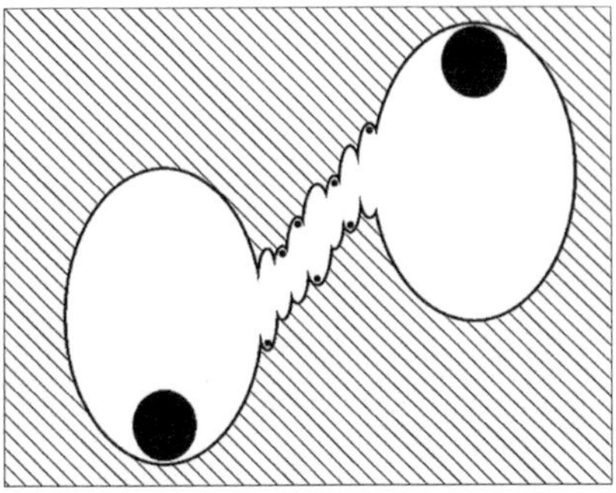

Fig. 27 Shear band with secondary voids between two primary voids

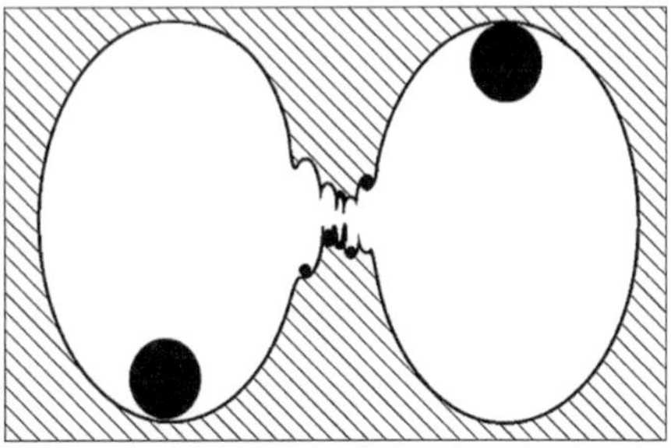

Fig. 28 Plastic collapse of the material bridge through secondary voids between two primary voids

2.3.1 Influence of the Stress Multiaxiality on Void Coalescence

Numerous studies show that void growth and void coalescence depend on the multiaxial stress state [119]. Since the coalescence process occurs in a quite small time interval (sometimes unstable) and material volume, it is difficult to examine it experimentally. However, the coalescence process can often be concluded from the shape of the dimples on the fracture surface. For high multiaxial stress states, as they are observed for example inside a necked round tensile bar, spherical or ellipsoide voids develop. On fracture surface, almost equiaxed dimples are found, see Fig. 29. At loads which are almost uniaxial, with fracture parallel to the greatest shear stress, the dimples are strongly defomed in the direction of shear, see Fig. 30. Finally pure shear stress leads to extremely distorted, squashed and elongated dimples [283], see Fig. 31. Baechem [24] even describes 14 possible honeycomb shapes.

Metallographic examinations show that the stress multiaxiality has a direct influence on the mechanism of void coalescence and the formation of secondary voids. Bandstra et al. [16] examine the ductile failure behaviour of a HY-100 steel with elongated manganese sulphides perpendicular to the loading direction. At higher multiaxiality $\sigma^{m}/\sigma^{vM} > 1$ the authors mainly find void coalescence perpendicular to the main direction of loading with the formation of small secondary voids, whereas they observed a shear failure with secondary voids between the large primary voids at lower multiaxiality. These results are confirmed by Bron & Besson [58] for an aluminium alloy AL2024. The influence of stress multiaxiality on the formation of secondary voids/ dimples is described by Besson et al. and Tanguy et al. [44, 267]. The authors show for steel X100 that the number of small secondary dimples on the fracture surface increases with decreasing stress multiaxiality.

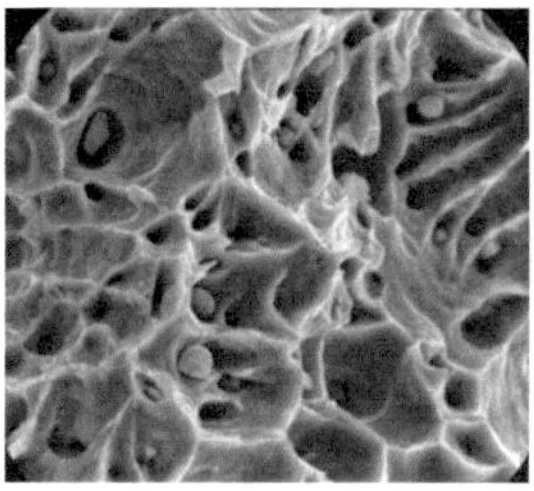

Fig. 29 equiaxed dimples, material copper [MPA archive]

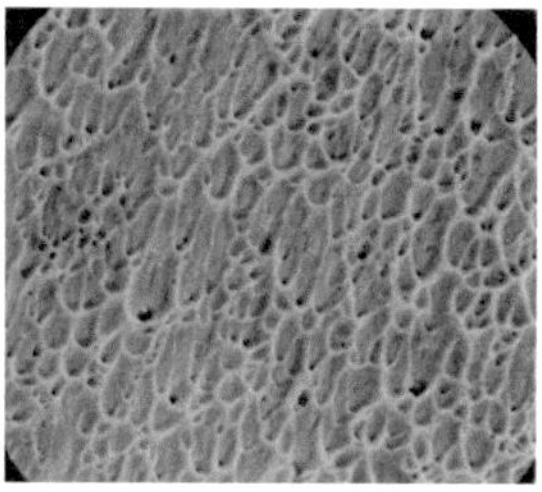

Fig. 30 dimples from tension-shear, material austenite [MPA archive]

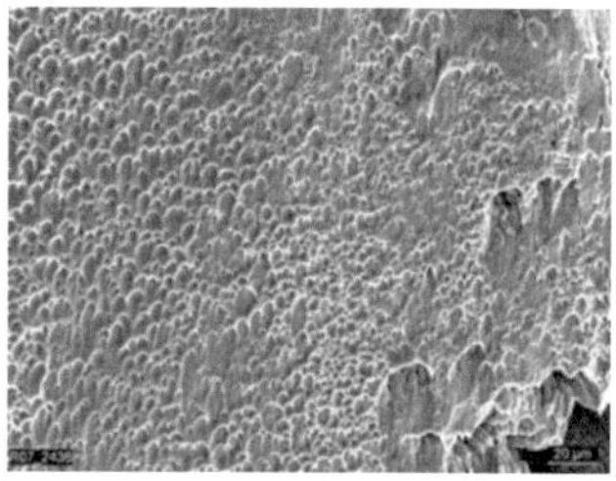

Fig. 31 dimples from pure shear, material HDT1200M [MPA archive]

2.3.2 Effect of Void Formation on the Void Coalescence

In general, if the void distance (perpendicular to the loading direction) is small, a failure by plastic collapse of the material bridges occurs [25]. If the voids are oriented rather under 45° with larger distances in between, the probability of shear band formation increases [25]. In elongated voids it seems as if orientation and rotation of the voids during deforming affect the coalescence mechanism [28]. Systematic studies on the influence of the void arrangement on the failure mechanism are hardly feasible because due to the manufacturing process the voids have a random position. Samples with artificially laser-drilled voids in the form of holes [284–287] offer a possibility to study the impacts more systematically. To analyze the void arrangement, Weck investigates two different types of tensile specimens a) two holes arranged perpendicular (90°) to the loading direction and b) two holes shifted under 45°. For the 90° arrangement Weck shows that coalescence takes place by a failure of the material bridge perpendicular to the loading direction, seeFig. 32. The specimen with the 45° shifted holes fails by shear band formation, see Fig. 33. On both fracture surfaces 'secondary' dimples were found.

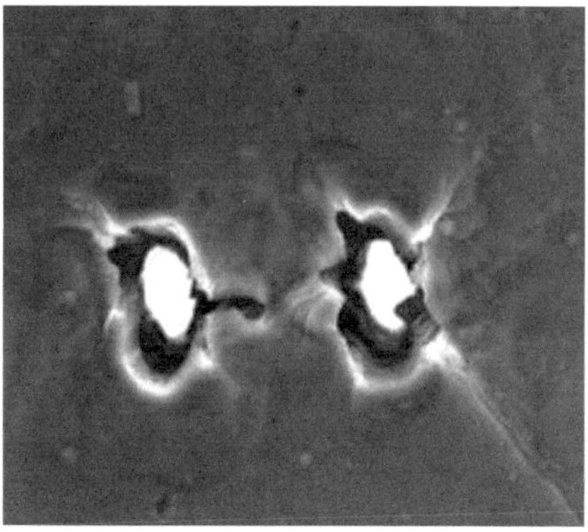

Fig. 32 Void coalescence of two voids arranged under 90° to loading direction, material Al 5052 [285]

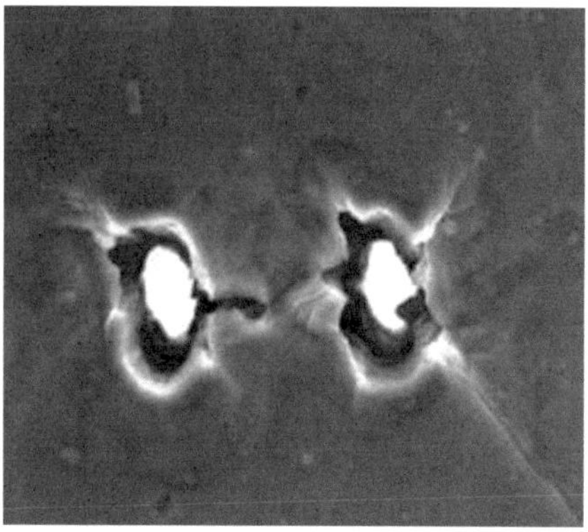

Fig. 33 Void coalescence of two voids arranged under 45° to loading direction, material Al 5052 [285]

2.3.3 Influence of the Materials on the Coalescence

The microstructure of the materials as well as size and composition of the particles affect the coalescence mechanism. Cox and Low [75] examine the failure behaviour of four different high-strength steels:

1. A commercial steel type AISI 4340 with large manganese sulphides (∅ ≈ 7.5 μm) and much smaller iron carbides. The volume fraction of manganese sulfides is 0.14 %.
2. A high purity version of the steel type AISI 4340 with slightly smaller manganese sulfides (∅ ≈ 4.2 μm) and also smaller iron carbides. The volume fraction of manganese sulfides here is only 0.06 %.
3. A commercial maraging steel (18Ni, 200 grade) with titanium carbonitrides (∅ ≈ 8.6 μm) and much smaller particles of an intermetallic phase. The volume fraction of titanium carbonitrides is 0.21 %.
4. A high purity version of maraging steel 18Ni, 200 grade with smaller titanium carbonitrides (∅ ≈ 3.0 μm) and much smaller particles of an intermetallic phase. The volume fraction of titanium carbonitrides is 0.09 %.

All four steels have a comparable yield strength of about 1400 MPa. While the void coalescence in the maraging steel mainly takes place by direct merging of the voids, in the AISI 4340 versions shear bands with secondary voids are observed. In the AISI 4340 steels, the small iron carbides are involved in void coalescence, while the small intermetallic phases have no direct influence on the convergence of the maraging steel.

3 Continuum Mechanical Models for Failure by Void Initiation, Growth and Coalescence [244]

To predict the macroscopic deformation and failure behaviour (crack initiation, crack growth and instability) of components and assembly groups, macroscopic continuum mechanical approaches are needed. Calculations on the level of microstructure or even on the atomic level which simulate single voids and/ or details from microstructure [13, 203, 206, 214, 225, 245, 251, 252], see Figs. 34 and 35, are not applicable to real components because of the huge computation time and time consuming modelling.

Based on the derivation of the macroscopic materials models it can be distinguished between empirical and micromechanical-based models:

- Empirical models approximate the experimentally observed macroscopic behaviour. These approaches are also called phenomenological or heuristic models. Due to the number of introduced material-dependent variables a more or less complex material behaviour can be approximated. Drawback of these models is that the used material parameters have no direct reference to

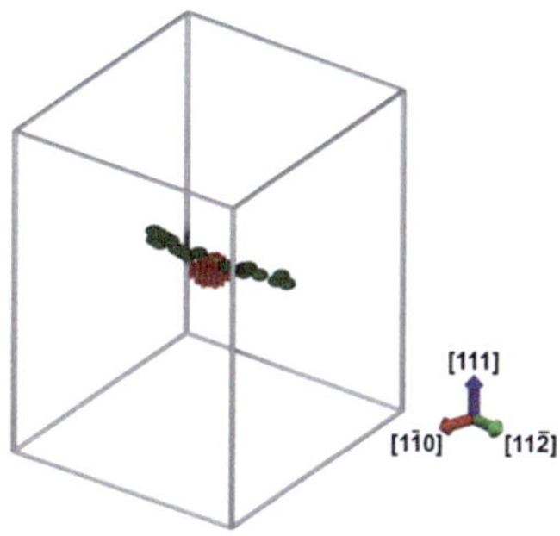

Fig. 34 Blocked dislocation by a copper precipitation (∅ = 1 nm) in cbc iron, atomistic simulation [172]

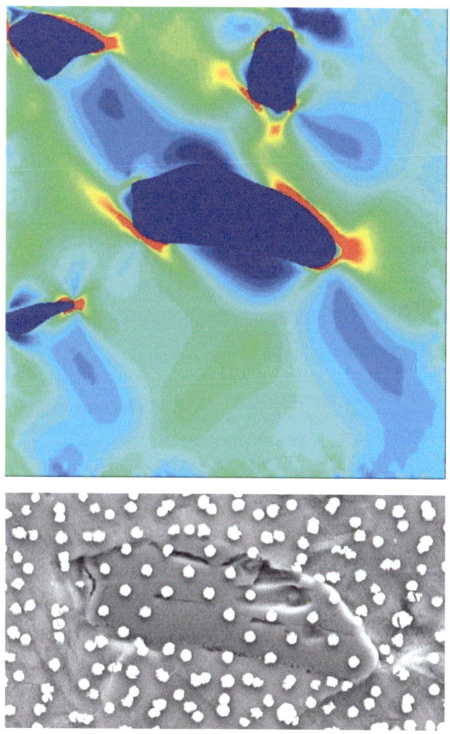

Fig. 35 Strain distribution in a dual phase Al-Al_2O_3 alloy; plane simulation [251]

material-physical parameters. The transferability of the empirical parameters to other load cases and materials is not given a priori. For example the following models are referred to:

- Cockcroft and Latham [73]
- Oyane [185]
- Gao et al. [106],
- Chaouadi et al. [71]
- Bai and Wierzbicki [15].

A discussion of this kind of models is found in [233].

- Another approach to describe the mechanical behaviour of materials is the use of so-called micromechanical-based models. These models try to describe the discontinuous micromechanical processes on a macroscopic level with mechanical and/ or thermo-mechanical approaches. For this purpose, the discontinuous stress and strain field is homogenized and described with continuum mechanical approaches. The advantage of this class of models is that a transferability of the material law and the used parameters to other loading situations is more likely. [86, 181, 182, 204]. Severe disadvantages are the simplifications needed for the derivation, for example ideal plastic material behaviour, axially symmetric voids and so on. As a result the transferability is limited.

In the following the focus of interest will be on the micromechanical-based material models. In case of dimple failure these models have to describe the microscopic processes

1. void initiation
2. void growth
3. void coalescence which leads to the formation of a micro crack

by means of continuum mechanical approaches. In general, each phase is described by a separate model. However, there are models which describe two or all three phases simultaneously. In this chapter, the classic models from the late 1960s, 1970s and 1980s are presented. Later published damage models are based almost exclusively on these classic formulations. Recent systematic reviews of Chaboche et al. [69] and Besson [46] confirm this.

3.1 Models Describing Void Initiation

The void initiation models described below simulate the formation of a void by decohesion (detachment) of a particle from the surrounding matrix. Micromechanical models that explicitly describe the fracture of a particle are not very common. An exception is for example the model of Huber et al. [125]. As being described later the decohesion models can often similarly be used to describe particle fracture. In principle, the decohesion models can be divided into three groups [6, 240], such as:

- stress criteria
- strain criteria
- energy criteria

If assuming void initiation by decohesion the normalized void volume f_0 is usually set equal to the normalized particle volume ${}^{inc}f$:

$$f_0 = {}^{inc}f = \frac{{}^{inc}V}{V_0} \tag{1}$$

In the following, the normalized void volume f_0 and the normalized particle volume ${}^{inc}f$ will be called void and particle volume.

In void formation by particle fracture the formed void volume f_0 is much smaller than the corresponding particle volume ${}^{inc}f$.

$$f_0 \ll {}^{inc}f = \frac{{}^{inc}V}{V_0} \tag{2}$$

3.1.1 Void Initiation Model of Tanaka et al.

Tanaka et al. [269] derived an energy based void initiation criterion. They assumed an elastic particle with radius R (in cm) in a plastic matrix. They derived a critical strain ε^c above which a void will initiate. Due to simplification they assumed that plastic strain in loading direction is higher than 1 % and that the macroscopically applied stress $\tilde{\sigma}$ is less than E/1000. If the elastic modulus of the particle is smaller than that of the matrix ε^c can be calculated as:

$$\varepsilon^c > \beta\sqrt{\frac{1}{R}} \tag{3}$$

The advantage of the Tanaka et al. approach is that a solution can also be found for particles with a higher elastic modulus than the matrix:

$$\varepsilon^c \geq \beta\sqrt{\frac{1}{\alpha R}} \quad \text{mit} \quad \alpha = \frac{{}^{inc}E}{{}^{mat}E} \tag{4}$$

R describes the radius of an inclusion and α the ratio between the elastic modulus of the inclusion and the matrix. β is a material dependent constant which can be calculated as follows:

$$\beta = \sqrt{2}\sqrt{\frac{48 \times 10^{-9}\{(7 - 5^{mat}\nu)(1 + {}^{inc}\nu) + \alpha(1 + {}^{mat}\nu)(8 - 10^{mat}\nu)\}}{(7 - 5^{mat}\nu)^2\{2(1 - 2^{inc}\nu) + \alpha(1 + {}^{mat}\nu)\}}} \\ * \sqrt{(7 - 5^{mat}\nu)(1 - {}^{inc}\nu) + 5\alpha(1 - {}^{mat}\nu^2)} \tag{5}$$

3.1.2 Void Initiation Model of Argon, Im and Safoglu

Argon, Im and Safoglu [6, 8] derived a stress-based criterion to predict void initiation by particle-matrix decohesion. For the initiation of voids by decohesion, sufficient energy for the creation of new surfaces must be available and the critical stress ${}^{\mathrm{deb}}\sigma^{\mathrm{c}}$ necessary for the debonding must be reached. The derived debonding stress ${}^{\mathrm{deb}}\sigma$ can be calculated with the following equation:

$$ {}^{\mathrm{deb}}\sigma \approx \mathrm{k}\tilde{\sigma}^{\mathrm{vM}} + \tilde{\sigma}^{\mathrm{m}} \geq {}^{\mathrm{deb}}\sigma^{\mathrm{c}} \tag{6} $$

In this equation $\tilde{\sigma}^{\mathrm{vM}}$ describes the macroscopic equivalent stress and $\tilde{\sigma}^{\mathrm{m}}$ the macroscopic hydrostatic stress. The constant k characterises the particle shape. For spherical particles k became 1.

3.1.3 Void Initiation Model of Gurson

The derivation of Gurson's void initiation model [115] is mainly based on the experimental work of Gurland [114] and the theoretical work of Argon et al. [6]. Gurland [114] observed in an uniaxially deformed steel with 1.05 % carbon content and coagulated cementite that the number of initiated voids depends approximately linear on the plastic equivalent strain. From this observation Gurson derives his strain based void initiation criterion in which the void initiation rate $\dot{f}^{\mathrm{nuc}}$ is proportional to the plastic equivalent strain rate $\dot{\varepsilon}^{\mathrm{p^v}}$. In addition, Gurson slightly modified the stress based void initiation criterion by Argon et al. [6]. Finally he defined a criterion that takes into account both strain and stress induced void formation:

$$ \dot{f}^{\mathrm{nuc}} = \mathrm{M}_1 \dot{\varepsilon}^{\mathrm{p^v}} + \mathrm{M}_2 \frac{\dot{\tilde{\sigma}}^{\mathrm{m}}}{(1-\mathrm{f})} \tag{7} $$

M_1 and M_2 are material dependent functions which should describe the interaction of particles. This model suggested by Gurson is of pure empirical nature.

3.1.4 Void Initiation Model of Goods and Brown

Goods and Brown derived a strain based micromechanical model to describe the void initiation [111]. By superposition of tension and a hydrostatic stress field as well as further simplification, the following equation for the local plastic limit strain $\varepsilon^{\mathrm{p^c}}$ was found in dependence on the particle radius R:

$$ \varepsilon^{\mathrm{p^c}} \geq \mathrm{KR}\left({}^{\mathrm{deb}}\sigma^{\mathrm{c}} - \frac{3\tilde{\sigma}^{\mathrm{m}} - \tilde{\sigma}_{\mathrm{I}}}{3} \right)^2 \tag{8} $$

Factor K can be calculated from the void volume ${}^{inc}f$ und Burgers vector b_i. Due to their dislocation based approach the authors assumed that the criterion is valid for particles up to a diameter of 2 μm.

3.1.5 Void Initiation Model Acc. to Chu and Needleman

The model suggested by Chu and Needleman [72] is based on the theories of Gurson [115] and Argon et al. [6]. They defined an empirical strain based and a stress based initiation model.

For the definition of their strain based initiation criteria Chu & Needleman started with Eq. 9 published by Gurson [115]. While Gurson assumed a proportional relationship between void initiation rate $\dot{f}^{nuc}$ and plastic equivalent strain rate $\dot{\varepsilon}^{p^v}$, Chu and Needleman proposed a dependency in the form of a normal distribution. For the void initiation volume rate they obtained their often cited relationship [3, 12, 39, 66, 72]:

$$\dot{f}^{nuc} = C\dot{\varepsilon}^{p^v} \tag{9}$$

$$C = \frac{\psi}{s^{\varepsilon}\sqrt{2\pi}} e^{-\frac{1}{2}\left(\frac{\varepsilon^{p^v} - \varepsilon^N}{s^{\varepsilon}}\right)^2} \tag{10}$$

ε^N is the expected value of the equivalent strain at void initiation and s^{ε} the standard deviation of the function. ψ is determined in a manner that the resulting void initiation volume associates with the consistent materials specific value. The reason why they assume a Gaussian distribution is not discussed by the authors. At high stress multiaxiality the void initiation predicted with Eq. 10 approaches zero.

For the derivation of their stress based criterion Chu and Needleman used the Argon et al. [6] criterion. Based on the work of Gurson [115] the authors derived the following equation with the assumption of a normally distributed void initiation volume rate $\dot{f}^{nuc}$:

$$\dot{f}^{nuc} = K(\dot{\bar{\sigma}}^{v} + \dot{\bar{\sigma}}^{m}) \quad \text{with} \quad K = \frac{\kappa}{s^{\sigma}R_e\sqrt{2\pi}} e^{-\frac{1}{2}\left(\frac{\bar{\sigma}^v + \bar{\sigma}^h - \sigma^N}{s^{\sigma} R_e}\right)^2} \tag{11}$$

s^{σ} represents the standard deviation and R_e the the yield stress. σ^N is the expected value of the normal distribution. κ is again determined in a way that the resulting void initiation volume is consistent with the experimental value.

It is possible to describe the decohesion process with cell models calculations. In a later work Needleman [178] showed that it was not possible to describe this micromechanical process (simulated with cell model calculations) correctly with both criteria proposed by Chu and Needleman [72]. To take better account of the stress multiaxiality he introduced another material dependent constant c to consider the influence of stress multiaxiality [178]:

$$\sigma^{N} = \tilde{\sigma}^{vM} + c\,\tilde{\sigma}^{m} \tag{12}$$

3.1.6 Void Initiation Model of Beremin

In the derivation of their void initiation criterion, the research group Beremin assumed an elastic particle in an elastic-plastically deformable infinite matrix [34]. They used a self-consistent approach [121, 138], in which the matrix material has the properties of the entire material. In their definition the entire material is the matrix material with particles.

$$^{deb}\sigma = \tilde{\sigma}_{I} + \chi\left(\tilde{\sigma}^{vM} - R_{e}\right) \geq {}^{deb}\sigma^{c} \tag{13}$$

If the debonding stress $^{deb}\sigma$ reaches the critical value $^{deb}\sigma^{c}$, a void initiates. The stress $\tilde{\sigma}_{I}$ refers to the macroscopic largest principal stress, $\tilde{\sigma}^{vM}$ to the macroscopic equivalent stress and the parameter χ to the shape of the particle.

3.1.7 Void Initiation Model of Huber et al.

In contrast to the authors discussed so far, Huber et al. [125] defined a model for predicting particle fracture. Their void initiation criterion combines the micromechanical-based void initiation criterion of Beremin [34] with the simple empirical strain criterion of Gurson [115]. They assumed that the maximum principal stress triggers particle fracture and that there is a dependence of the fracture stress on the particle size. They justified this size dependence with an increasing number of defects in larger particles due to the larger volume. Their model is defined as follows:

- Void initiation starts if the criterion of the Beremin model (Eq. 13) is fulfilled for the largest particles:

$$^{disb}\sigma^{c} \leq {}^{disb}\sigma = \tilde{\sigma}_{I} + \chi\left(\tilde{\sigma}^{vM} - R_{e}\right) \tag{14}$$

  For initiation the corresponding value of equivalent plastic strain is denoted by $^{start}\varepsilon^{p^{v}}$.

- For $\varepsilon^{p^{v}} > {}^{start}\varepsilon^{p^{v}}$ a phase of continuous void formation follows. To describe this phase they modified the void initiation criterion of Gurson [115]:

$$\dot{f}^{nuc} = C(\varepsilon^{p^{v}})\dot{\varepsilon}^{p^{v}} \tag{15}$$

The constant C from the Gurson equation is defined by Huber et al. as a polynomial function of the equivalent plastic strain:

$$C(\varepsilon^{p^v}) = a_1(\varepsilon^{p^v})^6 + a_2(\varepsilon^{p^v})^4 + a_3(\varepsilon^{p^v})^2 + a_4 \tag{16}$$

- Void initiation is finished if the material dependent equivalent plastic strain ${}^{end}\varepsilon^{p^v}$ is reached.

The factors a_i are defined in such a way that $C({}^{start}\varepsilon^{p^v}) = C({}^{end}\varepsilon^{p^v}) = 0$ is valid. Simultaneously the derivatives at these points should be zero. These requirements docs not have a micromechanical background, but are justified by numerical advantages.

Void Initiation Model of Morgeneyer et al.

Starting point for the derivation of the void initiation model by Morgeneyer et al. [173] are experimental investigations on the failure behaviour of thin metal sheets. Several studies [44, 57, 58] showed that for shear fracture the number of secondary voids increases sharply. Morgeneyer et al. postulated that the formation of the secondary voids depends on the multiaxiality of the strain state. They assumed that the tendency for shear fracture can be described with the Lode angle $\mu_{\dot{\varepsilon}}$ [162]:

$$\mu_{\dot{\varepsilon}} = \frac{\dot{\varepsilon}_{II}}{\dot{\varepsilon}_{I} - \dot{\varepsilon}_{III}}, \tag{17}$$

where $\dot{\varepsilon}_I$, $\dot{\varepsilon}_{II}$ and $\dot{\varepsilon}_{III}$ describe the principal strain rates.

The authors assumed that the initiation rate of secondary voids is particularly high for a Lode angle close to 0 (pure shear). As Gurson and Chu & Needleman they accepted, that void initiation further depends on the equivalent plastic strain rate. Starting from the initiation equation proposed by Gurson, the authors proposed the following empirical formula to describe the initiation of secondary voids:

$$\dot{f}^{nuc} = A_0 e^{-\left(\frac{\mu_{\dot{\varepsilon}}}{\mu_{\dot{\varepsilon}}^0}\right)^2} \dot{\varepsilon}^{p^v} \quad \text{for } \varepsilon^{p^v} > {}^{start}\varepsilon^{p^v}, \tag{18}$$

where A_0 is a material dependent constant. $\mu_{\dot{\varepsilon}}^0$ describes the shape of the normal distribution of the void initiation. Void initiation is predicted only for equivalent plastic strains $\varepsilon^{p^v} > {}^{start}\varepsilon^{p^v}$.

3.1.8 Void Initiation Caused by Particle Fracture

Especially in materials with pronounced elongated and branched particles void formation is often caused by particle fracture. The previously presented void initiation models, with the exception of the Huber criterion, are all based on the assumption that the voids are formed by a separation of matrix and second phase particles. If non plastically deformable particles are assumed, then the results that have been achieved for failure by matrix detachment can be transmitted very easily to the failure mechanism of particle fracture.

Relevant for the fracture of a brittle particle is the major principal stress perpendicular to the respective cross section. If the particles are linear-elastic it can be assumed approximately that stresses and strains are constant in the particle, for example [88]. In this case, the first principal stress in the particle ${}^{inc}\sigma_{I}$ is equal to the debonding stress ${}^{deb}\sigma$, see for example [8, 195].

3.2 Models Describing Void Growth

Due to their derivation micromechanical-based models for describing void growth can be divided into two groups [200, 240]:

1. The growth of a cylindrical, spherical or ellipsoidal void in a finite or infinite matrix is explicitly described with continuum mechanics based formulations.
2. The behaviour of porous materials is described with thermodynamic and continuum mechanics laws. Within these approaches no single voids are examined and due to this the derivation of the model is not so clear but nevertheless the basic laws of mechanics and thermodynamics are fulfilled.

Another classification of void growth models, as it is used in this study, is whether the void growth has an influence on the macroscopic deformation behaviour or not:

1. **uncoupled models**: The void growth is determined by a constitutive relation as a function of stress, strain and internal state variables. However, there is no coupling between void volume growth with the macroscopic material behaviour. The mechanical behaviour of the material is still described with the von Mises yield criterion. The void growth does not affect the hardening and deformation behaviour of the material. The material softening caused by void initiation and growth cannot be described with this class of models. Consequently, strain localisation which is important for the failure process cannot be simulated.
2. **coupled models**: Here, the calculated void volume has a direct influence on the yield behaviour of the material. A high degree of damage leads to a reduced load bearing capacity of the material. Damage becomes an internal state variable in the constitutive equations and thus influences directly the strength and yield behaviour of the material. Strain localisation can be described with this class of models.

In the following some of the well-established uncoupled and coupled void growth models are presented:

3.2.1 Uncoupled Models

McClintock Model

McClintock's void growth model [169] is the first known void growth model. For the derivation of his model McClintock made the following assumptions:

- a cylindrical void with circular cross section in an infinite matrix
- the matrix material is rigid and perfectly plastic
- the material behaviour is described with the von Mises yield criteria
- the material is fully plastic in the unit cell
- 'generalized plane strain' conditions are assumed
- The infinite unit cell is loaded with an axisymmetric radial stress σ_r and with an uniaxial strain ε_z in axial direction.

McClintock derived the following void growth law:

$$\frac{\dot{R}}{R} = \frac{\sqrt{3}}{2}\dot{\tilde{\varepsilon}}_z \sinh\left[\sqrt{3}\frac{\tilde{\sigma}_r}{R_e}\right] - \frac{\dot{\tilde{\varepsilon}}_z}{2} \tag{19}$$

Following Gross and Seelig [112], Eq. 19 can be transformed into an evolution law which describes void volume growth:

$$\frac{\dot{f}}{f} = \sqrt{3}\dot{\tilde{\varepsilon}}^{v} \sinh\left[\sqrt{3}\left(\frac{\tilde{\sigma}^{m}}{R_e} - \frac{1}{3}\right)\right] \tag{20}$$

Rice and Tracey Model

The basis for the development of the void growth model acc. to Rice and Tracey [211] is a spherically shaped void in an infinite matrix. Similar to McClintock, the matrix material is assumed to be rigid and perfectly plastic.

Using these approximations, they obtained their well-known void growth model:

$$\frac{\dot{R}}{R} = \alpha\dot{\tilde{\varepsilon}}^{v} e^{\frac{3\tilde{\sigma}^{m}}{2R_e}} \quad \text{with } \alpha = 0.283 \tag{21}$$

To be able to compare the different void growth models, the radial growth $\dot{R}$ is converted into void volume growth $\dot{f}$:

$$\frac{\dot{f}}{f} = \alpha^{*}\dot{\tilde{\varepsilon}}^{v} e^{\frac{3\tilde{\sigma}^{m}}{2R_e}} \quad \text{with } \alpha^{*} = 0.849 \tag{22}$$

The original Rice and Tracey model is not able to describe a strain hardening material behaviour. Perfect plastic material behaviour is assumed in the derivation of their model. In practice, the yield strength R_e is often replaced by the current yield stress σ_0 or by the equivalent stress σ^v respectively to take material hardening into account. However, the micromechanical background does not cover this assumption. Only for visco-plastic material behaviour a solution of the problem is known [64].

Several other authors which applied the model to experimental data achieved different results for the factor α: The research group Beremin [35] found $\alpha = 0.5$, Shi [246] $\alpha = 0.6 - 0.7$, Pardoen et al. [190] $\alpha = 0.4$ and $\alpha = 0.34$, Maire et al. [165] confirmed the original factor of Rice and Tracey $\alpha = 0.283$, Marini et al. [168] recognized a dependence of α on the volume fraction f_0 and Bandstra et al. [16] recognized a dependence of α on stress multiaxiality.

In the following, the extension by Huang [124] of the Rice and Tracey will be discussed.

Huang also solved the mechanical problem of a spherical void in an infinite matrix. However, compared to Rice and Tracey, he assumed much more complex shape functions for describing the stress field in the matrix. Due to the complexity he solved the problem only numerically. For stress multiaxiality $\sigma^m/\sigma^{vM} \geq 1$ he received a value of $\alpha = 0.427$. This value is close to the experimentally determined Beremin value.

For smaller multiaxiality $\frac{1}{3} \leq \frac{\tilde{\sigma}^m}{R_e} < 1$ Huang suggested the following equation:

$$\frac{\dot{R}}{R} = 0.427\,\dot{\tilde{\varepsilon}}^{v} \left(\frac{\tilde{\sigma}^m}{R_e}\right)^{\frac{1}{4}} e^{\frac{3\tilde{\sigma}^m}{2R_e}} \tag{23}$$

3.2.2 Coupled Models

Lemaitre Type Models

The basis of Lemaitre's models [153–159] are the works of Kachanov [134] and Rabotnov [208]. When calculating the macroscopic stresses $\tilde{\sigma}_{ij}$ these authors took into account the decrease of the loaded cross section caused by voids, see Fig. 36. This results in an increase of the averaged stresses acting in the matrix, the so called effective stress:

$${}^{mat}\sigma_{ij} = \frac{\tilde{\sigma}_{ij}}{1 - D} \tag{24}$$

For the derivation of his model Lemaitre made the following assumptions and simplifications:

- The increase in volume caused by void growth is neglected.
- In the elastic range there is a linear relationship between stresses and strains for the matrix material.

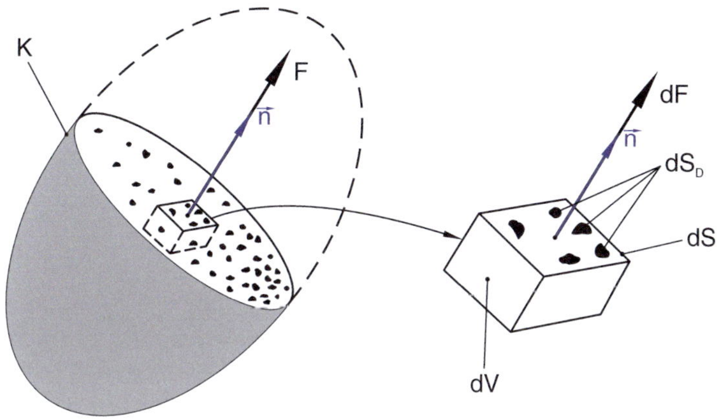

Fig. 36 Volume element with voids, acc. to [159]

- Damage D is coupled with the elastic strains.
- The elastic and plastic strains do not depend directly on each other.
- The damage D and the plastic strains do not depend directly on each other.

For the derivation of his damage law Lemaitre followed thermo-mechanical approaches. For the description of the material state he selected the internal variable r to describe the plastic equivalent strain and D to describe the material damage. The associated state variable R is related to the internal variable r. R describes the material hardening in dependence of r. The state variable associated with D is called Y which is defined in such a way that the product $Y\dot{D}$ is equal to the dissipated energy caused by fracture.

To describe the yield limit Lemaitre used the von Mises yield criterion. Derogating from the original model he replaced the equivalent stress in the total material by the equivalent stress in the matrix, where the function R(r) described the hardening of the matrix material:

$$\varphi = \frac{\bar{\sigma}^{\mathrm{vM}}}{1-\mathrm{D}} - \mathrm{R_e} - \mathrm{R(r)} = 0 \tag{25}$$

To deduce the damage evolution via the normality rule Lemaitre selected a non-associated flow rule. With this, the damage evolution is calculated as:

$$\dot{D} = -\frac{\mathrm{Y}}{\mathrm{A}}\dot{\tilde{\varepsilon}}^{\mathrm{p^v}}\,\mathrm{H}(\tilde{\varepsilon}^{\mathrm{p^v}} - {}^{\mathrm{c}}\varepsilon^{\mathrm{p^v}}), \tag{26}$$

where A is a material dependent scalar. By means of the Heaviside step function H the damage evolution starts when reaching the critical strain ${}^{\mathrm{c}}\varepsilon^{\mathrm{p^v}}$. The associated state variable Y can be calculated with:

$$Y = -\frac{(\tilde{\sigma}^{vM})^2}{2E(1-D)^2}\left[\frac{2}{3}(1+\nu) + 3(1-2\nu)\left(\frac{\tilde{\sigma}^{m}}{\tilde{\sigma}^{vM}}\right)^2\right] \tag{27}$$

In addition, Lemaitre often indicated the following extended evolution equation for damage [153, 156, 157]:

$$\dot{D} = \left(\frac{-Y}{A}\right)^{s0} \dot{\tilde{\varepsilon}}^{p^v} H(\tilde{\varepsilon}^{p^v} - {}^{c}\varepsilon^{p^v}), \tag{28}$$

where s0 is an additional material dependent parameter.

In the following two modifications of the Lemaitre approach will be discussed.

Bonora [50] modified the evolution equation proposed by Lemaitre in the following way:

$$\begin{aligned} \dot{D} &= \alpha \frac{(D_c - D_0)^{\frac{1}{\alpha}}}{\ln\left(\frac{{}^{end}\varepsilon}{{}^{start}\varepsilon}\right)} f\left(\frac{\tilde{\sigma}^{m}}{\tilde{\sigma}^{vM}}\right)(D_c - D)^{\frac{\alpha-1}{\alpha}} \frac{\dot{\tilde{\varepsilon}}^{pv}}{\tilde{\varepsilon}^{pv}} \quad \text{with} \\ f\left(\frac{\tilde{\sigma}^{m}}{\tilde{\sigma}^{vM}}\right) &= \frac{2}{3}(1+\nu) + 3(1-2\nu)\left(\frac{\tilde{\sigma}^{m}}{\tilde{\sigma}^{vM}}\right)^2 \end{aligned} \tag{29}$$

The damage evolution starts with the value D_0 and grows up to the critical value D^c at which failure is predicted. ${}^{start}\varepsilon$ denotes the strain at the beginning of damage and ${}^{end}\varepsilon$ the fracture strain (both under the assumption of uniaxial loading).

In contrast to most of the other models Bonora et al. [48, 50, 51] describe explicitly how to determine the material dependent parameters.

Chaboche et al. [69] modified the Lemaitre approach to take into account the volume growth induced by void growth. Lemaitre neglected this volume change in his model. In the new formulation acc. to Chaboche et al. hydrostatic stresses can induce plastic hydrostatic strains which induce a change in volume.

Gurson Model

The aim of Gurson was to derive a yield criterion and a flow rule for a ductile material containing voids [115, 116]. His model takes into account the influence of the hydrostatic stress on void growth and on the plastic deformation behaviour. The new defined yield criterion represents an upper limit for yielding.

For the derivation of his yield criterion Gurson made the following assumptions for simplicity:

- Gurson defined a unit cell which contains a single void and derived a yield criterion for a spherical void in a spherical matrix.
- The assumed material behaviour of the matrix is rigid/perfectly plastic and is described by the von Mises yield criterion.

With a brilliant derivation Gurson obtained the following upper bound approximation for the yield function:

$$\varphi = \left(\frac{\tilde{\sigma}^{vM}}{R_e}\right)^2 + 2f \cosh\left(\frac{3\tilde{\sigma}^m}{2R_e}\right) - 1 - f^2 = 0, \tag{30}$$

where R_e is the yield strength of the matrix material and not that of the whole unit cell.

With a similar derivation strategy Gurson [115] succeeded to transform approximately the Rice and Tracey model [211], see Eq. 21, into a yield function, that means, into a coupled model:

$$\varphi = \tilde{\sigma}^{vM} + 2\,\alpha\, f\, R_e e^{\frac{3\tilde{\sigma}^m}{2R_e}} - R_e = 0 \quad \text{mit} \quad \alpha = 0.283 \tag{31}$$

The number of published modifications and extensions of the Gurson model is quite high. This depends among other things on the fact, that the original, unmodified Gurson model cannot describe correctly void growth in strain hardening materials [199, 270, 277, 279].

Good, but not exhaustive overviews can be found for example in [29, 32, 46, 232, 235]. Some of the major extensions of the Gurson model will be discussed in the following:

Modification of the Gurson yield function for strain hardening materials

To take strain hardening into account Gurson [115] suggested to replace the yield strength R_e of the perfectly plastic matrix material by the averaged current yield stress σ_0 in the unit cell. To calculate this averaged yield stress σ_0, which is dependent on the material hardening, he suggested the following relationship:

$$\sigma_0 \dot{\varepsilon}^{p^v} = \frac{\tilde{\sigma}_{ij}\dot{\tilde{\varepsilon}}_{ij}^{p}}{1 - f} \tag{32}$$

With this assumption, the Gurson model loses much of his micromechanical background. Pardoen and Besson [32] indicated that this simplification is only approximately permitted if the hardening exponent is less than 0.2.

A much broader suggestion to describe the material hardening is made by Leblond, Perrin and Devaux [152] with her LPD model. They developed an analytical model for the behaviour of a spherical void in a spherical hardening matrix material.

Modification of the Gurson model of Tvergaard

Starting point for the development of the so-called Gurson Tvergaard model (GT model) [277] was the experimental work of Weinrich and French on shear band mechanisms [96, 97, 288]. Tvergaard attempted to simulate the shear band formation mechanisms with plane cell model calculations. For his cell model simulations with finite elements he assumed regularly arranged voids in a hardening elastic-plastic matrix. Due to the assumption of plane strain condition, he simulated the voids as infinitely long cylinders with circular cross sections. To simulate the behaviour of this cell model with a continuum mechanical macroscopic model, Tvergaard used the Gurson model which was actually derived for spherical voids. His results showed that the Gurson model overestimates the results from the cell model calculations. To get a better match with his cell model calculations, Tvergaard introduces three empirical constants q_1, q_2 and q_3:

$$\varphi = \left(\frac{\tilde{\sigma}^{vM}}{\sigma_0}\right)^2 + 2fq_1 \cosh\left(q_2 \frac{3\tilde{\sigma}^{m}}{2\sigma_0}\right) - 1 - q_3 f^2 = 0 \tag{33}$$

Due to the introduction of the 'adjusting parameters' [277] the micromechanical background of the original Gurson model becomes questionable.

For the best parameter combination to describe the behaviour of cylindrical voids in copper and brass he found the following values:

$$q_1 = 1,5, \; q_2 = 1,0 \quad \text{and} \quad q_3 = q_1^2 \tag{34}$$

In the same paper [277] Tvergaard mentions that the q-parameters are dependent on the hardening exponent of the material. Numerous studies show that the q-parameters can be influenced by:

- stress multiaxiality
- material hardening
- void shape
- void arrangement
- void volume
- kinematic hardening
- plastic strains

A more detailed discussion of the factors can be found in [244].

Additional modifications of the Gurson model

In literature, numerous modifications of the original Gurson or GT model can be found. Some examples are mentioned in the following:

- Pan et al. [186] extended the Gurson Tvergaard (GT) model for the description of viscous material behaviour.
- Gologanu et al. [110, 264] derived a void growth model for ellipsoidal voids in an ellipsoidal unit cell.
- Feucht et al. [91] and Ockewitz and Sun [183] tried to improve the model for low stress multiaxiality by coupling the GT model with the Johnson-Cook model [132, 133].
- An alternative suggestion to improve the failure prediction in the shear range is given by Nahshon et al. [177]. Nahshon et al. took the yield function of the GT

model without modifications. To improve the results in the shear range, they expanded the evolution law for the calculation of the void volume growth with an additional term:

$$\dot{f} = (1-\mathrm{f})\dot{\tilde{\varepsilon}}_{ii}^{p} + \mathrm{k}_{\omega}\mathrm{f}\omega(\tilde{\sigma}_{kl})\frac{\tilde{s}_{ij}\dot{\tilde{\varepsilon}}_{ij}^{p}}{\sigma^{vM}} \tag{35}$$

In Eq. 36 the function $\omega(\tilde{\sigma}_{kl})$ depends on the third invariant of the stress deviator.

The kinematic and mixed isotropic kinematic approaches referred to in literature rely almost exclusively on the work of Gurson, Tvergaard, and Gologanu. An overview of these models can be found in Besson and Guillemer-Neel [42].

Rousselier Model

Based on the thermo-mechanical approaches by Lemaitre and Chaboche [154] and own early works [223], Rousselier derived a model for simulating ductile damage. His model [220, 224] describes the elastic-plastic deformation and failure behaviour of a porous material, i.e. a material with voids. Rousselier defined a continuum mechanical yield condition with an associated flow rule:

$$\varphi = \frac{\tilde{\sigma}^{vM}}{\rho} + D\,\sigma^{k}\mathrm{f}\exp\left(\frac{\tilde{\sigma}^{m}}{\rho\,\sigma^{k}}\right) - \sigma_0\left(\tilde{\varepsilon}^{p^{v}}\right) \tag{36}$$

Here σ_0 describes the hardening behaviour of the overall material. D and σ^k are two integration constants resulting from the derivation. The original Rousselier yield criterion was derived for isothermal behaviour and small strains. In [220] Rousselier discusses these two points. He indicates that mainly the yield curve and the material-dependent constant σ^k depend on the temperature. The second integration constant D is assumed to be material independent D = 2. In the form presented here, it is assumed that $\rho \approx 1-\mathrm{f}$:

$$\varphi = \frac{\tilde{\sigma}^{vM}}{1-\mathrm{f}} + \sigma^{k}(\mathrm{T})\mathrm{f}\,\mathrm{D}\,\mathrm{e}^{\frac{\tilde{\sigma}^{m}}{(1-\mathrm{f})\,\sigma^{k}(\mathrm{T})}} - \sigma_0(\tilde{\varepsilon}^{p^{v}},\ \mathrm{T}) \tag{37}$$

The assumption $\sigma^k = \sigma^k(\mathrm{T})$ is not in contrast to the derivation of the original model. Whether D is temperature dependent or not is still an open point in literature.

Sainte Catherine et al. [229] and Poussard et al. [204] extended the Rousselier model for strain-rate dependent material behaviour. They selected a strain rate-dependent flow curve and extended the Rousselier model by defining σ^k as a function of strain rate:

$$\varphi = \frac{\tilde{\sigma}^{vM}}{1 - f} + \sigma^k(\dot{\varepsilon}) f\, D\, e^{\frac{\tilde{\sigma}^m}{(1-f)\sigma^k(\dot{\varepsilon})}} - \sigma_0(\tilde{\varepsilon}^{p^v}, \dot{\varepsilon}) \tag{38}$$

These authors do not violate any assumptions made in the derivation. The theoretical nature of Rousselier the yield function is not affected by their modification.

In contrast to the extensions shown before Tanguy and Besson [265, 266] modified the yield condition more extensively. For the definition of their visco-plastic material law the authors introduced the effective equivalent stress σ^{eff}. In their formulation σ^{eff} describes the behaviour of the matrix material and not of the overall material.

Lorentz et al. [163] showed that the constitutive equations of the Rousselier model are not convex with regard to all variables and therefore the solutions obtained do not have to be unique. To ensure a robust solution they proposed, among other things, to calculate the void growth not only from the plastic hydrostatic strains, but also to take into account the elastic parts. In addition they defined the yield condition and the constitutive equations with Cauchy stresses. Future applications will demonstrate whether and in which cases the new yield criterion provides comparable or better results compared to experiments.

Rousselier et al. [221, 222] formulated a polycrystalline damage model. Such a polycrystalline model allows the simulation of the orientation of crystal lattice and the resulting slip systems. This should improve the description of anisotropic and cyclic material behaviour. Rousselier et al. indicated that the needed computing time is not significantly higher than for other advanced material models and thus macroscopic structures can be simulated.

3.2.3 Discussion of the Void Growth Models

The micromechanical-based void growth models can be compared on the basis of the predicted void growth [2]. Only the empirical Lemaitre model, which uses a damage parameter, cannot be compared directly to the other models. For the comparison of the models the parameters and equations given in the original publications, see Table 1, are used.

As postulated in the derivations perfectly plastic behaviour of the matrix material is assumed. For the comparison of the models, it is postulated that failure occurs when a critical void volume f_c is reached. If the plastic equivalent failure strain is plotted as a function of the stress multiaxiality [10, 237] so called limit strain curves result. With the assumption of a critical void volume fraction of $f_c = 0.05$ the limit strain curves calculated with the models are shown in Fig. 37.

It can be seen that all the predicted curves are in a relatively narrow scatter band. This is not surprising, since all models are based on very similar basics.

Table 1 Comparison of void growth models

Model name	Equation
McClintock	$\frac{\dot{f}}{f} = \sqrt{3}\dot{\tilde{\varepsilon}}^{v} \sinh\left[\sqrt{3}\left(\frac{\tilde{\sigma}^{m}}{R_e} - \frac{1}{3}\right)\right]$
Rice and Tracey	$\frac{\dot{f}}{f} = 0.849\,\dot{\tilde{\varepsilon}}^{v}\,e^{\frac{3\sigma^{m}}{2R_e}}$
Rousselier[a]	$\varphi = \frac{\tilde{\sigma}^{vM}}{1-f} + \sigma^{k}\ f\ D\ e^{\frac{\tilde{\sigma}^{m}}{(1-f)\sigma^{k}}} - \sigma_0(\varepsilon^{p^{v}}) = 0$
Gurson	$\varphi = \left(\frac{\tilde{\sigma}_{ij}}{R_e}\right)^2 + 2\ f \cosh\left(\frac{3\tilde{\sigma}^{m}}{2R_e}\right) - 1 - f^2 = 0$
Rice and Tracey modell extended by Gurson	$\varphi = \tilde{\sigma}^{vM} + 2 \cdot 0.283\ f\ R_e e^{\frac{3\tilde{\sigma}^{m}}{2R_e}} - R_e = 0$

[a]As mentioned in [224] the Rousselier parameters σ^k und D are presumed to be $\sigma^k = 2/3R_e$ and $D = 2$

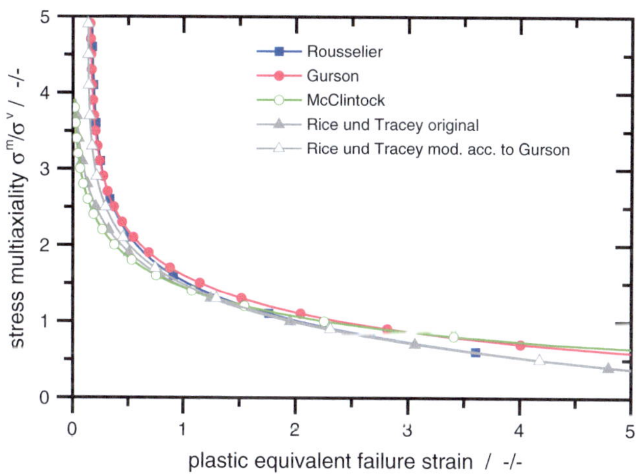

Fig. 37 Limit strain curves determined with different void growth laws, $f_c = 0.05$

At high multiaxialities the models of Rousselier, Gurson and Rice and Tracey (in the formulation of Gurson) produce comparable results. Lower fracture strains are only predicted by the original Rice and Tracey and McClintock model.

For low multiaxialities the Rousselier and the Rice and Tracey model (both formulations) give similar results. In comparison to this, the Gurson and McClintock model predicts much higher fracture strains.

3.3 *Models to Describe Void Coalescence*

The mechanism of void coalescence depends on the one hand on the microstructure of the materials, on the other hand on the external loads, see Sect. 2.3. However, which mechanism occurs and how to model it numerically has been investigated so far least of the three failure stages (initiation, growth, coalescence) [36, 187, 190].

Most authors assumed [137, 294] for their derivation that void coalescence can microscopically be described by plastic collapse of the material bridges between the single voids. Only the model proposed by Brown and Embury [61] describes the formation of shear bands. None of the models can predict which one of the two mechanisms is activated. Several approaches to model void coalescence are discussed in the following.

3.3.1 Coalescence When Reaching a Critical Void Volume, a Critical Void Growth or a Critical Damage Condition

The simplest and most often used approach is to assume the occurrence of void coalescence when a critical void volume f_c [240, 281, 292] or a critical void growth $(R/R_0)_c$ [35, 36, 167, 246, 247] is reached.

$$f \geq f_c \tag{39}$$

Based on experiments, Lemaitre also suggested that a critical damage [155] describes the final fracture. All these approaches are based on fractographic observations finding a void growth nearly independent of the multiaxiality of the stress state, e.g. [246].

The law defined by Tvergaard and Needleman [281], enabling the calculation of an accelerated void growth during coalescence, can be assigned to this model category as well. Since the original Gurson model predicts a, compared to experiments, too small void growth when considering the state of advanced void growth, Tvergaard and Needleman replaced the void volume fraction f by an empirical function $f^*(f)$:

$$\varphi = \left(\frac{\tilde{\sigma}^{vM}}{\sigma_0}\right)^2 + 2f^* q_1 \cosh\left(\frac{3\,\tilde{\sigma}^m}{2\sigma_0}\right) - 1 - (q_1 f^*)^2 = 0 \tag{40}$$

For $f^*(f)$, the following growth function is assumed:

$$f^* = \begin{cases} f & \text{for} \quad f \leq f_c \\ f_c + \kappa(f - f_c) & \text{for} \quad f > f_c \end{cases}$$
$$\text{with the acceleration coefficient } \kappa = \frac{f_u^* - f_c}{f_f - f_c} \tag{41}$$

f_c is the void volume at which void coalescence is starting. f_f is the void volume at final fracture of the material. f_u^* can be calculated with the relation $f_u^* = 1/q_1$. This approach simulates a continuous failing of the material. The empirical assumption of continuous formation has the advantage of resulting in less convergence problems in a finite element computation than a discontinuous formulation of the damage evolution.

In their studies, many authors find a more or less large dependence of the critical void volumes on the multiaxiality of the stress state [9, 35, 36, 136, 247, 257, 275], on the Lode angle [105, 235] and on the void shape [9]. This is contrary to experimental works of Shi et al. [247] finding a rather minor dependence on multiaxiality. These contradictions can possibly be solved by the volume portion of the voids. With cell model calculations, Kim et al. [136] showed that the influence of stress multiaxiality on the critical void volume is only minor when considering small initial void volumes ($f_0 < 0.001$). This observation is also described by Scheyvaerts and Pardoen [234].

When using any model of this category it has to be taken care that metallographic meaningful values are used for the critical void volumes and void growth rates.

3.3.2 Coalescence Triggered by Formation of Shear Bands Between Voids

Brown and Embury [61] assume that the voids coalesce due to the formation of shear bands between the single voids. As criterion for the critical strain $\tilde{\varepsilon}^c$ between void initiation and void coalescence, they found the following relation:

$$\tilde{\varepsilon}^c = \ln\left(\sqrt{\frac{\pi}{6f}} - \sqrt{\frac{2}{3}}\right) \tag{42}$$

This theory is supported by several experimental and numerical studies [149, 213, 245].

3.3.3 Plastic Limit Load-Model by Thomason for the Calculation of Void Coalescence

The best-known models describing the plastic collapse of material bridges between voids are the coalescence criteria by Thomason [271, 272, 275]. Thomason derived stress-based criteria for the description of void coalescence for different loading conditions and void geometries. In his derivations, he assumed cubic primitive arranged unit cells having one void at the center each, see Fig. 38. The material deformation behaviour of the matrix between the voids is assumed to be rigid/perfectly-plastic.

For the derivation of a three-dimensional coalescence criterion [272, 273] he assumed periodically arranged cuboidal voids with quadratic cross sections. The principal load direction is perpendicular to the quadratic base. The distance of the voids perpendicular to the principal load direction is 2d, the void height is 2a and the side length of the quadratic voids is 2b, see Fig. 39. For the localisation zone between the single voids, Thomason assumed simple displacement rate fields.

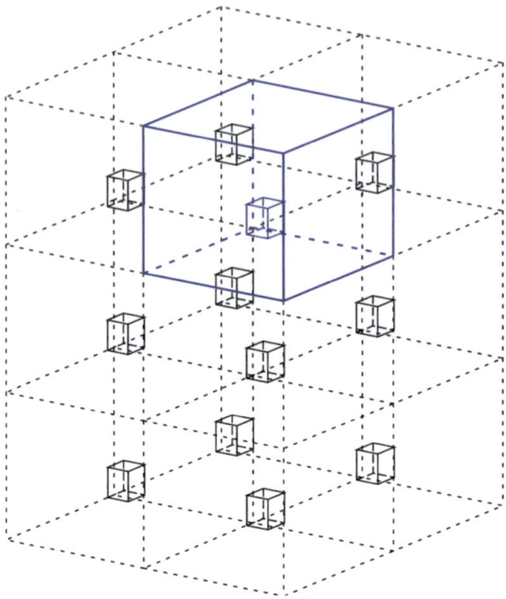

Fig. 38 Cubic primitive arrangement of unit cells

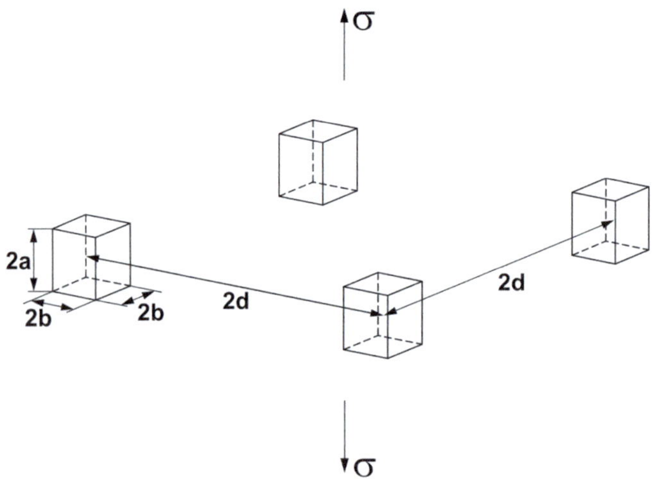

Fig. 39 Sizes and distances of cuboidal voids with cubic primitive arrangement [272]

Doing so, he obtained an upper limit for the load [47]. Thomason approximated his complex solutions for the upper limit of the ultimate load with the empirically found relation:

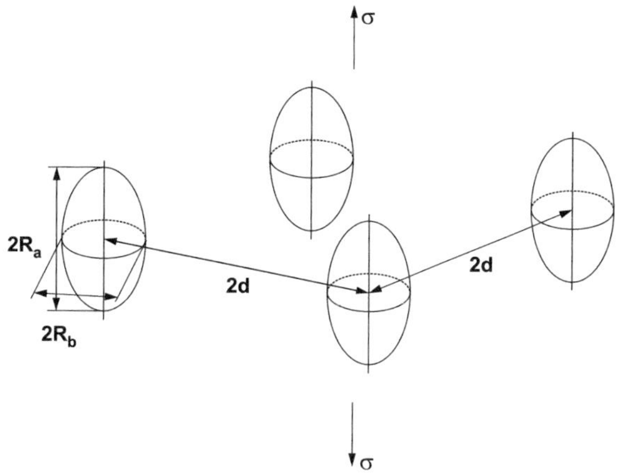

Fig. 40 Sizes and distances of ellipsoidal voids [272]

$$\tilde{\sigma}_I^c = {}^{mat}R_e\left(1-\xi^2\right)\left(0.1\left(\frac{\xi^{-1}-1}{\kappa}\right)^2+1.2\,\xi^{-\frac{1}{2}}\right) \quad \text{with } \xi=\frac{b}{d} \text{ and } \kappa=\frac{a}{b} \tag{43}$$

Here ${}^{mat}R_e$ is the yield strength of the matrix material. In order to be able to apply Eq. 43 to ellipsoidal voids as well, Thomason assumed that he can set the semi-axes of the ellipsoid (R_a und R_b) equal to the length of the sides of the cuboidal voids (a, b) [272, 273], see Fig. 40. He claimed this approximation to be valid as long as the void volume is smaller than 0.2. Analogous to Eq. 43 he obtained an upper limit for the ultimate stress acting macroscopically at the unit cell:

$$\sigma_I^c = {}^{mat}R_e\left(1-\xi^2\right)\left(0.1\left(\frac{\xi^{-1}-1}{\kappa}\right)^2+1.2\xi^{-\frac{1}{2}}\right) \quad \text{with } \xi=\frac{R_b}{d} \text{ and } \kappa=\frac{R_a}{R_b} \tag{44}$$

The Thomason-criterion is also used to compute a critical void volume f_c that depends on the state of strain or stress. In this case, f_c is not a material constant any more but a variable [293].

In order to incorporate material hardening, Pardoen et al. [187] enhanced the coalescence criterion derived by Thomason, Eq. 44. For the matrix surrounding the void, they assumed the following material law:

$${}^{mat}\sigma_0 = \left(1+\frac{E\varepsilon^p}{R_e}\right)^n R_e \quad \text{for } {}^{mat}\sigma_0 > R_e \tag{45}$$

Equation 44 is replaced by the empirical formulation:

$$\sigma_I^c = {}^{mat}\sigma_0\left(1-\xi^2\right)\left(\alpha(n)\left(\frac{\xi^{-1}-1}{\kappa}\right)^2+\beta(n)\xi^{-\frac{1}{2}}\right) \quad \text{with } \xi=\frac{R_b}{d} \text{ and } \kappa=\frac{R_a}{R_b} \tag{46}$$

The values of α and β depend on the hardening exponent n. In the range of $0 \leq n \leq 0.3$ the authors found $\alpha(\mathrm{n}) = 0.1 + 0.217\mathrm{n} + 4.83\mathrm{n}^2$ and $\beta(\mathrm{n}) = 1.24$.

3.3.4 Yield Criterion to Describe Material Behaviour in the Case of Plastic Collapse

To describe the material behaviour during void coalescence for arbitrary stress states, Benzerga [26] introduced an empirical yield condition based on the works by Pardoen and Hutchinson [187]:

$$\varphi = \tilde{\sigma}^{\mathrm{vM}} + \frac{3}{2}|\tilde{\sigma}^{\mathrm{m}}| - \frac{3}{2}\tilde{\sigma}_{\mathrm{I}}^{\mathrm{c}} = 0 \tag{47}$$

Seen in the principal stress space, his yield surface has the form of a double cone with the symmetry axis being on the hydrostatic stress axis, see Fig. 41.

The fact that the stress state around a void changes drastically during coalescence, as known from cell model calculations, is accounted for by the transition to the new yield surface. This approach models void coalescence as a continuous process.

3.3.5 Simulation of Void Coalescence Using Void Growth Models

Void growth models like the Rousselier- or the Gurson model also implicitly predict void coalescence. When the void volume fraction is about to reach 1 in the Rousselier- and Gurson model, respectively q_1^{-1} in the GT model, the calculated stresses go to zero. Values of about one for the void volume however are unrealistically high seen from a metallographic point of view. Likewise the predicted stress decreasing for high void volumes is too slow. The stresses in the Rousselier

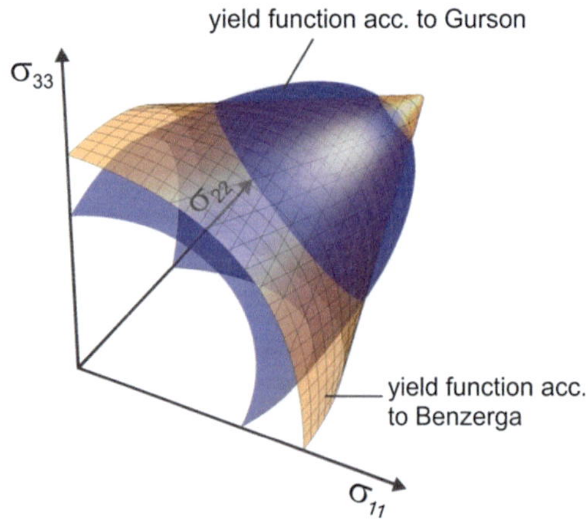

Fig. 41 Benzerga yield surface in the principal stress space

model approach zero only asymptotically. Due to these deficiencies, these void growth laws are in practice almost exclusively used in combination with an additional coalescence criterion.

3.3.6 Discussion of the Models Describing Void Coalescence

The models describing coalescence processes differ in their mechanisms. It is therefore difficult to compare them directly. Pardoen et al. [190] studied the growth and coalescence processes of voids in technical pure copper with different material hardening for different stress multiaxialities. They compared the following coalescence criteria:

- critical void growth
- shearing of material bridges (Brown & Embury criterion)
- plastic collapse of material bridges (Thomason-criterion)

They concluded that none of the models can reproduce the whole spectrum of investigated multiaxial stress states independently.

3.4 Common Combinations of Damage Models and a Comparison

To describe the whole process of dimple fracture, model combinations describing all three phases (void initiation, void growth and void coalescence) are needed. In principle it is possible to combine any models depending on the used material behaviour. However, in practice certain model combinations have become established and are successfully applied by a variety of users in research and industry.

3.4.1 Gurson, Tvergaard and Needleman (GTN) Model Combinations

The certainly most common combination of models describing void initiation, void growth and coalescence is the so-called Gurson-Tveergaard-Needleman model combination (GTN model) [39, 56, 90, 164, 176, 182, 238, 258, 259, 262, 263]. Although it is a combination of three independent damage mechanics models, the GTN combination is very often just called GTN model. Following models are combined in the GTN model combination:

- Void initiation by Chu and Needleman [72]
- Model of void growth by Gurson und Tvergaard [277]
- Void coalescence criterion by Tvergaard and Needleman [281]

Besides the initial void volume f_0, the characteristic length l_c (see Sect. 3.5) and the flow curve of the whole material, up to 10 additional material-depending parameters are needed for the GTN model:

$$s^{\varepsilon},\ \varepsilon^{N},\ \psi,\ \kappa,\ \sigma^{N},\ s^{\sigma},\ q_1,\ q_2,\ f_c \text{ and } \kappa \tag{48}$$

The large number of parameters that are in addition usually hard to determine presents a disadvantage [49, 51, 275] of the GTN model. From literature research it is known that different parameter sets can yield the same global response [1]. Especially the empirical parameters, having no actual physical meaning, cannot be identified independently. This problem will be discussed in Sect. 3.6.

3.4.2 Rousselier Seidenfuss (RS) Model Combination

Since the Rousselier model is not able to model the steep stress decrease occurring in void coalescence with sufficient accuracy, Seidenfuss proposed [146, 241] to combine the Rousselier model with a critical void volume for modelling void coalescence. This so-called RS model combination is often used in literature to describe the failure behaviour of different materials [67, 79, 81, 142–145, 191, 205].

- Modelling of void initiation. The above mentioned model combination assumes that an initial void volume f_0 exists, respectively, that the volume is created right after the yield stress is exceeded. These assumptions have been confirmed by a variety of authors, [8, 23, 59, 87, 89, 113, 125, 146, 190, 213, 243, 266, 274, 293].
- Void growth model of Rousselier [224]
- Void coalescence when reaching a critical void volume [240]

Besides the initial void volume f_0, the characteristic length l_c and the flow curve of the material mix, only 3 more model dependent material parameters are needed when using the RS model combination.

Many practical applications show that the GTN and RS model combinations yield similar results [20, 31, 175, 191, 241].

3.4.3 Gologanu, Leblond and Devaux (GLD) Model with Thomason-Criterion

The GLD model for the simulation of material behaviour with void growth of nonspherical voids has been combined very often with the Thomason coalescence criterion in form of a yield function [26, 188, 189]. Additionally, different model combinations in connection with the GLD model have been used, e.g. [105].

- Modelling of void initiation. In a first approximation, it is also assumed that the whole initial void volume f_0 exists from the beginning on.
- Model of void growth by Gologanu et al. [110, 264]:
- Model of void coalescence by Thomason [272] and Benzerga [26]

3.5 *Mesh Dependency of Results and Definition of a Characteristic Length*

Caused by increasing material damage resulting from the initiation and growth of voids, the material sooner or later starts to soften. If the material softens in a certain limited volume, strain localisation resulting in a shear band can occur. Therefor material models being able to capture material softening must be capable of predicting the finite dimension of strain and damage localization.

All so far presented material models are local models. Local means here that e.g. the stresses at a material point A are only dependent on the local state variables at that point. Neighbouring material points do not influence the stresses at point A. However, such a behaviour would only be valid for a perfectly isotropic and homogeneous material. Macroscopically seen, real metals and metal alloys fulfill these requirements only in a coarse approximation since they have a discrete microstructure resulting in inhomogeneity on microscopic scale. Material constituents have a finite size and influence each other. A void growing at point A can influence the growth of a neighbouring void at point B. This reciprocal influence is not considered with local material models.

Real mechanical problems can usually not be solved analytically, numerical approximations like e.g. the method of finite elements are used. If material softening is simulated with a local material model in combination with finite elements, the ellipticity of the initial-boundary value problem is lost and a bifurcation problem results. This means that a homogeneous strain respectively damage field will get unstable against a strongly localized one [209]. Since the method of finite elements approximates the displacement field volume by volume, the strains respectively damages cannot localize in an infinitely thin band due to the mathematical definition. The width of the localization zone is coupled with the size of the elements. This effect leads to the very often in literature discussed pathologic mesh dependency of results. The localization problem can only be solved by introducing an additional parameter, the characteristic length l_c.

Using finite element computations, the localisation problem is solved in practice very often by introducing a constant material-specific element size in areas where material softening can occur [18, 19, 39, 46, 90, 182, 224, 240]. Very often authors assume that the width of localization zones is directly related with the distance of the primary, failure causing voids [18, 62, 100, 107, 128, 150, 188, 199, 224, 240, 262, 292].

Numerous practical applications [45, 62, 86, 182, 216, 240, 241, 262] show that the problem can be solved satisfactorily (seen engineering-wise) this way. For steel materials, element sizes in the range of a few tenth of a millimeter result [20, 62, 100, 107, 135, 146, 199, 224, 240, 266].

A more general approach to solve the localisation problem and to avoid the mesh dependency of results is provided by the so-called nonlocal damage models as discussed in detail in Sect. 4.

3.6 *Determination of the Material Dependent Parameters*

When using damage models for simulating specimen and component behaviour it is not only necessary to select appropriate models but also crucial to determine the needed parameters reliably and uniquely. Although the majority of the presented models is in use for over 20 years now, no verified standardised procedures for the determination of the used parameters are available.

Determination procedures described in literature and/ or used in practice:

- Metallographical and fractographic determination of the parameters out of the microstructure of the material,
- direct or iterative determination out of macroscopically measured values from simple specimen or
- adaption to results of cell model computations are described shortly in the following sections.

3.6.1 Determination of the Parameters Out of the Microstructure of the Material

Since the models have been derived from micromechanical theories, some of the needed material dependent parameters have a direct relation to the microstructure. Examples are the initial void volume fraction f_0, the characteristic length l_c as well as the critical void volume fraction f_c. The determination can be done with metallographical cuts and/or with tomographic methods. The advantage of this method is that parameters can be determined isolated and conclusively without other parameters interfering. Drawbacks are an often occurring strong scattering of measured material constituents values as well as a relatively large fuzziness in the determination of the parameters [240].

3.6.2 Direct or Iterative Determination Out of Macroscopically Measured Values from Simple Specimens

Various parameters resulting from the derivations of the models or having been introduced as adjusting parameters do not have a direct relation to the microstructure and can therefore not be determined with metallographic methods. These parameters can partly be determined **directly** out of the load-deflection behaviour of a specimen or **indirectly** through a numerical adaption to experimental results.

The **indirect** determination is also called numerical calibration. Hereto, the deformation behaviour of a selected specimen is being simulated with finite elements. The parameter to be determined is varied in the simulation until a satisfying accordance between simulation and experiment is found.

The geometry of the specimen has to be chosen in a way that allows a preferably independent identification of the single parameters. The variation of a parameter should have a preferably large influence on the calculated macroscopic specimen behaviour. The numerical calibration has been tested in an European round robin test with only little success [31]. Values determined by the different participators turned out to have significant differences.

Springmann and Kuna [253, 255] present a method allowing an automatized determination of damage-mechanical parameters. Their approach is based on a nonlinear optimisation algorithm. With their method they tried to determine the material-dependent parameters out of the load-deformation behaviour of a simple specimen (notched flat tensile specimen, notched round tensile specimen, C(T)-specimen). However, it turned out that basically only one parameter can be determined with certainty. The other parameters have to be known. If more than one parameter is optimised with only one experiment, the parameters cannot be determined conclusively.

An advanced approach for a numerical parameter identification is presented by Springmann in [254]. Within this approach not only the macroscopic specimen length change is measured but additionally the displacement field on the surface with an optical measurement method [5]. Using a nonlinear optimisation algorithm several parameters can be adapted simultaneously. The authors claims that a maximum of 4 parameters can be identified with the adaption to the displacement field at the same time. The method could certainly be improved by using not only the result of one experiment but several specimen with different multiaxial stress states simultaneously.

3.6.3 Adaption to Results of Cell Model Computations

Often, damage mechanical parameters are identified by adapting them to results of cell model calculations [136, 137, 141]. The results of cell model computations are strongly influenced by the chosen boundary conditions [141, 244]. The quantitative identification of parameters for real materials is therefore not or only with strong restrictions possible, additionally, the validity has to be scrutinized very critically.

3.7 Concluding Remarks Considering Damage-Mechanical Models

The different damage models have often been compared among each other with respect to their ability to predict the failure behaviour in comparison to experiments. Especially the GTN model and the Rousselier model—in its original as well as in the RS formulation—have been investigated quite often. However, it is in the nature of such comparisons that different authors come to different conclusions:

- For fracture mechanics, some authors conclude that the models yield similar results [175, 241], while others see an advantage in the Rousselier model, e.g. [20].
- Pineau [200] calculates the behaviour of notched round tensile specimen with the models of Rice & Tracey, Rousselier and GTN. Compared to experiments, all three models deliver similar results, however, the Rousselier model has a slight advantage.
- An European ring round robin test also showed that the RS model delivers better results than the GTN model when compared to experiments [31] although less material-dependent parameters are needed.
- When considering cyclic loading in both tension and compression, the GTN model surely will deliver better predictions since the Rousselier model cannot describe a decrease of void volume when loading in compression occurs. Research done by e.g. Steglich et al. [256] shows that the LPD model which is based on the Gurson model can basically describe the processes occurring in cyclic plasticization.
- However, when considering loading with low multiaxiality, the original Rousselier model seems to have an advantage compared to the GTN model as Besson et al. [41, 43] shoed using the example of fractures with shear lips. Due to the basic assumptions in the derivation of the discussed models none of the models will give satisfactory results for very low multiaxialities like pure shear.
- Perrin and Leblond [194] on the other hand see advantages in the GTN model. In a theoretical study connected with a self-consistent unit cell they find disadvantages in the mathematical formulation of the Rousselier model. A restriction of this study is the used ‘Rousselier’ yield condition in a modified form. It is not stated how or from where it was derived.

A disadvantage of the micromechanical-based damage models based on the employed approach is the disability to correctly model a sharp crack tip occurring in a structure. The damage-mechanical approaches model the growth of the crack by a decrease of stress in the damaged elements. Since an element has a finite volume, the crack thus always has the width of this volume. In practice this circumstance is explained with a damage area growing in the vicinity of the crack tip. Since the damaged elements are usually not deleted from the structure stiffness matrix, such elements can lead to numerical problems. Suggestions on how to transform the volume damage to a crack path can for example be found in [80, 170, 249].

4 Nonlocal Damage Models

In experiments that are carried out until specimen’s failure, i.e. the complete loss of integrity of the material, technical steels typically show the formation of a process zone: while large parts of the specimen are deformed purely elastically, micromechanical processes such as plasticity or damage take place locally limited. The size of the process zone is determined by the micromechanical structure of the material and thus represents a material property.

4.1 Localization

The increase of strain or stress caused by the nucleation of microcracks or voids in the vicinity of microstructural defects or inclusions can lead to the localization of strain [226] in the process zone. This results in an increased evolution of damage [278, 279] and therefore to a further narrowing of the micromechanical active region. With increasing damage the load bearing capacity of the material reduces; with increasing strain a local decrease of stress is observed as local softening [109]. The force-displacement-curves of tensile tests with porous materials with hardening matrix materials show a maximum, which is followed by a region with a negative slope. This is caused by geometrical and material softening.

As shown in Sect. 3, in the framework of continuum mechanics, material softening can be reproduced by the consideration of damage in a plasticity model. With increasing damage lower stresses lead to plastic flow, the yield surface and thus the elastic region decrease.

The static equilibrium of a simple softening plastic continuum is unstable [180] and leads to strain localization. Mathematically, this means the loss of ellipticity [55] of the underlying system of differential equations, which is now of hyperbolic type.

The strain localization at the loss of ellipticity theoretically takes place in an infinite small zone. If a discretization of the continuum under investigation is carried out prior to the solution of the boundary value problem, the size of the localization zone is determined by the discretization size [55, 197]. When using the finite element method this effect is called mesh sensitivity: for meshes of smaller element sizes, the size of the localization zone reduces. As a result, the energy dissipated by damage converges to zero for infinitesimally small elements. This contradicts the experimental observation that both the size of the localization zone as well as the energy dissipated by the failure of the material are material properties [109].

4.2 Regularization Methods, Non-local Formulations

As shown in the previous section, the modeling of ductile damage within the framework of continuum mechanics of simple materials in combination with the numerical solution of the corresponding system of partial differential equations results in two fundamental properties of the considered problem: First, the material softening leads to instability of the material behavior. Second, material instability and strain localization are directly connected due to the strong interactions between increase of strain and evolution of damage for simple continua. For the numerical treatment of the boundary value problem this results in a size of the localization zone that depends on the discretization.

One possibility for setting a fixed size of the localization zone when using the FEM is to define a characteristic element size [38, 39, 184, 239]. In this method strain localization and material softening will take place within a certain element, element row or element layer, depending on the dimension of the problem. The element edge length becomes a model parameter. Thus, the accuracy of the solution cannot be judged by conventional convergence criteria. Furthermore, considerable numerical disadvantages arise in the treatment of problems with different length scales.

With the help of regularization approaches material softening and strain localization can occur independently. The use of so-called non-local formulations thus differs fundamentally from the pragmatic method above: the loss of ellipticity and the influence of discretization are eliminated [53, 196, 198, 259, 260].

The reduction of the mesh dependence which occurs when continuum damage models are used numerically was the subject of numerous physically motivated and phenomenological approaches. Reviewing articles can be found in [21, 85, 130, 193].

The fundamental approach of non-local formulations to reduce localization effects caused by discretization is to consider a finite region of surrounding material in the underlying equations

$$f(\omega) \Rightarrow f(\omega, \bar{\omega}), \tag{49}$$

where ω and $\bar{\omega}$ are local and non-local variables, respectively. This can be done efficiently by averaging a variable within an area characterized by a certain length parameter.

Depending on how averaging is done non-local formulations can be divided into non-local integral types [22, 196], explicit [54] and implicit gradient methods [84, 192]. Other classification may distinguish whether internal variables of a constitutive model or stresses and strains are averaged. Finally non-local formulations may be divided into weakly non-local and strongly non-local formulations [21, 130].

4.2.1 Nonlocal Integral Types

When using non-local integral type approaches a local variable ω is replaced by its non-local average $\bar{\omega}$ which is calculated by integrating the weighted local variable

$$\bar{\omega}(x) = \frac{1}{B} \int_B G(\xi)\omega(x + \xi) dB. \tag{50}$$

The influence of the surrounding material characterized by the distance vector ξ on the considered material point is given by the definition of the weight function $G(\xi)$. Here, typically GAUSSIAN or similar bell-shaped functions are used, which must at least satisfy the condition

$$\frac{1}{B} \int_B G(\xi) dB = 1. \tag{51}$$

Integral type approaches are strongly non-local methods, as the integration is carried out over the entire body under consideration. They are numerically robust methods and can in principle be applied to every type of constitutive models. In the treatment of complex geometries, the weight functions $G(\xi)$ must meet high standards, which requires the use of numerically robust algorithms in three dimensions.

4.2.2 Explicit Gradient Formulations

A formulation of Eq. 50 using gradients can be derived using Taylor series at the material point under consideration

$$\begin{aligned}\omega(x+\xi) = \omega(x) + \nabla\omega(x)\cdot\xi + \frac{1}{2!}\nabla(\nabla\omega(x))\\ : \xi\otimes\xi + \frac{1}{3!}\nabla^{(3)}\omega(x)\cdots\xi\otimes\xi\otimes\xi + \cdots\end{aligned} \quad (52)$$

Assuming isotropy and substituting back in Eq. 50, Eq. 52 gives

$$\bar{\omega}(x) = \omega(x) + c\nabla^2\omega(x) + d\nabla^4\omega(x) + \cdots \quad (53)$$

and after neglecting gradients of higher order an explicit gradient formulation of Eq. 50 can be written in reduced form as

$$\bar{\omega}(x) = \omega(x) + c\nabla^2\omega(x). \quad (54)$$

Explicit gradient methods are weakly non-local methods, since only the infinitesimal neighborhood of a material point is taken into account. Compared with nonlocal integral type approaches explicit gradient formulations possess significant numerical disadvantages [131, 193]. The constant $\sqrt{c}$ is introduced as an length paramcter that controls thc influence of the surrounding material.

4.2.3 Implicit Gradient Formulations

Differentiating Eq. 53 twice and reordering gives

$$\nabla^2\omega(x) = \nabla^2\bar{\omega}(x) - c\nabla^4\omega(x) - d\nabla^6\omega(x) - \cdots, \quad (55)$$

and after substituting Eq. 55 back into Eq. 53

$$\bar{\omega}(x) - c\nabla^2\bar{\omega}(x) = \omega(x) + (d - c^2)\nabla^4\omega(x) + \cdots \quad (56)$$

Thus, after neglecting higher gradients as done in Eq. 54 one finds an implicit gradient method to carry out the non-local averaging as

$$\bar{\omega}(x) - c\nabla^2\bar{\omega}(x) = \omega(x). \tag{57}$$

As well as non-local integral type approaches, implicit gradient formulations are strongly non-local formulations and are much easier to implement numerically compared to explicit methods. However, the partial differential equation of the HELMHOLTZ type Eq. 57 represents an additional field equation which necessitates the formulation of additional, possibly non-physical boundary conditions. A critical discussion of possible boundary conditions can be found in [193].

4.2.4 Non-local Formulations of Ductile Damage Models

The development of non-local formulations of ductile damage models has been the subject of numerous publications. Non-Local modifications of the GTN-model (Sect. 3.4.1) can be found in [82, 129, 151, 160, 210, 280]. Samal and coworkers [231] developed an implicit gradient formulation of the ROUSSELIER model (Sect. 3.4.2, [224]. From the above-mentioned modifications, the implicit gradient formulations used in [209, 230] spatially average the rate of the (modified) void volume fraction in a non-local sense

$$\dot{d} - c\nabla^2\dot{d} = \dot{f}. \tag{58}$$

The non-local modification of the GTN-model by Linse et al. [126, 160] is based on the micro-dilatational approach [95] and replaces the dilatational part of the plastic strain ε_p by its non-local spatial average

$$\bar{\varepsilon}_p - c\nabla^2\bar{\varepsilon}_p = \varepsilon_p, \tag{59}$$

where the rate of the non-local plastic strain enters the evolution equation for the growth of existing voids.

5 Combination of Damage Models in the Brittle-Ductile Transition Region

In the brittle and brittle-ductile transition region of ferritic steels cleavage fracture initiates at microcracks, while microvoids nucleate as a result of plastic deformation in the upper shelf. A consistent and independent description of these micromechanical processes is needed if the entire toughness region of ferritic steels is to be analyzed.

5.1 Beremin Model—Uncoupled Probabilistic Model for Cleavage Fracture

For ferritic steels, the brittle and transition region is characterized by a large scatter of fracture toughness values that results from the statistical distribution of microcracks. Using the WEIBULL theory, the BEREMIN-model was developed to describe cleavage of ferritic steels [37, 174]. Here, three fundamental assumptions are made:

- microcracks are created during plastic deformation; the probability density function for the size of the microcracks follows a constant power law
- the critical stress of each microcrack is determined by the GRIFFITH criterion
- weakest-link: the propagation of one single crack leads to the failure of the whole structure.

Under these assumptions, the probability of failure for a specific load level L is derived by

$$P_f(L) = 1 - \exp\left\{-\left(\frac{\sigma_W(L)}{\sigma_u}\right)^m\right\}. \quad (60)$$

The WEIBULL-stress σ_W is calculated as

$$\sigma_W(L) = \sqrt[m]{\frac{1}{V_0}\int\limits_{V_{pl}} (\sigma_I(L))^m dV}, \quad (61)$$

where σ_I is the maximum principal stress and V_{pl} the plastic volume of the structure. The reference volume V_0 must be large enough to represent the microstructure of the material as well as small enough to fulfill the requirements of the Griffith criterion. If the reference volume is chosen constant, the BEREMIN model uses two model parameters: the WEIBULL reference stress σ_u and the WEIBULL-modulus m.

Bordet et al. [52] showed that most of the problems that arise in engineering applications of the model are a result of a oversimplified description of local cleavage in the BEREMIN-model. Most modifications of the BEREMIN-model [37, 100–104, 139] change the calculation of the WEIBULL-stress. Bernauer et al. [40] proposed a modification that takes into account that the nucleation of voids is promoted by the presence of carbide particles. Consequently, the number of cleavage initiation points is reduced with the increase of the nucleated void volume fraction.

Note that in the sense of the classification of damage models given in Sect. 3.2 the BEREMIN-model and its modifications must be termed uncoupled, since cleavage is not modeled for each individual microcrack but rather by quantifying the cleavage fracture probability of many microcracks.

In principle, the BEREMIN-model for the calculation of the probability of cleavage fracture can be used together with a ductile damage model to account for ductile crack growth preceding cleavage in the transition region [4, 227]. In conjunction with the Griffith criterion the stress fields must be resolved very accurate which requires the use non-local ductile damage models in the presence of high stress gradients. However, following the weakest-link assumption, the model cannot predict cleavage crack arrest and thus most likely underestimates the fracture toughness values in the case of such effects as pop-in or crack arrest, see e.g. [127, 161]. In [161] fracture toughness values were predicted by numerical simulation of fracture mechanics tests using a combination of a non-local GTN-model together with the BEREMIN model. The calculated values agree well with experimental results in the brittle region and in the ductile region. However, in the brittle-ductile transition region, the predicted fracture toughness values are much smaller than the experimental values.

5.2 *Coupled Models for Cleavage Softening*

Among simplified strip yield models, e.g. [98], softening of metals by cleavage is mostly modeled in a coupled way by means of cohesive zone elements, see e.g. [140]. Here, softening initiates when the maximum principal stress exceeds the cohesive strength and the work of cohesive separation can be correlated to the fracture toughness.

The combination of cleavage softening with ductile damage models was realized numerically using computational cell simulations, see e.g. [99, 179, 248]. Here, discrete volumes of material (cells) are removed on a the basis of a stress criterion for cleavage, ductile damage is modelled in a continuous way. The results strongly depend on the discretization, since not only the accuracy of the computed stress and strain fields depend on the chosen mesh size, but the properties of the microstructure are directly correlated to the chosen cell size.

In [127] a consistent continuum formulation covering both cleavage softening and ductile damage is formulated using a non-local modification of the GTN model [126, 160] together with a cohesive zone model [219] for cleavage. The developed formulation captures many effects known from experiments such as the constraints sensitivity, cleavage initiation, cleavage crack propagation and crack arrest (pop-ins), a size independent lower-bound toughness and the possibility of stable cleavage crack propagation. The simulations were performed with homogeneous material properties and a high sensitivity with respect to small deviations of the material parameters was obtained in the transition region. However, in contrast to the application of probabilistic models, statistical predictions on the failure of structures cannot be obtained with this type of model.

6 Conclusions

Within this article the micromechanical processes leading to dimple fracture are discussed in detail. It is shown that for this kind of fracture mode the processes leading to material failure are similar for very different metals and metal alloys. Especially the processes leading to void initiation and void coalescence are complex and depend to a great extent on the microstructure of the selected materials. To describe the three phases of dimple fracture at high stress multiaxialities a large number of different mathematical approaches are available. The special focus of this paper is on so-called micromechanical-based models.

For the micromechanical derivation of the void initiation models, many simplifications must be made concerning the microstructure, the material behaviour and the initiation mechanisms. Hence, to select an adequate model for the considered material metallographic examinations are essential. It is also recommended to calibrate the material specific parameters on the basis of metallographic observations when using such micromechanical-based models. To take all the parameters only from literature or to use only numerical calibration procedures is not recommended at all.

Since the basics for the derivation of the void growth models are similar, the received results are comparable when taking into account the assumptions made in the derivations of the models. However, when these assumptions are not valid for a given material, e.g. when observing strain hardening, this can lead to different results of the models. In conclusion it can be said that none of the discussed void growth models is able to describe exactly the micromechanical processes in a strain hardening material in a wide range of stress multiaxiality. The Gurson model in the GT, LDP and GLD formulation has probably the largest potential in describing void growth.

The particular void coalescence models describe different kinds of merging mechanisms. Therefore a direct comparison is difficult. Pardoen et al. [190] compared different coalescence criteria. They came to the conclusion that none of the models is able to describe independently the whole area of the examined multiaxiality range. To simulate the different micromechanical processes during void coalescence, probably several models must be coupled.

In summary it can be said that the all models should be selected with respect to material and microstructure. If possible, the material-dependent parameters needed for the models should be determined metallographically or the numerically adjusted parameters should be verified with metallographic values. It is surely not recomended to regard the parameter determination as pure adaptation procedure to experimental values. Thus, there is a risk that the micromechanical background of the models is lost and the models are reduced to pure 'fitting models' [275].

The greater the number of material-dependent parameters is, the more difficult a unique determination is. Especially for the GTN-model, the large number of parameters partly difficult to access is often felt as a disadvantage [49, 51].In particular, the empirical parameters which have no physical background, cannot be determined independently. The following quote summarises the problem in determining the material-dependent constants:

> The limited success that has been achieved with the dilational-plastic models of ductile fracture is mainly the result of the large number of adjustable parameters that have been incorporated in the models. These models can now act in an analogous manner to a polynomial curvefitting technique, and the parameters (q_1, q_2, f_c, f_F, f_u) can be suitably adjusted to give a reasonable fit to any particular set of experimental results [275].

While basically applicable for the prediction of cleavage failure probabilities, uncoupled BEREMIN-type models most likely overestimate cleavage fracture in the presence of ductile damage. For the case of using coupled cleavage softening models, it remains open to perform (numerically expensive) Monte-Carlo-simulations for statistical failure predictions. In both cases, non-local formulations of ductile damage models are advantageous.

References

1. Abendroth, M. (2004) Identifikation elastoplastischer und schädigungsmechanischer Materialparameter aus dem Small Punch Test. Dissertation, Technische Universität Bergakademie Freiberg
2. Aboutayeb, S. M. (2000) Comportement a l'endommagement des materiaux metalliques heterogenes: Simulation et experience. These de doctorat, Universite des Sciences et Technologies de Lille
3. Acharyya S, Dhar S (2008) A complete GTN model for prediction of ductile failure of pipe. J Mater Sci 43(6):1897–1909
4. Anderson TL, Stienstra D, Dodds R (1994) A theoretical framework for addressing fracture in the ductile-brittle transition region. ASTM STP 1207:186
5. ARAMIS 2M (2007) Benutzerinformation - Hardware, Gesellschaft für optische Messtechnik. Braunschweig
6. Argon A (1975) Cavity formation from inclusions in ductile fracture. Metall Trans A 6A:825–837
7. Argon A (1975) Seperation of second phase particles in spheroidized 1045 steel, Cu-0.6Pct Cr alloy, and maragin steel in plastic straining. Metall Trans A 6A:839–851
8. Argon A (1976) Formation of cavities from nondeformable second-phase particles in low temperature ductile fracture. J Eng Mater Technol 60–68 (pub. by the American Society of Mechanical Engineers)
9. Arndt J, Majedi H, Dahl W (1996) Influence of strain history on ductile failure of steel. Le Journal de Physique IV 06, C6:23–32
10. Arndt J (1997) Experimentelle und rechnerische Untersuchungen zur Schädigung von Baustählen bei duktilem Versagen. Dissertation, IEHK, RWTH Aachen
11. Ashby M, Gandhi C, Taplin D (1979) Fracture-mechanism maps and their construction for f.c.c. metals and alloys. Acta Metall 27:699–729
12. Aurich D, Gerwien P, Häcker R, Hünecke J, Klingbeil D, Krafka H, Künecke G, Ohm K, Veith H, Wossidlo P (1997) Experimentelle und numerische Untersuchungen des statischen und dynamischen Rißwiderstandsverhaltens verschiedener höherfester Baustähle im Temperaturbereich von 20 °C bis 350 °C. 23. MPA-Seminar: Sicherheit und Verfügbarkeit in der Anlagentechnik 1, Universität Stuttgart 14.1–14.24
13. Baaser H, Gross D (2003) Analysis of void growth in a ductile material in front of a crack tip. Comput Mater Sci 26:28–35

14. Babout L (2004) Damage initiation in model metallic materials: X-ray tomography and modeling. Acta Mater 52(8):2475–2487
15. Bai Y, Wierzbicki T (2008) A new model of metal plasticity and fracture with pressure and Lode dependence. Int J Plast 24(6):1071–1096
16. Bandstra J, Goto D, Koss D (1998) Ductile failure as a result of a void-sheet instability: experiment and computational modeling. Mater Sci Eng, A 249:46–54
17. Barnby JT (1967) The initiation of ductile failure by fractured carbides in an austenitic stainless steel. Acta Metall 15:903–909
18. Batisse R (1987) Ductile fracture of a 508 Cl 3 steel in relation with inclusion content: the benefit of the local approach of fracture and continuum damage mechanics. Nucl Eng Des 105:113–120
19. Batisse R (1988) Contribution à la modelisation de la rupture ductile des aciers. Dissertation, Université de Technologie de Compiegne
20. Bauvineau L, Burlet H, Eripret C, Pineau A (1996) Modelling ductile stable crack growth in a C-Mn steel with local approaches. Le J de Physique IV 06(C6):33–42
21. Bažant Z, Jirásek M (2002) Nonlocal integral formulations of plasticity and damage: survey of progress. J Eng Mech 128(11):1119–1149
22. Bažant ZP (1984) Imbricate continuum and its variational derivation. J Eng Mech 110(12):1693–1712
23. Bažant Z, Pijaudier-Cabot G (1988) Nonlocal continuum damage, localization instability and convergence. J Appl Mech 55:287–293
24. Beachem CD (1973) The effects of crack tip plastic flow directions upon microscopic dimple shapes. Metall Trans A 6A:377–383
25. Becker R, Needleman A, Richmond O, Tvergaard V (1988) Void growth and failure in notched bars. J Mech Phys Solids 36(3):317–351
26. Benzerga A (2002) Micromechanics of coalescence in ductile fracture. J Mech Phys Solids 50:1331–1362
27. Benzerga A, Besson J, Pineau A (2004) Anisotropic ductile fracture—part I: experiments. Acta Mater 52(15):4623–4638
28. Benzerga A, Besson J, Pineau A (2005) How much input is needed from the microstructure to model ductile fracture? Int Conf Fract 11
29. Benzerga AA, Leblond J-B (2010) Ductile fracture by void growth to coalescence. Adv Appl Mech 44:169
30. Benzerga A, Besson J, Pineau A (1999) Coalescence-controlled anisotropic ductile fracture. J Eng Mater Technol 121(2):221–229
31. Bernauer G, Brocks W (2002) Micro-mechanical modelling of ductile damage and tearing—results of a European numerical round robin. Fatigue Fract Eng Mater Struct 25:363–384
32. Berdin C, Besson J, Bugat S (2004) Local approach to fracture. Paris: les Presses de l'Ecole des mines – ISBN 2-911762-55-X
33. Berg CA (1969) Plastic dilation and void interaction, inelastic behavior of solids. McGraw-Hill Book Company, New York
34. Beremin F (1981) Cavity formation from inclusions in ductile fracture of A508 steel. Metall Trans A 12A:723–731
35. Beremin F (1981) Study of fracture criteria for ductile rupture of A508 steels. In: 5th international conference on fracture, Cannes, pp 809–816
36. Beremin FM (1981) Experimental and numerical study of the different stages in ducitle rupture: application to crack initiation and stable crack growth, three-dimensional constitutive relations and ductile fracture. North Holland Publishing Company, pp 185–205
37. Beremin FM (1983) A local criterion for cleavage fracture of a nuclear pressure vessel steel. Metall Trans A 14A:2277–2287
38. Bernauer G (1997) Einsatz mikromechanischer schädigungsmodelle im sprödduktilen Übergangsbereich. Dissertation, Universität Karlsruhe

39. Bernauer G, Brocks W, Muehlich U, Steglich D, Werwer M (1999) Hinweise zur Anwendung des Gurson-Tvergaard-Needleman-Modells. Technical note gkss/wmg/99/10, GKSS-Forschungszentrum Geesthacht
40. Bernauer G, Brocks W, Schmitt W (1999) Modifications of the Beremin model for cleavage fracture in the transition region of a ferritic steel. Eng Fract Mech 64:305–325
41. Besson J, Steglich D, Brocks W (2001) Modeling of crack growth in round bars and plane strain specimens. Int J Solids Struct 38:8259–8284
42. Besson J, Guillemer-Neel C (2003) An extension of the green and gurson models to kinematic hardening. Mech Mater 35:1–18
43. Besson J, Steglich D, Brocks W (2003) Modeling of plane strain ductile rupture. Int J Plast 19:1517–1541
44. Besson J, Shinohara Y, Morgeneyer T, Madi Y (2008) Effect of prestrain on ductility of a X100 pipeline steel. In: Proceedings of the 17th european conference on fracture, pp 757–764
45. Besson J (2009) Damage of ductile materials deforming under multiple plastic or viscoplastic mechanisms. Int J Plast 25(11):2204–2221
46. Besson J (2010) Continuum models of ductile fracture: a review. Int J Damage Mech 19 (1):3–52
47. Bishop J, Hill R (1951) A Theory of the plastic distortion of a polycrystalline aggregate under combined stresses. Philos Mag Ser 7:414–427
48. Bonora N, Gentile D, Pirondi A (2004) Identification of the parameters of a non-linear continuum damage mechanics model for ductile failure in metals. J Strain Anal 39(6): 639–651
49. Bonora N, Gentile D, Pirondi A, Newaz G (2005) Ductile damage evolution under triaxial state of stress: theory and experiments. Int J Plast 21(5):981–1007
50. Bonora N (1997) A nonlinear CDM model for ductile failure. Eng Fract Mech 58(1/2):11–28
51. Bonora N (1999) Identification and measurement of ductile damage parameters. J Strain Anal Eng Des 34(6):463–478
52. Bordet S, Karstensen A, Knowles D, Wiesner C (2005) A new statistical local criterion for cleavage fracture in steel. Part I: model presentation. Eng Fract Mech 72:435–452
53. de Borst R (1991) Simulation of strain localization: a reappraisal of the cosserat continuum. Eng Comp 8:317–332
54. de Borst R, Mühlhaus HB (1992) Gradient-dependent plasticity: formulation and algorithmic aspects. Int J Numer Meth Eng 35:521–539
55. de Borst R, Sluys L, Mühlhaus HB, Pamin J (1993) Fundamental issues in finite element analyses of localization of deformation. Eng Comput 10(2):99–121
56. Brocks W, Cornec A, Scheider I (2003) Computational aspects of nonlinear fracture mechanics. Compr Struct Integrity 3:127–209
57. Bron F (2004) Ductile rupture in thin sheets of two grades of 2024 aluminum alloy. Mater Sci Eng, A 380(1-2):356–364
58. Bron F, Besson J (2006) Simulation of the ductile tearing for two grades of 2024 aluminum alloy thin sheets. Eng Fract Mech 73(11):1531–1552
59. Brooksbank D, Andrews KW (1968) Thermal expansion of some inclusions found in steels and relation to tessellated stresses. J Iron Steel Inst 206:595–599
60. Broek D (1972) The role of inclusions in ductile fracture and fracture toughness, symposium on fracture and fatigue on the school of engineering and applied science. George Washington University, Washington, pp 55–65
61. Brown LM, Embury JD (1973) The initiation and growth of voids at second phase particles. In: Proceedings of the third international conference on the strength of metals and alloys 1
62. Brocks W, Klingbeil D, Künecke G, Sun D-Z (1995) Application of the gurson model to ductile tearing resistance. Constraint Eff Fract Theory Appl 2:232–252
63. Brunet JC, Bellot J (1973) Deformation of MnS inclusions in steel. J Iron Steel Inst 211: 511–512

64. Budiansky B, Hutchinson J, Slutsky S (1982) Void growth and collapse in viscous solids. Mechanics of solids—the rodney hill 60th anniversary volume
65. Burghard H (1974) The influence of precipitate morphology on microvoid growth and coalescence in tensile fractures. Metall Trans 5:2083–2094
66. Butcher C, Chen Z, Worswick M (2006) A lower bound damage-based finite element simulation of stretch flange forming of Al–Mg alloys. Int J Fract 142(3-4):289–298
67. Büttner M, Seidenfuß M, Krätschmer D, Roos E (2011) Experimentelle und schädigungsmechanische Analyse der Rissentwicklung in Mischnähten. 37. MPA-Seminar, Universität Stuttgart, pp 39.1–39.21
68. Calhoun C (1970) The effects of particles on fracture processes in Magnesium alloys. Metall Trans 1:997–1006
69. Chaboche JL, Boudifa M, Saanouni K (2006) A CDM approach of ductile damage with plastic compressibility. Int J Fract 137(1-4):51–75
70. Chao H (1964) Deformation and fracture of MnS crystals. Trans ASME 386–398
71. Chaouadi R, de Meester P, Scibetta M (1996) Micromechanical modeling of ductile fracture initiation to predict fracture toughness of reactor pressure vessel steels. Le J de Physique IV 06(C6):53–64
72. Chu C (1980) Void nucleation effects in biaxially stretched sheets. J Eng Mater Technol 102:249–256
73. Cockcroft M, Latham D (1968) Ductility and the workability of metals. J Inst Met 96:33–39
74. Cottrell, A. (1959) Theoretical aspects of fracture. Department of Metallurgy, University of Cambridge: 20–53
75. Cox TB (1974) An investigation of the plastic fracture of AISI 4340 and 18 nickel–200 grade maraging steels. Metall Trans 5:1457–1470
76. Curran D (1987) Dynamic failure of solids. Physic Rep 147(5 & 6):253–388
77. Decamp K, Bauvineau L, Besson J, Pineau A (1997) Size and geometry effects on ductile rupture of notched bars in a C-Mn steel: experiments and modeling. Int J Fract 88:1–18
78. Deimel P, Sattler E (1998) Non-metallic inclusions and their relation to the J-Integral, Ji, phys, at physical crack initiation for different steels and weld metals. J Mater Sci 33:1723–1736
79. Eckstein J, Roos E, Roll K, Ruther M, Seidenfuß M (2007) Experimental and numerical investigations to extend the process limits in self-pierce riveting. In: 10th ESAFORM conference on material forming, pp 279–286
80. Eckstein J (2009) Numerische und experimentelle Erweiterung der Verfahrensgrenzen beim Halbhohlstanznieten hochfester Bleche. Dissertation, Institut für Materialprüfung, Werkstoffkunde und Festigkeitslehre, Universität Stuttgart
81. Eisele U, Seidenfuss M, Pitard-Bouet JM (1996) Comparison between fracture mechanics and local approach models for the analysis of shallow cracks. J de Physique IV:C6-75–C6-89
82. Enakoutsa K, Leblond J, Perrin G (2007) Numerical implementation and assessment of a phenomenological nonlocal model of ductile rupture. Comput Methods Appl Mech Eng 196(13–16):1946–1957
83. Engel L (1982) Rasterelektronenmikroskopische Untersuchungen von Metallschäden. 2. neubearbeitete Auflage, Carl Hanser Verlag München. ISBN 3-446-13416-6
84. Engelen RAB, Geers MGD, Baaijens FPT (2003) Nonlocal implicit gradientenhanced elasto-plasticity for the modelling of softening behaviour. Int J Plast 19(4):403–433
85. Engelen R, Fleck N, Peerlings R, Geers M (2006) An evaluation of higher-order plasticity theories for predicting size effects and localisation. Int J Solids Struct 43(7–8):1857–1877
86. Eripret C, Rousselier G (1994) First spinning cylinder test analysis using a local approach to fracture. Nucl Eng Des 152:11–18
87. Ervasti E, Stahlberg U (2005) Void initiation close to a macro-inclusion during single pass reductions in the hot rolling of steel slabs: a numerical study. J Mater Process Technol 170(1-2):142–150
88. Eshelby JD (1957) The determination of the elastic field of an ellipsoidal inclusion, and related problems. Math Phys Sci 241(1226):376–396

89. Faleskog J, Gao X, Shih C (1998) Cell model for nonlinear fracture analysis—I. Micromech calibration. Int J Fract 89:355–373
90. Fesich T, Mohan P, Marzougui D, Kan CD (2008) A study of the gurson damage model and numerical simulation of ductile failure in LS-DYNA. 7. LS-DYNA Anwenderforum, Bamberg 2008, Crash III. Versagen, Barrieren
91. Feucht M, Sun DZ, Erhart T, Frank T (2006) Recent development and applications of the Gurson model. 5. LS-DYNA Anwenderforum, Ulm 2006, Material II—Metalle, pp D-II-21–D-II-32
92. Feucht M (1998) Ein gradientenabhängiges Gursonmodell zur Beschreibung duktiler Schädigung mit Entfestigung. Dissertation, TU Darmstadt
93. Fisher JR, Gurland J (1981) Void nucleation in spheroidized carbon steels—part 1: experimental. Metal Sci 15(5):185–192
94. Fisher JR, Gurland J (1981) Void nucleation in spheroidized carbon steels—part 2: model. Metal Sci 15(5):193–202
95. Forest S, Sievert R (2006) Nonlinear microstrain theories. Int J Solids Struct 43(24):7224–7245
96. French I, Weinrich P (1975) The influence of hydrostatic pressure on the tensile deformation and fracture of copper. Metall Trans A 6A:785–790
97. French I, Weinrich P (1976) The shear mode of ductile fracture in materials with few inclusions. Metall Trans A 7A:1841–1845
98. Freund L, Lee Y (1990) Observations on high strain rate crack growth based on a strip yield model. Non-linear fracture. Springer, Berlin, pp 261–276
99. Gao X, Shih C, Tvergaard V, Needleman A (1996) Constraint effects on the ductile-brittle transition in small scale yielding. J Mech Phys Solids 44(8):1255–1282
100. Gao X, Ruggieri C, Dodds RH Jr (1998) Calibration of weibull stress parameters using fracture toughness data. Int J Fract 92:175–200
101. Gao X, Dodds R Jr, Tregoning R, Joyce A, Link R (1999) A Weibull stress model to predict cleavage fracture in plates containing surface cracks. Fatigue Fract Eng Mater Struct 22:481–493
102. Gao X, Dodds RH Jr, Tregoning RL, Joyce JA, Link RE (1999) A weibull stress model to predict cleavage fracture in plates containing surface cracks. Fatigue Fract Eng Mater Struct 22:481–493
103. Gao X, Zhang G, Srivatsan T (2005) Prediction of cleavage fracture in ferritic steels: a modified Weibull stress model. Mater Sci Eng, A 394:210–219
104. Gao X, Zhang G, Srivatsan T (2006) A probabilistic model for prediction of cleavage fracture in the ductile-to-brittle transition region and the effect of temperature on model parameters. Mater Sci Eng A 415:264–272
105. Gao X, Kim J (2006) Modeling of ductile fracture: significance of void coalescence. Int J Solids Struct 43:6277–6293
106. Gao X, Zhang G, Roe C (2010) A study on the effect of the stress state on ductile fracture. Int J Damage Mech 19(1):75–94
107. Gao X, Faleskog J, Shih C, Dodds R Jr (1998) Ductile tearing in part-through cracks: experiments and cell-model predictions. Eng Fract Mech 59(6):761–777
108. Gardner R, Pollock T, Wilsdorf H (1977) Crack initiation at dislocation cell boundaries in the ductile fracture of metals. Mater Sci Eng 169–174
109. Geers M (1997) Experimental analysis and computational modelling of damage and fracture. PhD thesis, Eindhoven University of Technology
110. Gologanu M, Leblond J, Devaux J (1994) Approximate models for ductile metals containing nonspherical voids—case of axisymmetric oblate ellipsoidal cavities. J Eng Mater Technol 116:290–297
111. Goods SH, Brown LM (1979) The nucleation of cavities by plastic deformation. Acta Metall 27:1–15
112. Gross D, Seelig T (2007) Bruchmechanik: Mit einer Einführung in die Mikromechanik. 4. bearb. Auflage, Berlin [u.a.]: Springer, Berlin. ISBN 3-540-37113-3

113. Gurland J (1963) The mechanism of dutile rupture of metals containing inclusions. Trans ASME 56:443–454
114. Gurland J (1972) Oberservation on the fracture of cementile particles in spheroidized 1,05 % C steel deformed at room temperature. Acta Metall 20(5):735–741
115. Gurson AL (1975) Plastic flow and fracture behaviour of ductile materials: incorporating void nucleation, growth, and interaction. Thesis, Brown University, Rhode Island
116. Gurson A (1977) Continuum theory of ductile rupture by void nucleation and growth: part I yield criteria and flow rules for porous ductile media. J Eng Mat Technol 99(1):2−15
117. Hancock JW (1976) On the mechanisms of ductile failure in high-strength steels subjected to multi-axial stress states. J Mech Phys Solids 24:147–169
118. Helms R (1977) Theoretische und Experimentelle Untersuchungen zum Mechanismus des duktilen Bruches metallischer Werkstoffe. Archiv für das Eisenhüttenwesen 48, Nr. 5:297–302
119. Henry G, Horstmann D (1979) De ferri metallographia V : Fraktographie und Mikrofraktographie. Düsseldorf, London: Verlag Stahleisen, Heyden. ISBN 3-514-00215-0
120. Hentrich M, Veit P, Stroppe H (1981) Der duktile Bruch von Materialien mit Einschlüssen. Wissenschaftliche Zeitschrift der Technischen Hochschule Otto von Guerike Magdeburg, Heft 2
121. Hill R (1965) A self-consistent mechanics of composite materials. J Mech Phys Solids 13:213–222
122. Hor A, Lebrun J-L, Morel F (2009) Experimental study and local approach modelling of ductile damage in steels over a wide temperature range. In: 7th EUROMECH solid mechanics conference
123. Hosseini SB, Temmel C, Karlsson B, Ingesten N-G (2007) An in-situ scanning electron microscopy study of the bonding between mns inclusions and the matrix during tensile deformation of hot-rolled steels. Metall Mater Trans A 38A:982–989
124. Huang Y (1991) Accurate dilatation rates for spherical voids in triaxial fields. J Appl Mech 58:1084–1086
125. Huber G, Brechet Y, Pardoen T (2005) Predictive model for void nucleation and void growth controlled ductility in quasi-eutectic cast aluminium alloys. Acta Mater 53:2739–2749
126. Hütter G, Linse T, Mühlich U, Kuna M (2013) Simulation of ductile crack initiation and propagation by means of a non-local gurson-model. Int J Solids Struct 50(5):662–671
127. Hütter G, Linse T, Roth S, Mühlich U, Kuna M (2013) A modeling approach for the complete ductile-brittle transition region: cohesive zone in combination with a non-local gurson-model. Int J Fract 185(1-2):1–25
128. Ishikawa N, Parks D, Kurihara M (2000) Micromechanism of ductile crack initiation in structural steels based on void nucleation and growth. ISIJ Int 40(5):519–527
129. Jackiewicz J, Kuna M (2003) Non-local regularization for fe simulation of damage in ductile materials. Comput Mater Sci 28(3-4):684–695
130. Jirásek M (1998) Nonlocal models for damage and fracture: comparison of approaches. Int J Solids Struct 35(31-32):4133–4145
131. Jirásek M, Rolshoven S (2009) Localization properties of strain-softening gradient plasticity models. part I: strain-gradient theories. Int J Solids Struct 46:2225–2238
132. Johnson, G., W. Cook (1983) A constitutive model and data for metals subjected to large strains, high strain rates and high temperatures. In: Proceedings of 7th international symposium on ballistics, pp 541–547
133. Johnson G, Cook W (1985) Fracture characteristics of three metals subjected to various strains, strain rates, temperatures and pressures. Eng Fract Mech 21(1):31–48
134. Kachanov LM (1999) Rupture time under creep conditions (1958). Int J Fract 97:11–18
135. Kanvinde AM (2006) Void growth model and stress modified critical strain model to predict ductile fracture in structural steels. Struct Eng 1907–1918
136. Kim J, Gao X, Srivatsan T (2004) Modeling of void growth in ductile solids: effects of stress triaxiality and initial porosity. Eng Fract Mech 71:379–400

137. Koplik J, Needleman A (1988) Void growth and coalescence in porous plastic solids. Int J Solids Struct 24(8):835–853
138. Kröner E (1961) Zur plastischen Verformung des Vielkristalls. Acta Metall 9:155–161
139. Kroon, M., Faleskog J. (2002) A probabilistic model for cleavage fracture with a length scale-influence of material parameters and constraint. Int J Fract 99–118
140. Kroon M, Faleskog J (2005) Micromechanics of cleavage fracture initiation in ferritic steels by carbide cracking. J Mech Phys Solids 53(1):171–196
141. Kuna M, Sun D (1996) Three-dimensional cell model analyses of void growth in ductile materials. Int J Fract 81:235–258
142. Kussmaul K, Eisele U, Seidenfuss M (1993) On the Applicability of local approaches for the determination of the failure behavior of ductile steels. J Pressure Vessel Technol 115:214–220
143. Kussmaul K, Seidenfuss M (1993) On the transferability of micro-mechanical damage models to specimens of different size and geometry. In: The 6th German-Japanese joint seminar on structural strength and NDE Problems in nuclear engineering
144. Kussmaul K, Seidenfuss M, Eisele U (1993) On the applicability of damage models for the description of the failure behaviour of ductile steels. In: 8th international confernce on fracture
145. Kussmaul K, Eisele U, Seidenfuss M (1995) On the applicability of local approach models for the determination of the failure behaviour of steels of different toughness. Fatigue Fract Mech Press Vessel Pip ASME PVP 304:17–25
146. Kußmaul K, Seidenfuß M, Elsäßer K, Mayer U, Zies G (1995) Reaktorsicherheitsforschung - Vorhaben-Nr. 1500 913: Experimentelle und numerische Untersuchungen zur Beschreibung des Versagensverhaltens von Stählen unterschiedlicher Zähigkeit mit Hilfe von Schädigungsmodellen. MPA Universität Stuttgart
147. Lange G (1983) Mikroskopische und makroskopische Erscheinungsformen des duktilen Gewaltbruches (Gleitbruch), Systematische Beurteilung techn. Schadensfälle, DGM, pp 79–87
148. Lautridou J (1981) Crack initiation and stable crack growth resistance in A508 steels in relation to inclusion distribution. Eng Fract Mech 15:55–71
149. Le Roy G, Embury JD, Edwards G, Ashby M (1981) A model of ductile fracture based on the nucleation and growth of voids. Acta Metall 29:1509–1522
150. Leblond J-B, Mottet G (2008) A theoretical approach of strain localization within thin planar bands in porous ductile materials. C R Mécanique 336(1−2):176−189
151. Leblond J, Perrin G, Devaux J (1994) Bifurcation effects in ductile metals with nonlocal damage. J Appl Mech 61:236
152. Leblond J, Perrin G, Devaux J (1995) An improved Gurson-type model for hardenable ductile metals. Eur J Mech A/Solids 14:499–527
153. Lemaître J, Desmorat R (2005) Engineering damage mechanics: ductile, creep, fatigue and brittle failures. Springer, Berlin. ISBN 3-540-21503-4
154. Lemaitre J, Chaboche JL (1978) Aspects phénoménologique de la rupture par endommagement. J de Mécanique appliquée 2(3):317–365
155. Lemaitre J (1984) How to use damage mechanics. Nucl Eng Des 80:233–245
156. Lemaitre J (1985) Coupled elasto-plasticity and damage constitutive equations. Comput Methods Appl Mech Eng 51:31–49
157. Lemaitre J (1985) A continuous damage mechanics model for ductile fracture. J Eng Mater Technol 107:83–89
158. Lemaitre J (1990) Micro-mechanics of crack initiation. Int J Fract 42:87–99
159. Lemaitre J (1996) A course on damage mechanics, 2nd revised and enlarged edition. Springer, Berlin. ISBN 3-540-60980-6
160. Linse T, Hütter G, Kuna M (2012) Simulation of crack propagation using a gradient-enriched ductile damagemodel based on dilatational strain. Eng Fract Mech 95:13–28

161. Linse T, Kuna M, Viehrig HW (2014) Quantification of brittle-ductile failure behavior of ferritic reactor pressure vessel steels using the small-punch-test and micromechanical damage models. Mater Sci Eng, A 614:136–147
162. Lode W (1926) Versuche über den Einfluß der mittleren Hauptspannung auf das Fließen der Metalle Eisen. Kupfer und Nickel. Zeitschrift für Physik 36(11–12):913–939
163. Lorentz E, Besson J, Cano V (2008) Numerical simulation of ductile fracture with the Rousselier constitutive law. Comput Methods Appl Mech Eng 197(21–24):1965–1982
164. Luong Dung N, Appeltauer J (1992) Repräsentative Bruchkriterien in der Kaltmassivumformung metallischer Werkstoffe. Forschung im Ingenieurwesen-Eng Res 58 (3):54–60
165. Maire E, Bouaziz O, Di Michiel M, Verdu C (2008) Initiation and growth of damage in a dual-phase steel observed by X-ray microtomography. Acta Mater 56:4954–4964
166. Margolin H (1978) Void formation, void growth and tensile fracture in Ti-6Al-4V. Metall Trans A 9A:781–790
167. Marini B (1985) Ductile rupture of A508 steel under non-radial loading. Eng Fract Mech 22 (3):375–386
168. Marini B, Mudry F, Pineau A (1985) Experimental study of cavity growth in ductile rupture. Eng Fract Mech 22(6):989–996
169. McClintock F (1968) A criterion for ductile fracture by the grwoth of holes. J Appl Mech 363–371
170. Mediavilla J, Peerlings R, Geers M (2004) Application of a gradient ductile damage model to metal forming processes including crack propagation and mesh adaptivity. In: 11th international conference on sracture
171. Menzemer C, Srivatsan T, Al-Hajri M, Ortiz R (2000) The impact toughness and tensile properties of 8320 steel. Mater Sci Eng, A 289:198–207
172. Molnar D (2011) Private communications. IMWF Universität Stuttgart
173. Morgeneyer T, Proudhon H, Besson J (2010) Study of the flat to slant crack transition in ductile thin sheet material: Simulations and experiments. In: Proceedings of the 18th European conference on fracture
174. Mudry F (1987) A local approach to cleavage fracture. Nucl Eng Des 105(1):65–76
175. Münstermann S, Langenberg P, Seidenfuss M (2005) Numerische Bestimmung von duktilen Rissinitiierungskennwerten unter Berücksichtigung der Mikrostruktur. DVM-Bericht 237 "Technische Sicherheit, Zuverlässigkeit und Lebensdauer":85–99
176. Münstermann, S. (2006) Numerische Beschreibung des duktilen Versagensverhalten von hochfesten Baustählen unter Berücksichtigung der Mikrostruktur. Dissertation, IEHK, RWTH-Aachen
177. Nahshon K, Hutchinson J (2008) Modification of the Gurson model for shear failure. Eur J Mech A Solids 27:1–17
178. Needleman A (1987) Continuum model for void nucleation by inclusion debonding. J Appl Mech 54(3):525–531
179. Needleman A, Tvergaard V (1995) Analysis of a brittle-ductile transition under dynamic shear loading. Int J Solids Struct 32(17):2571–2590
180. Nikitin L (1996) Softening solids: reality or misinterpretation? Technische Mechanik 16 (1):86–96
181. Nonn A (2009) Experimentelle und numerische Analyse des Schädigungsverhaltens von Hybridlaserschweißverbindungen. Dissertation, Rheinisch-Westfälische Technische Hochschule Aachen
182. Nonn A, Kalwa C (2010) Modelling of damage behaviour of high strength pipeline steel. In: 18th European conference on fracture
183. Ockewitz A, Sun D-Z, Klamser H, Malcher D (2006) Damage modelling of automobile components of aluminium materials under crash loading. 5. LS-DYNA Anwenderforum, Ulm 2006, pp 1–12
184. Ortiz M, Leroy Y, Needleman A (1987) A finite element method for localized failure analysis. Comput Methods Appl Mech Eng 61(2):189–214

185. Oyane M (1972) Criteria of ductile fracture strain. Jpn Soc Mech Eng 15(90):1507–1513
186. Pan J, Saje M, Needleman A (1983) Localization of deformation in rate sensitive porous plastic solids. Int J Fract 21:261–278
187. Pardoen T, Hutchinson J (2000) An extended model for void growth and coalescence. J Mech Phys Solids 48:2467–2512
188. Pardoen T, Hutchinson J (2003) Micromechanics-based model for trends in toughness of ductile metals. Acta Mater 51:133–148
189. Pardoen T (2006) Numerical simulation of low stress triaxiality ductile fracture. Comput Struct 84:1641–1650
190. Pardoen T (1998) Experimental and numerical comparison of void growth models and void coalescence criteria for the prediction of ductile fracture in copper bars. Acta Mater 46(2):541–552
191. Pavankumar T, Samal M, Chattopadhyay J, Dutta B, Kushwaha H, Roos E, Seidenfuss M (2005) Transferability of fracture parameters from specimens to component level. Int J Press Vessels Pip 82:386–399
192. Peerlings R, de Borst R, Brekelmans W, de Vree J (1996) Gradient enhanced damage for quasi-brittle materials. Int J Numer Meth Eng 39(19):3391–3403
193. Peerlings R, Geers M, de Borst R, Brekelmans W (2001) A critical comparison of nonlocal and gradient-enhanced softening continua. Int J Solids Struct 38(44–45):7723–7746
194. Perrin G, Leblond J (1990) Analytical study of a hollow sphere made of plastic porous material and subjected to hydrostatic tension-application to some problems in ductile fracture of metals. Int J Plast 6:677–699
195. Pettermann H (2000) Numerical simulations of a compositionally graded structure using a hierarchical approach. Mat Sci Eng, A276 276:277−282
196. Pijaudier-Cabot G, Bazant ZP (1987) Nonlocal damage theory. J Eng Mech 113(10):1512–1533
197. Pijaudier-Cabot G, Baant ZP, Tabbara M (1988) Comparison of variousmodels for strain-softening. Eng Comput 5(2):141–150
198. Pijaudier-Cabot G, Benallal A (1993) Strain localization and bifurcation in a nonlocal continuum. Int J Solids Struct 30(13):1761–1775
199. Pineau A., Joly P (1991) Local versus global approaches to elastic-plastic fracture mechanics. Application to ferritic steels and a cast duplex stainless steel, components—fundamentals and applications. In: ESIS/EGF9, pp 381–414
200. Pineau A (1997) Modelling of scatter and size effects in ductile and brittle fracture. In: Transactions of the 14th international conference on SMiRT
201. Pineau A (2006) Development of the local approach to fracture over the past 25 years: theory and applications. Int J Fract 138(1):139–166
202. Pospiech J (1995) Ductile fracture of carbon steels: a review. J Mater Eng Perform 4(1):82–89
203. Potirniche G, Horstemeyer M, Wagner G, Gullett P (2006) A molecular dynamics study of void growth and coalescence in single crystal nickel. Int J Plast 22:257–278
204. Poussard C, Sainte–Catherine C, Galon P, Forget P (2002) Finite element simulations of sub-size charpy tests and associated transferability to toughness results. In: Francois D, Pineau A (eds) Charpy to present impact testing. Elsevier science, London, pp 469–478
205. Poussard C, Seidenfuss M (1997) On the simulation of ductile crack growth using the Rousselier model. In: Transactions of the 14th international conference on structural mechanics in reactor technology, pp 673–680
206. Prahl U, Papaefthymiou S, Uthaisangsuk V, Bleck W, Sietsma J, van der Zwaag S (2007) Micromechanics-based modelling of properties and failure of multiphase steels. Comput Mater Sci 39:17–22
207. Puttik K (1959) Ductile fracture in metals. Phil Mag 4:964–969
208. Rabotnov Y (1968) Creep rupture. Applied mechanics. In: Proceedings of the twelfth international congress of applied mechanics, pp 342–349

209. Reusch F (2003) Entwicklung und Anwendung eines nicht-lokalen Materialmodells zur Simulation duktiler Schädigung in metallischen Werkstoffen. Dissertation, Universität Dortmund. ISBN 249591204
210. Reusch F, Svendsen B, Klingbeil D (2003) A non-local extension of Gursonbased ductile damagemodeling. Comput Mater Sci 26:219–229
211. Rice J, Tracey D (1969) On the ductile enlargement of voids in triaxial stress fields. J Mech Phys Solids 17:201–217
212. Rivalin F, Pineau A, Di Fant M, Besson J (2001) Ductile tearing of pipeline-steel wide plates: I. Dynamic and quasi-static experiments. Eng Fract Mech 68:329–345
213. Roberts W, Lehtinen B, Easterling KE (1976) An in situ SEM study of void development around inclusions in steel during plastic deformation. Acta Metall 24:745–758
214. Roos E, Seebich HP, Seidenfuss M, Schmauder S, Kizler P (2005) Effect of variation of microstructure on fracture mechanics parameters. In: 3rd Indo-German seminar on “advances in structural integrity & safety”, Paper J03: S. mechanics
215. Roos E, Eisele U, Lammert R, Restemeyer D, Schuler X, Seebich H-P, Seidenfuß M, Silcher H, Stumpfrock L (2006) Reaktorsicherheitsforschung – Vorhaben-Nr. 1501 240: Kritische Überprüfung des Masterkurve-Ansatzes im Hinblick auf die Anwendung bei deutschen Kernkraftwerken. MPA Universität Stuttgart
216. Roos E, Seidenfuss M, Krämer D, Krolop S, Eisele U, Hindenlang U (1991) Application and evaluation of different numerical methods for determining crack resistance curves. Nucl Eng Des 130:297–308
217. Rösch L (1969) Relationship between precipitation and dimple fracture in an 18 percent nickel maraging steel. Electron microfractography, In: ASTM STP453, pp 3–32
218. Rosenfield A (1972) Fracture of steels containing pearlite. Metall Trans 3:2797–2804
219. Roth S, Kuna M. (2011) Numerical study on interfacial damage of sprayed coatings due to thermo-mechanical fatigue. In: Proceedings of the XI international conference on computational plasticity
220. Rousselier G (2001) Dissipation in porous metal plasticity and ductile fracture. J Mech Phys Solids 49:1727–1746
221. Rousselier G, Pastor J, Bilger N, Leclercq S (2004) Recent results on ductile fracture modeling at the macro and microscales. In: 11th international conference on fracture
222. Rousselier G, Leclercq S (2006) A simplified “polycrystalline” model for viscoplastic and damage finite element analyses. Int J Plast 22:685–712
223. Rousselier G (1981) Finite deformation constitutive relations including ductile fracture damage. Three-dimensional constitutive relations and ductile fracture. North-Holland Publishing Company, Holland, pp 331–355
224. Rousselier G (1987) Ductile fracture models and their potential in local approach of fracture. Nucl Eng Des 105:97–111
225. Rudd RE, Belak JF (2002) Void nucleation and associated plasticity in dynamic fracture of polycrystalline copper: an atomistic simulation. Comput Mater Sci 24:148–153
226. Rudnicki, J. W., Rice J.R. (1975) Conditions for the localization of deformation in pressure-sensitive dilatant materials. J Mech Phys Solids 23(6):371–394
227. Ruggieri C, Dodds R Jr (1996) A transferability model for brittle fracture including constraint and ductile tearing effects: a probabilistic approach. Int J Fract 79:309–340
228. Sabirov I, Kolednik O (2005) The effect of inclusion size on the local conditions for void nucleation near a crack tip in a mild steel. Scripta Mater 53:1373–1378
229. Sainte–Catherine C, Poussard C, Vodinh J, Schill R, Hourdequin N, Galon P, Forget P (2002) Finite element simulations and empirical correlation for Charpy-V and subsize Charpy tests on an unirradiated low-alloy RPV ferritic steel. small specimen test techniques 4. In: ASTM STP 1418, pp 107–136
230. Samal, MK (2007) Nonlocal damage models for structural integrity analysis. Dissertation, Universität Stuttgart
231. Samal M, Seidenfuss M, Roos E, Dutta B, Kushwaha H (2008) Finite element formulation of a new nonlocal damagemodel. Finite Elem Anal Des 44(6–7):358–371

232. Samal M, Seidenfuss M, Roos E, Dutta B, Kushwaha H (2008) A mesh-independent Gurson–Tvergaard–Needleman damage model and its application in simulating ductile fracture behavior. Proc IMechE. Part C: J Mech Eng Sci 223:283–292
233. Schiffmann R (2001) Experimentelle Bestimmung und modellmässige Beschreibung der Schädigung beim Gleitbruch von Stählen. Dissertation, IEHK, RWTH-Aachen
234. Scheyvaerts F, Pardoen T, Onck P (2010) A new model for void coalescence by internal necking. Int J Damage Mech 19:95–126
235. Scheyvaerts F, Onck P, Tekoglu C, Pardoen T (2011) The growth and coalescence of llipsoidal voids in plane strain under combined shear and tension. J Mech Phys Solids 59:373–397
236. Scheil E, Schnell R (1952) Die Verformbarkeit von Schlackeneinschlüssen im Stahl und ihre Bedeutung für die Beurteilung von Schmiedestücken. Stahl und Eisen
237. Schlüter N (1997) Einfluss der Beanspruchung und des Gefüges auf die lokale Schädigung beim Gleitbruch von Baustählen. Dissertation, IEHK, RWTH Aachen
238. Schmitt W, Keim E, Sun D-Z, Blauel J, Nagel G (1999) Load-carrying capacity and crack resistance of a cladding by the Sigma-oscillating wire technique. Nucl Eng Des 190:149–158
239. Seebich H-P (2007) Mikromechanisch basierte Schädigungsmodelle zur Beschreibung des Versagensablaufs ferritischer Bauteile. dissertation, Universität Stuttgart
240. Seidenfuß M (1992) Untersuchungen zur Beschreibung des Versagensverhaltens mit Hilfe von Schädigungsmodellen am Beispiel des Werkstoffes 20 MnMoNi 5 5. Dissertation, Fakultät Energietechnik, Universität Stuttgart
241. Seidenfuß M, Steglich D, Heerens J (1998) Beschreibung des Verhaltens von Al_3Ti-partikelverstärktem Aluminium durch zwei verschiedene Schädigungsmodelle. DVM-Bericht 230:59–72
242. Seidenfuß M, Elsäßer K (1999) MPA/VGB Forschungsvorhaben 5.4.1: Analytische Ermittlung des Bruchverhaltens austenitischer Rohrleitungen - Local Approach – Teilschritt 2 : Teilbericht 5.4.1-2: Bestimmung der Schädigungsparameter für den austenitischen Rohrleitungswerkstoff 1.4541. MPA Universität Stuttgart
243. Seidenfuß M, Turan K, Didra H-P, Kuna M (1999) MPA/VGB Forschungsvorhaben 5.4.1: Analytische Ermittlung des Bruchverhaltens austenitischer Rohrleitungen - Local Approach – Teilschritt 1 : Teilbericht 5.4.1-1: Nachrechnung Schlitzversuch, Nachrechnung Oberflächenkerbe. MPA Universität Stuttgart
244. Seidenfuß M (2014) Schädigungsmechanische Modelle zur Beschreibung des Versagensablaufs in metallischen Bauteilen. Habilitationsschrift, Techn.-wiss. Ber., MPA Universität Stuttgart
245. Seppälä E, Belak J, Rudd R (2004) Onset of void coalescence during dynamic fracture of ductile metals. Phys Rev Lett 93:24
246. Shi Y (1989) Critical void growth for ductile rupture of steel welds. Eng Fract Mech 34 (4):901–907
247. Shi Y, Barnby J, Nadkarni A (1991) Void Growth at ductile crack initiation of a structural steel. Eng Fract Mech 39(1):37–44
248. Shterenlikht A, Howard I (2006) The cafe model of fracture—application to a tmcr steel. Fatigue Fract Eng Mater Struct 29(9–10):770–787
249. Simatos A (2010) Methode XFEM pour la modelisation de grandes propagations de fissure en dechirure ductile: transition d'un milieu continu vers une fissure via un modele de zone cohesive pour le modele de Rousselier. Dissertation, Mecanique, Energétique, Génie Civil, Acoustique, L'Institut National des Sciences Appliquées de Lyon
250. Soppa E, Schmauder S, Fischer G (2004) Particle cracking and debonding criteria in Al/Al_2O_3 composites. In: Proceedings of the sixth international conference for mesomechanics, pp 312–317
251. Soppa E, Fischer G, Seidenfuß M, Lammert R, Wackenhut G, Diem H (2008) Deformation and damage in Al based composites, FE simulations and experiments. Aluminium Alloys—Their Phys Prop 2:1225–1231

252. Soppa E, Nellesen J, Romanova V, Fischer G, Crostack H-A, Beckmann F (2010) Impact of 3D-model thickness on FE-simulations of microstructure. Mater Sci Eng, A 527(3):802–811
253. Springmann M, Kuna M (2003) Identification of material parameters of the Rousselier model by non-linear optimization. Comput Mater Sci 26:202–209
254. Springmann M (2005) Identifikation von Materialparametern schädigungsmechanischer Gesetze unter Einbeziehung der Dehnungslokalisierung. Dissertation, Technische Universität Bergakademie Freiberg
255. Springmann M, Kuna M (2005) Identification of material parameters of the Gurson-Tvergaard-Needleman model by combined experimental and numerical techniques. Comput Mater Sci 33:501–509
256. Steglich D, Pirondi A, Bonora N, Brocks W (2005) Micromechanical modelling of cyclic plasticity incorporating damage. Int J Solids Struct 42:337–351
257. Steglich D, Wafai H, Besson J (2010) Interaction between anisotropic plastic deformation and damage evolution in Al 2198 sheet metal. Eng Fract Mech 77(17):3501−3518
258. Steglich D (1999) Bestimmung von mikrostrukturellen Parametern in Schädigungsmodellen für duktile Metalle. Dissertation, Technische Universität Berlin
259. Steinmann P (1999) Formulation and computation of geometrically non-linear gradient damage. Int J Numer Meth Eng 46(5):757–779
260. Steinmann P, Larsson R, Runesson K (1997) On the localization properties of multiplicative hyperelasto-plastic continua with strong discontinuities. Int J Solids Struct 34(8):969–990
261. Stroppe H (1981) Ermittlung der Bruchzähigkeit duktiler Werkstoffe aus Parametern der Mikrostruktur und der Fließkurve. Neue Hütte, Heft 12:446–448
262. Sun D-Z, Siegele D, Voss B, Schmitt W (1989) Application of local damage models to the numerical analysis of ductile rupture. Fatigue Fract Eng Mater Struct 12(3):201–212
263. Sun D-Z, Hönig A, Böhme W, Schmitt W (1995) Application of micromechanical models to the analysis of ductile fracture under dynamic loading. Fracture mechanics. In: ASTM STP 1220, vol 25, pp 343–356
264. Suquet P (1997) Continuum micromechanics. Springer, Wien, New York, pp 61-130. ISBN 3-211-82902-4
265. Tanguy B, Besson J (2002) An extension of the Rousselier model to viscoplastic temperature dependent materials. Int J Fract 116:81–101
266. Tanguy B, Besson J, Piques R, Pineau A (2005) Ductile to brittle transition of an A508 steel characterized by Charpy impact test: Part II: modeling of the Charpy transition curve. Eng Fract Mech 72:413–434
267. Tanguy B, Luu T, Perrin G, Pineau A, Besson J (2008) Plastic and damage behaviour of a high strength X100 pipeline steel: experiments and modeling. Int J Press Vessels Pip 85:322–335
268. Tanaka J (1970) Fractographic analysis of the low energy fracture of an aluminium alloy. Review of developments in plane strain fracture toughness testing. In: ASTM STP 463, pp 191–215
269. Tanaka K, Mori T, Nakamura T (1970) Cavity formation at the interface of a spherical inclusion in a plastically deformed matrix. Phil Mag 21(170):267–279
270. Thomson C, Worswick M, Pilkey A, Lloyd D (2003) Void coalescence within periodic clusters of particles. J Mech Phys Solids 51:127–146
271. Thomason P (1968) A theory for ductile fracture by internal necking of cavities. J Inst Metals 96:360–365
272. Thomason P (1985) A three-dimensional model for ductile fracture by the growth and coalescence of microvoids. Acta Metall 33(6):1087–1095
273. Thomason P (1985) Three-dimensional models for the plastic limit-loads at incipient failure of the intervoid matrix in ductile porous solids. Acta Metall 33(6):1079–1085
274. Thomason P (1990) Ductile fracture of metals, 1st edn. Pergamon Press, Oxford. ISBN 0-08-040178-3
275. Thomason P (1998) A view on ductile-fracture modeling. Fatigue Fract Eng Mater Struct 21:1105–1122

276. Tipper C (1948) The fracture of metals. Metallurgia: Br J Metals 133–137
277. Tvergaard V (1981) Influence of voids on shear band instabilities under plane strain conditions. Int J Fract 17(4):389–407
278. Tvergaard V (1982) Material failure by void coalescence in localized shear bands. Int J Solids Struct 18(8):659–672
279. Tvergaard V (1982) On localization in ductile materials containing spherical voids. Int J Fract 18:237–252
280. Tvergaard V, Needleman A (1995) Effects of nonlocal damage in porous plastic solids. Int J Solids Struct 32(8-9):1063–1077
281. Tvergaard V, Needleman A (1984) Analysis of the cup-cone fracture in a round tensile bar. Acta Metall 32(1):157–169
282. Van Stone R, Cox TB, Low J Jr, Psioda J (1985) Microstructural aspects of fracture by dimpled rupture. Int Metals Rev 30:157–179
283. Verein Deutscher Eisenhüttenleute (1996) Erscheinungsformen von Rissen und Brüchen metallischer Werkstoffe, 2nd edn. Verlag Stahleisen GmbH, Düsseldorf
284. Weck A, Wilkinson D, Toda H, Maire E (2006) 2D and 3D visualization of ductile fracture. Adv Eng Mater 8:469–472
285. Weck A (2007) The role of coalescence on ductile fracture. Dissertation, McMaster University, Hamilton Ontario
286. Weck A, Segurado J, LLorca J, Wilkinson D, Böhm H (2008) Numerical simulations of void linkage in model materials using a nonlocal ductile damage approximation. Int J Fract 148:205–219
287. Weck A, Wilkinson D (2008) Experimental investigation of void coalescence in metallic sheets containing laser drilled holes. Acta Mater 56:1774–1784
288. Weinrich P, French I (1976) The influence of hydrostatic pressure on the fracture mechanisms of sheet tensile specimens of copper and brass. Acta Metall 24:317–322
289. Wilson D (1971) Effects of second-phase particles on formability at room temperature, Effect of second-phase particles on the mechanical properties of steel. Scarborough GB 28–36
290. Wilsdorf H (1975) Void initiation, growth, and coalescence in ductile fracture of metals. J Electron Mater 4(5):791–809
291. Wood I (1963) Fracture and deformation of sulfide inclusions in steel. Trans ASME 56:770–773
292. Xia L, Shih CF (1995) Ductile crack growth—I. A numerical study using computational cells with microstructurally-based length scales. J Mech Phys Solids 43(2):233–259
293. Zhang ZL, Thaulow C, Odegard J (2000) A complete Gurson model approach for ductile fracture. Eng Fract Mech 67:155–168
294. Zhang ZL, Skallerud B (2010) Void coalescence with and without prestrain history. Int J Damage Mech 19:153–174

Recent Trends in the Development of Gurson's Model

Jacek Jackiewicz

Abstract The original Gurson model for porous materials has undergone numerous modifications in order to improve its adequacy with experimental or numerical results. In this chapter various modifications of Gurson's model and models created on the basis of the idea of Gurson's model are presented. This chapter includes the following issues: (i) development of Gurson's model, (ii) development of models for nucleation, growth and coalescence of voids and (iii) modification of Gurson's model for failure prediction under shear deformation.

1 Introduction

The year 2015 marks the 40th anniversary of the publication of a PhD thesis of Gurson, in which yield criteria and flow rules for porous materials were proposed, centering around the effect of void nucleation and growth, as observed during ductile fracture. Along the past 40 years, the original formulation of Gurson's model, established in a micro-mechanical context, has been continuously developed by many scientists. This model has been subjected to many studies, various improvements, as well as criticisms. However, the research interests in Gurson's model do not decline. Based on Web Search Engines, which index the full text of scholarly literature across an array of publishing formats and disciplines the fundamental ideas of Gurson's model have appeared in over 400 publications each year in the last five years. The present chapter summarizes the recent trends in the development of Gurson's model. Emphasis on description of particular results is put in order to direct the reader's attention to valuable new applications and promising or interesting ways of solving problems in ductile fracture of polycrystalline metals.

J. Jackiewicz (✉)
Faculty of Mathematics, Physics & Technical Sciences, Kazimierz Wielki University, ul. Chodkiewicza 30, 85–064 Bydgoszcz, Poland
e-mail: jaceksnd@aim.com

G. Hütter and L. Zybell, *Recent Trends in Fracture and Damage Mechanics*, DOI 10.1007/978-3-319-21467-2_17

Among others, the following main topics are explored with respect to Gurson's model: Sect. 2 reviews the development of Gurson's model; Sect. 3 presents the general development of models for nucleation, growth and coalescence of voids; the modification of Gurson's model for failure prediction under shear deformation is discussed in Sect. 4; and finally, some concluding remarks are given in Sect. 5.

2 Development of Gurson's Model

Since experiments show that the ductile fracture of metals involves nucleation, growth and coalescence of voids, the study of the constitutive behavior of the porous material has become the basis of the ductile damage and the ductile fracture analysis. McClintock [1] considered the deformation of a cylindrical cavity in an infinite, perfectly plastic matrix material subjected to axial and transverse stresses. The same problem but for the growth of an initially spherical void in an infinite matrix was analyzed by Rice and Tracey [2]. Needleman [3] performed finite element analysis of large flow in two-dimensional models of porous materials.

The standard methodology in homogenization theories pioneered in the works of Mandel [4] and Hill [5] is to connect macroscopic quantities through volume averaging of their microscopic counterparts by means of generic cell models, which are applied to accommodate for the micro-macro coupling. However, Gurson [6] was the first who developed a damage-based plastic yield criterion and flow rules for a porous ductile material based on the micro-macro approach of limit analysis about the hollow sphere model with a von Mises solid matrix and subjected to conditions of homogeneous boundary strain rate. More precisely, based on a rigid-perfectly plastic analysis of spherically symmetric deformation around a spherical cavity, Gurson [6] established a macroscopic yield condition for porous metals in the form:

$$\left(\frac{\Sigma_{\mathrm{eq}}}{\sigma_0}\right)^2 + 2f_\upsilon \cosh\left(\frac{3\Sigma_{\mathrm{h}}}{2\sigma_0}\right) - 1 - f_\upsilon^2 = 0, \tag{1}$$

where f_υ is the porosity (void volume fraction), σ_0 = const. represents the tensile yield stress of the ideal plastic matrix material, Σ_{h} denotes the macroscopic hydrostatic stress and Σ_{eq} is the macroscopic von Mises equivalent stress. The plastic matrix is assumed to fulfill the following von Mises microscopic yield criterion:

$$f(\sigma) = \sigma_{\mathrm{eff}} - \sigma_0 = 0 \tag{2}$$

with

$$\sigma_{\mathrm{eff}} \overset{\mathrm{def}}{=} \sqrt{\frac{3}{2}\boldsymbol{\sigma}^{\mathrm{dev}} : \boldsymbol{\sigma}^{\mathrm{dev}}}, \tag{3}$$

where $\boldsymbol{\sigma}$ denotes the microscopic stress tensor, $\boldsymbol{\sigma}^{dev}$ its deviatoric part and σ_{eff} the microscopic von Mises effective stress. Generalizations to include hardening matrix material were also made.

For a porous representative volume element Ω containing voids that occupy volume ω, a basic microstructural variable consists of the void volume fraction $f_\upsilon \stackrel{def}{=} \omega/\Omega$. Due to the plastic incompressibility of the material surrounding the void (i.e., the matrix material) and the neglect of its elastic deformations, the change in material volume may only originate from the variation in the volume of the voids. Hence, $\dot{\omega} = \dot{\Omega}$ so that the change in porosity during plastic deformation is given by the evolution equation:

$$\dot{f}_\upsilon = \frac{d}{dt}\left(\frac{\omega}{\Omega}\right) = \left(1 - \frac{\omega}{\Omega}\right)\frac{\dot{\Omega}}{\Omega}. \tag{4}$$

The above equation may be restated as

$$\dot{f}_\upsilon = (1 - f_\upsilon)\,\mathrm{tr}\,\mathbf{D}^p, \tag{5}$$

where $\mathbf{D}^p$ is the tensor of plastic macrostrain rate. Other evolution equations, which take into account nucleation and growth of voids, have been considered.

Note that the strain additive decomposition is used: $\mathbf{D} = \mathbf{D}^e + \mathbf{D}^p$, where $\mathbf{D}$ is the tensor of total strain rate, whereas $\mathbf{D}^e$ is the tensor of elastic strain rate, which components are significantly smaller than components of $\mathbf{D}^p$ for finite strains. In addition the rate of the Eulerian finite strain tensor $\dot{\mathbf{E}} = \mathbf{F}^T \cdot \mathbf{D} \cdot \mathbf{F} \cong \mathbf{D}$ because, in the case of application of an updated Lagrangian finite element method, it is assumed that $\mathbf{F} = \mathbf{I} + \delta\mathbf{F}$, where $\mathbf{F}$ is the deformation gradient tensor and $\mathbf{I}$ is the unit tensor.

Based on the approximation of Koplik and Needleman [7] for the elastic dilatation, the porosity change due to the elastic deformation of material can be stated as

$$f_\upsilon = 1 - (1 - \mathrm{f}_{\upsilon 0})\frac{\mathrm{V}_0}{V}\left[1 + \frac{3(1 - 2\mathrm{v})}{\mathrm{E}}\Sigma_\mathrm{h}\right], \tag{6}$$

where V_0 and V are the initial and current volumes of the cell, respectively; $\mathrm{f}_{\upsilon 0}$ is the initial porosity, v is Poisson's ratio and E is Young's modulus of the matrix material.

If the matrix is rigid plastic with isotropic strain hardening, the flow effective stress of the matrix material σ_e relates to the effective plastic strain ε_e^p as $\sigma_e = \sigma_e(\varepsilon_e^p)$, the macroscopic yield surface obtained from Gurson's model is identical to Eq. (1) with σ_e substituting σ_0.

According to the equivalence between the rate of macroscopic plastic work $\Sigma_{ij}\dot{E}_{ij}^p$ and the rate of plastic work dissipated in the matrix material inside the same representative volume element $(1 - f_\upsilon)\sigma_e\dot{\varepsilon}_e^p$, what can be mathematically written as

$$\Sigma_{ij}\dot{E}_{ij}^{p} = (1-f_{\upsilon})\sigma_{e}\dot{\varepsilon}_{e}^{p}, \tag{7}$$

the rate of the effective yield stress of the matrix takes the form:

$$\dot{\sigma}_{e} = \frac{h\,\Sigma_{ij}\dot{E}_{ij}^{p}}{(1-f_{\upsilon})\sigma_{e}}, \tag{8}$$

where $h = d\sigma_{e}/d\varepsilon_{e}^{p}$.

Gurson's analysis consists in the use of the kinematic approach of limit analysis in order to obtain an upper bound to the macroscopic criterion of the spherically porous material, at least in the sense of the celebrated composite sphere assemblage of Hashin and Shtrikman [8–10]. In conformity with upper bound theories, the velocity field of the unit cell model needs to meet both the incompressibility conditions and the kinematical boundary conditions at the inner and outer radius of the hollow sphere. Since the Gurson model is based on upper bound plasticity theory, this model will overestimate the material strength and underestimate the porosity in the material by restricting void growth. The overestimation of the material strength might result from the assumption of the fully plastic flow fields.

The original Gurson criterion (1) for porous materials has undergone numerous modifications in order to improve its adequacy with experimental or numerical results. To improve agreement between predictions of Gurson's yield criterion and experimental data, Tvergaard [11] introduced three additional calibration parameters q_1, q_2 and q_3 in the structure of this criterion. He suggested the values $q_1 \approx 1.5$, $q_2 \approx 1$ and $q_3 \approx q_1^2$ based on considerations of bifurcation in shear of a square array of cylindrical holes. Tvergaard's criterion takes the form:

$$\left(\frac{\Sigma_{eq}}{\sigma_{e}}\right)^{2} + 2q_{1}f_{\upsilon}\cosh\left(q_{2}\frac{3\Sigma_{h}}{2\sigma_{e}}\right) - \left(1 + q_{3}f_{\upsilon}^{2}\right) = 0. \tag{9}$$

Further, Tvergaard [12] gave the meaning of q_1. Assuming that the ultimate value of void volume fraction, denoted by $f_{\upsilon U}$, is the porosity at which the macroscopic stress carrying capacity vanishes, Eq. (9) reduces to:

$$2q_{1}f_{\upsilon U} - \left(1 + q_{3}f_{\upsilon U}^{2}\right) = 0. \tag{10}$$

Note for the case where $q_3 \approx q_1^2$, the ultimate void volume fraction is calculated as $f_{\upsilon U} = 1/q_1$.

It is worth to stress that voids coalesce before they grow to the point, at which they touch. Brown and Embury [13] revealed that coalescence occurs between two voids when the length of the voids is approximately equivalent to the spacing. Therefore, a critical void volume fraction, $f_{\upsilon C}$, is used in addition to the ultimate void volume fraction, $f_{\upsilon U}$. Quite often it is assumed that $f_{\upsilon C}$ is approximately 0.15. For most engineering alloys, however, following the localization of plastic deformation, coalescence occurs at a relatively low void volume fraction ($f_{\upsilon} < 0.1$). Once the voids begin to coalesce, there is an enhancement in the change of void volume fraction during an increment of deformation. Several methods were introduced to include the void coalescence in Gurson's model. The most widely used one is the method proposed by Tvergaard and Needleman [14]. They proposed a two-part yield condition, the first part producing the original yield loci for void volume fractions less than $f_{\upsilon C}$ and the second part including the enhancement occurring for void volume fractions greater than $f_{\upsilon C}$. As a result of substituting in Eq. (9) a function $f_\upsilon^*(f_\upsilon)$ in place of the variable f_υ, the two-part yield condition takes the form:

$$\Phi_{\mathrm{GTN}}\left(\boldsymbol{\Sigma}, \sigma_e, f_\upsilon^*\right) \stackrel{\text{def}}{=} \left(\frac{\Sigma_{eq}}{\sigma_e}\right)^2 + 2q_1 f_\upsilon^* \cosh\left(q_2 \frac{3\Sigma_h}{2\sigma_e}\right) - 1 - \left(q_1 f_\upsilon^*\right)^2 = 0. \qquad (11)$$

According to the above condition, the yield loci depends on the function $f_\upsilon^*(f_\upsilon)$ specified by:

$$f_\upsilon^*(f_\upsilon) = \begin{cases} f_\upsilon, & \text{for } f_\upsilon \le f_{\upsilon C} \\ f_{\upsilon C} + \frac{f_{\upsilon U} - f_{\upsilon C}}{f_{\upsilon F} - f_{\upsilon C}}(f_\upsilon - f_{\upsilon C}), & \text{for } f_\upsilon > f_{\upsilon C} \end{cases} \qquad (12)$$

where $f_{\upsilon F}$ is the void volume fraction at fracture (when material load-carrying capacity vanishes). Based on the results of a numerical model analysis by Andersson [15] for initially spherical voids in a rigid-perfectly plastic matrix subjected to a highly triaxial stress state, Tvergaard and Needleman recommended $f_{\upsilon F} = 0.25$.

The modified Gurson model, which consists of Eqs. (11) and (12), is often referred in literature as the Gurson-Tvergaard-Needleman (GTN) model. In conditions prevail to promote ductile fracture, the material porosity is usually very small until just prior to fracture, when the stress carrying capacity of the material having ductile properties is reducing rapidly.

Koplik and Needleman [7] stated that the parameters q_i $(i = 1, 2, 3)$ are arbitrary constants required to ensure the relevant dependence of the equivalent plastic strain E^p_{eq} on the porosity growth Δf_υ for various values of the imposed macroscopic stress triaxiality ratio, $T \stackrel{\text{def}}{=} \Sigma_h/\Sigma_{eq}$. Based on extensive comparisons with FE cell model results, Koplik and Needleman [7] showed that using values of $q_1 \approx 1.25$ and $q_2 \approx 1$ in the range $1 \le T \le 3$, the GTN model (11) captures well, after the onset of localization, the very little growth of the macroscopic equivalent plastic strain E^p_{eq}, the rapid breaking down of the macroscopic Mises stress Σ_{eq}, as well as the rapid increase of the void volume fraction f_υ, which marks the development of coalescence. It was also shown that the failure strain is significantly reduced for increasing stress triaxiality. However, for $T = 1/3$ and other low triaxialities the GTN model overestimates the rate of void growth and, moreover, works slightly worse for high stress triaxiality cases. They also noted that q_1 should increase with decreasing strain hardening.

The good performance of the GTN model can be rationalized by considering the rate of increase of porosity $\dot{f}_\upsilon$. For rate-independent plasticity, Eq. (5) becomes

$$\dot{f}_\upsilon = (1 - f_\upsilon)\dot{\Lambda}\frac{\partial \Phi_{\mathrm{GTN}}}{\partial \Sigma_\mathrm{h}}\left(\frac{\partial \Sigma_\mathrm{h}}{\partial \Sigma_{11}} + \frac{\partial \Sigma_\mathrm{h}}{\partial \Sigma_{22}} + \frac{\partial \Sigma_\mathrm{h}}{\partial \Sigma_{33}}\right) = (1 - f_\upsilon)\dot{\Lambda}\frac{\partial \Phi_{\mathrm{GTN}}}{\partial \Sigma_\mathrm{h}}, \tag{13}$$

where the property of macroscopic normality is used and $\dot{\Lambda}$ is the rate of the plastic multiplier. By elimination of the plastic multiplier $\dot{\Lambda} = D^\mathrm{p}_\mathrm{eq}/\left(\partial \Phi_{\mathrm{GTN}}/\partial \Sigma_\mathrm{eq}\right)$, Eq. (13) can be written as:

$$\frac{\dot{f}_\upsilon}{1 - f_\upsilon} = \left(\frac{\partial \Phi_{\mathrm{GTN}}}{\partial \Sigma_\mathrm{h}} \Big/ \frac{\partial \Phi_{\mathrm{GTN}}}{\partial \Sigma_\mathrm{eq}}\right) D^\mathrm{p}_\mathrm{eq}. \tag{14}$$

Combining Eqs. (11) and (14) one gets:

$$\frac{\dot{f}^*_\upsilon}{1 - f^*_\upsilon} = \frac{3}{2}\mathrm{q}_1 \mathrm{q}_2 f^*_\upsilon \frac{\sigma_\mathrm{e}}{\Sigma_\mathrm{eq}} \sinh\left(\mathrm{q}_2 \frac{3\Sigma_\mathrm{eq}}{2\sigma_\mathrm{e}} T\right) D^\mathrm{p}_\mathrm{eq}, \tag{15}$$

which shows the exponential effect of stress triaxiality on void growth, a well known trend found by Rice and Tracey [2].

In addition, Koplik and Needleman [7], and Zhang [16] suggested on the basis of FE cell model studies that $\mathrm{f}_{\upsilon C}$ and $\mathrm{f}_{\upsilon F}$ vary slowly with stress triaxiality and with matrix strain hardening, but depend strongly on the initial porosity $\mathrm{f}_{\upsilon 0}$.

Perrin and Leblond [17] derived an estimate for q_1 assuming $\mathrm{q}_2 = 1$ based on an analytical solution of a hollow sphere with a porous matrix loaded hydrostatically. Using a self-consistent scheme they obtained $\mathrm{q}_1 = 4/e = 1.47$, which is close to the value of 1.5 proposed by Tvergaard [12].

FE cell model studies indicated that the values assigned to q_1 and also q_2, significantly vary with geometry and loading conditions [7, 18]. Such variations revealed that the introduction of the q_i (i = 1, 2, 3) factors may not be simply associated with a physical interaction effect, but are likely to reflect some inaccuracies in Gurson's model.

As stated in [19] and [20] the q_i factors may depend on different features, such as the elastic and inelastic material parameters, as well as the void volume fraction f_υ. Trying to adjust the q_i factors to the computed data it was found in [21] that there is not a set of constant-value q_i factors that are able to predict all of the calculated asymptotic stress responses of FE cell models. In addition, it was found that for $\mathrm{q}_2 = 1$ the GTN model cannot reproduce all actual shapes of macroscopic yield surfaces established by FE cell models with a sufficient accuracy. In particular, the shape transition of macroscopic yield surfaces from the convex surfaces with regular continuous bending field for $f_\upsilon \approx 0.1$ % to smooth convex surfaces for high porosities ($f_\upsilon \geq 10$ %) is rather not possible.

Mear and Hutchinson [22] were the first to introduce linear kinematic hardening into the Gurson yield function. They incorporated the effects of anisotropic (kinematic) hardening by replacing in Eq. (1): Σ_eq of $\boldsymbol{\Sigma}^\mathrm{dev}$ with Σ_eq of $(\boldsymbol{\Sigma} - \boldsymbol{\alpha})^\mathrm{dev}$; Σ_h of $\boldsymbol{\Sigma}$ with Σ_h of $\boldsymbol{\Sigma} - \boldsymbol{\alpha}$, where $\boldsymbol{\alpha}$ is the back stress defining the center of the yield surface in the stress space; and σ_0 with $\sigma_\upsilon \overset{\mathrm{def}}{=} b\, \sigma_0 + (1 - b)\, \sigma_\mathrm{e}$, where σ_0 is the initial yield stress of the matrix, σ_e the current yield stress, $\text{б} = 0$ represents isotropic hardening and $\text{б} = 1$ kinematic hardening.

To take account for the development of description of shear bands for the yielding process, Richmond and Smelser [23] have also proposed a modified form of Gurson's yield criterion. They introduced the concept of an effective void volume fraction in the framework of Gurson's yield criterion. Their modification of Gurson's yield criterion for porous materials with isotropic matrices can be stated as

$$\left(\frac{\Sigma_{eq}}{\sigma_e}\right)^2 + 2f_\upsilon^m \cosh\left(\frac{3m}{2}\frac{\Sigma_h}{\sigma_e}\right) - 1 - f_\upsilon^{2m} = 0. \tag{16}$$

The main reason for this modification was that a cubic array of spherical voids may yield in shear at lower stress than predicted by Gurson's model, which gives an effective volume fraction proportional to f_υ^m (with m = 2/3 for a non-hardening material and m = 1 for a linearly hardening material). They also suggested that m = (2 + n)/3 for a material obeying the stress-strain power law $\sigma_e = K\left(\varepsilon_e^p\right)^n$, where K and n are the strength coefficient and strain-hardening exponent, respectively.

Afterwards, Sun and Wang [24] were the first who constructed the statically admissible stress field around the spherical void and determined the lower bound yield function for porosity as follows:

$$\left(\frac{\Sigma_{eq}}{\sigma_e}\right)^2 + \frac{f_\upsilon\left[\beta_1 \sinh\left(\frac{3\Sigma_h}{2\sigma_e}\right) + \beta_2 \cosh\left(\frac{3\Sigma_h}{2\sigma_e}\right)\right]}{\sqrt{1 + \beta_4 f_\upsilon^2 \sinh^2\left(\frac{3\Sigma_h}{2\sigma_e}\right)}} - \beta_3 = 0 \tag{17}$$

where the coefficients β_j ($j = 1, 2, \ldots, 4$) for $f_\upsilon \leq 0.3$ can be established via numerical analysis as

$$\beta_1 = 0, \quad \beta_2 = 2 - \frac{1}{2}\ln f_\upsilon, \quad \beta_3 = 1 + f_\upsilon(1 + \ln f_\upsilon), \tag{18a}$$

$$\beta_4 = \left(\frac{\beta_2}{\beta_3}\right)^2 \coth^2\left(\frac{3\Sigma_h^0}{2\sigma_e}\right) - \frac{1}{f_\upsilon^2 \sinh^2\left(\frac{3\Sigma_h^0}{2\sigma_e}\right)} \quad \text{with } \Sigma_h^0 \stackrel{\text{def}}{=} -0.65\sigma_e \ln f_\upsilon. \tag{18b}$$

Assuming an incompressible isotropic hardening matrix material Leblond et al. [25] derived another yield function (extended Leblond-Perrin-Devaux model, acronymed as the LPD model) based on the analysis of a spherical void in a spherical volume element:

$$\left(\frac{\Sigma_{eq}}{\sigma_{e_1}}\right)^2 + 2q_1 f_\upsilon^* \cosh\left(q_2 \frac{3\Sigma_h}{2\sigma_{e_2}}\right) - 1 - \left(q_1 f_\upsilon^*\right)^2 = 0, \tag{19}$$

where the quantities σ_{e_1} and σ_{e_2} result from a re-calculation of the homogenisation problem for hardening material. Usually, the σ_{e_1} and σ_{e_2} in Eq. (19) are identified with a single parameter, the yield stress σ_0 in the case of an ideal plastic matrix material, or with an weighted effective flow stress $\sigma_e\left(\left\langle \varepsilon_e^p \right\rangle_r\right)$ related to some weighted effective plastic strain $\left\langle \varepsilon_e^p \right\rangle_r$ through the hardening law of the matrix material. Moreover, the σ_{e_1} and σ_{e_2} were introduced to overcome some drawbacks of Gurson's model: (i) the Gurson model cannot simultaneously determine with the same precision yield stresses under pure shear and pure hydrostatic tension for hardenable matrices, (ii) for a hollow sphere in hydrostatic tension made of a rigid hardening material, predictions determined by Gurson's model are incompatible with the analytical solution, and (iii) the Gurson model cannot reproduce the dependency of the porosity growth rate on the hardening function of the matrix.

To avoid the discrepancies mentioned above, Leblond et al. [25] performed a homogenization of the boundary value problem of a spherical void in a hardening matrix and derived the following expressions:

$$\sigma_{e_1} = \frac{1}{b^3 - a^3} \int_{a^3}^{b^3} \sigma_e\left(\left\langle \varepsilon_e^p \right\rangle_r\right) dr^3, \tag{20a}$$

$$\sigma_{e_2} = \frac{1}{\ln(b^3/a^3)} \int_{a^3}^{b^3} \sigma_e\left(\left\langle \varepsilon_e^p \right\rangle_r\right) \frac{dr^3}{r^3} \tag{20b}$$

with

$$\left\langle \varepsilon_e^p \right\rangle_r = \frac{2}{3}\left[\sinh^{-1}(u) - \frac{\sqrt{1+u^2}}{u}\right]_{u=\frac{2}{3}E_{kk}^p/\left(E_{eq}^p r^3\right)}^{u=\frac{2}{3}E_{kk}^p/\left(E_{eq}^p \mathrm{R}^3\right)}, \tag{21a}$$

$$b^3 = \exp\left(E_{kk}^p\right), \tag{21b}$$

$$b^3 - a^3 = 1 - f_{\upsilon 0} \quad \therefore \quad a^3 = \exp\left(E_{kk}^p\right) + f_{\upsilon 0} - 1, \tag{21c}$$

$$\mathrm{R}^3 = r^3 - b^3 + 1, \tag{21d}$$

where the hollow sphere (i.e. the representative volume element—RVE) has a current, internal radius a (i.e. the radius of the pore) and a current, external radius b, whereas R ($\mathrm{A} \le \mathrm{R} \le \mathrm{B}$) and r ($a \le r \le b$) correspond to the initial and current radial distances of the RVE point from a fixed origin, respectively. The initial radii of the hollow sphere are A and B (where B is assumed to be equal 1.0 for reasons of computational simplicity).

Comparing two different expressions (20a) and (20b) for σ_{e_1} and σ_{e_2} respectively, it is clear that σ_{e_1} and σ_{e_2} cannot be identical due to a different weighting factor for the region in the vicinity of the pore, where the stresses, as well as strains are naturally greater even in the case of the overall plastic deformation of the matrix material.

In the GTN model the parameter q_2 is usually used to correct the void-growth overestimation, which is generated by the implementation of the original Gurson model. For the case of complete softening of the material that has completely lost cohesion and cannot carry any stresses, Hommel [26] introduced the scaling factor $\ln(q_1 f_\upsilon^*)/\ln(f_\upsilon)$ to reset the argument of the cosh function of the LPD model, when the value of $q_1 f_\upsilon^*$ tends to 1. Thus, the LPD model takes the form [27]:

$$\left(\frac{\Sigma_{eq}}{\sigma_{e_1}}\right)^2 + 2q_1 f_\upsilon^* \cosh\left(\frac{\ln(q_1 f_\upsilon^*)}{\ln(f_\upsilon)} q_2 \frac{3\Sigma_h}{2\sigma_{e_2}}\right) - 1 - (q_1 f_\upsilon^*)^2 = 0. \tag{22}$$

For the purpose of the description of plastic anisotropy resulting from the non-spherical void evolution, the effects of void aspect ratio should be accounted. Since the void aspect ratio S significantly affects void coalescence, even when the voids are initially spherical in the porous material, an evolution law for S should also be derived.

Constitutive models incorporating void shape changes were proposed, among others, by Gologanu et al. [28, 29], and Ponte Castañeda and Zaidman [30]. Employing a micromechanical analysis to both prolate and oblate spheroidal voids, Gologanu et al. [31] extended the Gurson model, which is known in the literature as the Gologanu-Leblond-Devaux (GLD) model. The GLD model is based on the analysis of the growth of a spheroidal void in a finite, perfectly plastic solid with an outer confocal surface subjected to homogeneous straining. For mathematical convenience, the GLD model is expressed in terms of the inner and outer eccentricities, e_1 and e_2, respectively, which are uniquely related to f_υ and S by the following equations.

$$e_1 = \sqrt{1 - \exp(-2|S|)}, \tag{23}$$

$$\left(\frac{1-e_1^2}{1-e_2^2}\right)^{n^*} = f_\upsilon \left(\frac{e_1}{e_2}\right)^3 \quad \text{with } n^* \stackrel{\text{def}}{=} \begin{cases} 1, & \text{for prolate shape: 'p'} \quad (S \geq 0) \\ 1/2, & \text{for oblate shape: 'o'} \quad (S < 0) \end{cases}. \tag{24}$$

Pardoen and Hutchinson [32] developed the GLD model and the evolution equation of void aspect ratio S for the axisymmetric problems of strain-hardening in the material matrix, what can be mathematically written as

$$\begin{aligned} C^* \left(\frac{\Sigma_z - \Sigma_r + \eta\Sigma_h}{\sigma_e}\right)^2 + 2q(g+1)(g+f_\upsilon)\cosh\left(\kappa\frac{\Sigma_h}{\sigma_e}\right) + \\ - (g+1)^2 - q^2(g+f_\upsilon)^2 = 0, \end{aligned} \tag{25}$$

$$\dot{S} = (1 + h_S h_T h_\upsilon)(\dot{E}_z^p - \dot{E}_r^p) + h_{S\upsilon}\dot{E}_{kk}^p, \tag{26}$$

and

$$\Sigma_{\mathrm{h}} = 2\alpha_2 \Sigma_{\mathrm{r}} + (1 - 2\alpha_2)\Sigma_{\mathrm{z}}. \tag{27}$$

The phenomenological energy balance, expressed by Eq. (7), for plastic work originally proposed by Gurson [6] is retained by Pardoen and Hutchinson [32]. However, this energy balance equation is not used in the model of Gologanu et al. [31], developed on the basis of perfect plasticity of the material matrix.

The parameter g in Eq. (25), which is defined as

$$\text{'p':} \quad g \stackrel{\text{def}}{=} 0 \quad \text{and} \quad \text{'o':} \quad g \stackrel{\text{def}}{=} e_2^3 / \sqrt{1 - e_2^2} \tag{28}$$

can be interpreted as a correction (shape) factor for porosity of the real material with spheroidal voids referring to the fictitious material with ideal spherical voids. Note the Gurson model, given by Eq. (1), is recovered from Eq. (25) when $S = 0$. The following parameters and internal functions used in Eqs. (25)–(27) emerge from the micromechanical analysis carried out by Pardoen and Hutchinson [32] and are expressed in terms of the e_1 and e_2:

$$\text{'p':} \quad \alpha_2 = \frac{1 + e_2^2}{3 + e_2^4}, \quad \text{'o':} \quad \alpha_2 = \frac{(1 - e_2^2)(1 - 2e_2^2)}{3 - 6e_2^2 + 4e_2^4}, \tag{29}$$

$$\text{'p':}\ \kappa^{-1} = \frac{1}{\sqrt{3}} + \frac{1}{\ln(f_\upsilon)}\left[\left(\sqrt{3} - 2\right)\ln\left(\frac{e_1}{e_2}\right) + \frac{1}{\sqrt{3}}\ln\left(\frac{3 + e_2^2 + 2\sqrt{3 + e_2^4}}{3 + e_1^2 + 2\sqrt{3 + e_1^4}}\right) + \ln\left(\frac{\sqrt{3} + \sqrt{3 + e_1^4}}{\sqrt{3} + \sqrt{3 + e_2^4}}\right)\right], \tag{30a}$$

$$\text{'o':} \quad \kappa^{-1} = \frac{2}{3} + \frac{\frac{2}{3}(g_\upsilon - g_1) + \frac{2}{5}\left(g_\upsilon^{5/2} - g_1^{5/2}\right)\left(\frac{4}{3} - g_\upsilon^{5/2} - g_1^{5/2}\right)}{\ln(g_\upsilon / g_1)}, \tag{30b}$$

$$\text{with 'o':} \quad g_\upsilon \stackrel{\text{def}}{=} \frac{g}{g + f_\upsilon} \quad \text{and} \quad g_1 \stackrel{\text{def}}{=} \frac{g}{g + 1}, \tag{30c}$$

$$\eta = \frac{\kappa q(1 - f_\upsilon)(g + 1)(g + f_\upsilon)\mathrm{sh}}{(g + 1)^2 + q^2(g + f_\upsilon)^2 + 2q(g + 1)(g + f_\upsilon)[\kappa(\alpha_1 - \alpha_2)\mathrm{sh} - \mathrm{ch}]}, \tag{31a}$$

$$\text{with} \quad \mathrm{sh} \stackrel{\text{def}}{=} \sinh(2\kappa(\alpha_1 - \alpha_2)) \quad \text{and} \quad \mathrm{ch} \stackrel{\text{def}}{=} \cosh(2\kappa(\alpha_1 - \alpha_2)), \tag{31b}$$

$$\text{'p':} \quad \alpha_1 = \frac{e_1 - (1 - e_1^2)\tanh^{-1}(e_1)}{2e_1^3}, \tag{32a}$$

$$\text{'o':} \quad \alpha_1 = \frac{-e_1(1 - e_1^2) + \sqrt{1 - e_1^2}\sin^{-1}(e_1)}{2e_1^3}, \tag{32b}$$

$$C^* = -\frac{\kappa q(g+1)(g+f_\upsilon)\mathrm{sh}}{\eta[1-f_\upsilon+2\eta(\alpha_1-\alpha_2)]}, \tag{33}$$

$$h_S = \frac{9\alpha_1 - \alpha_1^G}{21 - 3\alpha_1}, \tag{34a}$$

$$\text{with 'p':} \quad \alpha_1^G \stackrel{\text{def}}{=} \frac{1}{3-e_1^2} \quad \text{and} \quad \text{'o':} \quad \alpha_1^G \stackrel{\text{def}}{=} \frac{1-e_1^2}{3-2e_1^2} \tag{34b}$$

$$h_\upsilon = \left(1-\sqrt{f_\upsilon}\right)^2, \tag{35}$$

$$h_{S\upsilon} = \frac{1-3\alpha_1}{f_\upsilon} + 3\alpha_2 - 1. \tag{36}$$

In addition, for the factor q that reflects the GLD model limitations to account for interactions between cavities, Pardoen and Hutchinson [32] obtained by calibrating numerical results in the domain $f_\upsilon < 5 \times 10^{-2}$ and $S > -2$ (and thus not for penny-shaped cracks) the following expression:

$$q = \left|\frac{b-1}{\pi}\right| \tan^{-1}(4(2.5-T)) + \frac{b}{2} + \frac{1}{2} \tag{37a}$$

$$\text{with} \quad b \stackrel{\text{def}}{=} 1 + \left(0.655 - 1.75\mathrm{n} - 0.533\sqrt[4]{f_\upsilon}\right)\left(\frac{1}{2} + \frac{\tan^{-1}(2(1.2-S))}{\pi} + 0.0288\exp(-1.08(0.2+S))\right) \tag{37b}$$

Equations (37a) and (37b) represents a good fit of values of q in order to adjust the void growth rates predicted by the GLD model to the void growth rates predicted by FE cell model computations in the early stages of straining. As indicated in Eqs. (37a) and (37b), the necessity of decreasing the value of q for increasing n had already been recognized by several authors (e.g., Koplik and Needleman [7], Søvik and Thaulow [33]). The effect of the stress triaxiality ratio T on the factor q, that is included in Eq. (37a), was also examined.

Pardoen and Hutchinson [32] adjusted the parameter h_T appearing in Eq. (26) to give the best predictions for the void shape rates $\dot{S}$ (valid only for $T < 4$) at the two following strain-hardening levels:

$$\mathrm{n} = 0.1: \quad h_T = 1 - 0.555T^2 - 0.045T^4 + 0.002T^6, \tag{38a}$$

$$\mathrm{n} = 0.3: \quad h_T = 1 - 0.54T^2 + 0.034T^4 - 0.00124T^6. \tag{38b}$$

The form of the function, defined by Eq. (38a), in which the dependent variable is h_T for n = 0.1 is close to the one proposed by Gologanu et al. [31] for perfect plasticity.

Note, based upon Eq. (37a), elongated prolate voids ($S \geq 0$) give rise to a value of $q \approx 1$ for all values of n, which is consistent with the Gurson model for cylindrical voids. One of the possible extensions of the GLD model to arbitrary multiaxial stress states can be found in [31].

Besides the classical Gurson model [6], the Gurson Hashin–Shtrikman upper bound [34] and the Garajeu–Suquet upper bound [35], it is worth paying attention to a micromechanics-based modification of Gurson's criterion proposed by Monchiet et al. [36] for porous media subjected to arbitrary loadings, which also has the advantage to preserve the upper bound character. The modification of Gurson's criterion developed by Monchiet et al. [37] was derived in the framework of limit analysis. The obtained formulation is the result of the consideration of Eshelby-like exterior point trial velocity fields for the determination of the macroscopic dissipation. Such an approach was implemented for perfectly plastic rigid von Mises matrix containing spherical voids. Based on the minimization procedure, for which the use of the Eshelby-like trial velocity fields was required, Monchiet et al. [37] derived the following two-field estimate of the macroscopic yield function:

$$\left(\frac{\Sigma_{\mathrm{eq}}}{\sigma_0}\right)^2 + 2f_\upsilon \cosh\left(\sqrt{\frac{9}{4}\frac{\Sigma_{\mathrm{h}}^2}{\sigma_0^2} + \frac{2}{3\rho(f_\upsilon)}\frac{\Sigma_{\mathrm{eq}}^2}{\sigma_0^2}}\right) - 1 - f_\upsilon^2 = 0, \tag{39}$$

where $\rho(f_\upsilon) \stackrel{\mathrm{def}}{=} 1 - 4f_\upsilon\left(1 - f_\upsilon^{2/3}\right)^2/(1 - f_\upsilon)$. Note $\rho(f_\upsilon) \approx 1$ for small values of the porosity f_υ.

Preserving the accuracy of the classical Gurson model for high stress triaxialities, the above criterion contains the equivalent stress Σ_{eq} in the argument of the hyperbolic cosine function that makes the significant difference in comparison with the criterion (1). However, the most important feature of the modification of Gurson's criterion is that for low stress triaxialities the derived yield surface is close to the Hashin-Shtrikman bound established by Ponte-Castañeda [38] in the framework of the non-linear homogenization applied to a RVE of isotropic porous media.

More recently, Leblond et al. [39] criticized the prediction of Gurson's model for predictions of the porosity increase rate $\dot{f}_\upsilon$. They noted that Gurson [6] considered only the approximation of the overall plastic dissipation, based on the first-order Taylor expansion of one term appearing in the integral defining it, which led to his well-known homogenized criterion. The most important result of their research is that the correction of the predicted porosity increase rate $\dot{f}_\upsilon$ brought by the second- and third-order approximations to the first-order one, represented by $\dot{f}_{\upsilon(\mathrm{Gurson})}$, are quite significant as indicated in Fig. 1. For a fixed porosity $f_\upsilon = 10^{-2}$, Fig. 1 shows

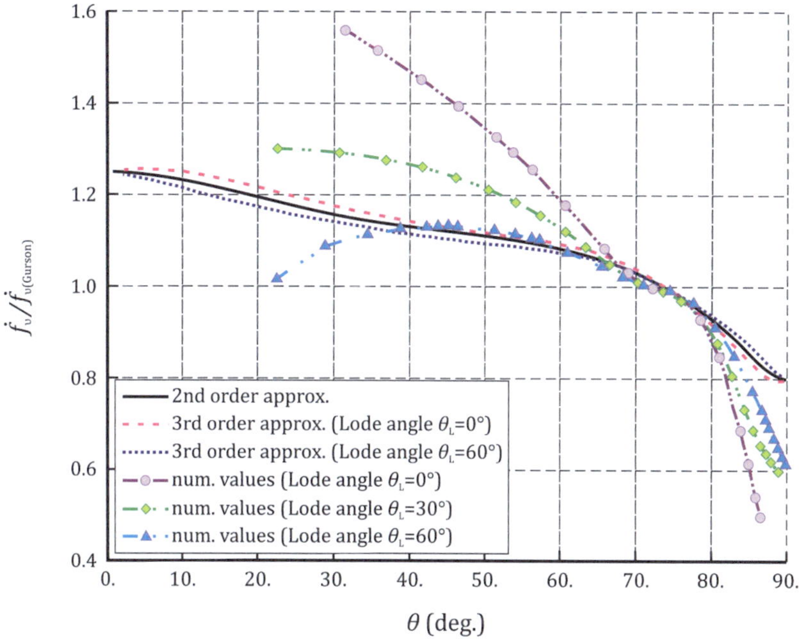

Fig. 1 Comparison of values of $\dot{f}_\upsilon/\dot{f}_{\upsilon(\text{Gurson})}$ as a function of the angle θ ($f_\upsilon = 10^{-2}$)

that the porosity increase rate, predicted by the second- and third-order approximations and normalized by results of Gurson's model, $\dot{f}_\upsilon/\dot{f}_{\upsilon(\text{Gurson})}$, as a function of the angle $\theta \stackrel{\text{def}}{=} \arctan(T)$, as well as normalized numerical values of $\dot{f}_\upsilon/\dot{f}_{\upsilon(\text{Gurson})}$, obtained by FE cell model computations, versus the angle θ.

Since all values of $\dot{f}_\upsilon/\dot{f}_{\upsilon(\text{Gurson})}$ achieved by FE cell model calculations, except those corresponding to the second- and third-order approximations, are sensitive to the value of Lode's angle, three kinds of loadings, corresponding to the following Lode angles θ_L: (i) 0° (axisymmetric load with major axial stress), (ii) 30° (pure shear with superposed hydrostatic tension) and (iii) 60° (axisymmetric load with major lateral stress), were considered by Leblond et al. [39]. The Lode angle is defined in the work [39] as

$$\theta_\text{L} \stackrel{\text{def}}{=} \frac{1}{3}\arccos\left(\frac{27}{2}\Sigma_1^{\text{dev}}\Sigma_2^{\text{dev}}\Sigma_3^{\text{dev}}\Big/\Sigma_{\text{eq}}^3\right). \tag{40}$$

Madou and Leblond [40] made further progress in extending the famous Gurson model to deal with general (non-spheroidal) ellipsoidal cavities, through approximate homogenization of some representative elementary porous cells. They performed a limit-analysis of such a cell, which was represented by an ellipsoidal volume made of some rigid-ideal-plastic von Mises material and containing a confocal ellipsoidal void, loaded arbitrarily under conditions of homogeneous boundary strain rate. Their analysis provided an estimate of the overall plastic dissipation based on a family of trial incompressible velocity fields discovered by Leblond and Gologanu [41], satisfying conditions of homogeneous strain rate on all ellipsoids confocal with the void and the outer boundary. The asymptotic behavior of the integrand in the expression of the global plastic dissipation was studied both far from and close to the void. The results obtained by Madou and Leblond [40] suggest approximations, leading to explicit approximate expressions of the overall dissipation and yield function. The important parameters of these expressions are estimated in [42].

The approximate analytic criterion proposed by Madou and Leblond [40, 42] for plastic porous materials containing arbitrary ellipsoidal voids was validated in [43] by comparing its predictions with the results of some numerical limit-analyses of elementary cells containing such voids. Moreover, Madou and Leblond [44] proposed an evolution equations for the length and orientation of the axes of the voids to complete their analytic criterion.

Pastor and Kondo [45] adapted and extended 3D static and kinematic FEM codes developed by Trillat and Pastor [46] for spherical cavities to the case of a von Mises matrix containing spheroidal (and confocal) cavities, subjected to axisymmetric loadings. The main objective of the research done by Pastor and Kondo [45] was to assess existing analytical criteria, by providing the numerical lower and upper bound limit analyses for various loading cases. The numerical developments required by such study were described in detail. Pastor and Kondo [45] conducted validation tests for a hollow sphere model with three values of porosity and models of spheroid confocal cavities. The test, which concerned the hollow sphere model, confirmed that the exact yield criterion of the homogenized material depends on the third invariant of the macroscopic stress tensor such, as was found in [47]. The subsequent tests, which concerned the models of spheroid confocal cavities, show that the Gologanu [48] criterion should be preferred to that of Gologanu and Leblond [28] and Gologanu et al. [29] for uniform strain rate on the boundary. Moreover, the Monchiet et al. [36] criterion appeared as a good estimate for uniform strain rate loading at the boundary, despite the fact that it is based on a kinematic approach under conditions, which can be interpreted as uniform stress boundary conditions.

Based on micromechanical considerations and non-linear homogenization, Stewart and Cazacu [49] obtained the macroscopic yield criterion incorporating anisotropy for a porous material. The form of this criterion is similar, in a certain sense, to the classical Gurson criterion. However, it was assumed that the matrix material is incompressible, anisotropic and displays tension–compression asymmetry. It should be emphasized that recently formulated yield criteria are quite often examined with respect to Gurson's model [50].

Benallal et al. [51] included the Lode angle in Gurson's approach, which allowed for analyzing the role of the third stress invariant in yielding of porous materials. They carried out a profound assessment of these effects, whereby for small porosities encountered on ductile fracture of metals, observed changes and effective roles of the Lode angle are rather small although from the qualitative point of view, non-symmetry of the yield locus and changes on its curvature are observed.

The elastic constants degrade with the evolution of the porosity in the material and contribute to additional material softening. From a practical perspective, the porosity in most metals is too small to be a significant factor, but should be accounted for in the model for completeness. Weng [52] derived the Mori-Tanaka [53] estimates of the effective shear and bulk moduli of a two-phase composite. Therefore, the degraded shear G and bulk K moduli of a voided material can be expressed as the following functions of their initial values and the porosity:

$$G = \frac{G_0(1 - f_\upsilon)}{1 + \frac{6(K_0 + 2G_0)}{9K_0 + 8G_0} f_\upsilon}, \quad K = \frac{4K_0G_0(1 - f_\upsilon)}{4G_0 + 3K_0 f_\upsilon}, \tag{41}$$

where G_0 and K_0 are the initial (undamaged) values of shear and compressibility modulus, respectively.

3 Development of Models for Nucleation, Growth and Coalescence of Voids

A number of methods used in the theoretical description of the mechanism of the nucleation of voids may be distinguished. Among these methods, there are two main approaches—the first ones, based on the theory of dislocations, are used to describe particles with small diameters (order of 1 μm or less than 1 μm), which interact with dislocations strongly—the second ones, specified by the theory of continuous media for particles with diameters greater than 1 μm, to characterize the interaction of these particles with the matrix material.

In metals containing large inclusions in the form of clusters of manganese sulfide or aluminum oxide, the nucleation of voids occurs as a result of fracture of inclusions, which can be easily cracked in the initial stage of plastic deformation. In this respect, it is assumed that nucleation of voids occurs almost imperceptibly at the beginning of plastic deformation and is characterized by an initial microporosity of the material, $f_{\upsilon 0}$.

Continuous models of nucleation of voids should be applied in the case of materials, in which voids nucleate from carbides or intermetallic phases. This nucleation may be dependent on the stress/strain state.

Various models of the nucleation of voids are sometimes even contradictory to each other, and could give contradictory estimates. Inconsistency results between nucleation models are probably due to the fact that each of these models has been developed for different types of alloy structures.

However, there are many common conclusions with respect to nucleation of microvoids in materials that have been drawn independently. Most researchers agree that larger second-phase particles lead to earlier nucleation than rather smaller

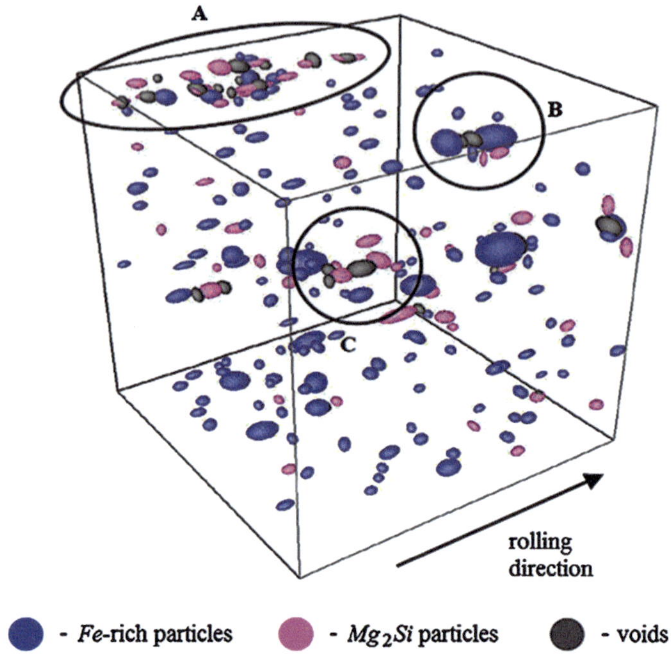

Fig. 2 3D tomographic image of the AA5182 sheet with dimensions 70 μm × 70 μm × 70 μm (**a** - particle/void cluster, **b** - fractured Fe-rich particles with voids, **c** - fractured Mg_2Si particles with voids) [61]

ones [54–57]. Larger particles of the second phase contain more surface defects in connections to the matrix material. Another common conclusion is that the equi-axed inclusions are susceptible to decohesion from the metal matrix, while the inclusions with different lengths of axis are prone to multiple cracking [56–60].

Scanning electron microscopy (SEM) coupled with energy dispersive spectrometry (EDS) can analyze the chemical composition of samples of material (qualitative, semiquantitative and quantitative analysis). It is also possible to perform distribution maps of particular chemical elements that may comprise the analyzed structure and the creation of three-dimensional, tomographic images of the samples. Figure 2 shows the microstructure of a sample of aluminum-alloy AA 5182, which was illuminated with the monochromatic beam of high photon flux for the acquisition of greyscale image with a voxel resolution of 0.7 μm at the European Synchrotron Radiation Facility in Grenoble [61].

The total rate of increase in porosity is determined by summing the nucleation and growth rate of new and existing voids, respectively

$$\dot{f}_{\upsilon} = \dot{f}_{\upsilon(\text{nucleation})} + \dot{f}_{\upsilon(\text{growth})} \tag{42}$$

The void nucleation rate component may differ depending on the nucleation criterion adapted. Chen et al. [62], in their study of AA5182 aluminum alloy, used a criterion suggested by Gurson [63], which assumes that voids nucleate at second phase particles and there is a normal distribution of nucleation strain for the total population of particles

$$\dot{f}_{\upsilon(\text{nucleation})} = \mathcal{A}\dot{\varepsilon}_{\text{eff}} \tag{43a}$$

$$\mathcal{A} = \frac{\upsilon_N}{s_N\sqrt{2\pi}}\exp\left[-\frac{1}{2}\left(\frac{\varepsilon^p_{\text{eff}} - \varepsilon_N}{s_N}\right)^2\right] \tag{43b}$$

where υ_N is the volume fraction of all second-phase particles with potential for microvoid nucleation, $\varepsilon^p_{\text{eff}}$ is the effective plastic strain, ε_N and s_N are the mean strain for void nucleation and its standard deviation, respectively.

The GTN model can describe the effect of material weakening due to the growth of existing microvoids in a ductile material, and thus may provide for their growth rate during the plastic deformation by Eq. (5).

Coalescence of voids in the ductile material is the final stage of damage development, and occurs after void nucleation at particles of the second material phase and after their growth. During this process, there is a phenomenon of localization of plastic deformation in the connecting layer between adjacent voids. Faleskog et al. [19] distinguished two coalescence mechanisms that occur through localizations of plastic deformation due to tension and shear.

A variety of models of microvoid coalescence was developed, and experimental observations of this phenomenon were also conducted. To describe the evolution of voids in ductile materials, the application of criteria of plastic flow for porous ductile materials is more effective than the use of strict models of porous microstructures, which are usually constructed by means of finite-element computational cells. Although the application of the initial yield and plastic flow criteria is more effective than the use of the finite-element microstructure models, their application requires, in addition, the incorporation of mathematical models for ductile microvoid coalescence.

Some of the most important models of microvoid coalescence are as follows: (a) model of Rosenfield and Hahn [64], (b) McClintock's model [1], (c) model of Hancock and McKenzie [65], (d) model of Budiansky, Hutchinson and Slutsky [66] and (e) Thomason's model [67–70] in particular.

Rosenfield and Hahn [64] introduced the term of the maximum sustainable size of void diameter D:

$$\sigma\sqrt{D} = \mathrm{K}, \tag{44}$$

where σ determines the level of stress in the ductile material and K is a material constant. McClintock [1] proposed a criterion according to which coalescence occurs, when the average size of voids reaches the average distance between voids. Hancock and McKenzie [65] obtained the formula defining the void volume fraction $f_{\upsilon L}$ required to occurrence the strain localization as a function of the triaxial stress state:

$$f_{\upsilon L} = \left[0.56 \exp\left(\frac{3\Sigma_h}{2\sigma_{eff}}\right)\right]^{-1}. \quad (45)$$

According to Eq. (45) a lower volume fraction of voids in the ductile material will be necessary to trigger the coalescence process, while the stress triaxiality coefficient, Σ_h/σ_{eff}, increases.

Budiansky et al. [66] studied the growth of voids in solids with viscoplastic properties and established estimates for strains, which are required for the appearance of coalescence:

$$\frac{\varepsilon_{CR}}{\mathfrak{c}} = \left(\frac{\dot{\omega}}{\dot{\varepsilon}\omega}\right)_F^{-1} \quad \text{when} \quad \left(\frac{\dot{\omega}}{\dot{\varepsilon}\omega}\right)_F \leq 3, \quad (46a)$$

$$\frac{\varepsilon_{CR}}{\mathfrak{c}} = \left[1 + \frac{2}{3}\left(\frac{\dot{\omega}}{\dot{\varepsilon}\omega}\right)_F\right]^{-1} \quad \text{when} \quad \left(\frac{\dot{\omega}}{\dot{\varepsilon}\omega}\right)_F > 3, \quad (46b)$$

where

$$\mathfrak{c} \stackrel{\text{def}}{=} \ln\left[\frac{(\mathscr{b}/\mathscr{w})_{CR}}{(\mathscr{b}/\mathscr{w})_0}\right]^2, \quad (46c)$$

ε_{CR} is a critical strain, $\dot{\varepsilon}$ strain rate, ω the volume of voids, $\mathscr{b}/\mathscr{w}$ describes the relative size of the representative void, which is related to the distance between two adjacent voids, $\mathscr{w}$, and indices '0' and 'CR' mean initial and critical values, respectively.

The plastic limit load criterion established by Thomason [67–70] can be used to determine the onset of the phenomenon of coalescence as follows:

$$\frac{\Sigma_1}{\sigma_e} - \left[\alpha_T\left(\frac{1}{\mathscr{r}} - 1\right)^2 + \frac{\beta_T}{\sqrt{\mathscr{r}}}\right](1 - \pi\mathscr{r}^2) = 0, \quad (47)$$

where $\alpha_T = 0.1$ and $\beta_T = 1.2$ are constants introduced by Thomason, $\mathscr{r}$ is the void space ratio and Σ_1 is the current maximum principal stress. The Thomason model developed by Pardoen and Hutchinson [32] concerns the localization of plastic strains in the arrangement of relatively narrow zones of concentrated deformations and simultaneously causes changes in forms of yield criteria and flow rules for porous (dilatant) ductile materials.

Zhang and Niemi [71] implemented Thomason's plastic limit load criterion to the Gurson model and found that the plastic limit load model is not only physically based but also reasonably accurate in comparison with cell model results obtained by Koplik and Needleman [7]. Quite recently, Benzerga and Leblond [72] derived an effective yield function for a porous ductile solid near a state of failure by microvoid coalescence. For this purpose, they used homogenization theory combined with limit analysis, wherein a problem, which was considered, is similar to that carried out by Thomason [69] for a square-prismatic cell containing a square-prismatic void. A cylindrical cell was taken to contain a coaxial cylindrical void of finite height. Plastic flow in the intervoid matrix was described by J_2 theory, while regions above and below the void remained rigid. Velocity boundary conditions were employed, which are compatible with an overall uniaxial straining of the cell, and consistent with post-localization kinematics that is omnipresent during the coalescence of neighboring microvoids in rate-independent solids. However such assumed boundary conditions are not of the uniform strain rate kind, as it is in the case for Gurson-like models. Their yield criterion, in closed form, for a porous material near a state of failure by void coalescence can be expressed as follows:

$$\Phi_{\mathrm{BL}}(\boldsymbol{\Sigma}, \sigma_{\mathrm{e}}, \chi, \mathcal{W}) \stackrel{\mathrm{def}}{=} \frac{|\Sigma_{33}|}{\sigma_{\mathrm{e}}} + \\ -\frac{1}{\sqrt{3}}\left[2-\sqrt{1+3\chi^4}+\ln\left(\frac{1+\sqrt{1+3\chi^4}}{3\chi^2}\right)+\frac{\chi^3-3\chi+2}{3\chi\mathcal{W}}\right]=0, \tag{48}$$

where $\mathcal{W} \stackrel{\mathrm{def}}{=} h_{\mathrm{V}}/R_{\mathrm{V}}$ is the void aspect ratio and $\chi \stackrel{\mathrm{def}}{=} R_{\mathrm{V}}/L_{\mathrm{C}}$ is the ligament parameter, wherein $2h_{\mathrm{V}}$ is the height of cylindrical void, R_{V} the void radius and L_{C} the cell radius. In Eq. (48) the absolute value of Σ_{33} permits to express the limit-loads in both tension and compression. The need for criteria of such type (48) in conjunction with the GTN model (11)–(12) is strongly motivated by cell model calculations [7].

Morin et al. [73] used numerical limit-analysis results to assess the coalescence criterion developed by Benzerga and Leblond [72], as well as two latter coalescence criteria, which are based on two different sets of kinematically admissible velocity fields consistent with nonuniform strain-rate conditions on the cell boundary. The first of two new upper-bound estimates is based on a generalized discontinuous velocity field, for which the shape of the rigid–plastic interface is free. The second one is based on a continuous velocity field. Both fields are consistent with the presence of rigid zones between the regions where the strain rate localizes. Both the new models give some improvement, compared to the model of Benzerga and Leblond [72].

4 Modification of Gurson's Model for Failure Prediction Under Shear Deformation

The original formulation of Gurson's model does not include shear effects, which excludes the possibility of predicting shear localization and fracture under conditions of low triaxiality. Under shear dominated loading conditions, the distortion of voids and inter-void linking promotes an effective increase of the material internal degradation and contributes to the material softening. In order to improve the predictive ability of all Gurson-type models, under both zero and low levels of stress triaxiality, certain researchers [74–80] suggested the introduction of effects of shear deformation, which can be based on geometrical or phenomenological assumptions. Simple formulations for the mechanism of shear deformation are included in the resulting evolution laws, which take into account the influence of the third invariant of the deviatoric stress tensor, the plastic strain tensor and its rate.

Equation (5) reveals that shear deformation does not trigger void growth and consequently, the GTN model does not predict failure under shear deformation. Nahshon and Hutchinson [76] claimed that f_υ should not be directly tied to the plastic volume change but rather it should be regarded as a damage parameter D_F, and introduced as an additional term in the evolution equation of f_υ to account for shear damage.

In order to incorporate the shear induced damage Xue [77] proposed a modification to the original GTN model, which is based upon the solution of McClintock [81] for the coalescence of holes in a shear band. Xue [77] developed his shear damage law with regard to the change of the void ligament of a unit cell model under simple shear deformation. This damage evolution law can be written as follows:

$$\dot{D}_F = q_1 \left(\dot{f}_{\upsilon(\text{nucleation})} + \dot{f}_{\upsilon(\text{growth})} \right) + q_4 f_\upsilon^{q_5} g_L \left(\theta_L^* \right) \varepsilon_{\text{eff}}^{P} \dot{\varepsilon}_{\text{eff}}^{P}, \tag{49}$$

where q_4 is the material parameter used to scale the shear damage growth rate, q_5 represents the additional parameter in Xue's model, $g_L\left(\theta_L^*\right) \stackrel{\text{def}}{=} 1 - 6\left|\theta_L^*\right|/\pi$ is the Lode angle-dependent function, and recall that θ_L^* denotes the Lode angle alternatively defined as $\theta_L^* \stackrel{\text{def}}{=} \arctan\left((2(\Sigma_2 - \Sigma_3)/(\Sigma_1 - \Sigma_3) - 1)/\sqrt{3}\right)$. It should also be paid attention to a new shear mechanism, which was proposed in [82] to improve the GTN model. The need for this type of modification aroused in order to reduce the substantial differences in results of the Gurson model and FE cell model computations at low levels of stress triaxiality, which are clearly shown in Fig. 1.

5 Conclusions

In this chapter the most important modifications and improvements of all kinds of Gurson's model have been presented with reference to their chronological development. In conclusion, it is worth paying attention to the genesis of these changes and whether the considerable effort in improving this model is already sufficient. Probably one of the reasons for failure in the application of Gurson-type models is an inappropriate selection of the application scope without awareness of limitations of these models. In this respect, it is worth to refer to Fig. 3 paying close attention to the area around the abscissa axis and including the possibility to fulfill Eq. (7), which expresses the equivalence between the rate of macroscopic plastic work

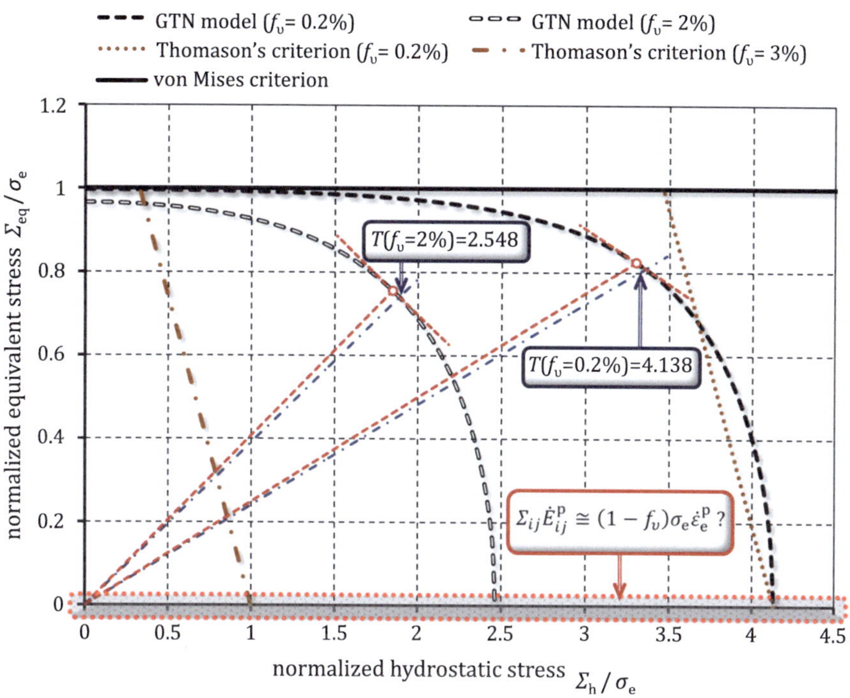

Fig. 3 Yield functions showing the reduction in macroscopic equivalent stress with increasing hydrostatic stress and porosity

$\Sigma_{ij}\dot{E}_{ij}^{p}$ and the rate of plastic work dissipated in the matrix material inside the same volume (RVE) $(1-f_\upsilon)\sigma_e\dot{\varepsilon}_e^{p}$. The basic assumption about the Gurson model should also be recalled. Namely, this model is based on a rigid-perfectly plastic analysis of spherically symmetric deformation around a spherical cavity. However, it is known from literature [83] that in a thick-walled spherical shell under hydrostatic tension there may be some form of plasticity called contained plasticity or constrained plasticity, when the elastic strains and the plastic strains are more or less of the same magnitude at the inner surface. In other words, the plastic region has high strains, and the surrounding elastic region has low strains. By virtue of the elastic region being outside the plastic region, as it is known on the boundary between the regions the circumferential strains must have the same value, i.e., there must be strain continuity on the elastic-plastic boundary. This means that the corresponding cases of elastic-plastic deformation patterns of the hollow sphere of the Gurson-type models cannot be considered, although this phenomenon could take place in some cases of loading.

Moreover, Williams and Schapery [84] studied fracture initiation due to a hydrostatic tensile stress field by considering an idealized solid containing a small spherical cavity. They found that for small void sizes, the stress, particularly at the void surface, becomes inordinately high, leading perhaps to unrealistic values of the fracture stress. In addition the weighting factor $1/r^3$ in Eq. (20b) must be noted. The situation is thus somewhat analogous to that occurring in the vicinity of the crack point in sheet materials where the elastic stresses become unbounded. Also in this case, the Gurson model cannot be applied directly. As indicated in [32], for stress triaxiality ratio T larger than 5, unstable void growth has been found and the distinction between void growth and void coalescence process becomes harder and harder to discern and the notion of critical porosity is no longer meaningful (see Fig. 1).

An attempt to identify borderline cases, when the Guson model cannot be applied, was taken in [85, 86] through rearrangement of the form of the GTN model. For this purpose also Jackiewicz [86] proposed a coalescence criterion based on the assumption that the close-to-zero value of a function of the continuity state triggers the coalescence of micro-voids in materials.

It should be highlighted that the identification of material model parameters of the Gurson-type models causes considerable difficulties [87], due to the large number of them. The material parameters of the Gurson-type models can be grouped into three following classes: (i) constitutive parameters (related to the Gurson yield surface), (ii) initial parameters (associated with the origin of the porosity) and (iii) critical parameters (related to the interaction between neighboring voids). The gradient-based methods (e.g., the methods of steepest descent) often used to determine the material model parameters consist in computing the steepest-descent direction associated with the gradient, and then the optimal stepsize along this direction. The local search technique of gradient-based methods can easily be trapped in local minima. Hence, there is no guarantee that a steepest descent algorithm finds an optimal solution and, therefore, the gradient-based methods are usually sensitive to initial guesses. This is consistent with results obtained by Springmann and Kuna [88, 89] and means that a different set of material parameters could reproduce numerically nearly the same stress–strain curves close to the experimental data. Multi-objective evolutionary algorithms [90] can be used instead to overcome such difficulties. To handle problems of multi-objective optimization, the Pareto optimal solutions can be used, while it is very important to be aware of the non- uniqueness in the constitutive parameters of Gurson-type models [91, 92]. A structural design is Pareto optimal, if there is no any other design that satisfies all of the objectives better. To this end a calibration procedure of material parameters proposed in [93] can also be used. However, this procedure is not, as efficient as, the gradient-based algorithm. In summary, the Gurson model seems to be a good analytic expression for description of porous materials with spherical cavities. Moreover, the latest extensions of this model allow to consider not only the void volume fraction as the parameter representing voids but also void size effects by means of the Taylor dislocation model [94] and the Fleck and Hutchinson strain gradient plasticity model [95], and even the description of the growth of submicron voids on the basis of interface stresses effects [96]. Nevertheless, research effort should be directed towards the creation of more efficient numerical cell models than to build more complex analytical models, such as the model of Pardoen and Hutchinson [32], which is described in this chapter.

References

1. McClintock FA (1968) A criterion for ductile fracture by the growth of holes. J Appl Mech 35:363–371
2. Rice JR, Tracey DM (1969) On the ductile enlargement of voids in triaxial stress fields. J Mech Phys Solids 17:201–217
3. Needleman A (1972) Void growth in an elastic-plastic medium. J Appl Mech Trans ASME 39:964–970
4. Mandel J (1964) Contribution théorique à l'étude de l'écrouissage et des lois de l'écoulement plastique. In: Proceedings of the 11th International Congress on Applied Mechanics. Springer, Munich, pp. 502–509
5. Hill R (1967) The essential structure of constitutive laws for metal composites and polycrystals. J Mech Phys Solids 15:79–95
6. Gurson A (1977) Continuum theory of ductile rupture by void nucleation and growth—part I, Yield criteria and flow rules for porous ductile media. J Eng Mater Trans ASME 99:2–15
7. Koplik J, Needleman A (1988) Void growth and coalescence in porous plastic solids. Int J Solids Struct 24:835–853
8. Hashin Z, Shtrikman S (1962) On some variational principles in anisotropic and nonhomogeneous elasticity. J Mech Phys Solids 10:335–342
9. Hashin Z, Shtrikman S (1962) A variational approach to the theory of the elastic behaviour of polycrystals. J Mech Phys Solids 10:343–352
10. Hashin Z, Shtrikman S (1963) A variational approach to the theory of the elastic behaviour of multiphase materials. J Mech Phys Solids 11:127–140
11. Tvergaard V (1981) Influence of voids on shear band instabilities under plane strain conditions. Int J Fract 17:389–407
12. Tvergaard V (1982) Material failure by void coalescence in localized shear bands. Int J Solids Struct 18:659–672
13. Brown LM, Embury JD (1973) The initiation and growth of voids at second-phase particles. In: Proceedings of 3rd International Conference on Strength of Metals and Alloys. Institute of Metals, London, pp. 164–169
14. Tvergaard V, Needleman A (1984) Analysis of cup-cone fracture in a round tensile bar. Acta Metall 32:157–169
15. Andersson H (1977) Analysis of a model for void growth and coalescence ahead of a moving crack tip. J Mech Phys Solids 25:217–233
16. Zhang ZL (2001) A complete Gurson model. In: Alibadi MH (ed) Nonlinear fracture and damage mechanics. WIT Press, Southampton, pp 223–248
17. Perrin G, Leblond J-B (1990) Analytical study of a hollow sphere made of plastic porous material and subjected to hydrostatic tension—application to some problems in ductile fracture of metals. Int J Plast 6:677–699
18. Gao X, Faleskog J, Shih CF (1998) Cell model for nonlinear fracture analysis—II, Fracture-process calibration and verification. Int J Fract 89:374–386
19. Faleskog J, Gao X, Shih C (1998) Cell model for nonlinear fracture analysis—I, Micromechanics calibration. Int J Fract 89:355–373
20. Kim J, Gao X, Srivatsan TS (2004) Modeling of void growth in ductile solids: effects of stress triaxiality and initial porosity. Eng Fract Mech 71:379–400
21. Fritzen F, Forest S, Böhlke T, Kondo D, Kanit T (2012) Computational homogenization of elasto-plastic porous metals. Int J Plast 29:102–119
22. Mear ME, Hutchinson JW (1985) Influence of yield surface curvature on flow localization in dilatant plasticity. Mech Mater 4:395–407
23. Richmond O, Smelser RE (1985) Alcoa technical center memorandum, March 7. In: Hom CL, McMeeking RM (eds) Void growth in elastic–plastic materials. J Appl Mech 56:309–317
24. Sun Y, Wang D (1989) A lower bound approach to the yield loci of porous materials. Acta Mech Sinica 5:399–406

25. Leblond J-B, Perrin G, Devaux J (1995) An improved Gurson-type model for hardenable ductile metals. Eur J Mech A Solid 14:499–527
26. Hommel J-H (2007) Mechanismenorientierte Simulation von Kurzzeitermüdung metallischer Werkstoffe und Strukturen. PhD thesis, Bochum, Germany. Download at http://www-brs.ub.rub.de/netahtml/HSS/Diss/HommelJanHendrik/
27. Kintzel O (2006) Modeling of elasto-plastic material behavior and ductile micropore damage of metallic materials at large deformations. PhD thesis, Bochum, Germany. Download at http://www-brs.ub.rub.de/netahtml/HSS/Diss/KintzelOlafE/
28. Gologanu M, Leblond J-B, Devaux J (1993) Approximate models for ductile metals containing nonspherical voids: case of axisymmetric prolate ellipsoidal cavities. J Mech Phys Solids 41:1723–1754
29. Gologanu M, Leblond J-B, Devaux J (1994) Approximate models for ductile metals containing nonspherical voids: case of axisymmetric oblate ellipsoidal cavities. J Eng Mater Trans ASME 116:290–297
30. Ponte Castañeda P, Zaidman M (1994) Constitutive models for porous materials with evolving microstructure. J Mech Phys Solids 42:1459–1497
31. Gologanu M, Leblond J-B, Perrin G, Devaux J (1995) Recent extensions of Gurson's model for porous ductile metals. In: Suquet P (ed) Continuum micromechanics. Springer, Heidelberg, pp 61–130
32. Pardoen T, Hutchinson JW (2000) An extended model for void growth and coalescence. J Mech Phys Solids 48:2467–2512
33. Søvik OP, Thaulow C (1997) Growth of spheroidal voids in elastic-plastic solids. Fatigue Fract Eng Mater Struct 20:1731–1744
34. Leblond J-B, Perrin G, Suquet P (1994) Exact results and approximate models for porous viscoplastic solids. Int J Plast 10:213–235
35. Garajeu M, Suquet P (1997) Effective properties of porous ideally plastic or viscoplastic materials containing rigid particles. J Mech Phys Solids 45:873–902
36. Monchiet V, Charkaluk E, Kondo D (2007) An improvement of Gurson-type models of porous materials by using Eshelby-like trial velocity fields. CR Mécanique 335:32–41
37. Monchiet V, Charkaluk E, Kondo D (2011) A micromechanics-based modification of the Gurson criterion by using Eshelby-like velocity fields. Eur J Mech A Solid 30:940–949
38. Ponte-Castañeda P (1991) The effective mechanical properties of nonlinear isotropic composites. J Mech Phys Solids 39:45–71
39. Leblond J-B, Morin L, Cazacu O (2014) An improved description of spherical void growth in plastic porous materials with finite porosities. Procedia Mater Sci 3:1232–1237
40. Madou K, Leblond J-B (2012) A Gurson-type criterion for porous ductile solids containing arbitrary ellipsoidal voids—I, Limit-analysis of some representative cell. J Mech Phys Solids 60:1020–1036
41. Leblond J-B, Gologanu M (2008) External estimate of the yield surface of an arbitrary ellipsoid containing a confocal void. CR Mécanique 336:813–819
42. Madou K, Leblond J-B (2012) A Gurson-type criterion for porous ductile solids containing arbitrary ellipsoidal voids—II, Determination of yield criterion parameters. J Mech Phys Solids 60:1037–1058
43. Madou K, Leblond J-B (2013) Numerical studies of porous ductile materials containing arbitrary ellipsoidal voids—I, Yield surfaces of representative cells. Eur J Mech A Solid 42:480–489
44. Madou K, Leblond J-B, Morin L (2013) Numerical studies of porous ductile materials containing arbitrary ellipsoidal voids—II, Evolution of the length and orientation of the void axes. Eur J Mech A Solid 42:490–507
45. Pastor F, Kondo D (2013) Assessment of hollow spheroid models for ductile failure prediction by limit analysis and conic programming. Eur J Mech A Solid 38:100–114
46. Trillat M, Pastor J (2005) Limit analysis and Gurson's model. Eur J Mech A Solid 24:800–819
47. Thoré P, Pastor F, Pastor J (2011) Hollow sphere models, conic programming and third stress invariant. Eur J Mech A Solid 30:63–71

48. Gologanu M (1997) Etude quelques problèmes de rupture ductile des métaux. Thèse de doctorat, Université Paris-6
49. Stewart JB, Cazacu O (2011) Analytical yield criterion for an anisotropic material containing spherical voids and exhibiting tension–compression asymmetry. Int J Solids Struct 48:357–373
50. Cazacu O, Revil-Baudard B (2015) New three-dimensional plastic potentials for porous solids with a von Mises matrix. CR Mécanique 343:77–94
51. Benallal A, Desmorat R, Fournage M (2014) An assessment of the role of the third stress invariant in the Gurson approach for ductile fracture. Eur J Mech A Solid 47:400–414
52. Weng GJ (1984) Some elastic properties of reinforced solids with special reference to isotropic ones containging spherical inclusions. Int J Eng Sci 22:845–856
53. Mori T, Tanaka K (1973) Average stress in matrix and average elastic energy of materials with misfitting inclusions. Acta Metall 21:571–574
54. Stone, van RH, Cox TB, Low JR Jr, Psioda JA (1985) Microstructural aspects of fracture by dimpled fracture. Int Met Rev 30:157–179
55. Tanaka K, Mori T, Nakamura T (1970) Cavity formation at the interface of a spherical inclusion in a plastically deformed matrix. Philos Mag 21:267–279
56. Gurland J (1972) Observations on the fracture of cementite particles in a spheroidized 1.05 % C steel deformed at room temperature. Acta Metall 20:735–741
57. Cox TB, Low JR Jr (1974) An investigation of the plastic fracture of AISI 4340 and 18 Nickel-200 grade maraging steels. Metall Trans 5:1457–1470
58. McMahon CJ Jr, Cohen M (1965) Initiation of cleavage in polycrystalline iron. Acta Metall 13:591–604
59. Argon AS, Im J (1975) Separation of second phase particles in spheroidized 1045 steel, Cu-0.6Pct Cr alloy, and maraging steel in plastic straining. Metall Trans A 6:839–851
60. Goods SH, Brown LM (1979) The nucleation of cavities by plastic deformation. Acta Metall 27:1–15
61. Maire E, Buffiere JY, Salvo L, Blandin JJ, Ludwig W, Letang JM (2001) On the application of X-ray tomography in the field of materials science. Adv Eng Mater 3:539–546
62. Chen ZT, Worswick MJ, Cinotti N, Pilkey AK, Lloyd DJ (2003) A linked FEM-damage percolation model of aluminum alloy sheet forming. Int J Plast 19:2099–2120
63. Gurson AL (1975) Plastic flow and fracture behaviour of ductile materials incorporating void nucleation, growth, and interaction. PhD thesis, Brown University, Providence, RI
64. Rosenfield AR, Hahn GT (1966) Numerical descriptions of the ambient low-temperature and high-strain rate flow and fracture behavior of plain carbon steel. ASM Trans 59:962–980
65. Hancock JW, Mackenzie AC (1976) On the mechanisms of ductile failure in high-strength steels subjected to multi-axial stress-states. J Mech Phys Solids 24:147–169
66. Budiansky B, Hutchinson JW, Slutsky S (1982) Void growth and collapse in viscous solids. In: Mechanics of solids—The Rodney Hill 60th anniversary volume. Pergamon Press, Oxford, pp. 13–45
67. Thomason PF (1968) A theory for ductile fracture by internal necking of cavities. J Inst Met 96:360–365
68. Thomason PF (1981) Ductile fracture and the stability of incompressible plasticity in the presence of microvoids. Acta Metall 29:763–777
69. Thomason PF (1985) A three-dimensional model for ductile fracture by the growth and coalescence of microvoids. Acta Metall 33:1087–1095
70. Thomason PF (1993) Ductile fracture by the growth and coalescence of microvoids of non-uniform size and spacing. Acta Metall Mater 41:2127–2134
71. Zhang ZL, Niemi E (1995) A new failure criterion for the Gurson-Tvergaard dilational constitutive model. Int J Fract 70:321–334
72. Benzerga AA, Leblond J-B (2014) Effective yield criterion accounting for microvoid coalescence. J Appl Mech 81:031009
73. Morin L, Leblond J-B, Benzerga AA (2015) Coalescence of voids by internal necking, Theoretical estimates and numerical results. J Mech Phys Solids 75:140–158

74. McVeigh C, Vernerey F, Liu WK, Moran B, Olson G (2007) An interactive micro-void shear localization mechanism in high strength steels. J Mech Phys Solids 55:225–244
75. Barsoum I, Faleskog J (2007) Rupture in combined tension and shear, Experiments. Int J Solids Struct 44:1768–1786
76. Nahshon K, Hutchinson J (2008) Modification of the Gurson model for shear failure. Eur J Mech A Solid 27:1–17
77. Xue L (2008) Constitutive modeling of void shearing effect in ductile fracture of porous materials. Eng Fract Mech 75:3343–3366
78. Butcher C, Chen Z, Bardelcik A, Worswick M (2009) Damage-based finite-element modeling of tube hydroforming. Int J Fract 155:55–65
79. Stoughton TB, Yoon JW (2011) A new approach for failure criterion for sheet metals. Int J Plast 27:440–459
80. Lecarme L, Tekoglu C, Pardoen T (2011) Void growth and coalescence in ductile solids with stage III and stage IV strain hardening. Int J Plast 27:1203–1223
81. McClintock FA (1968) A criterion for ductile fracture by the growth of holes. J Appl Mech 35:363–371
82. Malcher L, Andrade Pires FM, César de Sá JMA (2014) An extended GTN model for ductile fracture under high and low stress triaxiality. Int J Plast 54:193–228
83. Jones RM (2008) Deformation Theory of Plasticity. Bull Ridge Corporation
84. Williams ML, Schapery RA (1965) Spherical flaw instability in hydrostatic tension. Int J Fract Mech 1:64–72
85. Jackiewicz J, Kuna M (2003) Non-local regularization for FE simulation of damage in ductile materials. Comp Mater Sci 28:684–695
86. Jackiewicz J (2011) Use of a modified Gurson model approach for the simulation of ductile fracture by growth and coalescence of microvoids under low, medium and high stress triaxiality loadings. Eng Fract Mech 78:487–502
87. Bernauer G, Brocks W (2002) Micro-mechanical modelling of ductile damage and tearing—results of a European numerical round robin. Fatigue Fract Eng Mater Struct 25:363–384
88. Springmann M, Kuna M (2005) Identification of material parameters of the Gurson–Tvergaard–Needleman model by combined experimental and numerical techniques. Comp Mater Sci 32:544–552
89. Springmann M, Kuna M (2006) Determination of ductile damage parameters by local deformation fields, Measurement and simulation. Arch Appl Mech 75:775–797
90. Hariharan K, Chakraborti N, Barlat F, Lee M-G (2014) A novel multi-objective genetic algorithms-based calculation of Hill's coefficients. Metall Mater Trans A 45:2704–2707
91. Zhang ZL (1996) A sensitivity analysis of material parameters for the Gurson constitutive model. Fatigue Fract Eng Mater Struct 19:561–570
92. Kiran R, Khandelwal K (2014) Gurson model parameters for ductile fracture simulation in ASTM A992 steels. Fatigue Fract Eng Mater Struct 37:171–183
93. Jackiewicz J (2009) Calibration and evaluation of a combined fracture model of microvoid growth that may compete with shear in the polycrystalline microstructure by means of evolutionary algorithms. Comp Mater Sci 45:133–149
94. Wen J, Huang Y, Hwang KC, Liu C, Li M (2005) The modified Gurson model accounting for the void size effect. Int J Plast 21:381–395
95. Monchiet V, Bonnet G (2013) A Gurson-type model accounting for void size effects. Int J Solids Struct 50:320–327
96. Dormieux L, Kondo D (2010) An extension of Gurson model incorporating interface stresses effects. Int J Eng Sci 48:575–581

MIX
Papier aus verantwortungsvollen Quellen
Paper from responsible sources
FSC® C105338

If you have any concerns about our products,
you can contact us on
ProductSafety@springernature.com

In case Publisher is established outside the EU,
the EU authorized representative is:
Springer Nature Customer Service Center GmbH
Europaplatz 3, 69115 Heidelberg, Germany

Printed by Libri Plureos GmbH
in Hamburg, Germany